Digital
Fundamentals

Thomas L. Floyd

Digital
Fundamentals

Fourth Edition

**Merrill, an imprint of
Macmillan Publishing Company
New York**

**Collier Macmillan Canada, Inc.
Toronto**

**Maxwell Macmillan International Publishing Group
New York   Oxford   Singapore   Sydney**

Cover Photo: Joseph Drivas/The Image Bank

This book was set in Times Roman.

Administrative Editor: David Garza
Developmental Editor: Carol Thomas
Production Editor: Carol Driver
Art Coordinator: Ruth Ann Kimpel
Cover Designer: Cathy Watterson
Text Designer: Cynthia Brunk

Macmillan Publishing Company
866 Third Avenue, New York, NY 10022

Collier Macmillan Canada, Inc.

Library of Congress Catalog Number: 89–61807
International Standard Book Number: 0–675–21217–0
Printed in the United States of America
  4 5 6 7 8 9—93 92

Preface

While all of the popular features and the organization of the third edition have been retained, the coverage of many topics has been improved and expanded in this fourth edition. Also, several new features have been added.

In this book, as in most good texts, you will probably find more topics than can be covered because of time limitations in most programs or because of program emphasis. This selection of topics provides flexibility in tailoring the course to meet the needs of various program orientations. For example, some of the design-oriented and advanced topics may not be appropriate in certain programs, while in other programs less emphasis may be placed on troubleshooting. These topics are designed to be easily omitted without affecting the rest of the coverage.

Some of the improvements to this edition of *Digital Fundamentals* are:

1. An increase in end-of-chapter problems of approximately fifty percent.
2. An increase in illustrations of approximately thirty-two percent.
3. An increase in examples of approximately twenty percent.
4. A functional two-color format.
5. A glossary at the end of each chapter.
6. An improved and expanded coverage of memories in Chapter 10.
7. A revised coverage of interfacing in Chapter 11.
8. A completely new coverage of microprocessors in Chapter 12.
9. An expanded coverage of IC technologies in Chapter 13, which can be used as a ''floating'' chapter if desired.
10. An expanded and improved coverage of PLDs.
11. An expanded and improved coverage of troubleshooting, including more problems.
12. An increased emphasis on applications and systems concepts.
13. An improved coverage of Karnaugh maps.
14. Designation of troubleshooting and design-oriented problems for easy identification.
15. An expanded instructional aids package that now includes a test bank in both hard copy and computer disk, transparencies and transparency masters, two lab manuals, and an instructor's manual.

Content and Organization

There are thirteen chapters in this text, beginning with an introduction to digital concepts and logic elements in Chapter 1 and progressing through logic gates, Boolean algebra, combinational logic, sequential logic, memories, interfacing, and an introduction to microprocessors in Chapter 12. Chapter 13, a new chapter, covers inside-the-chip circuitry, operating characteristics, and data sheet parameters of digital IC families. Logic family interfacing and other practical considerations are also included in this chapter. A summary and glossary are included at the end of most chapters, with the glossary serving as the definition part of the summary.

Chapter 13, "Integrated Circuit Technologies," is optional. If used in whole or in part, it is designed to function as a "floating" chapter that can be covered at various points in the chapter sequence, depending on the individual program and on the preference of the instructor. This chapter is indicated by a tab-edge design for easy reference.

Within each chapter there is a review at the end of each section with answers at the end of that chapter. There is a self-test at the end of each chapter with all answers and/or solutions at the end of the book indicated by a tab mark. Also, there are sectionalized chapter problem sets with odd-numbered answers at the end of the book indicated by a tab mark. For quick identification and assignment, troubleshooting problems are contained in one section and are designated by an icon in the margin. All design-oriented problems are designated by a $\mathbb{D}$ in the margin.

The ANSI/IEEE Std. 91-1984 logic symbols with dependency notation are introduced gradually and conservatively at appropriate points while the use of the more traditional symbols is retained throughout. Although it is important for the student to become familiar with this newer standard, the symbols and notation represent a significant departure from many of the traditional logic symbols, and thus a gradual and limited introduction is appropriate. Omission of this coverage will not affect the rest of the material.

Suggestions for Use

In a two-term course sequence, this text may be divided into two parts: combinational logic (Chapters 1 through 6) and sequential logic (Chapters 7 through 12) with Chapter 13 inserted at an appropriate point if desired.

For programs having a one-term digital fundamentals course or for those courses where more time is spent on less material, the following topics are suggested for possible deletion. Final deletion decisions, of course, depend on program emphasis and other factors.

1. Chapters that may be considered for complete or partial deletion:
   Chapter 1: Digital Concepts
   Chapter 10: Memories and Programmable Logic Devices
   Chapter 11: System Interfacing
   Chapter 12: Introduction to Microprocessor-Based Systems
   Chapter 13: Integrated Circuit Technologies
2. Sections within other chapters that may be considered for deletion:
   Section 2–7: Binary Arithmetic with Signed Numbers
   Section 2–11: Digital Codes
   Section 3–7: Integrated Circuit Logic Families

Section 4–6: Simplification Using Boolean Algebra
Section 4–7: Simplification Using the Karnaugh Map
Section 5–2: Design of Combinational Logic
Section 6–3: Ripple-Carry versus Look-Ahead Carry Adders
Section 6–10: Parity Generators/Checkers
Section 7–4: Data Lock-Out Flip-Flops
Section 7–8: Astable Multivibrators and Timers
Section 8–4: Design of Sequential Circuits
Section 8–9: Logic Symbols with Dependency Notation
Section 9–7: Shift Register Counters
Section 9–10: Logic Symbols with Dependency Notation

Acknowledgments

We have strived to make this fourth edition of Digital Fundamentals the best yet. This final product is a result of the efforts of many people including the reviewers, the users of the third edition who made many valuable suggestions, and the editorial and production people at Merrill.

I want to express my appreciation to Carol Thomas, Steve Helba, Dave Garza, Carol Driver, and Ruth Kimpel at Merrill for their continued enthusiasm and dedication to quality, and to Ann Colowick for a fine job copy editing the manuscript.

Also, I am grateful to the following reviewers and accuracy checkers who contributed many valuable suggestions for improvement that helped me very much during the preparation of this new edition: Frank T. Gergeli, Metropolitan Technical Institute; Bruce Koller, Diablo Valley College; L. W. Heselton, Mohawk Valley Community College; Richard Copes, Indiana Vocational Technical College; Charles J. Abate, Onondaga Community College; Jack Braun, Houston Community College; Leonard J. Bundra, Lincoln Technical Institute; Dave Buchla, Yuba College; Jerry Stierwalt, Tulsa Vo-Tech School; Gary Farmer, DeVry Institute of Technology—Columbus; Francis Crozier, Rets Electronics Institute; David Hata, Portland Community College; Fitz Husbands, Lamar University—Orange; Ed Risinger, Electronic Computer Programming Institute; Bob Mobley, Knoxville State Technical College; Susan Garrod, Purdue University; Roy Siegel, DeVry Institute of Technology—Los Angeles; Gary Snyder, Yuba College; Kenneth Dennis, San Jacinto College; Nazar Karzay, Indiana Vocational Technical College; Roger Bertrand, Central Maine Vocational Technical Institute; Larry Ryan, Olympic College; Pete Chesebrough, Ferris State College; Bill Coyle, College of the Redwoods; Warren Foxwell, DeVry Institute of Technology—Chicago; Steve Yelton, Cincinnati Technical College; Gary House, DeVry Institute of Technology—Atlanta; Tim Lauck, College of the Redwoods; Stephen Harsany, Mt. San Antonio College; Francis Erazmus, CHI Institute; Terry Lewis, Computer Processing Institute—Cambridge; and Rick Schulmeister, Heald College.

The photographs used in this book are courtesy of Hewlett Packard, Tektronix, and Motorola. The data sheets are courtesy of Texas Instruments.

Finally, I want to thank my wife, Sheila, for her loving support during the many hours I have spent on this project.

Tom Floyd

Contents

7
FLIP-FLOPS AND RELATED DEVICES

313

8
COUNTERS

373

MERRILL'S INTERNATIONAL SERIES IN ELECTRICAL AND ELECTRONICS TECHNOLOGY

Digital electronics began in 1946 with an electronic digital computer called ENIAC, which was implemented with vacuum-tube circuits. The concept of the digital computer can be traced to Charles Babbage, who developed a mechanical computation device in the 1830s. The first digital computer was built in 1944 at Harvard University, but it was electromechanical, not electronic.

The term *digital* is derived from the way in which computers perform operations by counting *digits*. For years, applications of digital electronics were confined to computer systems. Today, digital techniques are applied in many diverse areas, such as telephone systems, data processing, radar, navigation, military systems, medical instruments, process control, and consumer products. Digital technology has progressed from vacuum-tube circuits to integrated circuits and microprocessors.

After completing this chapter, you will be able to

☐ appreciate the many diverse applications of digital electronics.

☐ explain the basic differences between digital systems and analog systems.

☐ explain the difference between positive and negative logic.

☐ show how voltage levels are used to represent digital quantities.

☐ measure various parameters of a pulse waveform such as rise time, fall time, pulse width, frequency, period, and duty cycle.

☐ explain the basic logic operations of NOT, AND, and OR.

☐ describe the basic functions of the comparator, adder, encoder, decoder, counter, register, multiplexer, and demultiplexer.

☐ understand how a complete digital system is formed from the basic functions in a simple application.

☐ identify digital integrated circuits according to their complexity and the type of circuit packaging.

☐ identify pin numbers on integrated circuit packages.

☐ recognize digital instruments and understand how they are used in troubleshooting digital circuits and systems.

1

Digital Concepts

DIGITAL ELECTRONICS

1–1

Before you begin studying the fascinating and rapidly growing field of digital electronics, it is important to understand some of the history of its development and to be aware of the diversity of its applications.

History

Early experiments in electronics involved electric currents flowing in glass tubes. One of the first to conduct such experiments was a German named Heinrich Geissler (1814–79). Geissler found that when he removed most of the air from a glass tube, the tube glowed when an electric voltage was applied to it.

About 1878 Sir William Crookes (1832–1919), a British scientist, experimented with tubes similar to those of Geissler. In his experiments Crookes found that the current in the tubes seemed to consist of particles.

Thomas Edison (1847–1931), while experimenting with the carbon-filament light bulb that he had invented, made another important finding. He inserted a small, positively charged metal plate in the bulb and observed that a current flowed from the filament to the plate. This device was the first thermionic diode, a forerunner of the modern semiconductor diodes used in digital circuits and other types of electronic circuits. Other work in the development of the vacuum diode was done by Sir John Fleming (1849–1945), a British scientist, in the early 1900s.

Major progress in electronics awaited the development of a device that could boost, or amplify, weak electric signals. Such a device was the audion, patented in 1907 by Lee De Forest (1873–1961), an American. It was a vacuum tube capable of amplifying small electrical signals. Great improvements were made on De Forest's invention by other researchers, including Langmuir, Schottky, and Tellegen.

Although crystal detectors used in early radios were the predecessors of modern semiconductor diodes, the era of solid state electronics actually began with the invention of the transistor in 1947 at Bell Labs. The inventors were Walter Brattain (b. 1902), John Bardeen (b. 1908), and William Shockley (b. 1910).

Transistor technology led to the development of the integrated circuit *(IC)* in the early 1960s. The integrated circuit incorporated many transistors, diodes, and other components on a single tiny chip of semiconductor material. Integrated circuit technology continues to develop and improve, allowing increasingly complex circuits to be created on smaller chips. The introduction of the microprocessor in the early 1970s created another electronics revolution: the entire processing portion of a computer was placed on a single small silicon chip.

The computer has probably had a greater impact on modern technology than any other type of electronic system. The first electronic digital computer was completed in 1946 at the University of Pennsylvania. It was called the Electronic Numerical Integrator and Computer (ENIAC). One of the most significant developments in computers was the *stored program* concept, developed in the 1940s by John Von Neumann (1903–57), an American mathematician. In the stored program concept, instructions to the computer for performing specified operations are stored in the computer's memory.

Applications

Computers One of the most important types of digital systems is the computer. Its applications are broad and diverse. For example, computer applications in business include record keeping, accounting payrolls, inventory control, market analysis, and financial projections.

Scientists use computers to process huge amounts of *data* (numbers and other information) and to perform complex and lengthy calculations. In industry, computers are used for controlling and monitoring intricate manufacturing processes. Communications, navigation, medicine, the military, and the home are a few of the other areas in which computers are used extensively.

Communications Digital communications encompasses a wide range of specialized fields. Included are space and satellite communications, telephone, data communications, radar, and military applications.

Automation Digital systems are employed extensively in the control of manufacturing processes. Control of ingredient mixes, operation of machine tools, control of automated assembly lines, industrial robots, and the control and distribution of power are some of the applications.

Medicine Digital techniques are applied widely in the medical field in instruments used for monitoring various body functions and in diagnostic equipment.

Entertainment and consumer products Electronic products used directly by consumers for entertainment, information, recreation, or work around the home use digital circuits to a great extent. Some examples are appliance controls, personal computers, electronic games, VCRs, compact disc players, and digital recording devices.

Digital systems are used in automobiles to control and monitor engine functions, control braking, provide entertainment, and display useful information to the driver.

**SECTION 1–1
REVIEW**

Answers are found at the end of the chapter.
1. List two major events in the development of digital electronics.
2. List five general areas in which digital electronics is important.

DIGITAL AND ANALOG QUANTITIES

1–2

The field of electronics can be divided into two broad categories: **digital** *and* **analog**. *Digital electronics involves quantities with discrete values, and analog electronics involves quantities with continuous values.*

Most things that can be measured quantitatively appear in nature in analog form. For example, the air temperature changes over a continuous range of values. During a given day, the temperature does not go from, say, 70 degrees to 71 degrees instantaneously;

it takes on all the infinite values in between. If you graphed the temperature on a typical summer day, you would have a smooth, continuous curve similar to Figure 1–1.

Rather than graphing the temperature on a continuous basis, suppose we just took a temperature reading every hour. Now we would have discrete values representing the temperature change over a period of time, as indicated in Figure 1–2. We would have converted an analog quantity to digital form.

Sound is another example of an analog quantity, and we can use electronics to process sound. For example, a linear amplifier that raises a very low-level audio signal to a level sufficient to drive a speaker is an illustration of analog electronics. The amplifier does not change the continuous nature of the sound wave; it simply increases its volume.

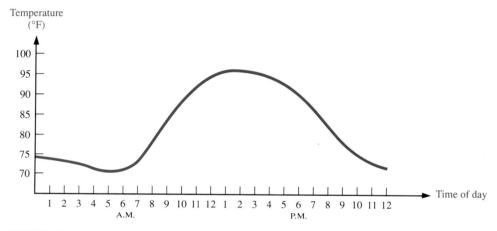

FIGURE 1–1
Graph of an analog quantity (temperature versus time).

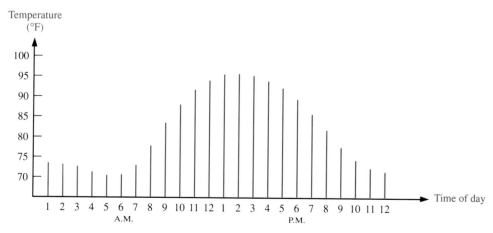

FIGURE 1–2
Discrete-value (digital) representation of an analog quantity.

An example of an electronic system in which both analog and digital electronics are used is the digital audiotape (DAT) recorder. The audio signal must be amplified and then converted to digital form for processing and storage on the tape. This process involves analog-to-digital conversion *(ADC),* in which the analog signal is converted to a series of coded numbers representing the digital values. For the stored digitized audio to be played, it must be converted back to analog form and amplified to drive a speaker. This process is called digital-to-analog conversion *(DAC).*

SECTION 1–2 REVIEW

Answers are found at the end of the chapter.
1. Define the term *analog.*
2. Define the term *digital.*

LOGIC LEVELS AND PULSE WAVEFORMS

1–3

Digital electronics involves circuits and systems in which there are only two possible states. Those states are usually represented by two different voltage levels: a HIGH and a LOW. The two states can also be represented by current levels, open and closed switches, or lamps turned on and off. In digital systems combinations of the two states, called **codes,** *are used to represent numbers, symbols, alphabetic characters, and other types of information. This two-state number system is called* **binary,** *and its two digits are 0 and 1. A binary digit is called a* **bit.**

Positive and Negative Logic

In digital systems two voltage levels represent the two binary digits, 1 and 0. If the more positive of the two voltages represents a 1 and the less positive voltage represents a 0, the system is called a *positive logic* system. If, on the other hand, the less positive voltage represents a 1 and the more positive voltage represents a 0, we have a *negative logic* system. To illustrate, suppose that we have $+5$ V and 0 V as our logic-level voltages. We will designate $+5$ V as the HIGH level and 0 V as the LOW level, so positive and negative logic can be defined as follows:

Positive logic	*Negative logic*
HIGH = 1	HIGH = 0
LOW = 0	LOW = 1

Both positive and negative logic are used in digital systems, but positive logic is more common. For this reason *we will use only positive logic in this text.*

Logic Levels

In practical applications each of the two logic levels is actually a range of values. In a digital circuit a HIGH can be any voltage between a specified minimum value and a specified maximum value. Likewise, a LOW can be any voltage between a specified minimum value and a specified maximum value.

Figure 1–3 illustrates the range of LOWs and HIGHs for a typical digital circuit. The variable $V_{H(\text{max})}$ represents the maximum HIGH value, and $V_{H(\text{min})}$ represents the minimum HIGH value. The maximum LOW value is represented by $V_{L(\text{max})}$, and the minimum LOW value by $V_{L(\text{min})}$. The range of voltages between $V_{L(\text{max})}$ and $V_{H(\text{min})}$ is a range of uncertainty. A voltage in that uncertain range can appear as either a HIGH or a LOW to a given circuit; we can never be sure. Therefore, the values in the uncertain range are prohibited values.

Pulse Waveforms

Pulses are very important in digital circuits and systems because voltage levels are normally changing back and forth between the HIGH and LOW states. Figure 1–4(a) shows that a single positive-going pulse is generated when the voltage (or current) goes from its normally LOW level to its HIGH level and then back to its LOW level. The negative-going pulse in Figure 1–4(b) is generated when the voltage goes from its normally HIGH level to its LOW level and back to its HIGH level.

As indicated in Figure 1–4, the pulse has two edges: a *leading edge* and a *trailing edge*. For a positive-going pulse, the leading edge is a rising edge, and the trailing edge is a falling edge. The pulses in Figure 1–4 are ideal because the rising and falling edges change in zero time (instantaneously). In practice, these transitions never occur instantaneously, although for most digital work we can assume ideal pulses. Figure 1–5 shows a nonideal pulse. The time required for the pulse to go from its LOW level to its HIGH level is called the *rise time* (t_r), and the time required for the transition from the HIGH

FIGURE 1–3

Logic level ranges for a typical digital circuit.

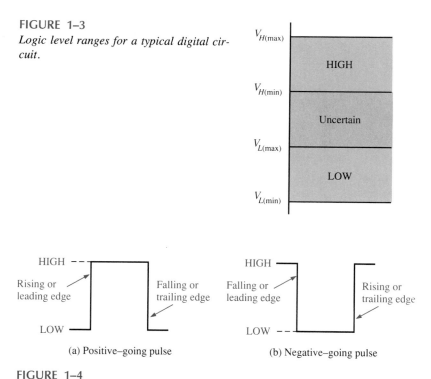

(a) Positive–going pulse (b) Negative–going pulse

FIGURE 1–4

Ideal pulses.

FIGURE 1–5

Nonideal pulse characteristics.

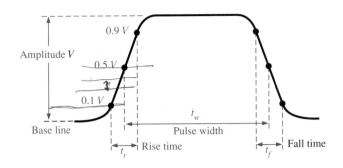

FIGURE 1–6

Examples of pulse waveforms.

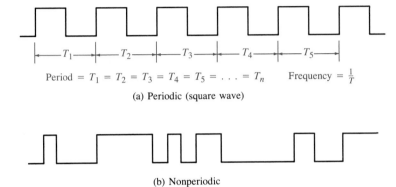

Period $= T_1 = T_2 = T_3 = T_4 = T_5 = \ldots = T_n$ Frequency $= \frac{1}{T}$

(a) Periodic (square wave)

(b) Nonperiodic

level to the LOW level is called the *fall time* (t_f). In practice, it is common to measure rise time from 10% of the pulse *amplitude* (height from baseline) to 90% of the pulse amplitude, and to measure the fall time from 90% to 10% of the pulse amplitude, as indicated in Figure 1–5. The reason for this practice is that nonlinearities commonly occur near the bottom and the top of the pulse.

The *pulse width* (t_W) is a measure of the duration of the pulse and is often defined as the time interval between the 50% points on the rising and falling edges, as indicated in Figure 1–5.

Most waveforms encountered in digital systems are composed of series of pulses and can be classified as either *periodic* or *nonperiodic*. A periodic pulse waveform is one that repeats itself at a fixed interval, called the *period* (T). The *frequency* (f) is the rate at which it repeats itself and is measured in pulses per second (pps) or Hertz (Hz). A nonperiodic pulse waveform, of course, does not repeat itself at fixed intervals and may be composed of pulses of differing pulse widths and/or differing time intervals between the pulses. An example of each type is shown in Figure 1–6.

The frequency of a pulse waveform (sometimes called the *pulse repetition rate*, or PRR) is the reciprocal of the period. The relationship between frequency and period is expressed as follows:

$$f = \frac{1}{T} \tag{1–1}$$

$$T = \frac{1}{f} \tag{1–2}$$

An important characteristic of a periodic pulse waveform is its *duty cycle*. The duty cycle is defined as the ratio of the pulse width (t_W) to the period (T) expressed as a percentage.

$$\text{Duty cycle} = \left(\frac{t_W}{T}\right) 100 \qquad\qquad \textbf{(1–3)}$$

EXAMPLE 1–1

A portion of a periodic pulse waveform is shown in Figure 1–7. The measurements are in milliseconds. Determine the duty cycle.

FIGURE 1–7

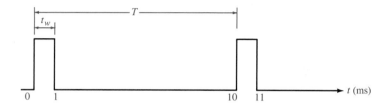

Solution

$$t_W = 1 \text{ ms} \qquad T = 10 \text{ ms}$$

$$\text{Duty cycle} = \left(\frac{t_W}{T}\right) 100 = \left(\frac{1 \text{ ms}}{10 \text{ ms}}\right) 100 = 10\%$$

Pulse Waveforms Carry Binary Information

Most binary information that is handled by digital systems appears as pulse waveforms. When the pulse waveform is HIGH, a 1 is present; when it is LOW, a 0 is present. (Remember that we are assuming positive logic. In a negative logic system, the reverse would be true.)

In digital systems all pulse waveforms are derived from and related to a basic timing waveform called the *clock*. The clock is a periodic waveform in which each pulse interval (period) is one *bit time*. When other pulse waveforms are shown in the proper time relationship with the clock or with each other, we have what is known as a timing diagram. An example is shown in Figure 1–8. Notice that each change in level of waveform *A* corresponds to a leading edge on the clock waveform. In some cases changes can occur on the trailing edges of the clock.

During each bit time of the clock in Figure 1–8, waveform *A* is either HIGH or LOW. Those HIGHs and LOWs represent the sequence of bits indicated. A group of several bits may represent a piece of binary information, such as a number or a letter.

Serial and Parallel Transfer

Binary information represented by pulse waveforms must be transferred within a digital system or from one system to another in order to be processed for a useful purpose. For example, numbers in the memory of a computer must be transferred to the central processor to be added together. The sum of this addition must then be transferred to the monitor for display.

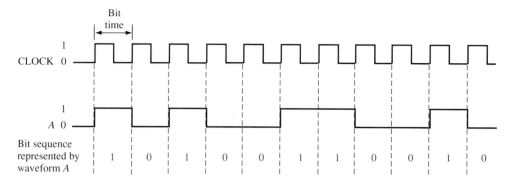

FIGURE 1–8
Example of a timing diagram and the waveform representation of a sequence of bits.

The transfer of a group of bits that make up a piece of binary information may be either *serial* or *parallel*. When transferred from system A to system B in serial form, the bits are sent on a single line one bit at a time, as illustrated in Figure 1–9(a). During the time interval from t_0 to t_1, the first bit appears on the line. During the time interval from t_1 to t_2, the second bit appears on the line, and so on. When transferred in parallel form, all the bits in a group are sent out on separate lines at the same time. There is one line for each bit, as shown in Figure 1–9(b) for the case of four bits.

As you can see, the advantage of serial transfer is that only one line is required. In parallel transfer four lines are required to transfer four bits, six lines are required to transfer six bits, and so on. The disadvantage of serial transfer is that a longer time is required to transfer a group of bits than is required with parallel transfer. For example,

FIGURE 1–9
Illustration of serial and parallel transfer of bits.

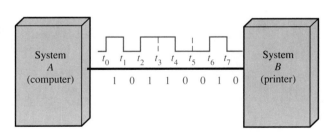

(a) Serial transfer of bits from A to B. Interval t_0 to t_1 is first.
Computer to printer is an example.

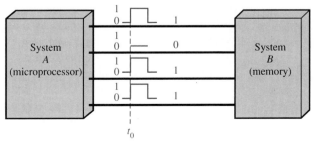

(b) Parallel transfer of bits from A to B. All four bits
are transferred simultaneously beginning at time t_0.
Microprocessor to memory is an example.

four bit times are required to transfer four bits of serial information. If each bit time is, say, 1 μs (1 microsecond, or 0.000001 second), then it takes 4 μs to transfer the four bits from one point to another. In the parallel form all four bits are transferred in only 1 μs because they are transferred simultaneously on four separate lines.

EXAMPLE 1–2

Determine the total time required to transfer the eight bits contained in the serial waveform (*A*) of Figure 1–10, and indicate the sequence of bits. The left-most bit is the first to be transferred. The 100 kHz clock is used as reference.

What is the total time to transfer the same eight bits in parallel?

FIGURE 1–10

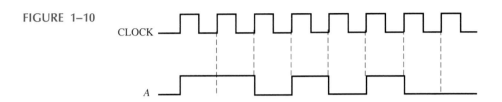

Solution Since the frequency of the clock is 100 kHz, the period is

$$T = \frac{1}{f} = \frac{1}{100 \text{ kHz}} = 10 \text{ μs}$$

It takes 10 μs to transfer each bit in the waveform. The total transfer time is

$$8 \times 10 \text{ μs} = 80 \text{ μs}$$

In parallel transfer it would take only 10 μs to transfer all eight bits.

To determine the sequence of bits, examine the timing diagram in Figure 1–10 during each bit time. If waveform *A* is HIGH during the bit time, a 1 is transferred. If waveform *A* is LOW during the bit time, a 0 is transferred. The bit sequence is illustrated in Figure 1–11.

FIGURE 1–11

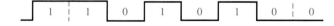

**SECTION 1–3
REVIEW**

Answers are found at the end of the chapter.

1. The rise time of a certain pulse is 5 μs. If the amplitude of the pulse is 5 V and its base line is 0 V, determine the actual change in voltage during the rise-time interval. (Use the definition of rise time.)
2. A certain periodic pulse waveform has a pulse width of 15 μs and a frequency of 16.67 kHz. Find the duty cycle.
3. The following bit sequence is transferred serially at a 10 MHz rate. What is the total time for transferring the bits?

1011000101110100

4. What is the main advantage of parallel transfer over serial transfer?

ELEMENTS OF DIGITAL LOGIC

1–4

*In its basic form, **logic** is the realm of human reasoning that tells us a certain proposition (declarative statement) is true if certain conditions are true. "The light is on" is an example of a proposition that can be only true or false. "The bulb is not burned out" and "The switch is on" are other examples of propositions that can be classified as true or false.*

Many situations, problems, and processes that we encounter in our daily lives can be expressed in the form of propositional, or logic, functions. Since such functions are true/false or yes/no statements, digital circuits, with their two-state characteristics, are extremely applicable.

Several propositions, when combined, form propositional, or logic, functions. For example, the propositional statement "The light is on" will be true if "The bulb is not burned out" is true and if "The switch is on" is true. Therefore, this logical statement can be made:

The light is on if and only if the bulb is not burned out *and* the switch is on.

In this example the first statement is true only if the last two statements are true. The first statement ("The light is on") is then the basic proposition, and the other two statements are the conditions on which the proposition depends.

A mathematical system for formulating logical statements with symbols, so that problems can be written and solved in a manner similar to ordinary algebra, was developed by the Irish logician and mathematician George Boole in the 1850s. *Boolean algebra,* as it is known today, finds application in the design and analysis of digital systems and will be covered in Chapter 4.

The term *logic* is applied to digital circuits used to implement logic functions. Several kinds of digital circuits are the basic elements that form the building blocks for such complex digital systems as the computer. We will now look at these elements and discuss their functions in a very general way. Later chapters will cover these circuits in full detail.

NOT

The first basic logic element is the *NOT* circuit. As illustrated in Figure 1–12, the primary function of this circuit is to change one logic level to the opposite logic level. If a HIGH level (1) is applied to the input of the NOT circuit, a LOW level (0) appears on the output. If a LOW level is applied to the input, a HIGH level appears on the output. A NOT circuit is commonly called an *inverter*.

FIGURE 1–12
The function of a NOT circuit.

AND

The second basic logic element is the *AND* circuit, which is a type of logic *gate*. The AND circuit can be viewed functionally as a switching circuit with two or more switches in series. Actually, logic circuits are electronic, but the switch analogy serves to illustrate their operation. The number of switches equals the number of conditions in a logic statement. A true condition is represented by a closed switch, and a false condition is represented by an open switch. A HIGH level on the output means that *all* conditions are true, and a LOW level means that at least one condition is false.

Figure 1–13 shows two switches, representing two conditions, and all the possible true/false combinations of the conditions. As you can see, the only time that the output is true (HIGH level) is when both conditions are true (both switches closed). The conditions can be thought of as *inputs* to the AND circuit.

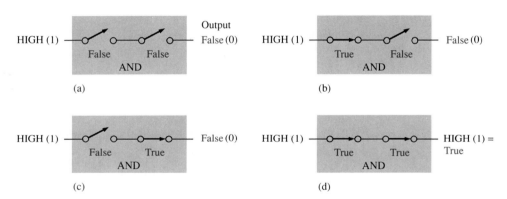

FIGURE 1–13
Switch equivalent of the AND function.

OR

The third basic logic element is the *OR* circuit, which is also a type of logic gate. The OR circuit can be viewed functionally as a switching circuit with two or more switches in parallel. The number of switches equals the number of conditions in a logic statement. As with the AND circuit, a true condition is represented by a closed switch and a false condition by an open switch. A HIGH level on the output means that one or more conditions are true, and a LOW level means that *all* conditions are false.

Figure 1–14 shows two switches, representing two conditions, and all the possible true/false combinations of the conditions. As you can see, whenever at least one of the conditions is true (switch closed), the output is true (HIGH level). The conditions can be thought of as inputs to the OR circuit.

Flip-Flop

The primary function of the *flip-flop* is to store, or "memorize," a binary digit for an indefinite period of time. The flip-flop is a basic memory element. It is distinguished from elements previously discussed by its ability to retain either logic level after input conditions have been removed. Actually, the flip-flop can be constructed from combi-

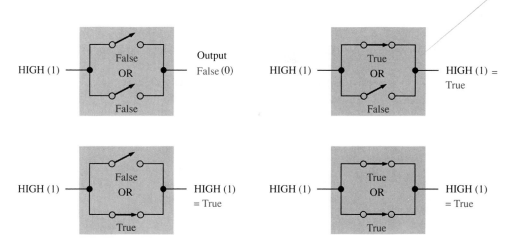

FIGURE 1–14
Switch equivalent of the OR function.

nations of basic gates, but it is treated as a distinct logic element because of its importance in digital systems. An important variation of the flip-flop is called the *latch*.

The basic elements—the inverter, the AND gate, the OR gate and the flip-flop—are used to construct more complex logic circuits, such as counters, registers, decoders, and memories. Those more complex logic functions are then combined to form complete digital systems designed to perform specified tasks.

SECTION 1–4 REVIEW

Answers are found at the end of the chapter.
1. List four basic logic elements.
2. A logic gate has HIGHs on both of its two inputs, and its output is HIGH. What type of gate is it?
   (a) AND gate   **(b)** OR gate   **(c)** either AND or OR gate
3. A logic gate has a LOW on one input and a HIGH on the other. Its output is LOW. What type of gate is it?

FUNCTIONS OF DIGITAL LOGIC

1–5

The inverter, the basic gates, and the flip-flop or latch combine to form more complex logic circuits that perform many operations and are used to build complete digital systems. Some of the more common logic functions are comparison, arithmetic, decoding, encoding, counting, memory, multiplexing, and demultiplexing. This section provides a general overview of these important functions so that you can begin to see how they form the building blocks of digital systems such as computers.

Comparison

The comparison function is performed by a logic circuit called a *comparator*. Its function is to compare two quantities and to indicate whether they are equal. For example, suppose we have two numbers and we wish to know if they are equal or not equal and, if unequal, which is greater. The comparison function is represented in Figure 1–15. One number is applied in binary form to input *P*, and the other to input *Q*. The outputs indicate the relationship of the two numbers by producing a HIGH level on the proper output line. Suppose that a binary representation of the number 2 is applied to input *P*, and a binary representation of the number 5 is applied to input *Q*. (We will discuss the binary representation of numbers and symbols in the next chapter.) A HIGH level will appear on the *P* < *Q* ("*P* is less than *Q*") output, indicating the relationship between the two numbers (2 is less than 5). The wide arrows represent a group of parallel lines that transfer the bits.

FIGURE 1–15
The comparator.

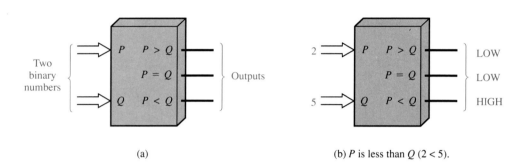

(a)   (b) *P* is less than *Q* (2 < 5).

Arithmetic

The addition function is performed by a logic circuit called an *adder*. Its function is to add two numbers (on inputs *P* and *Q*) with a *carry input* (CI) and generate a sum (Σ) and a *carry output* (CO), as shown in Figure 1–16(a). Figure 1–16(b) indicates the addition of 3 and 9. We all know that the sum is 12, and the adder indicates this result by producing 2 on the sum output and 1 on the carry output. We assume the carry input in this example to be 0.

Subtraction is the second arithmetic function that can be performed by digital circuits. A *subtractor* requires three inputs: the two numbers that are to be subtracted and a *borrow input*. The two outputs are the *difference* and the *borrow output*. When, for instance, 5 is subtracted from 8 with no borrow input, the difference is 3 with no borrow output. We will see in a later chapter how subtraction can actually be performed by an adder, because subtraction is simply a special case of addition.

Multiplication is the third arithmetic operation that can be performed by digital circuits. Since numbers are always multiplied two at a time, two inputs are required.

FIGURE 1–16
The adder.

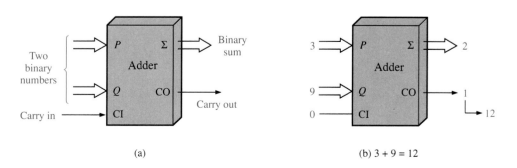

(a)   (b) 3 + 9 = 12

The output of the multiplier is the *product*. Since multiplication is simply a series of additions with shifts in the positions of the partial products, it can be performed by using an adder in conjunction with other circuits.

Division, the fourth type of arithmetic operation, is a series of subtractions, comparisons, and shifts, and thus it can also be performed by using an adder in conjunction with other circuits. Two inputs to the *divider* are required, and the outputs generated are called the *quotient* and the *remainder*.

Encoding

The *encoder* converts information, such as a decimal number or an alphabetic character, into some coded form. For example, a certain type of encoder converts each of the decimal digits, 0 through 9, to a binary code, as shown in Figure 1–17. A HIGH level on the input corresponding to a specific decimal digit produces the proper binary code on the output lines. The binary code and other codes are covered in the next chapter.

Decoding

The *decoder* converts coded information, such as a binary number, into another form, such as decimal form. For example, a particular type of decoder converts a four-bit binary code into the appropriate decimal digit, as illustrated in Figure 1–18.

FIGURE 1–17
Example of an encoder.

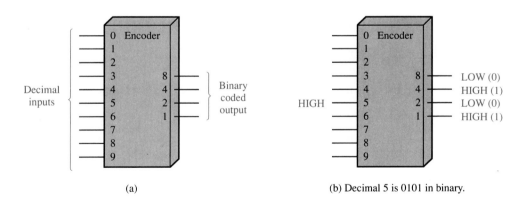

(a)

(b) Decimal 5 is 0101 in binary.

FIGURE 1–18
Example of a decoder.

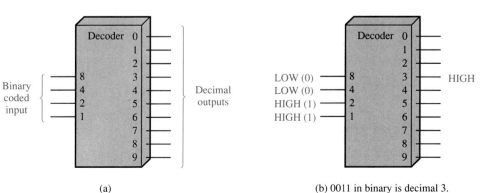

(a)

(b) 0011 in binary is decimal 3.

Counting

The counting operation is very important in digital systems. There are many types of digital *counters,* but their basic function is to count events represented by changing levels or pulses or to generate a particular code sequence. To count, the counter must "remember" the present number so that it can go to the next proper number in sequence. Therefore, memory capability is an important characteristic of all counters, and flip-flops are generally used to implement them. Figure 1–19 illustrates two simple counter applications.

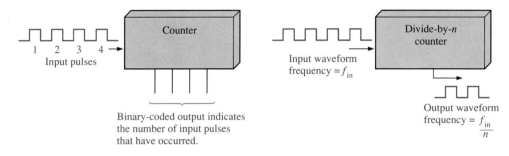

(a) Counter used to count events indicated by the occurrences of pulses.

(b) Counter used to divide an input frequency by a factor n. For example, if $f_{in} = 100$ kHz and $n = 10$, $f_{out} = 10$ kHz.

FIGURE 1–19
Examples of counter applications.

Registers (Memory)

Registers are digital circuits used for the temporary storage and shifting of information. For instance, a number in binary form can be stored in a register, and then its position within the register can be changed by shifting it one way or the other. Figure 1–20 illustrates simple forms of register operation. Latches and flip-flops are common memory elements used in registers.

Multiplexing and Demultiplexing

Multiplexing is an operation performed with digital logic circuits called *multiplexers.* Multiplexing allows information to be switched from several lines onto a single line in a specified sequence. A simple multiplexer can be represented by a switch operation that sequentially connects each of the input lines with the output, as illustrated in Figure 1–21. Assume that we have logic levels as indicated on the three inputs (a multiplexer can have any number of inputs). During time interval t_1, input A is connected to the output; during interval t_2, input B is connected to the output; and during interval t_3, input C is connected to the output. As a result of this multiplexing action, we have the three logic levels on the input lines appearing in sequence on the output line. Later you will learn how the switching action is controlled.

The inverse of the multiplexing function is called demultiplexing. In a *demultiplexer,* logic data from a single input line are sequentially switched onto several output lines, as shown in Figure 1–22.

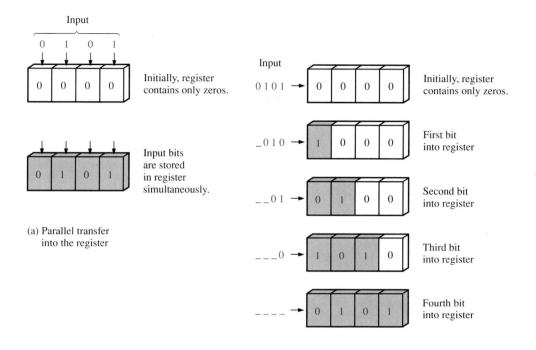

(a) Parallel transfer into the register

(b) Serial transfer into the register

FIGURE 1–20
Examples of simple shift register operations.

FIGURE 1–21
Example of simple multiplexer operation.

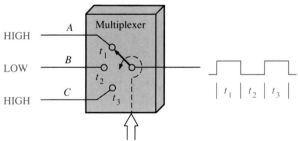

FIGURE 1–22
Example of simple demultiplexer operation.

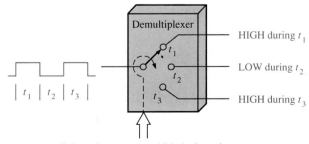

Digital circuits can be used to perform a large variety of tasks, limited only by the imagination. To give you a general picture of some aspects of digital logic, we have touched on only a few of the more basic and common functions, which can be used as building blocks for increasingly complex systems.

SECTION 1–5 REVIEW

Answers are found at the end of the chapter.

1. List seven primary functions of digital logic.
2. Identify the function that best performs each of the following operations:
   (a) Determining which of two numbers is greater.
   (b) Determining if two numbers are equal.
   (c) Producing the sum of two numbers.
   (d) Dividing the frequency of a pulse waveform by 4.

A SYSTEM APPLICATION

1–6

In this section an interesting but greatly simplified system application of the logic elements and functions that were discussed in the previous sections is presented. It is very important that an electronic technician or technologist understand how various digital functions can operate together as a total system to perform a specified task. It is also important to begin to think of system-level operation because, in practice, a large part of your work will involve systems rather than individual functions. Of course, to understand systems, you must first understand the basic elements and functions that make up a system.

The purpose of this section is to introduce you to the system concept. The example here will show you how logic functions can work together to perform a higher-level task and will get you started thinking at the system level. The specific system used here to illustrate the system concept is not necessarily the approach that would be used in practice, although it could be. In modern industrial applications like the one discussed here, instruments known as programmable controllers are often used.

A Hypothetical System

A pharmaceutical company uses the system shown in Figure 1–23 for automatically bottling tablets. The tablets are fed into a large funnel-like hopper. The narrow neck of the funnel allows only one table at a time to fall into a bottle on the conveyor belt below.

The digital system controls the number of tablets going into each bottle and displays a continually updated total near the conveyor line as well as at a remote location in another part of the plant. As you can see, this system utilizes all the basic logic functions that were introduced in the last section.

The general operation is as follows. An optical sensor at the bottom of the funnel neck detects each tablet that passes and produces an electrical pulse. This pulse goes to the counter and advances it by one count; thus, at any time during the filling of a bottle, the counter holds the binary representation of the number of tablets in the bottle. The binary count is transferred on parallel lines to the P input of the comparator. A preset binary number equal to the number of tablets that are to go into each bottle is placed

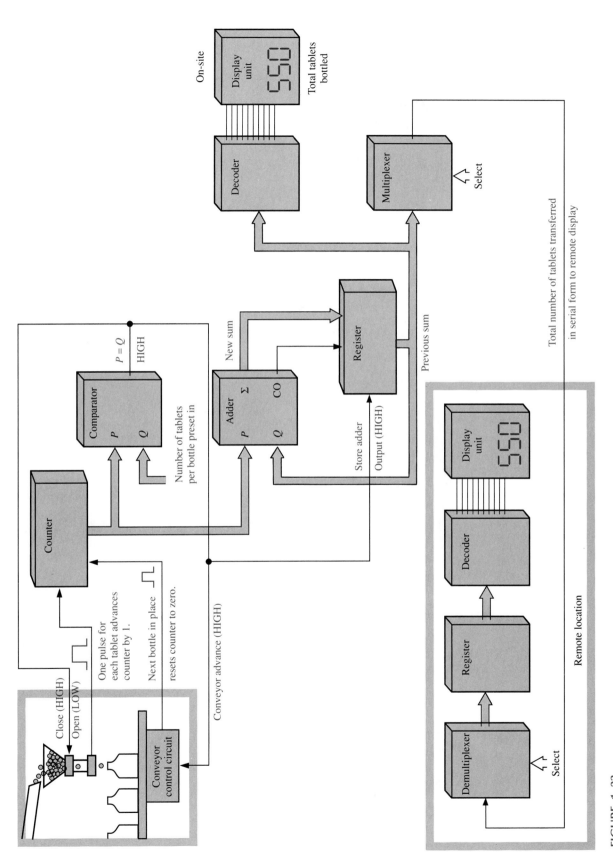

FIGURE 1–23
Example of the system concept.

21

on the Q input of the comparator. For example, suppose that each bottle is to hold fifty tablets. When the number in the counter reaches fifty, the $P = Q$ output of the comparator goes to its HIGH level, indicating that the bottle is full.

The HIGH output of the comparator immediately closes the valve in the neck of the funnel to stop the flow of tablets, and at the same time it activates the conveyor to move the next bottle into place under the funnel. When the next bottle is positioned properly under the neck of the funnel, the conveyor control circuit produces a pulse that resets the counter to zero. The $P = Q$ output of the comparator goes back to LOW, opening the funnel valve to restart the flow of tablets.

In the display portion of the system, the number in the counter is transferred in parallel to the P input of the adder. The Q input of the adder comes from a register that holds the total number of tablets bottled, up through the last bottle filled. For example, if ten bottles have been filled and each bottle holds fifty tablets, the register contains the binary representation for 500. Then, when the next bottle has been filled, the number 50 appears on the P input of the adder, and the number 500 is on the Q input. The adder produces a new sum of 550, which is stored in the register, replacing the previous sum of 500.

The content of the register is transferred in parallel to decoder A, which changes it from binary form to decimal form for display on a readout near the conveyor line. The content of the register is also transferred to a multiplexer and converted to serial form so that it can be transmitted along a single line to a remote location some distance away. (It is more economical to run a single line than to run several parallel lines when significant distances are involved.) At the other end of the single line, the serial information is converted back to parallel form by the demultiplexer and the register and is then decoded for display on the remote readout.

**SECTION 1–6
REVIEW**

Answers are found at the end of the chapter.
1. Explain the purpose of the comparator in the system in Figure 1–23.
2. What actions take place when the $P = Q$ output of the comparator goes to HIGH?
3. What is the content of the register at any given time?

DIGITAL INTEGRATED CIRCUITS

1–7

All the logic elements and functions that we have discussed—and many more—are available in integrated circuit (IC) form. Modern digital systems use ICs almost exclusively in their designs because of their small size, high reliability, low cost, and low power consumption. It is important to be able to recognize the IC packages and to know how the pin connections are numbered, as well as to be familiar with the way in which circuit complexities and circuit technologies determine the various IC classifications.

A monolithic integrated circuit is an electronic circuit that is constructed entirely on a single small chip of silicon. All the components that make up the circuit—transistors, diodes, resistors, and capacitors—are an integral part of that single chip.

FIGURE 1–24
Cutaway view of an IC package, showing the chip mounted inside, with leads to input and output pins.

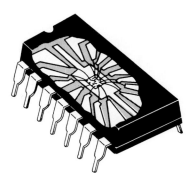

Figure 1–24 shows a cutaway view of one type of IC package, with the circuit chip shown within the package. Leads from the chip are connected to the package pins to allow input and output connections to the outside world.

Types of IC Packages

An IC package of the type shown in Figure 1–24 is called a dual in-line package *(DIP)*. This type of package has been the standard of the IC industry for many years and is still one of the most commonly used packages.

As integrated circuit technology advances, the circuits tend to become more complex and more compact. This development leads to smaller chips with more inputs and outputs, thus requiring more pins on a package. This situation has led to the development of an alternative packaging technique called surface-mount technology *(SMT)*, in which the IC packages are mounted right on the printed surface of the circuit board. The advantage of the SMT configurations is that the leads can be placed closer together than those on a DIP, because the DIP leads are designed to feed through holes in a circuit board. Those holes cannot be placed as close together as the connection pads for SMT devices. An SMT package is considerably smaller than the equivalent DIP because the pins are configured to allow closer spacing.

There are three basic types of SMT packages: the *SOIC* (small-outline IC), the *PLCC* (plastic-leaded chip carrier), and the *LCCC* (leadless ceramic chip carrier).

The SOIC resembles a small DIP with the leads bent out in a "gull wing" shape. The PLCC has leads that are turned up under the body in a J-like shape. The LCCC has no leads. Instead, it has metallic contacts molded into its ceramic body.

Figure 1–25 shows drawings of the four basic types of IC packages just discussed, with emphasis on their lead configurations and the way they are mounted on printed circuit (PC) boards.

Pin Numbering

Pin numbering for a 14-pin DIP and for a 20-pin chip carrier package are shown in Figure 1–26. Note the connections for the dc supply voltage (V_{CC}) and ground (GND). The other pins provide inputs and outputs for the logic circuits inside the packages. In some cases a few pins are not used.

Both DIP and SMT packages with more than 14 or 20 pins are used to house ICs of greater complexity.

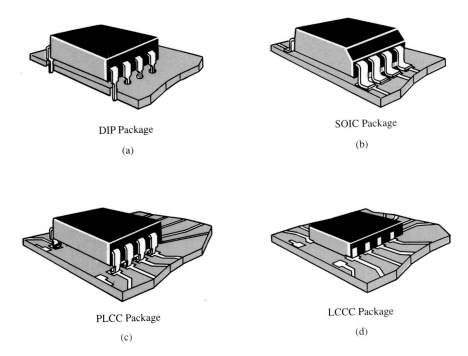

DIP Package

(a)

SOIC Package

(b)

PLCC Package

(c)

LCCC Package

(d)

FIGURE 1–25

Types of IC packages. The three main types of SMT packages (b, c and d) are surface mounted on the printed circuit board. The solder connections are made to pads on the top side (component side) of the board. The DIP (a) has leads that feed through to the other (bottom) side, where the solder connections are made. Cutaway views of printed circuit boards are shown, and drawings are not to scale.

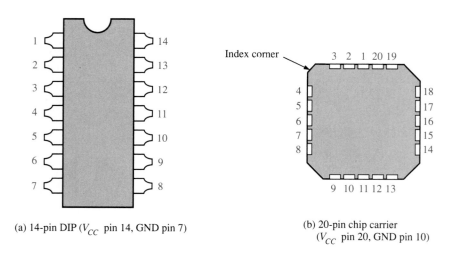

(a) 14-pin DIP (V_{CC} pin 14, GND pin 7)

(b) 20-pin chip carrier
(V_{CC} pin 20, GND pin 10)

FIGURE 1–26

Standard IC pin numbering. On the DIP, as viewed from the top, pin 1 is to the left of the notch. On the chip carrier, as viewed from the top, pin 1 is the middle pin to the right of the index corner. Pin numbers go in a counterclockwise sequence.

Circuit Complexity Classifications

Small-scale integration (SSI)   The least complex digital ICs are placed in the small-scale integration *(SSI)* category. These are circuits with up to 12 equivalent gate circuits on a single chip, and they include basic gate functions and flip-flops.

Medium-scale integration (MSI)   The next classification according to digital circuit complexity is called medium-scale integration *(MSI)*. These circuits have from 12 to 100 equivalent gates on a chip. They include the more complex logic functions, such as encoders, decoders, counters, registers, multiplexers, arithmetic circuits, small memories, and others.

Large-scale integration (LSI)   Circuits with complexities of 100 to 1000 equivalent gates per chip, including memories and some microprocessors, generally fall into the *LSI* category.

Very large-scale integration (VLSI)   Integrated circuits with complexities greater than 1000 equivalent gates per chip are generally considered VLSI. Very large memories, larger microprocessor systems, and single-chip computers are in this category.

The complexity figures stated here for SSI, MSI, LSI, and VLSI are generally accepted, but definitions may vary from one source to another.

The *microprocessor* is a VLSI device that can be programmed to perform arithmetic and logic operations and other functions in a prescribed sequence for the movement and processing of data. A typical microprocessor package is shown in Figure 1–27.

FIGURE 1–27
Microprocessor in a 40-pin dual in-line package. (Courtesy of Motorola)

The microprocessor is used as the central processing unit (CPU) in microcomputer systems, where it is connected with other ICs, such as memories and input/output interface circuits. Chapter 12 provides an introduction to microprocessors.

**SECTION 1–7
REVIEW**

Answers are found at the end of the chapter.
1. What is an integrated circuit?
2. Define the terms DIP, SMT, SSI, MSI, LSI, and VLSI.
3. Generally, in what classification does an IC with the following number of equivalent gates fall?
   (a) 75   **(b)** 500   **(c)** 10   **(d)** 10,000

DIGITAL TESTING AND TROUBLESHOOTING INSTRUMENTS

1–8

Troubleshooting *is the technique of systematically isolating, identifying, and correcting a fault in a circuit or system. A variety of special instruments are available for use in digital troubleshooting and testing. Some typical equipment is presented in this section.*

The Oscilloscope

Figure 1–28 shows a typical dual-channel oscilloscope. Pulse waveforms can be displayed on the screen, and parameters such as amplitude, rise and fall times, pulse width, period, and duty cycle can be measured. Also, abnormalities in the shape or characteristics of a pulse can be seen and analyzed. Two or more digital waveforms can be displayed simultaneously so that their time relationships can be determined and ana-

FIGURE 1–28
A typical oscilloscope. (Courtesy of Tektronix, Inc.)

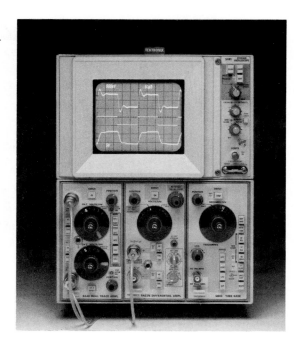

lyzed. Many specialized functions and performance levels are normally available with interchangeable plug-in modules. For these reasons the oscilloscope is considered to be one of the most versatile instruments for testing and troubleshooting.

The Logic Analyzer

Figure 1–29 shows a typical *logic analyzer*. This instrument is basically a multichannel oscilloscope with the ability to detect and display logic levels in several forms.

Most logic analyzers can display data in several formats. The timing diagram format displays several pulse waveforms with the proper time relationships. The bit format displays a bit pattern of 1s and 0s on the screen. Bit patterns from a functioning unit can be compared with the patterns of a faulty unit to detect errors.

A third format is a display using hexadecimal code. The hexadecimal code, as you will learn in the next chapter, is a convenient way to represent binary information. Some analyzers also provide an octal display and other code displays. Octal code is another way of representing binary information and is discussed in the next chapter.

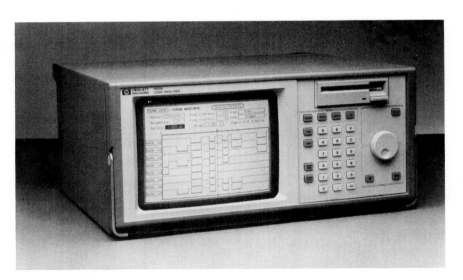

FIGURE 1–29
A typical logic analyzer. (Courtesy of Hewlett-Packard)

The Signature Analyzer

Another useful digital troubleshooting instrument is a *signature analyzer,* such as the one shown in Figure 1–30. Signature analysis is a troubleshooting technique that locates faults to the component level in microprocessor-based systems.

A signature analyzer reads a pattern of 1s and 0s at a given test point in the circuit under test, converts the pattern to hexadecimal code, and displays it. The hexadecimal code is called the *signature* for that particular point. By comparing the indicated signature with a known correct signature, a technician can identify a faulty device.

FIGURE 1–30
A typical signature analyzer. (Courtesy of Hewlett-Packard)

The Logic Probe

The *logic probe* provides a simple means of digital troubleshooting by detecting voltage levels or pulses at a given point. A typical logic probe is shown in Figure 1–31.

The logic probe uses a single lamp or multiple lamps to indicate the various states possible in a digital system, such as HIGHs, LOWs, single pulses, pulse trains, and open circuits. A typical indication format is as follows:

Lamp on:	HIGH
Lamp off:	LOW
Lamp dim:	Open or bad level
One flash:	Single pulse
Repetitive flashes:	Pulse train

Although the logic probe is a very useful instrument, it alone cannot solve all troubleshooting problems. Often it must be used in conjunction with oscilloscopes, logic analyzers, signature analyzers, and other types of probes.

The Current Tracer and Logic Pulser

These instruments are shown in Figure 1–32. Current tracing is a very effective troubleshooting technique in many cases. Isolating a bad element is particularly difficult when a given circuit node (point of connection) is stuck in one logic state and when several elements are common to the node. In such a situation, current tracing is very useful.

The hand-held *current tracer* has a one-lamp indicator that glows when its tip is held over a pulsing current path. The instrument can detect whether current is flowing at all and, most importantly, *where* the current is flowing. For instance, if a node is stuck in the LOW state because of a shorted input in one of the devices connected to the node, a very strong current exists between the circuit driving the node and the faulty component. For purposes of detection the current has to be pulsing; if it is not, a *logic pulser* can be used to force pulses on the node, and the current can then be traced to the bad component.

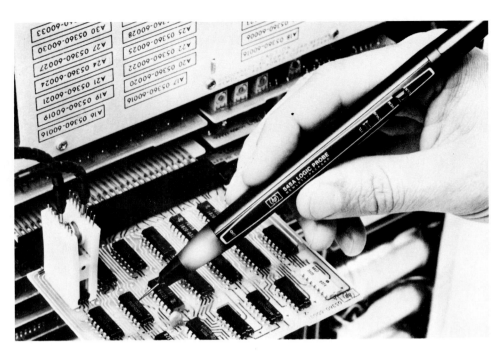

FIGURE 1–31
Using the logic probe. (Courtesy of Hewlett-Packard)

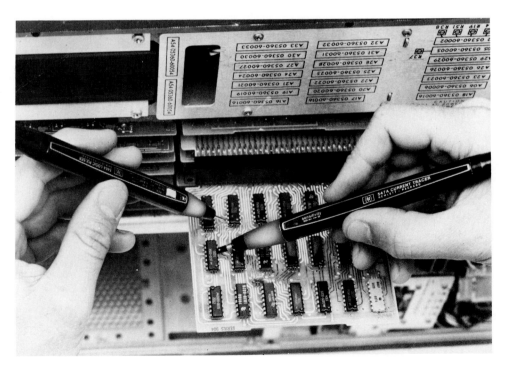

FIGURE 1–32
Using the logic pulser and the current tracer. (Courtesy of Hewlett-Packard)

FIGURE 1–33
Using the logic clip. (Courtesy of Hewlett-Packard)

The Logic Clip

A typical *logic clip* is shown in Figure 1–33. It is "clipped" to an integrated circuit, where it makes contact with each pin on the IC. The lamps on the clip then indicate the logic levels of the pins as follows:

Lamp on: HIGH
Lamp off: LOW
Lamp dim: Pulses

**SECTION 1–8
REVIEW**

Answers are found at the end of the chapter.
1. Which is the proper instrument for measuring the rise time of a pulse?
   (a) logic probe   (b) oscilloscope
   (c) logic analyzer   (d) pulser
2. List the conditions at a given point in a logic system that can be detected with a logic probe.

GLOSSARY

ADC Analog-to-digital conversion. The process of converting an analog signal to digital form.

Amplitude In a pulse waveform, the height or maximum value of the pulse as measured from the base line.

Analog Being continuous or having a continuous range of values, as opposed to having a set of discrete values.

AND A basic logic element in which a true (HIGH) output occurs if and only if all the input conditions are true (HIGH).

Binary   Having two values or states.

Bit   A binary digit, which can be either a 1 or a 0.

Boolean algebra   A mathematical system for formulating logical statements with symbols so that problems can be written and solved in a manner similar to ordinary algebra.

Circuit   An arrangement of electrical and/or electronic components interconnected in such a way as to perform a specified function.

Clock   The basic timing signal in a digital system.

Code   A combination of binary digits that represents such information as numbers, letters, and other symbols.

Comparator   A digital device that compares the magnitudes of two digital quantities and produces an output indicating the relationship of the quantities.

Computer   A digital electronic system that can be programmed to perform various tasks, such as mathematical computations, at very high speed and that can store large amounts of data.

Counter   A digital device capable of counting electronic events, such as pulses, by progressing through a sequence of binary states.

DAC   Digital-to-analog conversion. A process in which information in digital form is converted to analog form.

Data   Information in numeric, alphabetic, or other form.

Decoder   A digital device that converts coded information into another form.

Digital   Related to digits or discrete quantities. Having a set of discrete values as opposed to a continuous range of values.

DIP   Dual in-line package. A type of IC package whose leads must pass through holes to the other side of a circuit board.

Duty cycle   The ratio of pulse width to period.

Encoder   A digital device that converts information to a coded form.

Fall time   The time interval between the 90% point and the 10% point on the negative-going edge of a pulse.

Flip-flop   A basic memory element.

Frequency   The number of pulses in one second for a periodic waveform. The unit of frequency is the Hertz.

Gate   A logic circuit that performs a specified logic operation, such as AND or OR.

IC   Integrated circuit. A type of circuit in which all of the components are integrated on a single semiconductor chip of very small size.

Input   The signal or line going into a circuit. A signal that controls the operation of a circuit.

Inverter   A NOT circuit. A circuit that changes a HIGH to a LOW or vice versa.

Latch   A bistable digital device used for storing a bit.

LCCC   Leadless ceramic chip carrier. An SMT package that has metallic contacts molded into its body.

Leading edge   The first transition of a pulse.

Logic   In digital electronics, the decision-making capability of gate circuits, in which a HIGH represents a true statement and a LOW represents a false one.

LSI   Large-scale integration. A level of IC complexity in which there are 100–1000 equivalent gates per chip.

Magnitude   The size or value of a quantity.

Microprocessor   A VLSI device that can be programmed to perform arithmetic, logic, and other operations and to process data in a specified manner.

MSI   Medium-scale integration. A level of IC complexity in which there are 12–100 equivalent gates per chip.

Negative logic The system of logic in which the less positive voltage level represents a 1 and the more positive level represents a 0.

NOT A basic logic element that performs inversion.

OR A basic logic element in which a true (HIGH) output occurs when one or more of the input conditions are true (HIGH).

Parallel In digital systems, occurring simultaneously on several lines; transferring or processing several bits simultaneously.

Period The time required for a periodic waveform to repeat itself.

Periodic Repeating at fixed intervals.

PLCC Plastic-leaded chip carrier. An SMT package whose leads are turned up under its body in a J-like shape.

Positive logic The system of logic in which the more positive voltage level represents a 1 and the less positive level represents a 0.

Pulse A sudden change from one level to another, followed after a time by a sudden change back to the original level.

Pulse width The time interval between the 50% points on the leading and trailing edges of the pulse; the duration of the pulse.

Register A digital device capable of storing and shifting binary information.

Rise time The time required for the positive-going edge of a pulse to go from 10% of its full value to 90% of its full value.

Semiconductor A material used to construct such electronic devices as integrated circuits, transistors, and diodes. Silicon is the most common semiconductor.

Serial Having one element following another, as in a serial transfer of bits. Occurring, as pulses, in sequence rather than simultaneously.

SMT Surface-mount technology. An IC packaging technique in which the packages are smaller than DIPs and are mounted right on the printed surface of the circuit board.

SOIC Small-outline integrated circuit. An SMT package that resembles a small DIP but has its leads bent out in a ''gull wing'' shape.

SSI Small-scale integration. A level of IC complexity in which there are 12 or fewer equivalent gates per chip.

Storage The memory capability of a digital device. The process of retaining digital data for later use.

Trailing edge The second transition of a pulse.

Troubleshooting The technique of systematically isolating, identifying, and correcting a fault in a circuit or system.

VLSI Very large-scale integration. A level of IC complexity in which there are more than 1000 equivalent gates per chip.

SELF-TEST

Answers and solutions are found at the end of the book.

1. What are the binary digits?
2. What does a HIGH level represent in negative logic?
3. Define the term *bit*.
4. For a negative pulse, the leading edge corresponds to the _____ edge.
5. For a positive pulse, the trailing edge corresponds to the _____ edge.
6. Explain the difference between periodic and nonperiodic pulse waveforms.

7. Define *period*.
8. Define *frequency* and specify its unit.
9. For a given period, when the pulse width is increased, does the duty cycle increase or decrease?
10. A NOT circuit is also called an _____.
11. Explain the difference between an AND gate and an OR gate.
12. Define one characteristic that distinguishes a flip-flop from a gate.
13. Briefly describe the purpose of each of the eight functions of digital logic discussed in this chapter.
14. Define the term *DIP*.
15. Define the term *SMT*.
16. Define the term *troubleshooting*.
17. List at least five types of digital testing and troubleshooting instruments.

PROBLEMS

Answers to odd-numbered problems are found at the end of the book.

Section 1–1 Digital Electronics

1–1 Name five men who were instrumental in the development of the field of electronics.

1–2 Name a few things found in everyday life in which digital electronics is used.

Section 1–2 Digital and Analog Quantities

1–3 Explain the basic difference between analog and digital quantities.

1–4 Name an analog quantity other than temperature and sound.

Section 1–3 Logic Levels and Pulse Waveforms

1–5 Define the sequence of bits (1s and 0s) represented by each of the following sequences of levels in a positive logic system:
   (a) HIGH, HIGH, LOW, HIGH, LOW, LOW, LOW, HIGH
   (b) LOW, LOW, LOW, HIGH, LOW, HIGH, LOW, HIGH, LOW

1–6 Define the sequence of bits (1s and 0s) represented by each of the following sequences of levels in a negative logic system:
   (a) HIGH, LOW, HIGH, HIGH, HIGH, LOW, HIGH
   (b) HIGH, HIGH, HIGH, LOW, HIGH, LOW, LOW, HIGH

1–7 For the pulse shown in Figure 1–34, determine the following:
   (a) rise time   **(b)** fall time
   (c) pulse width   **(d)** amplitude

FIGURE 1–34

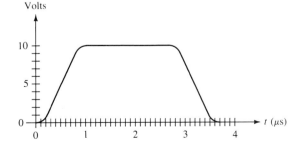

1–8   Determine the period of the waveform in Figure 1–35.

1–9   What is the frequency of the pulse waveform in Figure 1–35?

1–10   Is the pulse waveform in Figure 1–35 periodic or nonperiodic?

1–11   Determine the duty cycle in Figure 1–35.

1–12   Determine the bit sequence (positive logic) represented by the waveform in Figure 1–36. A bit time is 1 µs in this case.

1–13   What is the total serial transfer time for the eight bits in Figure 1–36? What is the total parallel transfer time?

FIGURE 1–35

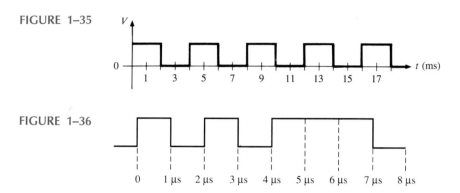

FIGURE 1–36

Section 1–4   Elements of Digital Logic

1–14   A basic logic element requires HIGHs on all its inputs to make the output HIGH. What type of logic circuit is it?

1–15   A basic two-input logic element has a HIGH on one input and a LOW on the other input, and the output is LOW. Identify the element.

1–16   A basic two-input logic element has a HIGH on one input and a LOW on the other input, and the output is HIGH. What type of logic element is it?

Section 1–5   Functions of Digital Logic

1–17   Name the function of each block in Figure 1–37.

FIGURE 1–37

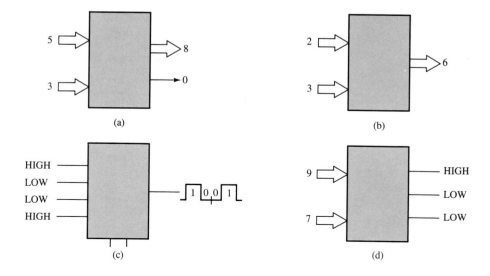

1–18 A pulse waveform with a frequency of 10 kHz is applied to the input of a divide-by-5 counter. What is the frequency of the output waveform?

1–19 Consider a register that can store eight bits. Assume that it has been reset so that it contains zeros in all positions. If we transfer four alternating bits (0101) serially into the register, beginning with a 1 and shifting to the right, what will the total content of the register be?

Section 1–6 A System Application

1–20 Define the term *system*.

1–21 In the system depicted in Figure 1–23, why are the multiplexer and demultiplexer necessary?

1–22 What action can be taken to change the number of tablets per bottle in the system of Figure 1–23?

Section 1–7 Digital Integrated Circuits

1–23 An integrated circuit chip has a complexity of 200 equivalent gates. How is it classified?

1–24 Explain the main difference between the DIP and SMT packages.

1–25 Label the pin numbers on the packages in Figure 1–38. Top views are shown.

FIGURE 1–38

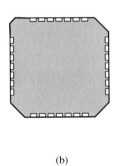

(a)   (b)

TBLS ▶

Section 1–8 Digital Testing and Troubleshooting Instruments

1–26 A pulse is displayed on the screen of an oscilloscope, and you measure the base line as 1 V and the top of the pulse as 8 V. What is the amplitude?

1–27 A logic probe is applied to a contact point on an IC that is operating in a system. The lamp on the probe flashes repeatedly. What does this indicate?

ANSWERS TO SECTION REVIEWS

Section 1–1 Review
1. digital computer, integrated circuit
2. computers, communications, medicine, automation, and entertainment

Section 1–2 Review
1. having a continuous set of values
2. having a discrete set of values

Section 1–3 Review
1. 4 V   2. 25%   3. 1.6 μs   4. speed

Section 1–4   Review

1. AND gate, OR gate, inverter, flip-flop
2. (c) either AND or OR gate
3. AND gate

Section 1–5   Review

1. comparison, arithmetic, encoding, decoding, counting, storage, multiplexing, and demultiplexing
2. **(a)** comparison   **(b)** comparison   **(c)** addition   **(d)** counting

Section 1–6   Review

1. The comparator determines when the tablet count reaches the preset number of tablets per bottle.
2. (1) The dispenser valve is closed. (2) The next bottle is positioned under the dispenser by advancement of the conveyor belt. (3) The new sum is stored in the register.
3. the total number of tablets bottled

Section 1–7   Review

1. a circuit with all components integrated on a single silicon chip
2. DIP: dual in-line package
   SMT: surface mount technology
   SSI: small-scale integration (up to 12 equivalent gates)
   MSI: medium-scale integration (12 to 100 equivalent gates)
   LSI: large-scale integration (100 to 1000 equivalent gates)
   VLSI: very large-scale integration (more than 1000 equivalent gates)
3. **(a)** MSI   **(b)** LSI   **(c)** SSI   **(d)** VLSI

Section 1–8   Review

1. (b) oscilloscope
2. HIGH, LOW, open, single pulse, pulse train

The binary number system and digital codes are fundamental to digital electronics. In this chapter the binary number system and its relationship to other number systems, such as the decimal system, the octal system, and the hexadecimal system, are studied. Arithmetic operations with binary numbers are covered, to provide a basis for understanding how computers and other digital systems operate. The following digital codes are introduced: BCD, Gray, Excess-3, and ASCII.

After completing this chapter, you will be able to

☐ count in the binary number system.

☐ convert from decimal form to binary form and from binary form to decimal form.

☐ add, subtract, multiply, and divide binary numbers.

☐ determine the 1's and 2's complements of a binary number.

☐ express signed numbers in binary form.

☐ carry out arithmetic operations with signed binary numbers.

☐ convert between the binary and octal number systems.

☐ convert between the binary and hexadecimal number systems.

☐ add numbers in hexadecimal form.

☐ express decimal numbers in binary coded decimal (BCD) form.

☐ add BCD numbers.

☐ convert between the binary system and the Gray code.

☐ express quantities in Excess-3 code.

☐ interpret the American Standard Code for Information Interchange (ASCII).

2

Number Systems and Codes

DECIMAL NUMBERS

2–1

We are all familiar with the decimal number system and use decimal numbers every day. Although decimal numbers are commonplace, many people do not understand their weighted structure. In this section we will review the structure of decimal numbers. This review will help you more easily understand the binary number system, which is so important in digital electronics.

Because human anatomy is characterized by four fingers and a thumb on each hand, it was only natural that our method of counting involve the use of ten digits, that is, a system with a base of ten. Each of the ten decimal digits, 0 through 9, represents a certain quantity. The ten symbols (digits) do not limit us to expressing only ten different quantities, because we use the various digits in appropriate positions within a number to indicate the magnitude of the quantity. We can express quantities up through nine before we run out of digits; if we wish to express a quantity greater than nine, we use two or more digits, and the position of each digit within the number tells us the magnitude it represents. If, for instance, we wish to express the quantity twenty-three, we use (by their respective positions in the number) the digit 2 to represent the quantity twenty and the digit 3 to represent the quantity three. Therefore, the position of each of the digits in the decimal number indicates the magnitude of the quantity represented and can be assigned a *weight*. The weights are powers of ten that increase from right to left, beginning with $10^0 = 1$.

The value of a decimal number is the sum of the digits after each has been multiplied by its weight. The following examples will illustrate.

EXAMPLE 2–1

$$23 = 2 \times 10^1 + 3 \times 10^0$$
$$= 2 \times 10 + 3 \times 1 = 20 + 3$$

The digit 2 has a weight of 10 (10^1), as indicated by its position, and the digit 3 has a weight of 1 (10^0), as indicated by its position.

EXAMPLE 2–2

$$568 = 5 \times 10^2 + 6 \times 10^1 + 8 \times 10^0$$
$$= 5 \times 100 + 6 \times 10 + 8 \times 1 = 500 + 60 + 8$$

The digit 5 has a weight of 100 (10^2), the digit 6 has a weight of 10 (10^1), and the digit 8 has a weight of 1 (10^0).

SECTION 2–1 REVIEW

Answers are found at the end of the chapter.
1. What weight does the digit 7 have in each of the following numbers?
   (a) 1370   (b) 6725   (c) 7051   (d) 75,898
2. Express each of the following decimal numbers as a sum of the products obtained by multiplying each digit by its appropriate weight:
   (a) 51   (b) 137   (c) 1492   (d) 10,658

BINARY NUMBERS

2–2

The **binary number system** *is simply another way to count. It is less complicated than the decimal system because it is composed of only two digits. It may seem more difficult at first because it is unfamiliar to you.*

Just as the decimal system with its ten digits is a base-ten system, the binary system with its two digits is a base-two system. The two binary digits (bits) are 1 and 0. The position of a 1 or 0 in a binary number indicates its weight, or value within the number, just as the position of a decimal digit determines the magnitude of that digit. The weight of each successively higher position (to the left) in a binary number is the next higher power of two.

Counting in Binary

To learn to count in the binary system, first look at how we count in the decimal system. We start at 0 and count up to 9 before we run out of digits. We then start another digit position (to the left) and continue counting 10 through 99. At this point we have exhausted all two-digit combinations, so a third digit is needed to count from 100 through 999.

A comparable situation occurs when we count in binary, except that we have only two digits. We begin counting, 0, 1. At this point we have used both digits, so we include another digit position and continue, 10, 11. We have now exhausted all combinations of two digits, so a third is required. With three digits we can continue to count, 100, 101, 110, and 111. Now we need a fourth digit to continue, and so on. A binary count of zero through thirty-one is shown in Table 2–1 on page 40.

An easy way to remember how to write a binary sequence as in Table 2–1 for a five-bit example is as follows:

1. The right-most position in the binary number begins with a 0 and changes with each number.
2. The next position begins with two 0s and changes after every two numbers.
3. The next position begins with four 0s and changes after every four numbers.
4. The next position begins with eight 0s and changes after every eight numbers.
5. The next position begins with sixteen 0s and changes after every sixteen numbers.

As you have seen, it takes at least five bits to count from zero to thirty-one. The following formula tells us how high we can count with n bits, beginning with zero:

$$\text{Highest decimal number} = 2^n - 1 \tag{2–1}$$

For instance, with two bits we can count from zero to three.

$$2^2 - 1 = 4 - 1 = 3$$

With four bits, we can count from zero to fifteen.

$$2^4 - 1 = 16 - 1 = 15$$

A table of powers of two (2^n) is given in Appendix C.

TABLE 2–1

Count	Decimal Number	Binary Number
zero	0	00000
one	1	00001
two	2	00010
three	3	00011
four	4	00100
five	5	00101
six	6	00110
seven	7	00111
eight	8	01000
nine	9	01001
ten	10	01010
eleven	11	01011
twelve	12	01100
thirteen	13	01101
fourteen	14	01110
fifteen	15	01111
sixteen	16	10000
seventeen	17	10001
eighteen	18	10010
nineteen	19	10011
twenty	20	10100
twenty-one	21	10101
twenty-two	22	10110
twenty-three	23	10111
twenty-four	24	11000
twenty-five	25	11001
twenty-six	26	11010
twenty-seven	27	11011
twenty-eight	28	11100
twenty-nine	29	11101
thirty	30	11110
thirty-one	31	11111

Binary-to-Decimal Conversion

A binary number is a weighted number, as mentioned previously. The value of a given binary number, expressed as its decimal equivalent, can be determined by multiplying each bit by its weight and adding the products. The right-most bit is the least significant bit (*LSB*) in a binary number and has a weight of $2^0 = 1$. The weights increase from right to left by a power of two for each bit. The left-most bit is the most significant bit (*MSB);* its weight depends on the size of the binary number. The method of converting a binary number to decimal form is illustrated by the following example.

EXAMPLE 2–3

Convert the binary number 1101101 to decimal form.

Solution

$$\begin{array}{llllllll}
\text{Binary weight:} & 2^6 & 2^5 & 2^4 & 2^3 & 2^2 & 2^1 & 2^0 \\
\text{Weight value:} & 64 & 32 & 16 & 8 & 4 & 2 & 1 \\
\text{Binary number:} & 1 & 1 & 0 & 1 & 1 & 0 & 1
\end{array}$$

$$
\begin{aligned}
& 1 \times 64 + 1 \times 32 + 0 \times 16 + 1 \times 8 + 1 \times 4 + 0 \times 2 + 1 \times 1 \\
= \ & 64 \ + \ 32 \ + \ 0 \ + \ 8 \ + \ 4 \ + \ 0 \ + \ 1 \\
= \ & 109
\end{aligned}
$$

This is the equivalent decimal value of the binary number. Notice that the weights corresponding to 1s in the binary number are added. The 0s are ignored.

The binary numbers so far in this book have been whole numbers. Fractional numbers can also be represented in binary by placing bits to the right of the *binary point,* just as fractional decimal digits are placed to the right of the decimal point.

The weights of the digits in a binary number are

$$2^{n-1} \ . \ . \ . \ 2^3 2^2 2^1 2^0 . 2^{-1} 2^{-2} \ . \ . \ . \ 2^{-n}$$
$$\underset{\text{binary point}}{\uparrow}$$

where n is the number of bits from the binary point. Thus, all the bits to the left of the binary point have weights that are positive powers of two, as previously discussed. All bits to the right of the binary point have weights that are negative powers of two, or fractional weights ($2^{-1} = 1/2^1 = 1/2 = 0.5$, etc.), as illustrated in the following example.

EXAMPLE 2–4

Determine the decimal value of the fractional binary number 0.1011.

Solution First determine the weight of each bit, and then sum the weights times the bits.

$$\begin{array}{lllll}
\text{Binary weight:} & 2^{-1} & 2^{-2} & 2^{-3} & 2^{-4} \\
\text{Weight value:} & 0.5 & 0.25 & 0.125 & 0.0625 \\
\text{Fractional binary number:} & 1 & 0 & 1 & 1
\end{array}$$

$$
\begin{aligned}
& 1 \times 0.5 + 0 \times 0.25 + 1 \times 0.125 + 1 \times 0.0625 \\
= \ & 0.5 \ + \ 0 \ + \ 0.125 \ + \ 0.0625 \\
= \ & 0.6875
\end{aligned}
$$

Remember that to determine the decimal value of a binary number, fractional or whole, we simply add the weights of the 1s and ignore the 0s, because the product of a 0 and its weight is 0.

Another method of evaluating a binary fraction is to determine the whole-number decimal value of the bits and divide by the total possible combinations of the number of bits appearing in the fraction. For instance, for the binary fraction in Example 2–4, we could use the following procedure:

1. If we ignore the binary point, the value of 1011_2 is 11 (eleven) in decimal.
2. For four bits, there are $2^4 = 16$ possible combinations.
3. Dividing, we obtain $11/16 = 0.6875$.

The following examples illustrate the evaluation of binary numbers in terms of their equivalent decimal values.

EXAMPLE 2–5

Determine the decimal value of the binary number 11101.011.

Solution

$$(16 + 8 + 4 + 1).(0.25 + 0.125) = 29.375$$

Using the alternative method to evaluate the fraction, we have $3/8 = 0.375$ (the same result).

EXAMPLE 2–6

Determine the decimal value of the binary number 110101.11.

Solution

$$(32 + 16 + 4 + 1).(0.5 + 0.25) = 53.75$$

Using the alternative method to evaluate the fraction gives us $3/4 = 0.75$ (the same result).

SECTION 2–2 REVIEW

Answers are found at the end of the chapter.
1. What is the weight of the 1 in the binary number 10000?
2. Convert each of the following binary numbers to decimal:
      (a) 1010   **(b)** 11010   **(c)** 10111101   **(d)** 110.101

DECIMAL-TO-BINARY CONVERSION

2–3

In Section 2–2 we discussed how to determine the equivalent decimal value of a binary number. Now you will learn two ways of converting from a decimal number to a binary number.

Sum-of-Weights Method

One way to find the binary number that is equivalent to a given decimal number is to determine the set of binary weight values whose sum is equal to the decimal number.

For instance, the decimal number 9 can be expressed as the sum of binary weights as follows:

$$9 = 8 + 1 = 2^3 + 2^0$$

By placing 1s in the appropriate weight positions, 2^3 and 2^0, and 0s in the 2^2 and 2^1 positions, we have the binary number for decimal 9:

$$
\begin{array}{cccc}
2^3 & 2^2 & 2^1 & 2^0 \\
1 & 0 & 0 & 1
\end{array} \quad \text{binary nine}
$$

EXAMPLE 2–7

Convert the following decimal numbers to binary form:
(a) 12 **(b)** 25 **(c)** 58 **(d)** 82

Solutions
(a) $12 = 8 + 4 = 2^3 + 2^2$ $\longrightarrow$ 1100
(b) $25 = 16 + 8 + 1 = 2^4 + 2^3 + 2^0$ $\longrightarrow$ 11001
(c) $58 = 32 + 16 + 8 + 2 = 2^5 + 2^4 + 2^3 + 2^1$ $\longrightarrow$ 111010
(d) $82 = 64 + 16 + 2 = 2^6 + 2^4 + 2^1$ $\longrightarrow$ 1010010

Repeated Division-by-2 Method

A more systematic method of converting from decimal to binary is the *repeated division-by-2* process. For example, to convert the decimal number 12 to binary, we begin by dividing 12 by 2. Then we divide each resulting quotient by 2 until there is a 0 quotient. The remainders generated by each division form the binary number. The first remainder to be produced is the least significant bit (LSB) in the binary number, and the last remainder to be produced is the most significant bit (MSB). This procedure is shown in the following steps:

EXAMPLE 2–8

Convert the following decimal numbers to binary:
(a) 19   **(b)** 45

Solutions

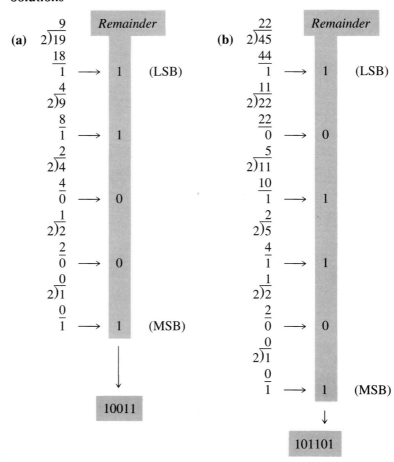

Converting Decimal Fractions to Binary

The previous examples demonstrated whole-number conversions. Now fractional conversions will be examined.

The sum-of-weights method can be applied to fractional decimal numbers, as shown in the following example:

$$0.625_{10} = 0.5 + 0.125 = 2^{-1} + 2^{-3} = 0.101_2$$

There is a 1 in the 2^{-1} position, a 0 in the 2^{-2} position, and a 1 in the 2^{-3} position.

As you have seen, decimal whole numbers can also be converted to binary by repeated division by 2. Decimal fractions can be converted to binary by repeated *multiplication* by 2. For example, to convert the decimal fraction 0.3125 to binary, we begin by multiplying 0.3125 by 2 and then multiplying each resulting fractional part of the product by 2 until the fractional product is zero. The carried digits, or *carries*, generated by the multiplications form the binary number. The first carry produced is the MSB, and the last carry is the LSB. This procedure is shown in the following steps:

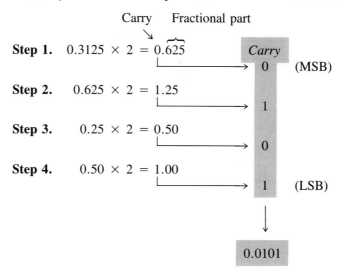

Carry   Fractional part

Step 1. $0.3125 \times 2 = 0.625$ *Carry*   0   (MSB)

Step 2. $0.625 \times 2 = 1.25$ 1

Step 3. $0.25 \times 2 = 0.50$ 0

Step 4. $0.50 \times 2 = 1.00$ 1   (LSB)

0.0101

SECTION 2–3
REVIEW

Answers are found at the end of the chapter.
1. Convert each decimal number to binary by using the sum-of-weights method.
    (a) 23   **(b)** 57   **(c)** 45.5
2. Convert each decimal number to binary by using the repeated division-by-2 method (repeated multiplication-by-2 for fractions).
    (a) 14   **(b)** 21   **(c)** 0.375

BINARY ARITHMETIC

2–4

Binary arithmetic is fundamental to all digital computers and to many other types of digital systems. To understand digital systems, you must know the basics of binary addition, subtraction, multiplication, and division. This section provides an introduction that will be expanded in later sections.

Addition

There are four basic rules for adding binary digits:

$$0 + 0 = 0$$
$$0 + 1 = 1$$
$$1 + 0 = 1$$
$$1 + 1 = 10 \quad \text{(0 with a carry of 1)}$$

Notice that three of the addition rules result in a single bit and that the addition of two 1s yields a binary two (10). When binary numbers are added, the last condition creates a sum of 0 in a given column and a carry of 1 over to the next higher column, as illustrated in the following addition:

$$
\begin{array}{r}
11\\
011\\
+001\\
\hline
100
\end{array}
$$

In the right column, $1 + 1 = 0$ with a carry of 1 to the next column to the left. In the middle column, $1 + 1 + 0 = 0$ with a carry of 1 to the next column to the left. In the left column, $1 + 0 + 0 = 1$

When there is a carry, we have a situation in which three bits are being added (a bit in each of the two numbers and a carry bit). This situation is illustrated as follows:

Carry bits—
$$
\begin{array}{lll}
1 + 0 + 0 = 01 & (1 \text{ with a carry of } 0)\\
1 + 1 + 0 = 10 & (0 \text{ with a carry of } 1)\\
1 + 0 + 1 = 10 & (0 \text{ with a carry of } 1)\\
1 + 1 + 1 = 11 & (1 \text{ with a carry of } 1)
\end{array}
$$

The following example will illustrate binary addition, with the equivalent decimal addition also shown for reference.

EXAMPLE 2–9

(a)
$$
\begin{array}{r}
11\\
+11\\
\hline
110
\end{array}
\qquad
\begin{array}{r}
3\\
+3\\
\hline
6
\end{array}
$$

(b)
$$
\begin{array}{r}
100\\
+10\\
\hline
110
\end{array}
\qquad
\begin{array}{r}
4\\
+2\\
\hline
6
\end{array}
$$

(c)
$$
\begin{array}{r}
111\\
+11\\
\hline
1010
\end{array}
\qquad
\begin{array}{r}
7\\
+3\\
\hline
10
\end{array}
$$

(d)
$$
\begin{array}{r}
110\\
+100\\
\hline
1010
\end{array}
\qquad
\begin{array}{r}
6\\
+4\\
\hline
10
\end{array}
$$

(e)
$$
\begin{array}{r}
1111\\
+1100\\
\hline
11011
\end{array}
\qquad
\begin{array}{r}
15\\
+12\\
\hline
27
\end{array}
$$

(f)
$$
\begin{array}{r}
11100\\
+10011\\
\hline
101111
\end{array}
\qquad
\begin{array}{r}
28\\
+19\\
\hline
47
\end{array}
$$

Subtraction

There are four basic rules for subtracting binary digits:

$$
\begin{array}{ll}
0 - 0 = 0 &\\
1 - 1 = 0 &\\
1 - 0 = 1 &\\
10 - 1 = 1 & (0{-}1 \text{ with a borrow of } 1)
\end{array}
$$

When subtracting numbers, we sometimes have to borrow from the next higher column. A borrow is required in binary only when we try to subtract a 1 from a 0. In this case, when a 1 is borrowed from the next higher column, a 10_2 is created in the column being subtracted, and the last of the four basic rules just listed must be applied. The following examples illustrate binary subtraction, with the equivalent decimal subtraction shown.

EXAMPLE 2–10

(a)
$$\begin{array}{rr} 11 & 3 \\ -01 & -1 \\ \hline 10 & 2 \end{array}$$

(b)
$$\begin{array}{rr} 11 & 3 \\ -10 & -2 \\ \hline 01 & 1 \end{array}$$

No borrows were required in this example. The binary number 01 is the same as 1.

EXAMPLE 2–11

$$\begin{array}{rr} 101 & 5 \\ -011 & -3 \\ \hline 010 & 2 \end{array}$$

Let us examine exactly what was done to subtract the two binary numbers.

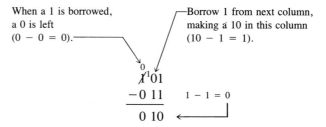

When a 1 is borrowed, a 0 is left $(0 - 0 = 0)$.

Borrow 1 from next column, making a 10 in this column $(10 - 1 = 1)$.

$$\begin{array}{r} \overset{0}{\cancel{1}}{}^{1}01 \\ -0\,11 \qquad 1 - 1 = 0 \\ \hline 0\,10 \end{array}$$

Multiplication

The following are four basic rules for multiplying binary digits:

$$0 \times 0 = 0$$
$$0 \times 1 = 0$$
$$1 \times 0 = 0$$
$$1 \times 1 = 1$$

Multiplication is performed with binary numbers in the same manner as with decimal numbers. It involves forming partial products, shifting each successive partial product left one place, and then adding all the partial products. A few examples will illustrate the procedure, with the equivalent decimal multiplication shown for reference.

EXAMPLE 2–12

(a)
$$\begin{array}{rr} 11 & 3 \\ \times 1 & \times 1 \\ \hline 11 & 3 \end{array}$$

(b)
$$\begin{array}{rr} 11 & 3 \\ \times 11 & \times 3 \\ \hline 11 & 9 \end{array}$$
partial products $\begin{cases} 11 \\ 11 \end{cases}$
$$\overline{1001}$$

(c)
$$\begin{array}{r} 111 \\ \times 101 \\ \hline 111 \\ 000 \\ 111 \\ \hline 100011 \end{array}$$
partial products

$$\begin{array}{r} 7 \\ \times 5 \\ \hline 35 \end{array}$$

(d)
$$\begin{array}{r} 1011 \\ \times 1001 \\ \hline 1011 \\ 0000 \\ 0000 \\ 1011 \\ \hline 1100011 \end{array}$$
partial products

$$\begin{array}{r} 11 \\ \times 9 \\ \hline 99 \end{array}$$

Division

Division in binary follows the same procedure as division in decimal, as the following example illustrates.

EXAMPLE 2–13

(a)
$$11\overline{)110}$$
$$\underline{11}$$
$$000$$

$$3\overline{)6}$$
$$\underline{6}$$
$$0$$

(b)
$$10\overline{)110}$$
$$\underline{10}$$
$$10$$
$$\underline{10}$$
$$00$$

$$2\overline{)6}$$
$$\underline{6}$$
$$0$$

(c)
$$100\overline{)1100}$$
$$\underline{100}$$
$$100$$
$$\underline{100}$$
$$000$$

$$4\overline{)12}$$
$$\underline{12}$$
$$0$$

(d)
$$110\overline{)1111.0}$$
$$\underline{110}$$
$$11\ 0$$
$$\underline{11\ 0}$$
$$00\ 0$$

$$6\overline{)15.0}$$
$$\underline{12}$$
$$3\ 0$$
$$\underline{3\ 0}$$
$$0$$

SECTION 2–4 REVIEW

Answers are found at the end of the chapter.
1. Perform the following binary additions:
   (a) 1101 + 1010   **(b)** 10111 + 01101
2. Perform the following binary subtractions:
   (a) 1101 − 0100   **(b)** 1001 − 0111
3. Perform the indicated binary operation:
   (a) 110 × 111   **(b)** 1100 ÷ 011

1'S AND 2'S COMPLEMENTS

2–5

The 1's complement and the 2's complement of a binary number are important because they permit the representation of negative numbers in digital systems. Although applications of both of these methods are found in practice, 2's complement arithmetic is the predominant method used in computers when operating with negative numbers. It is essential that you understand both of these methods.

Obtaining the 1's Complement of a Binary Number

The 1's complement of a binary number is found by simply changing all 1s to 0s and all 0s to 1s, as illustrated by a few examples:

Binary Number	1's Complement
10101	01010
10111	01000
111100	000011
11011011	00100100

1's Complement Subtraction

Subtraction of binary numbers can be accomplished by the direct method described in Section 2–4 or by the 1's complement method, which allows us to subtract using only addition.

For subtracting a smaller number from a larger number, the 1's complement method is as follows:

1. Determine the 1's complement of the smaller number.
2. Add the 1's complement to the larger number.
3. Remove the final carry and add it to the result. This step is called an *end-around carry*.

EXAMPLE 2–14

Subtract 10011 from 11001 by using the 1's complement method. Show direct subtraction for comparison.

Solution

Direct Subtraction	1's Complement Method	
11001	11001	
$-$ 10011	$+$ 01100	1's complement of 10011
00110	①00101	
	⤷→ +1	Adding end-around carry
	00110	Final answer

For subtracting a larger number from a smaller one, the 1's complement method is as follows:

1. Determine the 1's complement of the larger number.
2. Add the 1's complement to the smaller number.
3. There is no carry. The result has the opposite sign from the answer and is the 1's complement of the answer.
4. Change the sign and take the 1's complement of the result to get the final answer.

EXAMPLE 2–15

Subtract 1101 from 1001 by using the 1's complement method. Show direct subtraction for comparison.

Solution

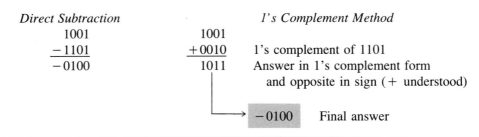

Direct Subtraction
```
  1001
− 1101
─────
− 0100
```

1's Complement Method
```
  1001
+ 0010    1's complement of 1101
─────
  1011    Answer in 1's complement form
             and opposite in sign (+ understood)
```

```
  → − 0100    Final answer
```

The 1's complement method is useful in arithmetic logic circuits because subtraction can be accomplished with an adder. Also, the 1's complement of a number is easily obtained by inverting each bit in the number.

Obtaining the 2's Complement of a Binary Number

The 2's complement of a binary number is found by adding 1 to the 1's complement. The following example shows how this is done.

EXAMPLE 2–16

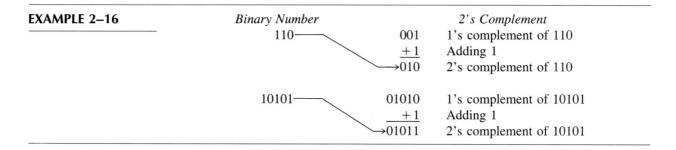

Binary Number
110
10101

2's Complement
```
  001     1's complement of 110
+ 1       Adding 1
────
→010     2's complement of 110
```

```
  01010   1's complement of 10101
+ 1       Adding 1
─────
→01011   2's complement of 10101
```

An alternative method of obtaining the 2's complement is demonstrated as follows for the binary number 1101011101000:

1. Start at the right and write the bits as they are up to and including the first 1.
2. Take the 1's complements of the remaining bits.

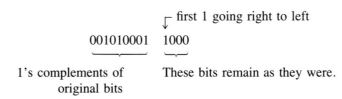

┌ first 1 going right to left
↓
001010001 1000

1's complements of   These bits remain as they were.
original bits

2's Complement Subtraction

For subtracting a smaller number from a larger one, the 2's complement method is applied as follows:

1. Determine the 2's complement of the smaller number.
2. Add the 2's complement to the larger number.
3. Discard the final carry (there is always a carry in this case).

EXAMPLE 2–17

Subtract 1011 from 1100 by using the 2's complement method. Show direct subtraction for comparison.

Solution

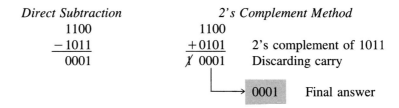

For subtracting a larger number from a smaller one, the 2's complement method is as follows:

1. Determine the 2's complement of the larger number.
2. Add the 2's complement to the smaller number.
3. There is no carry from the left-most column. The result is in 2's complement form and is negative.
4. To get the final answer, take the 2's complement of the result, and change the sign.

EXAMPLE 2–18

Subtract 11100 from 10011 using the 2's complement method.

Solution

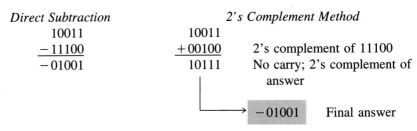

Note: In computers, negative numbers are stored in 2's complement form.

At this point, both the 1's and 2's complement methods of subtraction may seem excessively complex when compared with direct subtraction. However, as mentioned before, they both have distinct advantages when implemented with logic circuits, because they allow subtraction to be done by using only addition. Both the 1's and the 2's complements of a binary number are relatively easy to accomplish with logic circuits; the 2's complement has an advantage over the 1's complement in that an end-around carry operation does not have to be performed. You will see later how arithmetic operations are implemented with logic circuits.

SECTION 2–5 REVIEW

Answers are found at the end of the chapter.
1. Determine the 1's complement of each binary number.
   (a) 11010 **(b)** 10111 **(c)** 001101
2. Perform the following subtractions by using the 1's complement method:
   (a) 1101 − 0111 **(b)** 10110 − 11011
3. Determine the 2's complement of each binary number.
   (a) 10111 **(b)** 11111 **(c)** 010001
4. Perform the following subtractions by using the 2's complement method:
   (a) 1110 − 1001 **(b)** 1001 − 1110

BINARY REPRESENTATION OF SIGNED NUMBERS

2–6

The capability of handling both positive and negative numbers is a requirement of any digital system that does arithmetic operations. A signed number consists of both sign and magnitude information. The sign indicates whether a number is positive or negative, and the magnitude is the value of the number.

Sign and Magnitude

In binary systems the sign is represented by including an additional bit along with the magnitude bits. Conventionally, a 0 represents a positive sign and a 1 represents a negative sign. For example, in an eight-bit number the left-most bit is the sign bit, and the remaining seven bits are magnitude bits. This binary sign and magnitude representation is as follows for decimal values +107 and −20:

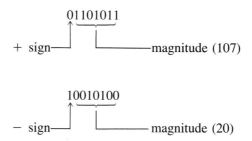

A *byte,* that is, eight bits, is a very common unit of information in computer systems. With an eight-bit sign-and-magnitude binary number, a range of values from 01111111 = +127 to 11111111 = −127 can be represented. Two or more bytes can be used to represent larger numbers. For example, a group of 16 bits includes one sign bit and 15 magnitude bits. With this format, a range of numbers from $0111111111111111 = +32,767$ to $1111111111111111 = -32,767$ can be represented.

1's Complement Form of Negative Numbers

Negative numbers are represented in 1's complement form by inverting all the magnitude bits and leaving the sign bit as is. For example,

$$-53 = 10110101 \longrightarrow 11001010$$
$$\text{true form} \qquad \text{1's complement form}$$

EXAMPLE 2–19

Convert each negative binary number to 1's complement form:
(a) 11000100 **(b)** 10001001 **(c)** 11110110 **(d)** 11000011

Solutions
(a) $11000100 \longrightarrow 10111011$ **(b)** $10001001 \longrightarrow 11110110$
(c) $11110110 \longrightarrow 10001001$ **(d)** $11000011 \longrightarrow 10111100$

2's Complement Form of Negative Numbers

Most digital systems use 2's complements to represent negative numbers for arithmetic operations. One method of obtaining the 2's complement of a negative number is to add 1 to the 1's complement, as previously discussed. Only the magnitude bits are complemented; the sign bit is left a 1. For example,

$$11101010 \longrightarrow 10010101 \qquad \text{1's complement}$$
$$\underline{+ \quad 1}$$
$$10010110 \qquad \text{2's complement}$$

A second method of obtaining the 2's complement has also been discussed. Again, only the magnitude bits are affected. The procedure is as follows: Starting with the right-most bit, the bits remain uncomplemented up to and including the first 1. The remaining magnitude bits are inverted; the sign bit remains a 1.

$$
\begin{array}{l}
\lceil \text{complemented} \\
\quad \lceil \text{uncomplemented} \\
10110110 \qquad \text{True form} \\
11001010 \qquad \text{2's complement}
\end{array}
$$

EXAMPLE 2–20

Convert each of the following negative binary numbers to 2's complement form by using both of the methods just discussed:
(a) 10010010 **(b)** 10110000 **(c)** 11110111 **(d)** 11000000

Solutions
(a) Taking the 1's complement and adding 1 yields

$$10010010 \longrightarrow 11101101 \qquad \text{1's complement}$$
$$\underline{+ \quad 1}$$
$$11101110 \qquad \text{2's complement}$$

Using the second method, we have

$$\underline{1\underline{0}010\underline{0}10} \longrightarrow 11101110$$

complemented⌐   └uncomplemented

(b)   Take the 1's complement and add 1:

$$10110000 \longrightarrow 11001111 \quad \text{1's complement}$$
$$\underline{+ \ 1}$$
$$11010000 \quad \text{2's complement}$$

Using the second method gives

$$10\underline{1\underline{1}0000} \longrightarrow 11010000$$

complemented⌐   └uncomplemented

(c)   Taking the 1's complement and adding 1, we obtain

$$11110111 \longrightarrow 10001000 \quad \text{1's complement}$$
$$\underline{+ \ 1}$$
$$10001001 \quad \text{2's complement}$$

Using the second method yields

$$\underline{1111011\underline{1}} \longrightarrow 10001001$$

complemented⌐   └uncomplemented

(d)   Take the 1's complement and add 1:

$$11000000 \longrightarrow 10111111 \quad \text{1's complement}$$
$$\underline{+ \ 1}$$
$$11000000 \quad \text{2's complement}$$

Using the second method gives

$$\underline{11000000} \longrightarrow 11000000$$

└uncomplemented

SECTION 2–6 REVIEW

Answers are found at the end of the chapter.

1. Determine the decimal equivalent of each of the signed binary numbers (all magnitudes are in true form).
    (a) 0011011   **(b)** 11011101   **(c)** 01010001
2. Express each decimal number in 2's complement.
    (a) -38 **(b)** -59 **(c)** -273

BINARY ARITHMETIC WITH SIGNED NUMBERS

2–7

Now that you know how negative numbers can be represented in complement form, we will look at how signed numbers are added, subtracted, multiplied, and divided. The processes introduced here are essential to the understanding of computers and microprocessor-based systems.

Binary Addition

There are four cases that must be considered when adding two numbers:

1. both numbers positive
2. positive number and smaller negative number
3. positive number and larger negative number
4. both numbers negative

We will take one case at a time. Eight bits (a byte) are used to represent the sign and magnitude of each number. The 2's complement is used to represent negative numbers.

Both numbers positive   In this case both sign bits are 0s, and a 2's complement is not required. To illustrate, add $+7$ and $+4$:

$$
\begin{array}{rl}
7 & \quad 00000111 \\
+\ 4 & \quad 00000100 \\
\hline
11 & \quad 00001011
\end{array}
$$

Positive number and smaller negative number   In this case the true binary form of the positive number is added to the 2's complement of the negative number. The sign bits are included in the addition, the final carry is discarded, and the result will be positive and in true form. To illustrate, add $+15$ and -6:

$$
\begin{array}{rl}
15 & \quad 00001111 \\
+\ -6 & \quad 11111010 \\
\hline
9 & \quad \cancel{1}\ \ 00001001
\end{array}
$$

discard carry

Notice that the sign of the sum is positive (0), as it should be.

Positive number and larger negative number   Again, the true binary form of the positive number is added to the 2's complement of the negative number. The sign bits are included in the addition, and the result will be negative. To illustrate, add $+16$ and -24:

$$
\begin{array}{rll}
16 & \quad 00010000 & \\
+\ -24 & \quad 11101000 & \\
\hline
-8 & \quad 11111000 & \text{2's complement of } -8
\end{array}
$$

Notice that the result automatically comes out in 2's complement form because it is a negative number.

Both numbers negative   In this case the 2's complements of both numbers are added, and of course, the sum is a negative number in 2's complement form. To illustrate, add -5 and -9:

$$
\begin{array}{rll}
-5 & \quad 11111011 & \\
+\ -9 & \quad 11110111 & \\
\hline
-14 & \quad \cancel{1}\ \ 11110010 & \text{2's complement of } -14
\end{array}
$$

discard carry

Overflow

When the number of bits in the sum exceeds the number of bits in each of the numbers added, *overflow* results, as illustrated by the following example.

EXAMPLE 2–21

$$
\begin{array}{cc}
\textit{Decimal} & \textit{Binary} \\
+9 & 01001 \\
+ \ \underline{+8} & + \ \underline{01000} \\
+17 & 10001
\end{array}
$$

sign incorrect————┘ └————magnitude incorrect

The overflow condition can occur only when both numbers are positive or both numbers are negative. It is indicated by an incorrect sign bit.

Summary of Signed Addition

The following is a summary of the four cases of signed binary addition.

Both numbers positive

1. Add both numbers in true (uncomplemented) form, including the sign bit.
2. The sign bit of the sum will be 0 (+).
3. Overflow into the sign bit is possible. The sign bit and the magnitude of the sum will be incorrect if an overflow occurs.

Both numbers negative

1. Take the 1's or 2's complements of the magnitudes of both numbers. Leave the sign bits as they are.
2. Add the numbers in their complement form, including sign-bits.
3. Add the end-around carry in the case of the 1's complement method; drop the carry in the case of the 2's complement method. The sign bit of the sum will be 1 (−), and the magnitude of the sum will be in complement form.
4. Overflow into the sign bit is possible. The sign bit and the magnitude of the sum will be incorrect if an overflow occurs.

Larger number positive, smaller number negative

1. Take the 1's or 2's complement of the magnitude of the negative number. Leave the sign bit as is. The positive number remains in true form.
2. Add the numbers, including the sign bits.
3. Add the end-around carry for the 1's complement method; drop the carry for the 2's complement method. The sign bit of the sum will be a 0 (+), and the magnitude will be in true form.
4. No overflow is possible.

Larger number negative, smaller number positive

1. Take the 1's or 2's complement of the magnitude of the negative number. Leave the sign bit as is. The positive number remains in true form.
2. Add the numbers, including the sign bits.
3. No carries will occur. The sum will have the proper sign bit, and the magnitude will be in complement form.
4. No overflow is possible.

Numbers Are Added Two at a Time

Now that we have examined the basic arithmetic processes by which *two* numbers can be added, let us look at the addition of a string of numbers, added two at a time. This can be accomplished by adding the first two numbers, then adding the third number to the sum of the first two, then adding the fourth number to this result, and so on. The addition of several numbers taken two at a time is illustrated in the following example. It is the basic way addition is accomplished in computer systems.

EXAMPLE 2–22

Add the numbers 2, 4, 6, 5, 8, and 9.

Solution

$$
\begin{array}{r r l}
  & 2 & \\
+ & 4 & \\
\hline
  & 6 & \text{First sum} \\
+ & 6 & \\
\hline
  & 12 & \text{Second sum} \\
+ & 5 & \\
\hline
  & 17 & \text{Third sum} \\
+ & 8 & \\
\hline
  & 25 & \text{Fourth sum} \\
+ & 9 & \\
\hline
  & 34 & \text{Final sum}
\end{array}
$$

Binary Subtraction

Subtraction is a special case of addition. For example, subtracting $+6$ (the subtrahend) from $+9$ (the minuend) is equivalent to adding -6 to $+9$.

Basically the subtraction operation changes the sign of the subtrahend and adds it to the minuend. The 2's complement method can be used in subtraction so that all operations require only addition. The four cases that were discussed in relation to the addition of signed numbers apply to the subtraction process because subtraction can be essentially reduced to an addition process. The following example will illustrate.

EXAMPLE 2–23

For each pair of decimal numbers, use the 2's complement method to perform the subtraction in binary. Show the equivalent decimal operation.
(a) $8 - 3$ **(b)** $12 - (-9)$ **(c)** $-25 - 8$
(d) $-58 - (-32)$ **(e)** $17 - 28$

Solutions

(a) $8 - 3 = 8 + (-3) = +5$

$$
\begin{array}{ll}
00001000 & \\
+11111101 & \text{2's complement of } -3 \\
\hline
\cancel{1}\ 00000101 & +5
\end{array}
$$

(b) $12 - (-9) = 12 + 9 = +21$

$$
\begin{array}{ll}
00001100 & \\
+00001001 & \\
\hline
00010101 & +21
\end{array}
$$

(c) $-25 - 8 = -25 + (-8) = -33$

$$
\begin{array}{ll}
11100111 & \text{2's complement of } -25 \\
+11111000 & \text{2's complement of } -\ 8 \\
\hline
11011111 & \text{2's complement of } -33
\end{array}
$$

(d) $-58 - (-32) = -58 + 32 = -26$

$$
\begin{array}{ll}
11000110 & \text{2's complement of } -58 \\
+00100000 & \\
\hline
11100110 & \text{2's complement of } -26
\end{array}
$$

(e) $17 - 28 = 17 + (-28) = -11$

$$
\begin{array}{ll}
00010001 & \\
+11100100 & \text{2's complement of } -28 \\
\hline
11110101 & \text{2's complement of } -11
\end{array}
$$

Binary Multiplication

In binary multiplication the two numbers are the *multiplicand* and the *multiplier,* and the result is the *product.* The magnitudes of the numbers must be in true form, and the sign bits are not used during the multiplication process.

The sign of the product depends on the signs of the two numbers that are being multiplied. If the two numbers have the same sign, either both positive or both negative, the product is positive. If the two numbers have different signs, the product is negative. This principle is illustrated with decimal numbers as follows:

$$(+3)(+5) = +15 \qquad (-4)(-5) = +20$$
$$(-10)(+3) = -30 \qquad (+6)(-8) = -48$$

In the multiplication process, the sign bits of the two numbers are checked, and the resulting sign of the product is stored before the actual multiplication occurs. Once the product sign is determined, the multiplicand and multiplier sign bits are discarded, leaving only the magnitudes to be multiplied.

The multiplication process Multiplication of binary numbers in digital systems is often accomplished by a series of additions and shifts similar to the way you multiply in longhand. So, as in subtraction, an adder can be used. The following example illustrates binary multiplication of two unsigned seven-bit numbers.

EXAMPLE 2–24

Multiply 1010011 (83) and 0111011 (59). The seven bits represent the magnitudes of the numbers and are given in true form. The sign bits are not included here.

Solution

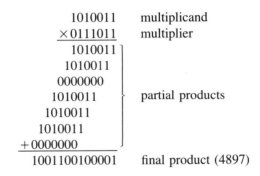

```
        1010011        multiplicand
      × 0111011        multiplier
        1010011  ⎫
        1010011  ⎪
        0000000  ⎪
        1010011  ⎬  partial products
        1010011  ⎪
        1010011  ⎪
      + 0000000  ⎭
   1001100100001        final product (4897)
```

Notice that the number of partial products equals the number of multiplier bits. Also, each partial product is shifted one bit to the left with respect to the previous partial product. The final product, which is the sum of the shifted partial products, can have up to twice the number of bits as in the original numbers.

Most computers can add only two numbers at a time. For that reason, each partial product is added to the sum of the previous partial products, and the sum of the partial products is accumulated. For example, the first partial product and the second partial product are added; then the third partial product is added to this sum; and so on.

Now we will go through the previous multiplication example as a typical computer might handle it.

EXAMPLE 2–25

Repeat the multiplication in Example 2–24, except this time keep a running total of the partial product sum so that only two numbers will have to be added at a time.

Solution

```
        1010011      Multiplicand
      × 0111011      Multiplier
        1010011      1st partial product
      + 1010011      2nd partial product
        11111001     Sum of 1st and 2nd
      + 0000000      3rd partial product
       011111001     Sum
```

+ 1010011	4th partial product
1110010001	Sum
+ 1010011	5th partial product
100011000001	Sum
+ 1010011	6th partial product
1001100100001	Sum
+ 0000000	7th partial product
1001100100001	Final product

The multiplier and the multiplicand must be in true form when multiplied. In a computer using the 2's complement system, any negative number is already in 2's complement form and must be converted to its true form before multiplying. A negative number is converted back to its true form by taking the 2's complement of the 2's complement. For example, let us convert a signed binary number from its true form to its 2's complement and then back to the true form:

$$10101110 \longrightarrow 11010010 \longrightarrow 10101110$$
$$\text{true} \qquad\qquad\quad \text{2's comp.} \qquad\qquad\quad \text{true}$$

To summarize the multiplication process in a typical 2's complement system, the following steps are listed:

1. Obtain the multiplicand and the multiplier.
2. If either number or both are negative and in complement form, convert them to true form.
3. Examine both sign bits, and store the sign of the product. (If both signs are the same, the product is positive; if signs are different, the product is negative.)
4. Examine each multiplier magnitude bit beginning with the right-most bit (LSB). When the multiplier bit is a 1, shift the partial product sum one place to the right, and add the multiplicand magnitude bits. This shift is, in effect, the same as shifting each new partial product one place to the left as in the previous example.

   If the multiplier bit is a 0, the partial product sum is shifted one place to the right, but no addition is performed. This procedure is effectively the same as adding an all-zeros partial product.
5. Attach the predetermined sign bit to the final product. If the sign is negative, put the product into 2's complement form.

The following example illustrates the steps in this process.

EXAMPLE 2–26

Multiply the following two signed binary numbers, which are equivalent to the decimal numbers $+107$ and -51. The negative number is in 2's complement form.

$$01101011 \times 11001101$$

Solution

Step 1. Put the negative number (-51) in true form:

$$11001101 \xrightarrow{\hspace{2cm}} 10110011$$
$$\text{2's complement} \qquad\qquad \text{true number}$$

Step 2. Determine the sign of the product. Because the signs are different, the product will have a negative sign bit (1).

Step 3. Multiply the magnitudes:

	1101011	Multiplicand magnitude (107)
	0110011	Multiplier magnitude (51)
1st multiplier bit $= 1$	1101011	1st partial product
	1101011	Shift right
2nd multiplier bit $= 1$	$+\,1101011$	2nd partial product
	101000001	Sum
	101000001	Shift right
3rd multiplier bit $= 0$	101000001	Shift right (3rd part. prod. $= 0$)
4th multiplier bit $= 0$	101000001	Shift right (4th part. prod. $= 0$)
5th multiplier bit $= 1$	$+\,1101011$	5th partial product
	11111110001	Sum
	11111110001	Shift right
6th multiplier bit $= 1$	$+\,1101011$	6th partial product
	1010101010001	Final product (magnitude)
		(7th partial product $= 0$)

Step 4. Attach the sign bit, and take the 2's complement of the final product because it is negative:

$$10101010101111 = -5457$$

Binary Division

In division, the two numbers are called the *dividend* and the *divisor*. The result is the *quotient*. The sign of the quotient is determined by the signs of the dividend and divisor. If both signs are the same, the quotient is positive. If the signs differ, the quotient is negative.

The division process Division can be accomplished by repeated subtraction. Since subtraction can be done with 2's complement addition, the division process can be accomplished with an adder. Keep in mind that the quotient is the number of times that the divisor goes into the dividend. The process is as follows:

1. Obtain the divisor and dividend. Initialize the quotient to zero.
2. Subtract the divisor from the dividend to get the partial remainder. If this remainder is positive, increment the quotient. (Add 1 to the quotient.) A positive partial remainder indicates that the divisor goes into the dividend (or

partial remainder). If the partial remainder is negative, the divisor does not go into the dividend (or partial remainder), and therefore the division is complete.

3. If the partial remainder is positive, repeat step 2 by subtracting the divisor from the partial remainder.

The following example will illustrate this process.

EXAMPLE 2–27

Divide 01100100 by 00110010. The decimal equivalent of this division is $+100$ divided by $+50$.

Solution

Step 1. Initialize the quotient to zero. Subtract the divisor from the dividend by using 2's complement addition.

$$
\begin{array}{ll}
01100100 & \text{Dividend} \\
+\ 11001110 & \text{2's complement of divisor} \\
\hline
00110010 & \text{Positive 1st partial remainder} \\
\end{array}
$$

$+$ sign ⤴

Step 2. A positive partial remainder indicates that the divisor goes into the dividend, and therefore the quotient is incremented by one. The new quotient is 00000001.

Step 3. Subtract the divisor from the 1st partial remainder:

$$
\begin{array}{ll}
00110010 & \text{1st partial remainder} \\
+\ 11001110 & \text{2's complement of divisor} \\
\hline
00000000 & \text{Positive 2nd partial remainder} \\
\end{array}
$$

$+$ sign ⤴

Step 4. Increment the quotient to 00000010.

Step 5. Subtract the divisor from the 2nd partial remainder:

$$
\begin{array}{ll}
00000000 & \text{2nd partial remainder} \\
+\ 11001110 & \text{2's complement of divisor} \\
\hline
11001110 & \text{Negative 3rd partial remainder} \\
\end{array}
$$

$-$ sign ⤴

The negative remainder indicates that the divisor will not go into the 2nd partial remainder. Therefore, the final quotient is $00000010 = 2$, and the division is complete.

In this particular example, the zero as the second partial remainder indicated that the division was complete. However, a zero remainder does not always occur, and therefore the check for a negative partial remainder is a more general test.

SECTION 2–7
REVIEW

Answers are found at the end of the chapter.
1. Add the following signed binary numbers (negative numbers are in 2's complement form):
   (a) 01101101 + 00001110   **(b)** 10011101 + 00111010
2. Perform the following signed binary subtractions (negative numbers are in 2's complement form):
   (a) 01110110 − 10111001   **(b)** 11011011 − 10111101
3. Multiply the following signed binary numbers:
   (a) 0101 × 0011   **(b)** 10100 × 01100

OCTAL NUMBERS

2–8

The primary application of octal numbers is representing binary numbers. It is very easy to convert from binary to octal, and it is much easier to read a number in octal form than in binary.

The *octal* number system is composed of eight digits, which are

$$0, 1, 2, 3, 4, 5, 6, 7$$

To count above 7, we begin another column and start over:

$$10, 11, 12, 13, 14, 15, 16, 17, 20, 21$$

and so on. Counting in octal is similar to counting in decimal, except that the digits 8 and 9 are not used. To distinguish octal numbers from decimal numbers, we will use the subscript 8 to indicate an octal number. For instance, 15_8 is equivalent to 13 in decimal.

Octal-to-Decimal Conversion

Since the octal number system has a base of eight, each successive digit position is an increasing power of eight, beginning in the right-most column with 8^0. The evaluation of an octal number in terms of its *decimal equivalent* is accomplished by multiplying each digit by its weight and summing the products, as illustrated here for 2374_8:

Weight:	8^3	8^2	8^1	8^0
Decimal value:	512	64	8	1
Octal number:	2	3	7	4

$$
\begin{aligned}
2374_8 &= 2 \times 8^3 + 3 \times 8^2 + 7 \times 8^1 + 4 \times 8^0 \\
&= 2 \times 512 + 3 \times 64 + 7 \times 8 + 4 \times 1 \\
&= 1024 + 192 + 56 + 4 \\
&= 1276
\end{aligned}
$$

Decimal-to-Octal Conversion

A method of converting a decimal number to an octal number is the *repeated division-by-8* method, which is similar to the method used in conversion of decimal numbers to binary. To show how it works, we convert the decimal number 359 to octal. Each successive division by 8 yields a remainder that becomes a digit in the equivalent octal number. The first remainder generated is the least significant digit (LSD).

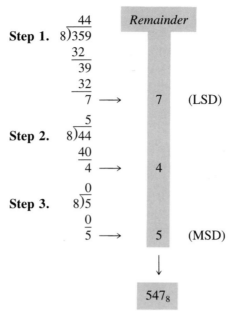

$$547_8$$

The octal numbers shown so far have been whole numbers. Fractional octal numbers are represented by digits to the right of the octal point.

The weights of an octal number are

$$8^{n-1} \ldots 8^3 8^2 8^1 8^0 \cdot 8^{-1} 8^{-2} 8^{-3} \ldots 8^{-n}$$

This display shows that all digits to the left of the octal point have weights that are positive powers of eight, as seen previously. All digits to the right of the octal point have fractional weights, or negative powers of eight, as illustrated in the following example.

EXAMPLE 2–28

Determine the decimal value of the octal fraction 0.325_8.

Solution First, determine the weight of each digit. Then multiply the weight by the digit. Finally, sum the products.

Octal weight:	8^{-1}	8^{-2}	8^{-3}
Decimal value:	0.125	0.015625	0.001953
Octal number:	0.3	2	5

$$
\begin{aligned}
0.325_8 &= 3(0.125) + 2(0.015625) + 5(0.001953) \\
&= 0.375 + 0.03125 + 0.009765 \\
&= 0.416015
\end{aligned}
$$

Octal-to-Binary Conversion

Because each octal digit can be represented by a three-bit binary number, it is very easy to convert from octal to binary and from binary to octal. The octal number system is often used in digital systems, especially for input/output applications.

Each octal digit is represented by three bits as indicated:

Octal Digit	Binary
0	000
1	001
2	010
3	011
4	100
5	101
6	110
7	111

To convert an octal number to a binary number, simply replace each octal digit with the appropriate three bits. This procedure is illustrated in the following example.

EXAMPLE 2–29

Convert each of the following octal numbers to binary:

(a) 13_8 (b) 25_8 (c) 47_8 (d) 170_8 (e) 752_8 (f) 5276_8
(g) 37.12_8

Solutions

(a)

(b)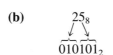

(c)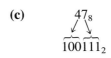

(d)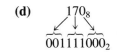

(e)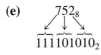

(f)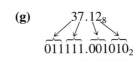

(g) $37.12_8 \rightarrow 011111.001010_2$

Binary-to-Octal Conversion

Conversion of a binary number to an octal number is also a straightforward process. Beginning at the binary point, simply break the binary number into groups of three bits, and convert each group to the appropriate octal digit. In binary whole numbers the binary point is understood to be immediately to the right of the LSB.

EXAMPLE 2–30

Represent each of the following binary numbers as its octal equivalent:

(a) 110101 (b) 101111001 (c) 1011100110 (d) 1001101.1011

Solutions

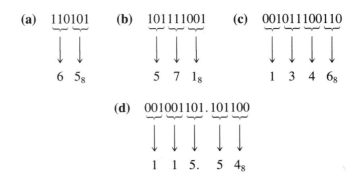

(a) 110101   **(b)** 101111001   **(c)** 001011100110

$6 \quad 5_8$ $5 \quad 7 \quad 1_8$ $1 \quad 3 \quad 4 \quad 6_8$

(d) 001001101.101100

$1 \quad 1 \quad 5. \quad 5 \quad 4_8$

In parts (c) and (d), additional zeros are included with the MSB and the LSB where needed to complete a group of three digits. These zeros are not in the original numbers, but they do not affect the values of the original numbers when included.

SECTION 2–8 REVIEW

1. Convert the octal numbers to decimal.
   (a) 73_8 **(b)** 125_8
2. Convert the decimal numbers to octal.
   (a) 98   **(b)** 163
3. Convert the octal numbers to binary.
   (a) 46_8 **(b)** 723_8 **(c)** 5624.37_8
4. Convert the binary numbers to octal.
   (a) 110101111   **(b)** 100110001.101111   **(c)** 10111111.0011

HEXADECIMAL NUMBERS

2–9

Like the octal system, the hexadecimal system provides a convenient way to express binary numbers and codes. It is used frequently in conjunction with computers and microprocessors to express binary quantities for input and output purposes and for easy readability.

The *hexadecimal* system has a base of sixteen; that is, it is composed of 16 digits and characters. Most digital systems process binary data in groups that are multiples of four bits, making the hexadecimal number very convenient because each hexadecimal digit represents a four-bit binary number (as listed in Table 2–2).

Ten numeric digits and six alphabetic characters make up the hexadecimal number system. The subscript 16 indicates a hexadecimal number.

TABLE 2–2

Decimal	Binary	Hexadecimal
0	0000	0
1	0001	1
2	0010	2
3	0011	3
4	0100	4
5	0101	5
6	0110	6
7	0111	7
8	1000	8
9	1001	9
10	1010	A
11	1011	B
12	1100	C
13	1101	D
14	1110	E
15	1111	F

Binary-to-Hexadecimal Conversion

Converting a binary number to hexadecimal is a very straightforward procedure. Simply break the binary number into four-bit groups, starting at the binary point, and replace each group with the equivalent hexadecimal symbol.

EXAMPLE 2–31

Convert the following binary numbers to hexadecimal:
(a) 1100101001010111 **(b)** 111111000101101001
(c) 1110011000.111

Solutions

(a)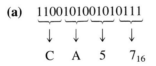
C   A   5 7_{16}

(b)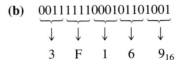
3   F   1   6 9_{16}

(c)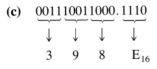
3   9   8 E_{16}

Zeros have been added where needed to complete a four-bit group.

Hexadecimal-to-Binary Conversion

To convert from a hexadecimal number to a binary number, reverse the process and replace each hexadecimal symbol with the appropriate four bits.

EXAMPLE 2–32   Determine the binary numbers for the following hexadecimal numbers:

(a) $10A4_{16}$ (b) $CF83_{16}$ (c) 9742_{16} (d) $D2E.8_{16}$

Solutions

(a)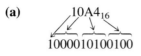

1000010100100

(b)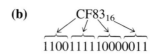

1100111110000011

(c)

1001011101000010

(d)

110100101110.1000

In part (a) the MSB is understood to have three zeros preceding it, thus forming a four-bit group.

The use of letters to represent numbers may seem strange at first, but keep in mind that *any* number system is only a set of sequential symbols. If we understand what quantities these symbols represent, then the form of the symbols themselves is unimportant once we get accustomed to using them.

It should be clear that it is much easier to write the hexadecimal number than the equivalent binary number, and since conversion is so easy, the hexadecimal system is a natural for representing binary numbers and digital codes.

Counting in Hexadecimal

How do we count in hexadecimal once we get to F? Simply start over with another column and continue as follows:

10, 11, 12, 13, 14, 15, 16, 17, 18, 19, 1A, 1B, 1C, 1D, 1E, 1F,
20, 21, 22, 23, 24, 25, 26, 27, 28, 29, 2A, 2B, 2C, 2D, 2E, 2F,
30, 31, . . .

With two hexadecimal digits, we can count up to FF_{16}, which is 255_{10}. To count beyond this, three hexadecimal digits are needed. For instance, 100_{16} is decimal 256_{10}, 101_{16} is decimal 257_{10}, and so forth. The maximum three-digit hexadecimal number is FFF_{16}, or 4095_{10}. The maximum four-digit hexadecimal number is $FFFF_{16}$, which is $65,535_{10}$.

Hexadecimal-to-Decimal Conversion

One way to find the decimal equivalent of a hexadecimal number is first to convert the hexadecimal number to binary and then to convert from binary to decimal. The following example illustrates this procedure.

EXAMPLE 2–33

Convert the following hexadecimal numbers to decimal:
(a) $1C_{16}$ **(b)** $A85_{16}$

Solutions

(a) $1C_{16}$

$\overline{0001} \ \overline{1100}$

$00011100 = 2^4 + 2^3 + 2^2$
$= 16 + 8 + 4$
$= 28$

(b) $A85_{16}$

$\overline{1010} \ \overline{1000} \ \overline{0101}$

$101010000101 = 2^{11} + 2^9 + 2^7 + 2^2 + 2^0$
$= 2048 + 512 + 128 + 4 + 1$
$= 2693$

Another way to convert a hexadecimal number to its decimal equivalent is to multiply the decimal value of each hexadecimal digit by its weight and then take the sum of these products. The weights of a hexadecimal number are increasing powers of 16 (from right to left). For a four-digit hexadecimal number, the weights are

$$16^3 \quad 16^2 \quad 16^1 \quad 16^0$$
$$4096 \quad 256 \quad 16 \quad 1$$

The following example shows this conversion method.

EXAMPLE 2–34

Convert the following hexadecimal numbers to decimal:
(a) $E5_{16}$ **(b)** $B2F8_{16}$

Solutions

(a) $E5_{16} = E \times 16 + 5 \times 1 = 14 \times 16 + 5 \times 1 = 224 + 5 = 229$

(b) $B2F8_{16} = B \times 4096 + 2 \times 256 + F \times 16 + 8 \times 1$
$= 11 \times 4096 + 2 \times 256 + 15 \times 16 + 8 \times 1$
$= 45,056 \qquad + 512 \qquad + 240 \qquad + 8$
$= 45,816$

Decimal-to-Hexadecimal Conversion

Repeated division of a decimal number by 16 will produce the equivalent hexadecimal number, formed by the remainders of the divisions. The first remainder produced is the LSD. This procedure is similar to repeated division by 2 for decimal-to-binary conversion and repeated division by 8 for decimal-to-octal conversion. The following example illustrates the procedure.

EXAMPLE 2–35

Convert the decimal number 650 to hexadecimal by repeated division by 16.

Solution

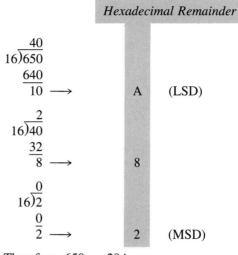

Therefore, $650 = 28A_{16}$.

Hexadecimal Addition

Addition can be done directly with hexadecimal numbers by remembering that the hexadecimal digits 0 through 9 are equivalent to decimal digits 0 through 9 and that hexadecimal digits A through F are equivalent to decimal numbers 10 through 15. When adding two hexadecimal numbers, use the following rules (decimal numbers are indicated by a subscript 10):

1. In any given column of an addition problem, think of the two hexadecimal digits in terms of their decimal value. For instance, $5_{16} = 5_{10}$ and $C_{16} = 12_{10}$.
2. If the sum of these two digits is 15_{10} or less, bring down the corresponding hexadecimal digit.
3. If the sum of these two digits is greater than 15_{10}, bring down the amount of the sum that exceeds 16_{10}, and carry a 1 to the next column.

EXAMPLE 2–36

Add the following hexadecimal numbers:
(a) $23_{16} + 16_{16}$ (b) $58_{16} + 22_{16}$ (c) $2B_{16} + 84_{16}$
(d) $DF_{16} + AC_{16}$

Solution

(a) $\begin{array}{r} 23_{16} \\ +16_{16} \\ \hline 39_{16} \end{array}$ right column: $3_{16} + 6_{16} = 3_{10} + 6_{10} = 9_{10} = 9_{16}$
   left column: $2_{16} + 1_{16} = 2_{10} + 1_{10} = 3_{10} = 3_{16}$

(b) $\begin{array}{r} 58_{16} \\ +22_{16} \\ \hline 7A_{16} \end{array}$ right column: $8_{16} + 2_{16} = 8_{10} + 2_{10} = 10_{10} = A_{16}$
   left column: $5_{16} + 2_{16} = 5_{10} + 2_{10} = 7_{10} = 7_{16}$

(c) $2B_{16}$ right column: $B_{16} + 4_{16} = 11_{10} + 4_{10} = 15_{10} = F_{16}$

  $+84_{16}$ left column: $2_{16} + 8_{16} = 2_{10} + 8_{10} = 10_{10} = A_{16}$

  $\overline{AF_{16}}$

(d) DF_{16} right column: $F_{16} + C_{16} = 15_{10} + 12_{10} = 27_{10}$

  $+AC_{16}$ $27_{10} - 16_{10} = 11_{10} = B_{16}$ with a 1 carry

  $\overline{18B_{16}}$ left column: $D_{16} + A_{16} + 1_{16} = 13_{10} + 10_{10} + 1_{10} = 24_{10}$

                          $24_{10} - 16_{10} = 8_{10} = 8_{16}$ with a 1 carry

Hexadecimal Subtraction Using 2's Complement Method

Since a hexadecimal number can be used to represent a binary number, it can also be used to represent the 2's complement of a binary number. For instance, the hexadecimal representation of 11001001_2 is $C9_{16}$. The 2's complement of this binary number is 00110111, which is written in hexadecimal as 37_{16}.

As you have learned, the 2's complement allows us to subtract by *adding* binary numbers. We can also use this method for hexadecimal subtraction, as the following example shows.

EXAMPLE 2–37

Subtract the following hexadecimal numbers:

(a) $84_{16} - 2A_{16}$ **(b)** $C3_{16} - 0B_{16}$

Solution

(a) $2A_{16} = 00101010$

  2's complement of $2A_{16} = 11010110 = D6_{16}$

  84_{16}

  $+D6_{16}$ Add.

  $\overline{\cancel{1}5A_{16}}$ Drop carry, as in 2's complement addition.

  The difference is $5A_{16}$.

(b) $0B_{16} = 00001011$

  2's complement of $0B_{16} = 11110101 = F5_{16}$

  $C3_{16}$

  $+F5_{16}$ Add.

  $\overline{\cancel{1}B8_{16}}$ Drop carry.

  The difference is $B8_{16}$.

**SECTION 2–9
REVIEW**

Answers are found at the end of the chapter.
1. Convert the binary numbers to hexadecimal.
    (a) 10110011 **(b)** 110011101000
2. Convert the hexadecimal numbers to binary.
    (a) 57_{16} **(b)** $3A5_{16}$ **(c)** $F80B_{16}$
3. Add the hexadecimal numbers directly.
    (a) $18_{16} + 34_{16}$ **(b)** $3F_{16} + 2A_{16}$
4. Subtract the hexadecimal numbers.
    (a) $75_{16} - 21_{16}$ **(b)** $94_{16} - 5C_{16}$

BINARY CODED DECIMAL (BCD)

2–10

*As you have learned, quantities can be represented by binary digits. Not only numbers, but also letters and other symbols, can be represented by 1s and 0s. In fact, **any** entity expressible as numbers, letters, or other symbols can be represented by binary digits and can therefore be processed by digital systems.*

Combinations of binary digits that represent numbers, letters, and symbols are digital codes. In many applications, special codes are used for such auxiliary functions as error detection and correction.

The 8421 Code

The 8421 code is a type of binary coded decimal *(BCD)* code. Binary coded decimal means that each decimal digit, 0 through 9, is represented by a binary code of four bits. The designation 8421 indicates the binary weights of the four bits (2^3, 2^2, 2^1, 2^0). The ease of conversion between 8421 code numbers and the familiar decimal numbers is the main advantage of this code. All you have to remember are the ten binary combinations that represent the ten decimal digits as shown in Table 2–3. The 8421 code is the predominant BCD code, and when we refer to BCD, we always mean the 8421 code unless otherwise stated.

TABLE 2–3
The 8421 BCD code.

8421 (BCD)	Decimal
0000	0
0001	1
0010	2
0011	3
0100	4
0101	5
0110	6
0111	7
1000	8
1001	9

You should realize that, with four bits, sixteen numbers (2^4) can be represented and that, in the 8421 code, only ten of these are used. The six code combinations that are not used—1010, 1011, 1100, 1101, 1110, and 1111—are invalid in the 8421 BCD code.

To express any decimal number in BCD, simply replace each decimal digit with the appropriate four-bit code, as shown by the following example.

EXAMPLE 2–38

Convert each of the following decimal numbers to BCD:

(a) 3   **(b)** 9.2   **(c)** 18   **(d)** 34.8   **(e)** 65   **(f)** 92
(g) 150   **(h)** 321   **(i)** 1472

Solutions

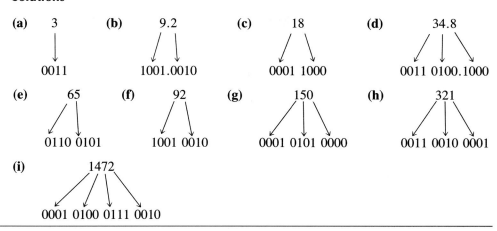

(a) 3 → 0011

(b) 9.2 → 1001.0010

(c) 18 → 0001 1000

(d) 34.8 → 0011 0100.1000

(e) 65 → 0110 0101

(f) 92 → 1001 0010

(g) 150 → 0001 0101 0000

(h) 321 → 0011 0010 0001

(i) 1472 → 0001 0100 0111 0010

It is equally easy to determine a decimal number from a BCD number. Start at the decimal point (to the right of the LSB in whole numbers), and break the code into groups of four bits. Then write the decimal digit represented by each four-bit group. An example illustrates.

EXAMPLE 2–39

Find the decimal numbers represented by the following BCD codes:

(a) 10000110   **(b)** 00110001   **(c)** 01010011   **(d)** 100101110100
(e) 0001100001100000.0111

Solutions

(a) 1000 0110
        ↓   ↓
        8   6

(b) 0011 0001
        ↓   ↓
        3   1

(c) 0101 0011
        ↓   ↓
        5   3

(d) 1001 0111 0100
        ↓   ↓   ↓
        9   7   4

(e) 0001 1000 0110 0000 . 0111
        ↓   ↓   ↓   ↓   ↓
        1   8   6   0 . 7

BCD Addition

BCD is a numerical code, and many applications require that arithmetic operations be performed. Addition is the most important operation because the other three operations (subtraction, multiplication, and division) can be accomplished by the use of addition. Here is how to add two BCD numbers:

1. Add the two BCD numbers, using the rules for binary addition in Section 2–4.
2. If a four-bit sum is equal to or less than 9, it is a *valid* BCD number.

3. If a four-bit sum is greater than 9, or if a carry out of the four-bit group is generated, it is an *invalid* result. Add 6 (0110) to the four-bit sum in order to skip the six invalid states and return the code to 8421. If a carry results when 6 is added, simply add the carry to the next four-bit group.

Several examples illustrate BCD addition for the case in which the sum of no four-bit column exceeds 9.

EXAMPLE 2–40

Add the following BCD numbers (the decimal number addition is shown for comparison):

(a)
```
  0011        3
+ 0100      + 4
  0111        7
```

(b)
```
  0110        6
+ 0010      + 2
  1000        8
```

(c)
```
  0010  0011       23
+ 0001  0101      + 15
  0011  1000       38
```

(d)
```
  1000  0110       86
+ 0001  0011      + 13
  1001  1001       99
```

(e)
```
  0100  0101  0000      450
+ 0100  0001  0111     + 417
  1000  0110  0111      867
```

(f)
```
  1000  0111  0011      873
+ 0001  0001  0010     + 112
  1001  1000  0101      985
```

Note that in each case the sum in any four-bit column does not exceed 9, and the results are valid BCD numbers.

Next, we deal with the case of an invalid sum (greater than 9 or a carry) by illustrating the procedure with several examples.

EXAMPLE 2–41

Add the following BCD numbers:

(a)
```
                    1001                                      9
                  + 0100                                    + 4
                    1101    Invalid BCD number (> 9)          13
                  + 0110    Add 6.
           0001     0011    Valid BCD number
             ↓        ↓
             1        3
```

(b)
```
                    1001
                  + 1001                                      9
          1        0010    Invalid because of carry         + 9
                  + 0110    Add 6.                            18
           0001     1000    Valid BCD number
             ↓        ↓
             1        8
```

(c)	0001	0110		16
	+0001	0101		+15
	0010	1011	Right group is invalid (> 9),	31
			left group valid. Add 6	
		+0110	to invalid code (add	
			carry, 0001, to next group).	
	0011	0001	Valid BCD number	
	↓	↓		
	3	1		

(d)		0110	0111		67
		+0101	0011		+53
		1011	1010	Both groups are invalid (> 9).	120
		+0110	+0110	Add 6 to both groups.	
	0001	0010	0000	Valid BCD number	
	↓	↓	↓		
	1	2	0		

SECTION 2–10 REVIEW

Answers are found at the end of the chapter.

1. What is the binary weight of each 1 in the following BCD numbers?
   (a) 0010   **(b)** 1000   **(c)** 0001   **(d)** 0100
2. Convert the following decimal numbers to BCD:
   (a) 6   **(b)** 15   **(c)** 273   **(d)** 84.9
3. What decimal numbers are represented by each BCD code?
   (a) 10001001   **(b)** 001001111000   **(c)** 00010101.0111

DIGITAL CODES

2–11

*In addition to the BCD code, there are many other special codes used in digital systems. They are used for a variety of purposes. Some types of codes are strictly numeric. Others are **alphanumeric**; that is, they are used to represent numbers, letters, symbols, and instructions. Three important codes—the Gray code, the Excess-3 code, and the ASCII code—are introduced in this section. For coverage of error detection and correction codes, which are more complex, refer to Appendix B.*

The Gray Code

The *Gray code* is an *unweighted* code; that is, there are no specific weights assigned to the bit positions. The important feature of the Gray code is that *it exhibits only a single bit change from one code number to the next*. This property is important in many applications, such as shaft position encoders, where error susceptibility increases with the number of bit changes between adjacent numbers in a sequence. The Gray code is not an arithmetic code.

Table 2–4 is a listing of four-bit Gray code numbers for decimal numbers 0 through 9. Like binary numbers, the Gray code can have any number of bits. Notice the single-bit change between successive code numbers. For instance, in going from decimal 3 to decimal 4, the Gray code changes from 0010 to 0110, while the binary code changes from 0011 to 0100, a change of three bits. The only bit change is in the third bit from the right in the Gray code; the others remain the same. Binary numbers are shown for reference.

TABLE 2–4
Four-bit Gray code.

Decimal	Binary	Gray
0	0000	0000
1	0001	0001
2	0010	0011
3	0011	0010
4	0100	0110
5	0101	0111
6	0110	0101
7	0111	0100
8	1000	1100
9	1001	1101

Binary-to-Gray conversion   By representing the ten decimal digits with a four-bit Gray code, we have another form of BCD code. The Gray code, however, can be extended to any number of bits, and conversion between binary code and Gray code is sometimes useful. First, we will discuss how to convert from a binary number to a Gray code number. The following rules apply:

1. The most significant bit (left-most) in the Gray code is the same as the corresponding MSB in the binary number.
2. Going from left to right, add each adjacent pair of binary code bits to get the next Gray code bit. Discard carries.

For example, let us convert the binary number 10110 to Gray code.

Step 1. The left-most Gray digit is the same as the left-most binary code bit.

1	0	1	1	0	Binary
↓					
1					Gray

Step 2. Add the left-most binary code bit to the adjacent one.

1 + 0	1	1	0	Binary
↓				
1 1				Gray

Step 3. Add the next adjacent pair.

$$\begin{array}{ccccccc} 1 & \boxed{0 \;+\; 1} & 1 & 0 & & \text{Binary} \\ & \downarrow & & & \\ 1 & 1 & 1 & & & \text{Gray} \end{array}$$

Step 4. Add the next adjacent pair and discard the carry.

$$\begin{array}{ccccccc} 1 & 0 & \boxed{1 \;+\; 1} & 0 & & \text{Binary} \\ & & \downarrow & & \\ 1 & 1 & 1 & 0 & & \text{Gray} \end{array}$$

Step 5. Add the last adjacent pair.

$$\begin{array}{ccccccc} 1 & 0 & 1 & \boxed{1 \;+\; 0} & & \text{Binary} \\ & & & \downarrow & \\ 1 & 1 & 1 & 0 & 1 & \text{Gray} \end{array}$$

The conversion is now complete; the Gray code is 11101.

Gray-to-binary conversion To convert from Gray code to binary, a similar method is used, but there are some differences. The following rules apply:

1. The most significant bit (left-most) in the binary code is the same as the corresponding bit in the Gray code.
2. Add each binary code bit generated to the Gray code bit in the next adjacent position. Discard carries.

For example, the conversion of the Gray code number 11011 to binary is as follows:

Step 1. The left-most bits are the same.

$$\begin{array}{ccccc} 1 & 1 & 0 & 1 & 1 \quad \text{Gray} \\ \downarrow & & & & \\ 1 & & & & \quad \text{Binary} \end{array}$$

Step 2. Add the last binary code bit just generated to the Gray code bit in the next position. Discard the carry.

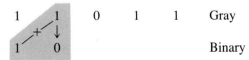

$$\begin{array}{ccccc} 1 & 1 & 0 & 1 & 1 \quad \text{Gray} \\ 1 & 0 & & & \quad \text{Binary} \end{array}$$

Step 3. Add the last binary code bit generated to the next Gray bit.

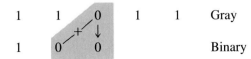

$$\begin{array}{ccccc} 1 & 1 & 0 & 1 & 1 \quad \text{Gray} \\ 1 & 0 & 0 & & \quad \text{Binary} \end{array}$$

Step 4. Add the last binary code bit generated to the next Gray bit.

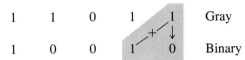

This completes the conversion. The final binary number is 10010.

EXAMPLE 2–42

(a) Convert the binary number 11000110 to Gray code.
(b) Convert Gray code 10101111 to binary.

Solutions

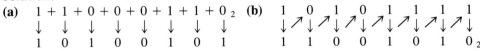

The Excess-3 Code

Excess-3 is a digital code related to BCD that is derived by adding 3 to *each* decimal digit and then converting the result of that addition to four-bit binary. Since no definite weights can be assigned to the four digit positions, Excess-3 is an unweighted code. For instance, the Excess-3 code for decimal 2 is

$$
\begin{array}{r}
2 \\
+3 \\
\hline
5 \rightarrow 0101
\end{array}
$$

The Excess-3 code for decimal 9 is

$$
\begin{array}{r}
9 \\
+3 \\
\hline
12 \rightarrow 1100
\end{array}
$$

The Excess-3 code for each decimal digit is found by the same procedure. The entire code is shown in Table 2–5.

Notice that ten of a possible 16 code combinations are used in the Excess-3 code. The six invalid combinations are 0000, 0001, 0010, 1101, 1110, and 1111.

TABLE 2–5
Excess-3 code.

Decimal	BCD	Excess-3
0	0000	0011
1	0001	0100
2	0010	0101
3	0011	0110
4	0100	0111
5	0101	1000
6	0110	1001
7	0111	1010
8	1000	1011
9	1001	1100

EXAMPLE 2–43

Convert each of the following decimal numbers to Excess-3 code:
(a) 13 **(b)** 35 **(c)** 87 **(d)** 159 **(e)** 430

Solutions First, add 3 to each digit in the decimal number, and then convert each resulting sum to its equivalent binary code.

(a)
```
   1    3
  +3   +3
   4    6
   ↓    ↓
 0100 0110    Excess-3
```

(b)
```
   3    5
  +3   +3
   6    8
   ↓    ↓
 0110 1000    Excess-3
```

(c)
```
   8    7
  +3   +3
  11   10
   ↓    ↓
 1011 1010    Excess-3
```

(d)
```
   1    5    9
  +3   +3   +3
   4    8   12
   ↓    ↓    ↓
 0100 1000 1100    Excess-3
```

(e)
```
   4    3    0
  +3   +3   +3
   7    6    3
   ↓    ↓    ↓
 0111 0110 0011    Excess-3
```

Alphanumeric Codes

In order to communicate, we need not only numbers, but also letters and other symbols. In the strictest sense, alphanumeric codes are codes that represent numbers and alphabetic characters (letters). Most such codes, however, also represent symbols and various instructions necessary for conveying information.

At a minimum, an alphanumeric code must represent 10 decimal digits and 26 letters of the alphabet, for a total of 36 items. This number requires six bits in each code combination because five bits are insufficient ($2^5 = 32$). There are 64 total combinations of six bits, so we have 28 unused code combinations. Obviously, in many applications, symbols other than just numbers and letters are necessary to communicate completely. We need spaces to separate words, periods to mark the ends of sentences or to serve as decimal points, instructions to tell the receiving system what to do with the information, and more. So with codes that are six bits long, we can handle decimal numbers, the alphabet, and 28 other symbols. This should give you an idea of the requirements for a basic alphanumeric code.

ASCII A standardized alphanumeric code called the American Standard Code for Information Interchange (ASCII) is the most widely used type. It is a seven-bit code in which the decimal digits are represented by the 8421 BCD code preceded by 011. The letters of the alphabet and other symbols and instructions are represented by other code combinations, shown in Table 2–6. For instance, the letter A is represented by 1000001 (41_{16}), the comma by 0101100 ($2C_{16}$), and ETX (end of text) by 0000011 (03_{16}).

ASCII is the most commonly used input/output code for computers. For example, the keystrokes on the keyboard of a computer are converted directly into ASCII for processing by the digital circuits.

EXAMPLE 2–44

Determine the codes that are entered into the computer's memory when the following BASIC program statement is typed on the keyboard by the operator:

20 PRINT "A = "; X

Solution The ASCII code for each character is found in Table 2–6.

Character	ASCII	Hexadecimal
2	0110010	32_{16}
0	0110000	30_{16}
Space	0100000	20_{16}
P	1010000	50_{16}
R	1010010	52_{16}
I	1001001	49_{16}
N	1001110	$4E_{16}$
T	1010100	54_{16}
Space	0100000	20_{16}
"	0100010	22_{16}
A	1000001	41_{16}
=	0111101	$3D_{16}$
"	0100010	22_{16}
;	0111011	$3B_{16}$
X	1011000	58_{16}

TABLE 2–6
American Standard Code for Information Interchange.

LSBs	MSBs							
	000 (0)	001 (1)	010 (2)	011 (3)	100 (4)	101 (5)	110 (6)	111 (7)
0000 (0)	NUL	DLE	SP	0	@	P	`	p
0001 (1)	SOH	DC$_1$	!	1	A	Q	a	q
0010 (2)	STX	DC$_2$	"	2	B	R	b	r
0011 (3)	ETX	DC$_3$	#	3	C	S	c	s
0100 (4)	EOT	DC$_4$	$	4	D	T	d	t
0101 (5)	ENQ	NAK	%	5	E	U	e	u
0110 (6)	ACK	SYN	&	6	F	V	f	v
0111 (7)	BEL	ETB	'	7	G	W	g	w
1000 (8)	BS	CAN	(	8	H	X	h	x
1001 (9)	HT	EM	)	9	I	Y	i	y
1010 (A)	LF	SUB	*	:	J	Z	j	z
1011 (B)	VT	ESC	+	;	K	[	k	{
1100 (C)	FF	FS	,	<	L	\	l	|
1101 (D)	CR	GS	–	=	M	]	m	}
1110 (E)	SO	RS	.	>	N	↑	n	~
1111 (F)	SI	US	/	?	O	___	o	DEL

Definitions of control abbreviations:

ACK	Acknowledge	GS	Group separator
BEL	Bell	HT	Horizontal tab
BS	Backspace	LF	Line feed
CAN	Cancel	NAK	Negative acknowledge
CR	Carriage return	NUL	Null
DC$_1$–DC$_4$	Direct control	RS	Record separator
DEL	Delete idle	SI	Shift in
DLE	Data link escape	SO	Shift out
EM	End of Medium	SOH	Start of heading
ENQ	Enquiry	SP	Space
EOT	End of transmission	STX	Start text
ESC	Escape	SUB	Substitute
ETB	End of transmission block	SYN	Synchronous idle
ETX	End text	US	Unit separator
FF	Form feed	VT	Vertical tab
FS	Form separator		

NOTE: The hexadecimal digit representing each bit pattern in shown in parentheses.

Answers are found at the end of the chapter.

SECTION 2–11 REVIEW

1. Convert the following binary numbers to the Gray code:
   (a) 1100   **(b)** 1010   **(c)** 11010   **(d)** 101110
2. Convert the following Gray codes to binary:
   (a) 1000   **(b)** 1010   **(c)** 11101   **(d)** 00110
3. Convert the following decimal numbers to Excess-3 code:
   (a) 3   **(b)** 15   **(c)** 87   **(d)** 349
4. What is the ASCII representation for each of the following characters? Express each as a bit pattern and in hexadecimal notation.
   (a) K   **(b)** r   **(c)** $ **(d)** +

SUMMARY

☐ A binary number is a weighted number in which the weight of each whole number digit is a positive power of two and the weight of each fractional digit is a negative power of two. The weights increase from right to left—from least significant digit to most significant.

☐ A binary number can be converted to a decimal number by summing the decimal values of the weights of all the 1s in the binary number.

☐ A decimal whole number can be converted to binary by using the sum-of-weights or the repeated division-by-2 method.

☐ A decimal fraction can be converted to binary by using the sum-of-weights or the repeated multiplication-by-2 method.

☐ The basic rules for binary addition are as follows:

$$0 + 0 = 0$$
$$0 + 1 = 1$$
$$1 + 0 = 1$$
$$1 + 1 = 10$$

☐ The basic rules for binary subtraction are as follows:

$$0 - 0 = 0$$
$$1 - 1 = 0$$
$$1 - 0 = 1$$
$$10 - 1 = 1$$

☐ The 1's complement of a binary number is derived by changing 1s to 0s and 0s to 1s.

☐ The 2's complement of a binary number can be derived by adding 1 to the 1's complement.

☐ Binary subtraction can be accomplished with addition by using the 1's or 2's complement method.

☐ A positive binary number is represented by a 0 sign bit.

☐ A negative binary number is represented by a 1 sign bit.

☐ For arithmetic operations negative binary numbers are represented in 1's complement or 2's complement form.

☐ In an addition operation an overflow is possible when both numbers are positive or when both numbers are negative. An incorrect sign bit in the sum indicates the occurrence of an overflow.

☐ The octal number system consists of eight digits, 0 through 7.

☐ A decimal whole number can be converted to octal by using the repeated division-by-8 method.

☐ Octal-to-binary conversion is accomplished by simply replacing each octal digit with its three-bit binary equivalent. The process is reversed for binary-to-octal conversion.

☐ The hexadecimal number system consists of 16 digits and characters, 0 through 9 and A through F.

☐ One hexadecimal digit represents a four-bit binary number, and its primary usefulness is in simplifying bit patterns and making them easier to read.

☐ A decimal number is converted to BCD by replacing each decimal digit with the appropriate four-bit binary code.

☐ The ASCII is a seven-bit alphanumeric code that is widely used in computer systems for input and output of information.

GLOSSARY

Addend   In addition, the number that is added to another number called the *augend*.

Alphanumeric   Consisting of numerals, letters, and other characters.

ASCII   American Standard Code for Information Interchange. The most widely used alphanumeric code.

Augend   In addition, the number to which the addend is added.

BCD   Binary coded decimal. A digital code in which each of the decimal digits, 0 through 9, is represented by a group of four bits.

Binary number system   A number system that has a base of two and utilizes 1 and 0 as its digits.

Byte   A group of *eight* bits.

Character   A symbol, letter, or numeral.

Difference   The result of a subtraction.

Dividend   In a division operation, the quantity that is being divided.

Divisor   In a division operation, the quantity that is divided into the dividend.

End-around carry   In 1's complement subtraction, the addition of the final carry to the result.

Error correction   The process of correcting bit errors occurring in a digital code.

Error detection   The process of detecting bit errors occurring in a digital code.

Excess-3   An unweighted digital code in which each of the decimal digits is represented by a four-bit code derived by adding 3 to the decimal digit and then converting to binary.

Gray code   An unweighted digital code characterized by a single bit change between adjacent code numbers in a sequence.

Hexadecimal   A number system with a base of 16.

LSB   Least significant bit. The right-most bit in a binary number or code.

Minuend   The number from which another number is subtracted.

MSB   Most significant bit. The left-most bit in a binary number or code.

Multiplicand   The number that is being multiplied by another number.

Multiplier   The number that multiplies the multiplicand.

Numeric   Related to numbers.

Octal   A number system with a base of eight.

Overflow   The condition that occurs when the number of bits in a sum exceeds the number of bits in each of the numbers added.

Product   The result of a multiplication.

Quotient   The result of a division.

Radix   The base of a number system. The number of digits in a given number system.

Subtrahend   The number that is being subtracted from the minuend.

Weight   The value of a digit in a number based on its position in the number.

Word   A group of bits representing a complete piece of information.

SELF-TEST

Answers and solutions are found at the end of the book.

1. What is the base of the decimal number system? This means that there are _____ digits in the system.

2. Break each decimal number into a sum of the products of the digits and their weights.
   (a) 28 (b) 389 (c) 1473 (d) 10,956

3. Write the sequence of binary numbers for the decimal numbers 32 through 42.

4. Find the decimal equivalent of each binary number.
   (a) 1101 (b) 100101 (c) 11011101
   (d) 1011.011 (e) 1110.1011

5. Express each of the following decimal numbers in binary:
   (a) 17 (b) 102 (c) 555 (d) 72.66

6. Perform the indicated binary operations.
   (a) $110 + 011$ (b) $11010 + 01111$
   (c) $110 - 010$ (d) $11011 - 10110$
   (e) 101×111 (f) $1110 \div 0111$

7. Convert each binary number to its 1's complement and also to its 2's complement.
   (a) 110101 (b) 0001101

8. Determine the decimal equivalent of each of the signed binary numbers. The left-most bit is the sign bit.
   (a) 1010 (b) 0111 (c) 11010101

9. In an addition operation, how are negative binary numbers handled?

10. Explain overflow, and list the conditions under which an overflow is possible.

11. Convert each decimal number to octal.
    (a) 16 (b) 45 (c) 380

12. Express each octal number in binary.
    (a) 7041_8 (b) 65315_8

13. Express the binary number 111000010110010 in octal.

14. What is the hexadecimal representation for $0011110100010000101 1_2$?

15. What binary number does $5A34F_{16}$ represent?

16. Convert each decimal number to BCD.
    (a) 23 (b) 79 (c) 718 (d) 954.61

17. What decimal number does the BCD sequence 0100100110000110 represent?

18. Convert 101011011_2 to Gray code.

19. What binary number does Gray code 1100100 represent?

20. Determine the ASCII representations for H, X, :, =, and %.

PROBLEMS

Answers to odd-numbered problems are found at the end of the book.

Section 2–1 Decimal Numbers

2–1 What is the weight of the digit 6 in each of the following decimal numbers?
   (a) 1386 (b) 54,692 (c) 671,920

2–2 Express each of the following decimal numbers as a power of ten.
   (a) 10 (b) 100 (c) 10,000 (d) 1,000,000

2–3 Give the value of each digit in the following decimal numbers.
   (a) 471 (b) 9356 (c) 125,000

2–4 How high can you count with four decimal digits?

Section 2–2 Binary Numbers

2–5 Convert the following binary numbers to decimal:
  (a) 11 **(b)** 100 **(c)** 111 **(d)** 1000
  (e) 1001 **(f)** 1100 **(g)** 1011 **(h)** 1111

2–6 Convert the following binary numbers to decimal:
  (a) 1110 **(b)** 1010 **(c)** 11100 **(d)** 10000
  (e) 10101 **(f)** 11101 **(g)** 10111 **(h)** 11111

2–7 Convert each binary number to decimal.
  (a) 110011.11 **(b)** 101010.01 **(c)** 1000001.111
  (d) 1111000.101 **(e)** 1011100.10101 **(f)** 1110001.0001
  (g) 1011010.1010 **(h)** 1111111.11111

2–8 What is the highest decimal number that can be represented by each of the following numbers of binary digits (bits)?
  (a) two **(b)** three **(c)** four **(d)** five
  (e) six **(f)** seven **(g)** eight **(h)** nine
  (i) ten **(j)** eleven

2–9 How many bits are required to represent the following decimal numbers?
  (a) 17 **(b)** 35 **(c)** 49 **(d)** 68
  (e) 81 **(f)** 114 **(g)** 132 **(h)** 205
  (i) 271

2–10 Generate the binary counting sequence for each decimal sequence.
  (a) 0 through 7 **(b)** 8 through 15 **(c)** 16 through 31
  (d) 32 through 63 **(e)** 64 through 75

Section 2–3 Decimal-to-Binary Conversion

2–11 Convert each decimal number to binary by using the sum-of-weights method.
  (a) 10 **(b)** 17 **(c)** 24 **(d)** 48
  (e) 61 **(f)** 93 **(g)** 125 **(h)** 186
  (i) 298

2–12 Convert each decimal number to binary by repeated division by 2.
  (a) 15 **(b)** 21 **(c)** 28 **(d)** 34
  (e) 40 **(f)** 59 **(g)** 65 **(h)** 73
  (i) 99

Section 2–4 Binary Arithmetic

2–13 Add the binary numbers.
  (a) 11 + 01 **(b)** 10 + 10 **(c)** 101 + 11
  (d) 111 + 110 **(e)** 1001 + 101 **(f)** 1101 + 1011

2–14 Use direct subtraction on the following binary numbers:
  (a) 11 − 1 **(b)** 101 − 100 **(c)** 110 − 101
  (d) 1110 − 11 **(e)** 1100 − 1001 **(f)** 11010 − 10111

2–15 Perform the following binary multiplications:
  (a) 11 × 11 **(b)** 100 × 10 **(c)** 111 × 101
  (d) 1001 × 110 **(e)** 1101 × 1101 **(f)** 1110 × 1101

2–16 Divide the binary numbers as indicated.
  (a) 100 ÷ 10 **(b)** 1001 ÷ 11 **(c)** 1100 ÷ 100

Section 2–5 1's and 2's Complements

2–17 Determine the 1's complement of each binary number.
  (a) 101 **(b)** 110 **(c)** 1010
  (d) 11010111 **(e)** 1110101 **(f)** 00001

2–18 Perform the following binary subtractions by using the 1's complement method:
    (a) $11 - 10$ **(b)** $100 - 11$ **(c)** $1010 - 111$
    (d) $1101 - 1010$ **(e)** $11100 - 1101$ **(f)** $100001 - 1010$
    (g) $1001 - 1110$ **(h)** $10111 - 11111$

2–19 Determine the 2's complement of each binary number.
    (a) 10 **(b)** 111 **(c)** 1001 **(d)** 1101
    (e) 11100 **(f)** 10011 **(g)** 10110000 **(h)** 00111101

2–20 Perform the following subtractions by using the 2's complement method:
    (a) $10 - 01$ **(b)** $111 - 110$ **(c)** $1101 - 1001$
    (d) $1111 - 1101$ **(e)** $10111 - 10011$ **(f)** $10001 - 11100$
    (g) $10101 - 10111$ **(h)** $1111000 - 1111111$

Section 2–6   Binary Representation of Signed Numbers

2–21 Convert each negatively signed true binary number to 1's complement form.
    (a) 10010001 **(b)** 11100111 **(c)** 10000101 **(d)** 11110111

2–22 Convert each negatively signed true binary number to 2's complement form.
    (a) 10000110 **(b)** 11111000 **(c)** 10101010 **(d)** 10001111

Section 2–7   Binary Arithmetic with Signed Numbers

2–23 Convert each decimal number to signed binary, and perform each addition, using the 2's complement method when appropriate. Leave negative results in 2's complement form.
    (a) $33 + 15$ **(b)** $56 + (-27)$ **(c)** $99 + (-75)$
    (d) $-46 + 25$ **(e)** $-110 + (-84)$

2–24 Perform the additions by using the 2's complement method. Each signed binary number is shown in true (uncomplemented) form. Leave negative results in 2's complement form.
    (a) $01001 + 10110$ **(b)** $00110 + 10011$ **(c)** $10100 + 10010$
    (d) $11101 + 10001$ **(e)** $11000 + 00111$ **(f)** $10101 + 01010$

2–25 Repeat Problem 2–24, using the 1's complement method.

2–26 Perform each subtraction by converting the decimal numbers to signed binary and using the 2's complement method. Leave negative results in 2's complement form.
    (a) $43 - 27$ **(b)** $20 - (-14)$ **(c)** $30 - (-45)$

2–27 Multiply the signed true binary numbers by using repeated addition.
                    Multiplicand: 011110110
                     Multiplier: 100000100

2–28 Divide 010000100 by 001100011. These are signed true binary numbers.

Section 2–8   Octal Numbers

2–29 Convert each octal number to decimal.
    (a) 12_8 **(b)** 27_8 **(c)** 56_8
    (d) 64_8 **(e)** 103_8 **(f)** 557_8
    (g) 163_8 **(h)** 1024_8 **(i)** 7765_8

2–30 Convert each decimal number to octal by repeated division by 8.
    (a) 15 **(b)** 27 **(c)** 46 **(d)** 70
    (e) 100 **(f)** 142 **(g)** 219 **(h)** 435
    (i) 791

2–31 Convert each octal number to binary.
    (a) 13_8 **(b)** 57_8 **(c)** 101_8
    (d) 321_8 **(e)** 540_8 **(f)** 4653_8
    (g) 13271_8 **(h)** 45600_8 **(i)** 100213_8
    (j) 103.45_8

2–32 Convert each binary number to octal.

(a)	111	**(b)**	10	**(c)**	110111
(d)	101010	**(e)**	1100	**(f)**	1011110
(g)	101100011001	**(h)**	10110000011	**(i)**	111111101111000
(j)	10011.011				

Section 2–9 Hexadecimal Numbers

2–33 Convert each hexadecimal number to binary.

(a)	38_{16}	**(b)**	59_{16}	**(c)**	$A14_{16}$	**(d)**	$5C8_{16}$
(e)	4100_{16}	**(f)**	$FB17_{16}$	**(g)**	$8A.9_{16}$		

2–34 Convert each binary number to hexadecimal.

(a)	1110	**(b)**	10	**(c)**	10111
(d)	10100110	**(e)**	1111110000	**(f)**	100110000010

2–35 Convert each hexadecimal number to decimal.

(a)	23_{16}	**(b)**	92_{16}	**(c)**	$1A_{16}$	**(d)**	$8D_{16}$
(e)	$F3_{16}$	**(f)**	EB_{16}	**(g)**	$5C2_{16}$	**(h)**	700_{16}

Convert each decimal number to hexadecimal.

2–36

(a)	8	**(b)**	14	**(c)**	33	**(d)**	52
(e)	284	**(f)**	2890	**(g)**	4019	**(h)**	6500

2–37 Perform the following additions.

(a)	$37_{16} + 29_{16}$	**(b)**	$A0_{16} + 6B_{16}$	**(c)**	$FF_{16} + BB_{16}$

2–38 Perform the following subtractions.

(a)	$51_{16} - 40_{16}$	**(b)**	$C8_{16} - 3A_{16}$	**(c)**	$FD_{16} - 88_{16}$

Section 2–10 Binary Coded Decimal (BCD)

2–39 Convert each of the following decimal numbers to 8421 BCD:

(a)	10	**(b)**	13	**(c)**	18
(d)	21	**(e)**	25	**(f)**	36
(g)	44	**(h)**	57	**(i)**	69
(j)	98	**(k)**	125	**(l)**	156

2–40 Convert each of the decimal numbers in Problem 2–39 to straight binary, and compare the number of bits required with that required for BCD.

2–41 Convert the following decimal numbers to BCD:

(a)	104	**(b)**	128	**(c)**	132
(d)	150	**(e)**	186	**(f)**	210
(g)	359	**(h)**	547	**(i)**	1051
(j)	2563				

2–42 Convert each of the BCD code numbers to decimal.

(a)	0001	**(b)**	0110	**(c)**	1001
(d)	00011000	**(e)**	11001	**(f)**	00110010
(g)	1000101	**(h)**	10011000	**(i)**	100001110000
(j)	011000011001				

2–43 Convert each of the BCD code numbers to decimal.

(a)	10000000	**(b)**	1000110111
(c)	1101000110	**(d)**	10000100001
(e)	11101010100	**(f)**	100000000000
(g)	100101111000	**(h)**	1011010000011
(i)	1001000000011000	**(j)**	0110011001100111

2–44 Add the following BCD numbers:

(a)	0010 + 0001	**(b)**	0101 + 0011
(c)	0111 + 0010	**(d)**	1000 + 0001
(e)	00011000 + 00010001	**(f)**	01100100 + 00110011
(g)	01000000 + 01000111	**(h)**	10000101 + 00010011

2–45 Add the following BCD numbers:
   (a) 1000 + 0110   **(b)** 0111 + 0101
   (c) 1001 + 1000   **(d)** 1001 + 0111
   (e) 00100101 + 00100111   **(f)** 01010001 + 01011000
   (g) 10011000 + 10010111   **(h)** 010101100001 + 011100001000

2–46 Convert each pair of decimal numbers to BCD, and add as indicated.
   (a) 4 + 3   **(b)** 5 + 2   **(c)** 6 + 4
   (d) 17 + 12   **(e)** 28 + 23   **(f)** 65 + 58
   (g) 113 + 101 **(h)** 295 + 157

Section 2–11 Digital Codes

2–47 In a certain application a four-bit binary sequence cycles from 1111 to 0000 periodically. There are four bit changes, and because of circuit delays, these changes may not occur at the same instant. For example, if the LSB changes first, the number will appear as 1110 during the transition from 1111 to 0000 and may be misinterpreted by the system. Illustrate how the Gray code avoids this problem.

2–48 Convert each binary number to Gray code.
   (a) 11011   **(b)** 1001010   **(c)** 1111011101110

2–49 Convert each Gray code to binary.
   (a) 1010   **(b)** 00010   **(c)** 11000010001

2–50 Convert each of the following decimal numbers to Excess-3 code:
   (a) 1   **(b)** 3   **(c)** 6
   (d) 10   **(e)** 18   **(f)** 29
   (g) 56   **(h)** 75   **(i)** 107

2–51 Convert each Excess-3 code number to decimal.
   (a) 0011   **(b)** 1001
   (c) 0111   **(d)** 01000110
   (e) 01111100 **(f)** 10000101

2–52 Decode the following ASCII coded message.

   1001000110010111011001101100110111101011110
   0100000100100011011111111011101000001100001
   1110010110010101000001111001110111111110101
   0111111

2–53 Write the message in Problem 2–52 in hexadecimal.

2–54 Convert the following computer program statement to ASCII:

   30 INPUT A, B

ANSWERS TO SECTION REVIEWS

Section 2–1 Review

1. **(a)** 10   **(b)** 100   **(c)** 1000   **(d)** 10,000
2. **(a)** $5 \times 10 + 1 \times 1$ **(b)** $1 \times 100 + 3 \times 10 + 7 \times 1$
   (c) $1 \times 1000 + 4 \times 100 + 9 \times 10 + 2 \times 1$
   (d) $1 \times 10,000 + 6 \times 100 + 5 \times 10 + 8 \times 1$

Section 2–2   Review

1. 16   **2.** **(a)** 10   **(b)** 26   **(c)** 189   **(d)** 6.625

Section 2–3   Review

1. **(a)** 10111   **(b)** 111001   **(c)** 101101.1
2. **(a)** 1110   **(b)** 10101   **(c)** 0.011

Section 2–4   Review

1. **(a)** 10111   **(b)** 100100
2. **(a)** 1001   **(b)** 0010
3. **(a)** 101010   **(b)** 100

Section 2–5   Review

1. **(a)** 00101   **(b)** 01000   **(c)** 110010
2. **(a)** 0110   **(b)** −00101
3. **(a)** 01001   **(b)** 00001   **(c)** 101111
4. **(a)** 0101   **(b)** −0101

Section 2–6   Review

1. **(a)** +27   **(b)** −93   **(c)** +81
2. **(a)** 1011010   **(b)** 1000101   **(c)** 1011101111

Section 2–7   Review

1. **(a)** 01111011   **(b)** 11010111
2. **(a)** 010111101   **(b)** 00011110
3. **(a)** 01111   **(b)** 1110000

Section 2–8   Review

1. **(a)** 59   **(b)** 85
2. **(a)** 142_8 **(b)** 243_8
3. **(a)** 100110   **(b)** 111010011   **(c)** 101110010100.011111
4. **(a)** 657_8 **(b)** 461.57_8 **(c)** 277.14_8

Section 2–9   Review

1. **(a)** $B3_{16}$ **(b)** $CE8_{16}$
2. **(a)** 01010111   **(b)** 001110100101
    (c) 1111100000001011
3. **(a)** $4C_{16}$ **(b)** 69_{16}
4. **(a)** 54_{16} **(b)** 38_{16}

Section 2–10   Review

1. **(a)** 2   **(b)** 8   **(c)** 1   **(d)** 4
2. **(a)** 0110   **(b)** 00010101   **(c)** 001001110011   **(d)** 10000100.1001
3. **(a)** 89   **(b)** 278   **(c)** 15.7

Section 2–11   Review

1. **(a)** 1010   **(b)** 1111   **(c)** 10111   **(d)** 111001
2. **(a)** 1111   **(b)** 1100   **(c)** 10110   **(d)** 00100
3. **(a)** 0110   **(b)** 01001000   **(c)** 10111010   **(d)** 011001111100
4. **(a)** 1001011 $\longrightarrow$ $4B_{16}$ **(b)** 1110010 $\longrightarrow$ 72_{16}
    (c) 0100100 $\longrightarrow$ 24_{16} **(d)** 0101011 $\longrightarrow$ $2B_{16}$

The emphasis in this chapter is on the logical operation, application, and troubleshooting of logic gates. The waveform relationship between the inputs and the output of the various types of gates is stressed by the use of timing diagrams.

The logic symbols used to represent the various types of gates are in accordance with ANSI/IEEE Standard 91–1984. This standard has been adopted by major corporations for use in their internal documentation as well as in their published literature. It has also been adopted by the military.

Because of the predominant use of integrated circuits (IC) in digital systems, the logic function of a device generally is of greater importance to the technician or technologist than the details of the component-level circuit operation buried within the package. Therefore, detailed coverage of the devices at the component level can be treated as an optional topic. For those who need it and have the time, a thorough coverage of digital IC technologies is available in Chapter 13. That "floating" chapter can be used after this chapter or at other appropriate points.

The following specific devices are introduced in this chapter: 7400, 7402, 7404, 7408, 7410, 7411, 7420, 7427, 7430, 7432, 7486, and 74133.

After completing this chapter, you will be able to

☐ describe the operation of the inverter, the AND gate, and the OR gate.

☐ describe the operation of the NAND gate and the NOR gate.

☐ describe the operation of the exclusive-OR and exclusive-NOR gates.

☐ recognize and use both the distinctive shape logic gate symbols and the rectangular outline logic gate symbols of ANSI/IEEE Standard 91–1984.

☐ construct timing diagrams showing the proper time relationships of inputs and outputs for the various logic gates.

☐ make basic comparisons between the two major IC technologies, TTL and CMOS.

☐ explain how the different series within the TTL and CMOS families differ from each other.

☐ define *propagation delay time, power dissipation, speed-power product,* and *fan-out,* in relation to logic gates.

☐ list specific devices that contain the various logic gates.

☐ use each logic gate in simple applications.

☐ troubleshoot logic gates for opens and shorts by using the logic pulser and probe or the oscilloscope.

3

Logic Gates

THE INVERTER

3–1

The inverter (NOT circuit) performs a basic logic function called **inversion** *or* **complementation.** *The purpose of the inverter is to change one logic level to the opposite level. In terms of bits, it changes a 1 to a 0 and a 0 to a 1.*

Standard logic symbols for the inverter are shown in Figure 3–1. Part (a) of the figure shows the *distinctive shape* symbols, and part (b) shows the *rectangular outline* symbols. In this text distinctive shape symbols are used; however, the rectangular outline symbols are found in many industry publications, and you should become familiar with them as well.

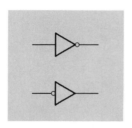

(a) Distinctive shape symbols
with negation indicators

(b) Rectangular outline symbols
with polarity indicators

FIGURE 3–1
Standard logic symbols for the inverter
(ANSI/IEEE Std. 91–1984).

The Negation and Polarity Indicators

The *negation indicator* is a "bubble" (○) appearing on the input or output of a logic element, as shown in Figure 3–1(a). When appearing on the input, the bubble means that a 0 is the active input state. When appearing on the output, the bubble means that a 0 is the active output state. The absence of a bubble on the input or output means that a 1 is the active state. Generally, inputs are on the left of a logic symbol and the output is on the right.

The *polarity indicator* is a "triangle" (◺) appearing on the input or output of a logic element, as shown in Figure 3–1(b). When appearing on the input, it means that a LOW level is the active input state. When appearing on the output, it means that a LOW level is the active output state.

Either indicator (bubble or triangle) can be used both on distinctive shape symbols and on rectangular outlines. The shaded portion of the figure, part (a), indicates the principal inverter symbol used in this text. Note that a change in the placement of the negation or polarity indicator does not imply a change in the way an inverter operates.

Because positive logic (HIGH = 1, LOW = 0) is used in this text, both indicators are equivalent and can be interchanged. However, we will use primarily the negation indicator (bubble) throughout this book to represent an inversion.

The placement of the bubble on the input or output of a logic element is determined by the *active state* of the signal (pulse or level). The active state is the state (1 or 0) when a signal is considered to be present. When the active state is a 0, the bubble is used. When the active state is a 1, the bubble is not used. When a signal is active, it is said to be asserted. When it is not active, it is unasserted.

Inverter Operation

When a HIGH level is applied to an inverter input, a LOW level will appear on its output. When a LOW level is applied to its input, a HIGH will appear on its output. This operation is summarized in Table 3–1, which shows the output for each possible input in terms of levels and bits. Such tables are called *truth tables*.

TABLE 3–1
Inverter truth tables.

Input	Output
LOW	HIGH
HIGH	LOW

Input	Output
0	1
1	0

Pulsed Operation

Figure 3–2 shows the output of an inverter for a pulse input, where t_1 and t_2 indicate the corresponding points on the input and output pulse waveforms. Note that when the input is LOW, the output is HIGH, and when the input is HIGH, the output is LOW, thereby producing an *inverted* output pulse.

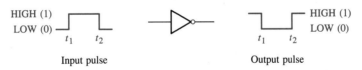

FIGURE 3–2
Inverter with pulse input.

EXAMPLE 3–1

A pulse waveform is applied to an inverter in Figure 3–3(a). Determine the output waveform corresponding to the input. According to the placement of the bubble, what is the active output state?

FIGURE 3–3

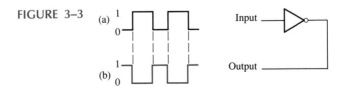

Solution The output waveform is exactly opposite to the input (inverted) at each point, as shown in Figure 3–3(b), which is a basic *timing diagram*. The active output state is 0.

SECTION 3–1
REVIEW

Answers are found at the end of the chapter.

1. When a 1 is on the input of an inverter, what is the output?

2. An active HIGH pulse (HIGH level when present, LOW level when absent) is required on an inverter input.

    (a) Draw the appropriate logic symbol, using the distinctive shape and the negation indicator, for the inverter in this application.

    (b) Describe the output signal when the pulse is present.

THE AND GATE

3–2

*The AND gate performs logical multiplication, more commonly known as the **AND** function. The mathematical aspects of this function are discussed in Chapter 4 on Boolean Algebra.*

The AND gate is composed of two or more inputs and a single output, as indicated by the standard logic symbols shown in Figure 3–4. Inputs are on the left, and the output is on the right in each symbol. Gates with two inputs are shown; however, an AND gate can have any number of inputs greater than one. Although examples of both distinctive shape symbols and rectangular outline symbols are shown, the distinctive shape symbol, shown shaded in part (a), is used predominantly in this book.

FIGURE 3–4

Standard logic symbols for the AND gate showing two inputs (ANSI/IEEE Std. 91–1984).

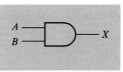

(a) Distinctive shape

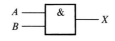

(b) Rectangular outline
with AND (&) qualifying symbol

Logical Operation of the AND Gate

The operation of the AND gate is such that the output is HIGH only when *all* of the inputs are HIGH. When *any* of the inputs are LOW, the output is LOW. Therefore, the basic purpose of an AND gate is to determine when certain conditions are simultaneously true, as indicated by HIGH levels on all of its inputs, and to produce a HIGH on its output to indicate that all these conditions are true. The inputs of the two-input AND gate in Figure 3–4 are labeled A and B, and the output is labeled X. We can express the gate operation with the following description:

If A AND B are HIGH, then X is HIGH. If A is LOW, or if B is LOW, or if both A and B are LOW, then X is LOW.

The HIGH level is the active output level for the AND gate. Figure 3–5 illustrates a two-input AND gate with all four possibilities of input combinations, and the resulting output for each.

FIGURE 3–5

All possible logic levels for a two-input AND gate.

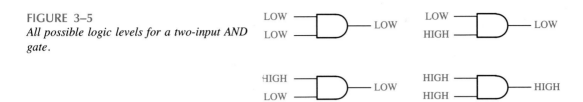

We generally express the logical operation of a gate with a table that lists all input combinations and the corresponding outputs. This table of combinations is also called a *truth table* and is illustrated in Table 3–2 for a two-input AND gate. The truth table can be expanded to any number of inputs.

TABLE 3–2

Truth table for a two-input AND gate.

Inputs		Output
A	*B*	*X*
0	0	0
0	1	0
1	0	0
1	1	1

NOTE: 1 ≡ HIGH, 0 ≡ LOW.

Although the terms HIGH and LOW tend to give a ''physical'' sense to input and output states, the truth table is shown with 1s and 0s, since a HIGH is equivalent to a 1 and a LOW is equivalent to a 0 in positive logic.

For any AND gate, regardless of the number of inputs, the output is HIGH *only* when *all* inputs are HIGH.

The total number of possible combinations of binary inputs is determined by the following formula:

$$N = 2^n \qquad\qquad \textbf{(3–1)}$$

where N is the total possible combinations and n is the number of input variables. To illustrate, the following calculations are made using Equation (3–1):

$$
\begin{aligned}
\text{For } two \text{ input variables:} &\quad N = 2^2 = 4 \\
\text{For } three \text{ input variables:} &\quad N = 2^3 = 8 \\
\text{For } four \text{ input variables:} &\quad N = 2^4 = 16
\end{aligned}
$$

This is how we determine the number of combinations for gates with any number of inputs.

EXAMPLE 3–2

(a) Develop the truth table for a three-input AND gate.

(b) Determine the total number of possible input combinations for a five-input AND gate.

Solutions

(a) There are eight possible input combinations for a three-input AND gate. The input side of the truth table (Table 3–3) shows all eight combinations of three bits. The output side is all 0s except when all three input bits are 1s.

TABLE 3–3

Inputs			Output
A	B	C	X
0	0	0	0
0	0	1	0
0	1	0	0
0	1	1	0
1	0	0	0
1	0	1	0
1	1	0	0
1	1	1	1

(b) $N = 2^5 = 32$

There are 32 possible combinations of inputs for a five-input AND gate.

Pulsed Operation

In a majority of applications, the inputs to a gate are not stationary levels but are voltages that change frequently between two logic levels and that can be classified as pulse waveforms. We will now look at the operation of AND gates with pulsed input waveforms. Keep in mind that an AND gate obeys the truth table operation regardless of whether its inputs are constant levels or pulsed levels.

In examining the pulsed operation of the AND gate, we will look at the inputs with respect to each other in order to determine the output level at any given time. For example, in Figure 3–6, the inputs are both HIGH (1) during the interval t_1, making the output HIGH (1) during this interval. During interval t_2, input A is LOW (0) and input B is HIGH (1), so the output is LOW (0). During interval t_3, both inputs are HIGH (1) again, and therefore the output is HIGH (1). During interval t_4, input A is high (1) and input B is LOW (0), resulting in a LOW (0) output. Finally, during interval t_5, input A is LOW (0), input B is LOW (0), and the output is therefore LOW (0). A diagram of input and output waveforms showing time relationships is called a *timing diagram*.

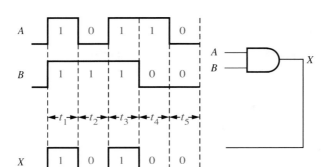

FIGURE 3-6
Example of pulsed AND gate operation.

EXAMPLE 3-3

If two waveforms, *A* and *B*, are applied to the AND gate inputs as in Figure 3–7(a), what is the resulting output waveform?

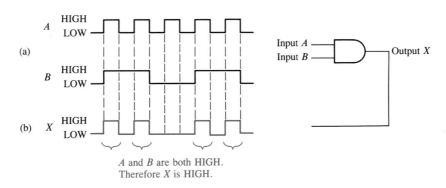

A and *B* are both HIGH.
Therefore *X* is HIGH.

FIGURE 3-7

Solution   See Figure 3–7(b). Waveform *X* is HIGH only when both *A* and *B* are HIGH.

It is very important, when analyzing the pulsed operation of logic gates, to pay careful attention to the time relationships of all the inputs to each other and to the output.

EXAMPLE 3-4

For the two input waveforms, *A* and *B*, in Figure 3–8(a), sketch the output waveform, showing its proper relation to the inputs for a two-input AND gate.

Solution   The output is HIGH only when both of the inputs are HIGH. See Figure 3–8(b).

FIGURE 3–8

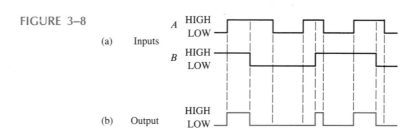

(a)   Inputs

(b)   Output

EXAMPLE 3–5

For the three-input AND gate in Figure 3–9(a), determine the output waveform in proper relation to the inputs.

FIGURE 3–9

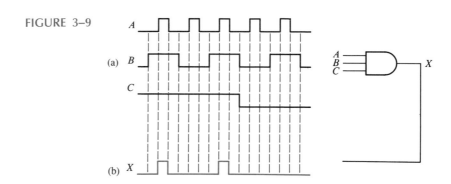

(a)

(b)

Solution   See Figure 3–9(b). The output of a three-input AND gate is HIGH only when all three inputs are HIGH.

SECTION 3–2
REVIEW

Answers are found at the end of the chapter.
1. When is the output of an AND gate HIGH?
2. When is the output of an AND gate LOW?
3. Develop the truth table for a four-input AND gate.

THE OR GATE

3–3

*The OR gate performs logical addition, more commonly known as the **OR function**. The mathematical aspects of this operation will be covered in Chapter 4, on Boolean algebra.*

An OR gate has two or more inputs and one output, as indicated by the standard logic symbols in Figure 3–10, where OR gates with two inputs are illustrated. An OR gate can have any number of inputs greater than one. Although both distinctive shape and rectangular outline symbols are shown for familiarization, the distinctive shape OR gate symbol will be used predominantly.

FIGURE 3–10
Standard logic symbols for the OR gate showing two inputs (ANSI/IEEE Std. 91–1984).

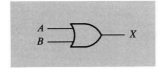

(a) Distinctive shape

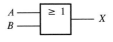

(b) Rectangular outline
with OR ($\geq$ 1) qualifying symbol

Logical Operation of the OR Gate

The operation of the OR gate is such that a HIGH on the output is produced when *any* of the inputs are HIGH. The output is LOW only when *all* of the inputs are LOW. Therefore, the purpose of an OR gate is to determine when one or more of its inputs are HIGH and to produce a HIGH on its output to indicate this condition. The inputs of the two-input OR gate in Figure 3–10 are labeled A and B, and the output is labeled X. We can express the operation of the gate as follows:

If either A OR B OR both are HIGH, then X is HIGH. If both A and B are LOW, then X is LOW.

The HIGH level is the active output level for the OR gate. Figure 3–11 illustrates the logic operation for a two-input OR gate for all four possible input combinations.

FIGURE 3–11
All possible logic levels for a two-input OR gate.

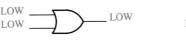

The logical operation of the two-input OR gate is described in Table 3–4. This truth table can be expanded for any number of inputs; but regardless of the number of inputs, the output is HIGH when *any* of the inputs are HIGH.

TABLE 3–4
Truth table for a two-input OR gate.

Inputs		Output
A	B	X
0	0	0
0	1	1
1	0	1
1	1	1

NOTE: 1 $\equiv$ HIGH, 0 $\equiv$ LOW.

Pulsed Operation

Let us now turn our attention to the operation of an OR gate with pulsed inputs, keeping in mind what we have learned about its logical operation.

Again, the important thing in analysis of gate operation with pulsed waveforms is the relationship of all the waveforms involved. For example, in Figure 3–12, inputs A

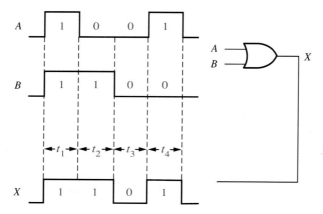

FIGURE 3–12
Example of pulsed OR gate operation.

and B are both HIGH (1) during interval t_1, making the output HIGH (1). During interval t_2, input A is LOW (0), but because input B is HIGH (1), the output is HIGH (1). Both inputs are LOW (0) during interval t_3, and we have a LOW (0) output during this time. During t_4, the output is HIGH (1) because input A is HIGH (1).

In this illustration, we have simply applied the truth table operation of the OR gate to each of the intervals during which the levels are nonchanging. A few examples further illustrate OR gate operation with pulse waveforms on the inputs.

EXAMPLE 3–6

If the two waveforms in Figure 3–13(a) are applied to the OR gate, what is the resulting output waveform?

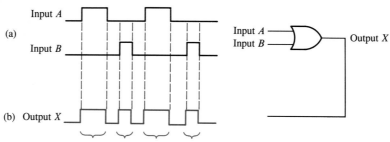

FIGURE 3–13

Solution See Figure 3–13(b). The output of a two-input OR gate is HIGH when either or both inputs are HIGH.

EXAMPLE 3–7

For the two input waveforms in Figure 3–14(a), sketch the output waveform, showing its proper relation to the inputs for a two-input OR gate.

FIGURE 3–14

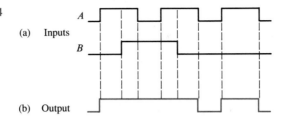

(a)   Inputs

(b)   Output

Solution   When either input or both inputs are HIGH, the output is HIGH. See Figure 3–14(b).

EXAMPLE 3–8

For the three-input OR gate in Figure 3–15(a), determine the output waveform in proper relation to the inputs.

FIGURE 3–15

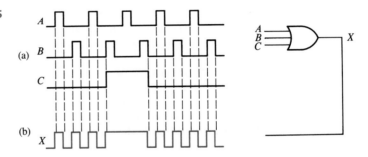

(a)

(b)

Solution   See Figure 3–15(b). The output is HIGH when any of the inputs are HIGH.

SECTION 3–3
REVIEW

Answers are found at the end of the chapter.
1. When is the output of an OR gate HIGH?
2. When is the output of an OR gate LOW?
3. Develop the truth table for a three-input OR gate.

THE NAND GATE

3–4

The NAND gate is a very popular logic element because it can be used as a universal function; that is, it can be used to construct an AND gate, an OR gate, an inverter, or any combination of these functions. In Chapter 5 we will examine this universal property of the NAND gate. In this chapter we are going to look at the logical operation of the NAND gate.

The term *NAND* is a contraction of *NOT-AND* and implies an AND function with a complemented (inverted) output. A standard logic symbol for a two-input NAND gate and its equivalency to an AND gate followed by an inverter are shown in Figure 3–16(a). A rectangular outline representation is shown in part (b).

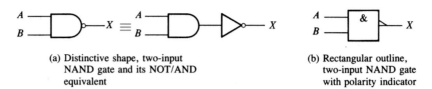

(a) Distinctive shape, two-input
NAND gate and its NOT/AND
equivalent

(b) Rectangular outline,
two-input NAND gate
with polarity indicator

FIGURE 3–16
Standard NAND gate logic symbols (ANSI/IEEE Std. 91–1984).

Logical Operation of the NAND Gate

The logical operation of the *NAND gate* is such that a LOW output occurs *only* when all inputs are HIGH. When any of the inputs are LOW, the output will be HIGH. For the specific case of a two-input NAND gate, as shown in Figure 3–16 with the inputs labeled *A* and *B* and the output labeled *X*, we can state the operation as follows:

If *A* AND *B* are HIGH, then *X* is LOW. If *A* is LOW or *B* is LOW, or if both *A* and *B* are LOW, then *X* is HIGH.

Note that this operation is opposite that of the AND as far as output is concerned. In a NAND gate, the LOW level (0) is the *active* output level, as indicated by the bubble on the output. Figure 3–17 illustrates the logical operation of a two-input NAND gate for all four input combinations, and the truth table summarizing the logical operation of the two-input NAND gate is Table 3–5.

FIGURE 3–17
Logical operation of a two-input NAND gate.

LOW / LOW → HIGH
LOW / HIGH → HIGH
HIGH / LOW → HIGH
HIGH / HIGH → LOW

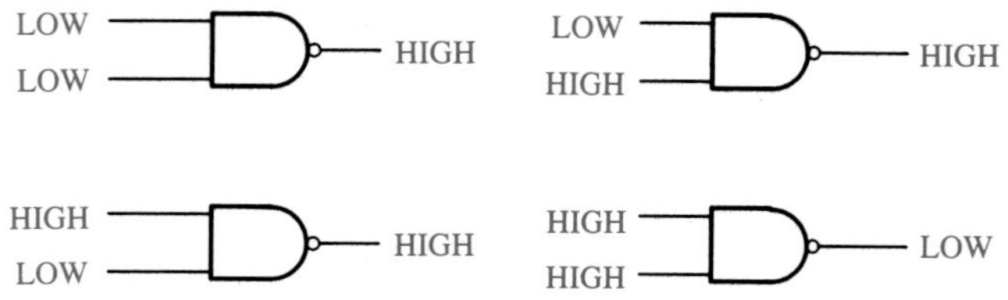

TABLE 3–5
Truth table for a two-input NAND gate.

Inputs		Output
A	*B*	*X*
0	0	1
0	1	1
1	0	1
1	1	0

NOTE: 1 ≡ HIGH, 0 ≡ LOW.

Pulsed Operation

We will now look at the pulsed operation of the NAND gate. Remember from the truth table that any time *all* of the inputs are HIGH, the output will be LOW, and this is the *only* time a LOW output occurs. The following examples illustrate pulsed operation.

EXAMPLE 3–9

If the two waveforms shown in Figure 3–18(a) are applied to the NAND gate, determine the resulting output waveform.

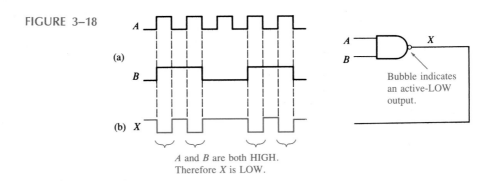

FIGURE 3–18

(a)

(b) *X*

A and *B* are both HIGH.
Therefore *X* is LOW.

Bubble indicates
an active-LOW
output.

Solution See Figure 3–18(b). Waveform *X* is LOW only when both *A* and *B* are HIGH.

EXAMPLE 3–10

Sketch the output waveform for the three-input NAND gate in Figure 3–19(a), showing its proper relationship to the inputs.

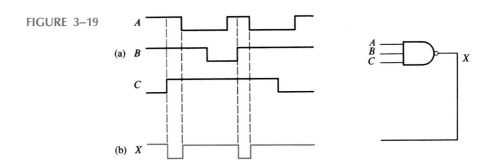

FIGURE 3–19

(a) *B*

C

(b) *X*

Solution The output is LOW only when all three inputs are HIGH. See Figure 3–19(b).

The NAND Gate as an OR Gate with Active-LOW Inputs

Inherent in the NAND gate's operation is the fact that one or more LOW inputs produce a HIGH output. If you look at Table 3–5, you will see that the output is HIGH (1) when any or all of the inputs are LOW (0). These conditions can be stated as follows:

If *A* is LOW OR *B* is LOW, OR if both *A* and *B* are LOW, then *X* is HIGH.

From this viewpoint, we have an OR operation that requires one or more LOW inputs to produce a HIGH output. This is referred to as *negative-OR*. The term *negative* in this context does not imply negative logic; it means only that the inputs are defined to be in the active state when LOW. When the NAND gate is looking for one or more LOWs on its inputs rather than for all HIGHs, it is *acting* as a negative-OR gate and is represented by the standard logic symbol in Figure 3–20(b). The two symbols in Figure 3–20 represent the same gate, but they also serve to define its role in a particular application, as illustrated by the following examples.

FIGURE 3–20
Standard symbols representing the two equivalent functions of the NAND gate.

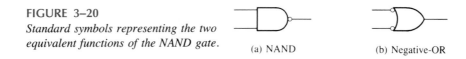

(a) NAND (b) Negative-OR

EXAMPLE 3–11

A manufacturing plant uses two tanks to store a certain liquid chemical that is required in a manufacturing process. Each tank has a sensor that detects when the chemical level drops to 25% of full. The sensors produce a 5 V level when the tanks are more than one-quarter full. When the volume of chemical in a tank drops to one-quarter full, the sensor puts out a 0 V level.

It is required that a single green lamp on an indicator panel show when *both* tanks are more than one quarter full. Show how a NAND gate can be used to implement this function.

FIGURE 3–21

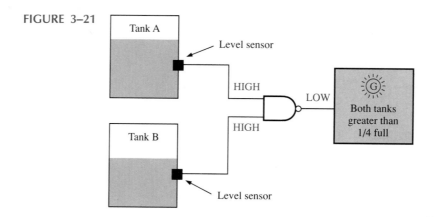

Solution Figure 3–21 shows a NAND gate with its two inputs connected to the tank level sensors and its output connected to the indicator panel.

As long as both sensor outputs are HIGH (5 V), indicating that both tanks are more than one quarter full, the NAND gate output is LOW (0 V). The green-lamp circuit is arranged so that a LOW-level voltage turns it on.

EXAMPLE 3–12

The supervisor of the manufacturing process described in Example 3–11 has decided that he would prefer to have a red lamp come on when at least one of the tanks falls to the quarter-full level rather than have the green lamp indicate when both are above one quarter. Show how this requirement can be implemented.

FIGURE 3–22

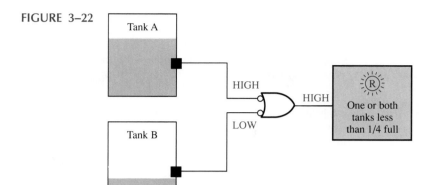

Solution Figure 3–22 shows the NAND gate operating as a negative-OR gate to detect the occurrence of at least one LOW on its inputs. A sensor puts out a low-level voltage if the volume in its tank goes to one-quarter full or less. When this happens, the gate output goes HIGH. Therefore, the red-lamp circuit in the panel must be arranged so that a HIGH-level voltage turns it on.

    Notice that, in this example and in Example 3–11, the same NAND gate is used, but the *way* it is used is different.

EXAMPLE 3–13

For the four-input NAND gate in Figure 3–23(a), operating as a negative-OR, determine the output with respect to the inputs.

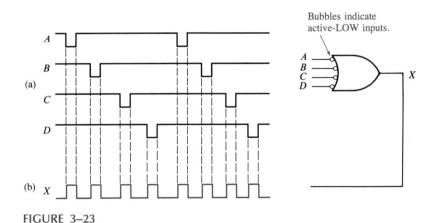

FIGURE 3–23

Solution The output is HIGH any time an input is LOW. See Figure 3–23(b).

SECTION 3–4 REVIEW

Answers are found at the end of the chapter.

1. When is the output of a NAND gate LOW?
2. When is the output of a NAND gate HIGH?
3. Describe the functional differences between a NAND gate and a negative-OR gate. Do they both have the same truth table?

THE NOR GATE

3–5

The NOR gate, like the NAND, is a very useful logic gate because of its universal property. We will examine the universal property of this gate in detail in Chapter 5.

The term *NOR* is a contraction of *NOT-OR* and implies an OR function with an inverted output. A standard logic symbol for a two-input NOR gate and its equivalent OR gate followed by an inverter are shown in Figure 3–24(a). A rectangular outline symbol is shown in part (b).

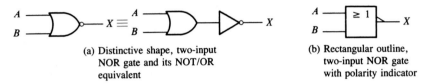

(a) Distinctive shape, two-input NOR gate and its NOT/OR equivalent

(b) Rectangular outline, two-input NOR gate with polarity indicator

FIGURE 3–24
Standard NOR gate logic symbols (ANSI/IEEE Std. 91–1984).

Logical Operation of the NOR Gate

The logical operation of the *NOR gate* is such that a LOW output occurs when *any* of its inputs are HIGH. Only when *all* of its inputs are LOW is the output HIGH. For the specific case of a two-input NOR gate, as shown in Figure 3–24 with the inputs labeled *A* and *B* and the output labeled *X,* we can state the operation as follows:

If *A* or *B* OR both are HIGH, then *X* is LOW. If both *A* and *B* are LOW, then *X* is HIGH.

This operation results in an output opposite that of the OR gate. In a NOR gate, the LOW output is the active output level as indicated by the bubble on the output. Figure 3–25 illustrates the logical operation of a two-input NOR gate for all four possible input combinations, and the truth table for the two-input NOR gate is Table 3–6.

FIGURE 3–25
Logical operation of a two-input NOR gate.

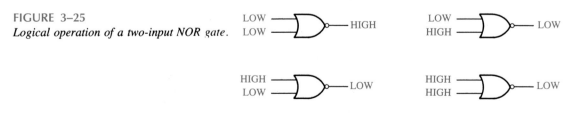

TABLE 3–6
Truth table for a two-input NOR gate.

Inputs		Output
A	*B*	*X*
0	0	1
0	1	0
1	0	0
1	1	0

NOTE: 1 ≡ HIGH, 0 ≡ LOW.

Pulsed Operation

The next three examples illustrate the logical operation of the NOR gate with pulsed inputs. Again, as with the other types of gates, we will simply follow the truth table operation to determine the output waveforms.

EXAMPLE 3–14

If the two waveforms shown in Figure 3–26(a) are applied to the NOR gate, what is the resulting output waveform?

FIGURE 3–26

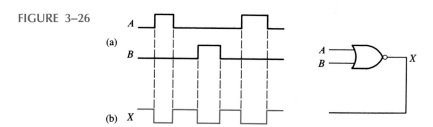

Solution   Whenever any input of a NOR gate is HIGH, the output is LOW. See Figure 3–26(b).

EXAMPLE 3–15

Sketch the output waveform for the three-input NOR gate in Figure 3–27(a), showing the proper relation to the inputs.

FIGURE 3–27

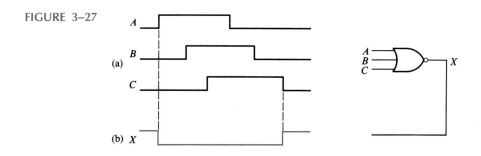

Solution   See Figure 3–27(b). The output is LOW when any input is HIGH.

The NOR Gate as an AND Gate with Active-LOW Inputs

The NOR gate, like the NAND, displays another aspect of its operation that is inherent in the way it logically functions. Table 3–6 shows that a HIGH is produced on the gate output only if *all* of the inputs are LOW. In reference to Figure 3–28(a), this aspect of NOR operation is stated as follows:

If both *A* AND *B* are LOW, then *X* is HIGH.

FIGURE 3–28
Standard symbols representing the two equivalent functions of the NOR gate.

(a) NOR

(b) Negative-AND

Here we have essentially an AND operation that requires all LOW inputs to produce a HIGH output. This is called *negative-AND* and is represented by the standard symbol in Figure 3–28(b). It is important to remember that the two symbols in this figure represent the same gate and serve only to distinguish between the two facets of the logical operation. The following two examples illustrate this.

EXAMPLE 3–16

A certain application requires that two lines be monitored for the occurrence of a HIGH level voltage on either or both lines. Upon detection of a HIGH level, the circuit must provide a LOW voltage to energize an alarm circuit. Sketch the operation.

Solution This application requires a NOR function, since the output must be active-LOW in order to give an indication of at least one HIGH on its inputs. The NOR symbol is therefore used to show the operation. See Figure 3–29.

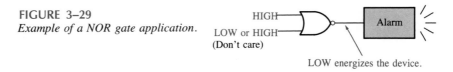

FIGURE 3–29
Example of a NOR gate application.

EXAMPLE 3–17

A device is needed to indicate when two LOW levels occur simultaneously on its inputs and to produce a HIGH output as an indication. Sketch the operation.

Solution Here, the negative-AND function is required, as shown in Figure 3–30.

FIGURE 3–30 LOW —◦⌐◯⌐— HIGH LOW —◦⌐

EXAMPLE 3–18

For the four-input NOR gate operating as a negative-AND in Figure 3–31(a), determine the output relative to the inputs.

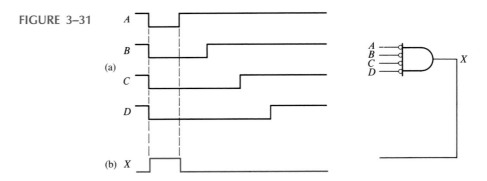

FIGURE 3–31

(a)

(b) X

Solution Any time *all* of the inputs are LOW, the output is HIGH. See Figure 3–31(b).

SECTION 3–5
REVIEW

Answers are found at the end of the chapter.
1. When is the output of a NOR gate HIGH?
2. When is the output of a NOR gate LOW?
3. Describe the functional difference between a NOR gate and a negative-AND gate. Do they both have the same truth table?

THE EXCLUSIVE-OR AND EXCLUSIVE-NOR GATES

3–6

These two gates are actually formed by a combination of other gates already discussed, as you will see in Chapter 5. However, because of their fundamental importance in many applications, these gates are treated as basic gates with their own unique symbols.

The Exclusive-OR Gate

Standard symbols for the exclusive-OR (XOR for short) gate are shown in Figure 3–32. The XOR gate has only two inputs. Unlike the other gates we have discussed, it never has more than two inputs.

The output of an *exclusive-OR* gate is HIGH *only* when the two inputs are at *opposite* levels. This operation can be stated as follows with reference to inputs A and B and output X:

If A is LOW and B is HIGH, OR if A is HIGH and B is LOW, then X is HIGH. If A and B are both HIGH or both LOW, then X is LOW.

The four possible input combinations and the resulting outputs for the exclusive-OR gate are illustrated in Figure 3–33. The HIGH level is the active output level and occurs only when the inputs are at opposite levels.

FIGURE 3–32
Standard logic symbols for the exclusive-OR gate.

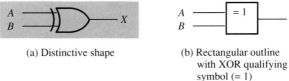

(a) Distinctive shape

(b) Rectangular outline with XOR qualifying symbol (= 1)

FIGURE 3–33
All possible logic levels for an exclusive-OR gate.

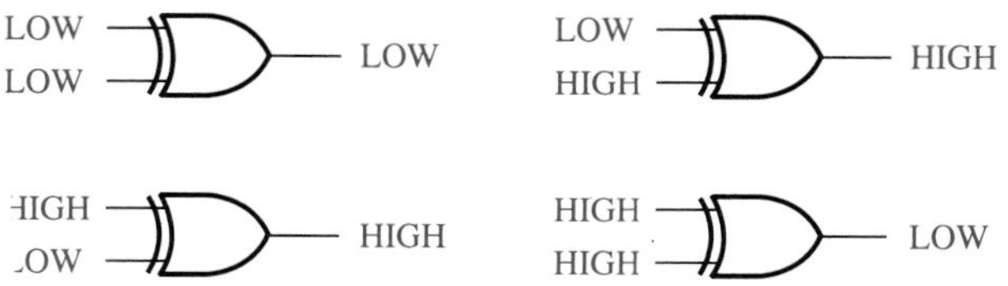

LOW
LOW — LOW

LOW
HIGH — HIGH

HIGH
LOW — HIGH

HIGH
HIGH — LOW

The logical operation of the XOR gate is summarized in the truth table shown in Table 3–7.

TABLE 3–7
Truth table for an exclusive-OR gate.

Inputs		Output
A	B	X
0	0	0
0	1	1
1	0	1
1	1	0

EXAMPLE 3–19

A certain system contains two identical circuits operating in parallel. As long as both are operating properly, the outputs of both circuits are always the same. If one of the circuits fails, the outputs will be at opposite levels at some time. Devise a way to detect a failure in one of the circuits.

FIGURE 3–34

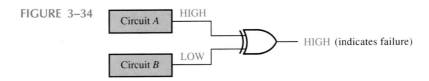

Solution The outputs of the circuits are connected to the inputs of an XOR gate as shown in Figure 3–34. A failure in one of the circuits causes the XOR inputs to be at opposite levels. This condition produces a HIGH on the output of the XOR gate, indicating a failure.

The Exclusive-NOR Gate

Standard symbols for the exclusive-NOR (XNOR) gate are shown in Figure 3–35. Like the XOR gate, the XNOR has only two inputs.

The bubble on the output of the XNOR symbol indicates that its output is opposite that of the XOR gate. When the two input levels are opposite, the output of the *exclusive-NOR* gate is LOW. The operation can be stated as follows:

If *A* is LOW and *B* is HIGH, OR if *A* is HIGH and *B* is LOW, then *X* is LOW. If *A* and *B* are both HIGH or both LOW, then *X* is HIGH.

FIGURE 3–35
Standard logic symbols for the exclusive-NOR gate.

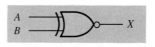

(a) Distinctive shape

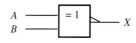

(b) Rectangular outline

The four possible input combinations and the resulting outputs for the XNOR gate are shown in Figure 3–36. The logical operation is summarized in Table 3–8. Notice that the output is HIGH when the *same* level is on both inputs.

TABLE 3–8
Truth table for an exclusive-NOR gate.

Inputs		Output
A	B	X
0	0	1
0	1	0
1	0	0
1	1	1

FIGURE 3–36
All possible logic levels for an exclusive-NOR gate.

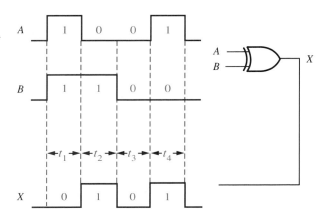

Pulsed Operation

As we have done with the other gates, we will examine the operation of the XOR and XNOR gates under pulsed input conditions. As before, we apply the truth table operation during each distinct time interval of the pulsed inputs, as illustrated in Figure 3–37 for an XOR gate. You can see that the input waveforms A and B are at opposite levels during time intervals t_2 and t_4. Therefore, the output X is HIGH during these two times. Since both inputs are the same level, either both HIGH or both LOW, during t_1 and t_3, the output is LOW during those times.

FIGURE 3–37
Example of pulsed exclusive-OR gate operation.

EXAMPLE 3–20

Determine the output waveforms for an XOR gate and for an XNOR gate, given the input waveforms in Figure 3–38(a).

FIGURE 3–38

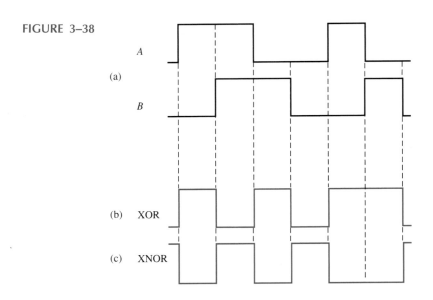

Solution   The XOR output is shown in Figure 3–38(b). Notice that it is HIGH only when both inputs are at opposite levels. The XNOR output is shown in Figure 3–38(c). Notice that it is HIGH only when both inputs are the same.

SECTION 3–6
REVIEW

Answers are found at the end of the chapter.
1. When is the output of an XOR gate HIGH?
2. When is the output of an XNOR gate HIGH?
3. How can you use an XOR gate to detect when two bits are different?

INTEGRATED CIRCUIT LOGIC FAMILIES

3–7

In the previous sections you learned about the logical operation of the various gates. In this section we will briefly introduce the two most widely used types of digital integrated circuits, TTL and CMOS, and will look at several specific devices in these IC families.

This coverage is important because it gives you information on the specific devices you will be using in the lab and in industry. A more detailed and thorough coverage of the circuitry, specifications, and parameters of IC families is provided in Chapter 13, which can be covered immediately after this chapter if desired. If you choose to cover Chapter 13, you may wish to omit all or part of this section.

TTL

The term *TTL* stands for *t*ransistor-*t*ransistor *l*ogic, which refers to the use of bipolar junction transistors in the circuit technology used to construct the gates at the chip level.

TTL consists of a series of logic circuits: standard TTL, low-power TTL, Schottky TTL, low-power Schottky TTL, advanced low-power Schottky TTL, and advanced Schottky TTL. The differences in these various types of TTL are in their performance characteristics, such as propagation delay times, power dissipation, and fan-out, which are explained later. The TTL family has a number prefix of 54 or 74, followed by a letter or letters that specify the series, as shown in Table 3–9. The prefix 54 indicates an operating temperature range of $-55°C$ to $125°C$ (generally for military use). The prefix 74 indicates a temperature range of $0°C$ to $70°C$ (for commercial use). We will use the prefix 74 throughout the book.

TABLE 3–9 *TTL series designations.*

TTL Series	Prefix Designation	Example of Device
Standard TTL	54 or 74 (no letter)	7400 (quad NAND gates)
Low-power TTL	54L or 74L	74L00 (quad NAND gates)
Schottky TTL	54S or 74S	74S00 (quad NAND gates)
Low-power Schottky TTL	54LS or 74LS	74LS00 (quad NAND gates)
Advanced low-power Schottky TTL	54ALS or 74ALS	74ALS00 (quad NAND gates)
Advanced Schottky TTL	54AS or 74AS	74AS00 (quad NAND gates)

CMOS

The term *CMOS* stands for *c*omplementary *m*etal *o*xide *s*emiconductor. Whereas TTL uses bipolar transistors in its circuit technology, CMOS uses field-effect transistors. Logic functions are the same, however, whether the device is implemented with TTL or CMOS technologies. The circuit technologies make a difference, not in logic function, but only in performance characteristics.

Several series of CMOS logic are available, but they fall basically into two process technology categories: metal-gate CMOS and silicon-gate CMOS. The older, metal-gate technology is the 4000 series. The newer, silicon-gate technology consists of the 74C, the 74HC, and the 74HCT. All of the 74 series CMOS devices are both pin compatible and function compatible with the TTL series. That is, a TTL IC and a CMOS IC of the same number have the same input, output, V_{CC}, or ground on the same pins, as well as the same logic gates. In addition, the 74HCT series is voltage-level compatible with TTL and requires no special interfacing as do the 74C and 74HC series. Other differences in the various types of 74 series CMOS are in their performance characteristics.

Performance Characteristics

Propagation delay time is a very important characteristic of logic circuits because it limits the speed (frequency) at which they can operate. The terms *low speed* and *high speed,* when applied to logic circuits, refer to the propagation delays; the shorter the propagation delay, the higher the speed of the circuit.

The propagation delay time of a gate is basically the time interval between the application of an input pulse and the occurrence of the resulting output pulse. There are two propagation delays associated with a logic gate:

t_{PHL}: The time between a specified reference point on the input pulse and a corresponding reference point on the output pulse, with the output changing from the HIGH level to the LOW level.

t_{PLH}: The time between a specified reference point on the input pulse and a corresponding reference point on the output pulse, with the output changing from the LOW level to the HIGH level.

EXAMPLE 3–21

Show the propagation delays of the inverter in Figure 3–39.

FIGURE 3–39

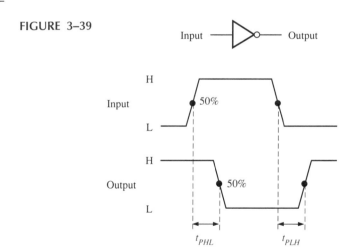

Solution The propagation delays, t_{PHL} and t_{PLH}, are indicated in the figure.

Power dissipation of a logic gate equals the dc supply voltage V_{CC} times the average supply current I_{CC}. Normally, the value of I_{CC} for a LOW gate output is higher than for a HIGH output. The manufacturer's data sheet usually specifies these values as I_{CCL} and I_{CCH}. The average I_{CC} is then determined, based on a 50% duty cycle operation of the gate (LOW half the time and HIGH half the time).

Fan-out of a gate is the maximum number of inputs of the same IC family series that the gate can drive while maintaining its output levels within specified limits. That is, the fan-out specifies the maximum load that a given gate is capable of handling.

For example, a standard TTL gate has a fan-out of 10 unit loads. This means that it can drive no more than 10 inputs of other standard TTL gates and still operate reliably. If the fan-out is exceeded, specified operation is not guaranteed. Figure 3–40 shows a gate driving 10 other gates.

FIGURE 3–40
The NAND gate output fans out to ten gate inputs.

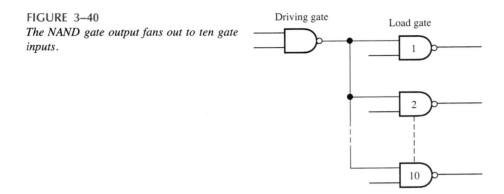

Speed-power product is sometimes specified by the manufacturer as a measure of the performance of a logic circuit. It is the product of the propagation delay time and the power dissipation at a specified frequency. The speed-power product is expressed in joules, symbolized J. For example, the speed-power product (SPP) of a 74HC CMOS gate at a frequency of 100 kHz is

$$SPP = (8 \text{ ns})(0.17 \text{ mW}) = 1.36 \text{ pJ}$$

Table 3–10 provides a comparison of some performance characteristics of CMOS and TTL.

TABLE 3–10
Comparison of performance characteristics of CMOS and TTL.

Technology	CMOS* (silicon-gate)	CMOS* (metal-gate)	TTL Std.	TTL LS	TTL S	TTL ALS	TTL AS
Device series	74HC	4000B	74	74LS	74S	74ALS	74AS
Power dissipation:							
Static	2.5 nW	1 μW	10 mW	2 mW	19 mW	1 mW	8.5 mW
@ 100 kHz	0.17 mW	0.1 mW	10 mW	2 mW	19 mW	1 mW	8.5 mW
Propagation delay time	8 ns	50 ns	10 ns	10 ns	3 ns	4 ns	1.5 ns
Fan-out (same series)			10	20	20	20	40

*Propagation delay is dependent on V_{CC}. Power dissipation and fan-out are a function of frequency.

Specific Devices

A wide variety of SSI (small-scale integration) logic gate configurations are available in both the TTL and the CMOS families. When specific devices in the SSI and MSI categories are referred to in this book, standard TTL will generally be used for illustrative purposes, although other types will occasionally be used. Keep in mind that most of the specific devices are also available in the other TTL series as well as in CMOS.

A selection of typical logic gate ICs is now presented. These devices are commonly housed in the dual in-line package (DIP) or a chip carrier (SMT) package. For simplicity, V_{CC} and ground connections to each gate are normally not shown in a logic diagram. In Figures 3–41 through 3–46, each device is represented by a distinctive shape logic diagram, with the pin numbers indicated in parentheses. Additionally, each device is shown as a rectangular outline logic symbol. The two representations are equivalent. Although the standard 74 series designation is used, most of these devices are also available in most if not all of the other TTL and CMOS series.

Hex inverter   The 7404 hex inverter is a standard TTL device consisting of six inverters in a 14-pin package, as shown in Figure 3–41.

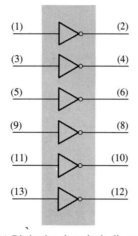

(a) Distinctive shape logic diagram

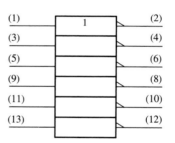

(b) Rectangular outline logic symbol with polarity indicators. The inverter qualifying symbol (1) appears in the top block and applies to all blocks below.

FIGURE 3–41
7404 hex inverter (pin numbers in parentheses).

AND gates   Several configurations of AND gates are available in IC form. The 7408 has four 2-input AND gates (it is called a quad 2-input AND); the 7411 has three 3-input AND gates (a triple 3-input AND); and the 7421 has two 4-input AND gates (a dual 4-input AND). These gates are shown in Figure 3–42.

NAND gates   A variety of NAND gates are available, including the 7400 with four 2-input gates, the 7410 with three 3-input gates, the 7420 with two 4-input gates, the

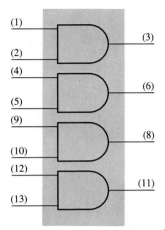

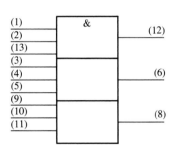

(a) 7408 Quad 2-input AND

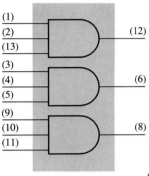

(b) 7411 Triple 3-input AND

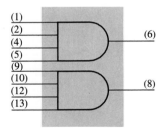

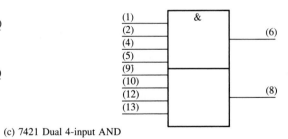

(c) 7421 Dual 4-input AND

FIGURE 3–42
AND gates.

FIGURE 3–43
NAND gates.

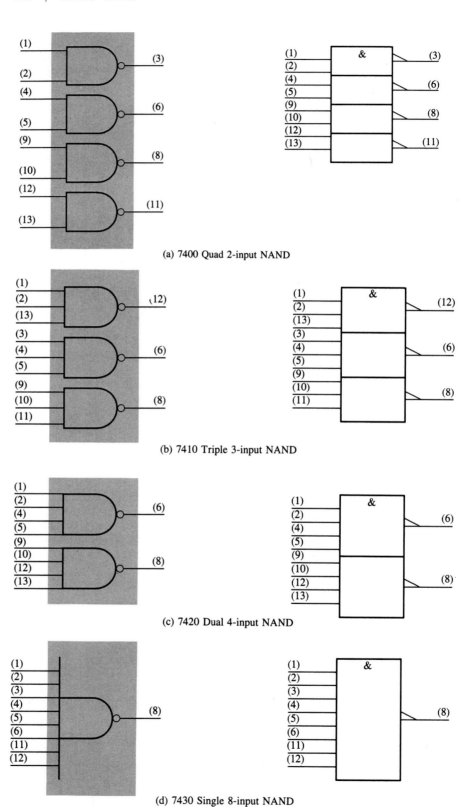

(a) 7400 Quad 2-input NAND

(b) 7410 Triple 3-input NAND

(c) 7420 Dual 4-input NAND

(d) 7430 Single 8-input NAND

FIGURE 3–43
(Continued)

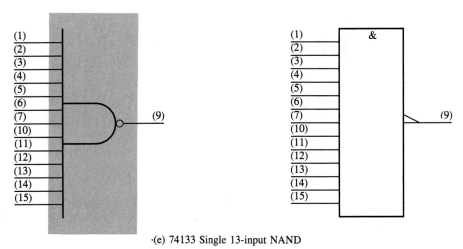

·(e) 74133 Single 13-input NAND

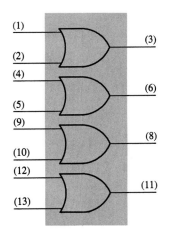

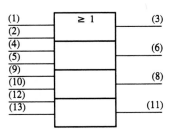

FIGURE 3–44
The 7432 Quad 2-input OR gates.

7430 with one 8-input gate, and the 74133 with one 13-input gate. These gates are shown in Figure 3–43. Notice that the 74133 requires 16-pins.

OR gates The 7432 has four 2-input OR gates, as shown in Figure 3–44.

NOR gates Examples of NOR gate configurations are shown in Figure 3–45. The 7402 has four 2-input gates, and the 7427 has three 3-input gates.

Exclusive-OR gates The 7486 has four exclusive-OR gates in the package, as shown in Figure 3–46.

FIGURE 3–45
NOR gates.

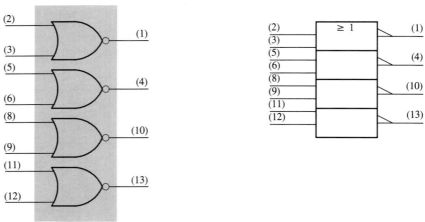

(a) 7402 Quad 2-input NOR

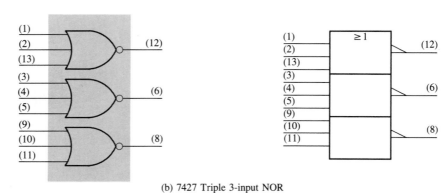

(b) 7427 Triple 3-input NOR

FIGURE 3–46
The 7486 quad exclusive-OR gates.

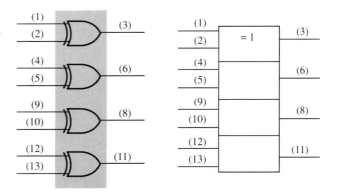

SECTION 3–7
REVIEW

Answers are found at the end of the chapter.

1. A positive pulse is applied to an inverter input. The time from the leading edge of the input to the leading edge of the output is 10 ns. The time from the trailing edge of the input to the trailing edge of the output is 8 ns. What are the values of t_{PLH} and t_{PHL}?

2. Define I_{CCL} and I_{CCH}.

3. Define *fan-out*.

4. Name the type and series of IC technology that exhibits each of the following performance characteristics:
  (a) fastest switching time (lowest propagation delay)
  (b) lowest power dissipation
  (c) highest fan-out
5. Calculate the speed-power product for an advanced Schottky (74AS) gate.

APPLICATIONS

3–8

In this section we will look at several simple examples of the application of logic gates to practical situations. You can probably think of several other useful ways in which the gates can be applied.

The AND Gate as an Enable/Inhibit Device

A common application of the AND gate is to *enable* (that is, to allow) the passage of a signal (pulse waveform) from one point to another at certain times and to *inhibit* (prevent) the passage at other times.

A simple example of this particular use of the AND gate is shown in Figure 3–47, where the AND gate controls the passage of a signal (waveform *A*) to a digital counter. The purpose of this circuit is to measure the frequency of waveform *A*. The Enable pulse has a width of precisely 1 second. When the Enable is HIGH, waveform *A* passes through the gate to the counter, and when the Enable is LOW, the signal is prevented from passing through (inhibited).

During the 1-second interval of the Enable pulse, a certain number of pulses in waveform *A* pass through the AND gate. The number of pulses passing through during 1 second is equal to the frequency of waveform *A*. For example, if 1000 pulses pass through the gate in the 1-second interval of the Enable pulse, there are 1000 pulses/s, or a frequency of 1000 Hz.

The counter counts the number of pulses per second and produces a binary output that goes to a decoding and display circuit to produce a readout of the frequency.

The Enable pulse repeats at certain intervals and a new updated count is made so that if the frequency changes, the new value will be displayed. Between Enable pulses,

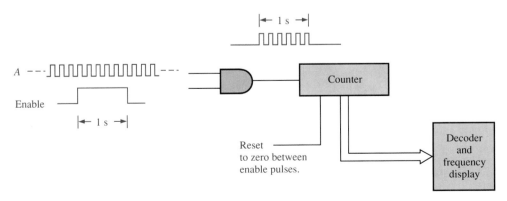

FIGURE 3–47
An AND gate performing an Enable/Inhibit function for a frequency counter.

the counter is reset so that it starts at zero each time an Enable occurs. Assume that the frequency count is held in a memory so that the display is unaffected by the resetting of the counter.

A Seat Belt Alarm System

In Figure 3–48 an AND gate is used in a simple automobile seat belt alarm system to detect when the ignition switch is on *and* the seat belt is unbuckled. If the ignition switch is on, a HIGH is produced on input *A* of the AND gate. If the seat belt is not properly buckled, a HIGH is produced on input *B* of the AND gate. Also, when the ignition switch is turned on, a timer is started that produces a HIGH on input *C* for 30 seconds.

If all three conditions exist—that is, if the ignition is on, the seat belt is unbuckled, and the timer is running—the output of the AND gate is HIGH, and an audible alarm is energized to remind the driver.

Before moving on to another application, let us look at the use of other types of gates in the seat belt alarm circuit. For example, if the ignition switch, the seat belt buckle, and the timer all produced LOW levels rather than HIGH levels when activated,

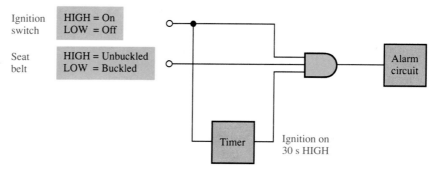

FIGURE 3–48
A simple seat belt alarm circuit using an AND gate.

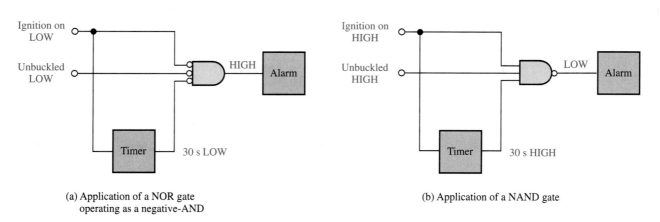

(a) Application of a NOR gate operating as a negative-AND

(b) Application of a NAND gate

FIGURE 3–49
Variations of the seat belt alarm system in Figure 3–48, using the gate symbols that best describe the circuit function.

FIGURE 3–50

A simplified intrusion detection system using an OR gate.

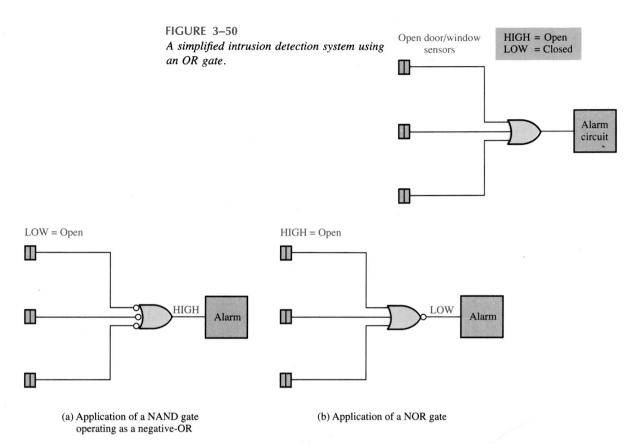

(a) Application of a NAND gate operating as a negative-OR

(b) Application of a NOR gate

FIGURE 3–51

Variations of the intrusion detection system, using the gate symbols that best describe the circuit function.

we could use a NOR gate functioning as a negative-AND, as indicated in Figure 3–49(a). As another case, suppose that the alarm circuit required a LOW to activate it rather than a HIGH, but the three inputs were all active-HIGH. In this situation a NAND gate would be required, as shown in Figure 3–49(b).

An Intrusion Detection System

A simplified portion of an intrusion detection and alarm system is shown in Figure 3–50. This system might be used for one room in a home—a room with two windows and a door. The sensors can be magnetic switches that produce a HIGH output when open and a LOW output when closed. As long as the windows and the door are secured, the switches are closed and all three of the OR gate inputs are LOW. When one of the windows or the door is opened, a HIGH is produced on that input to the OR gate and the gate output goes HIGH. It then activates an alarm circuit to warn of the intrusion.

Again, it is interesting to see how the requirements for the logic gate change when the active levels in the system are different. For example, if the window and door switches produced LOW levels when opened, we would have to use a NAND gate functioning as a negative-OR, as shown in Figure 3–51(a). If the inputs were still active-HIGH but the alarm circuit required a LOW to activate it, a NOR gate would have to be used, as indicated in Figure 3–51(b).

The Exclusive-OR Gate as a Two-Bit Adder

Recall from Chapter 2 that the basic rules for binary addition are as follows: $0 + 0 = 0$, $0 + 1 = 1$, $1 + 0 = 1$, and $1 + 1 = 10$. An examination of the truth table for the XOR gate will show you that its output is the binary sum of the two input bits. In the case where the inputs are both 1s, the output is the sum 0, but we lose the carry of 1. In chapter 6 we will see how XOR gates are combined to make complete adding circuits. Figure 3–52 illustrates the XOR gate used as a basic adder.

FIGURE 3–52
The XOR gate used to add two bits.

Input (bits)		Output (sum)
A	B	Σ
0	0	0
0	1	1
1	0	1
1	1	0 (without 1 carry)

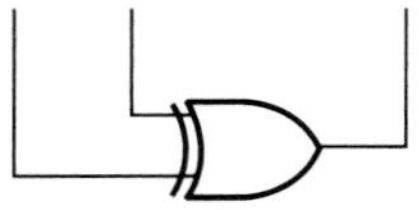

SECTION 3–8 REVIEW

Answers are found at the end of the chapter.
1. If a signal with a frequency of 1 MHz is applied to the AND gate in the frequency counter system of Figure 3–47, how many pulses will pass through the AND gate during the Enable pulse?
2. Show how you would alter the circuit in Figure 3–50 if the room had three windows and two doors.

TROUBLESHOOTING

3–9

Troubleshooting is the process of recognizing, isolating, and correcting a fault or failure in a circuit or system. To be an effective troubleshooter, you must understand how the circuit or system is supposed to work and be able to recognize incorrect performance. For example, to determine whether or not a certain logic gate is faulty, you must know what the output should be for given inputs.

At this point it may be helpful to review Section 1–8, on pulsers, probes, and oscilloscopes. Those are the most common test instruments, and most of the troubleshooting coverage in this text emphasizes their use.

Internal Failures of IC Logic Gates

Opens and shorts are the most common types of internal gate failures. These can occur on the inputs or on the output of a gate.

Open gate input This type of malfunction is generally caused by an open circuit component on the IC chip or by a break in the tiny wire lead connecting the IC chip to the pin connection. An open input prevents a signal on that input from getting to the gate circuitry and thus to the output, as illustrated in Figure 3–53. In a TTL gate an open input acts as a HIGH. In a CMOS gate an open input has an unpredictable effect.

Troubleshooting an open input Troubleshooting this type of failure can be accomplished with a logic pulser and a logic probe, as illustrated for a 7400 NAND gate in Figure 3–54.

The first step in troubleshooting an IC is to make sure the dc supply voltage (V_{CC}) and the ground are connected. Next, pulse one of the inputs (pin 13), and observe the output activity with the logic probe, as shown in Figure 3–54(a). If pulse activity is observed on the output, as indicated by the flashing logic probe, then that particular input is all right. Now pulse the other input (pin 12). No activity on the output indicates that the input is open, as shown in part (b). Notice that the input that is not being pulsed should be connected to a HIGH (+5 V).

FIGURE 3–53
The effect of an open input on a logic gate.

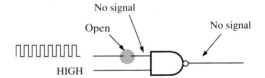

FIGURE 3–54
Troubleshooting an open gate input.

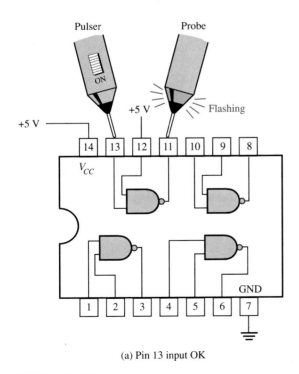

(a) Pin 13 input OK

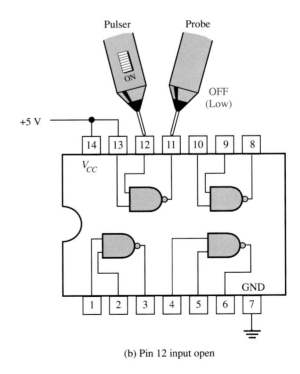

(b) Pin 12 input open

Open gate output An open gate output prevents a signal on either of the inputs from getting to the output, as shown in Figure 3–55. Since there is a break between the gate and the output pin on the IC, there is no voltage on the output pin.

Troubleshooting an open output Figure 3–56 illustrates the troubleshooting of an open output. In part (a) one of the inputs (pin 13) is pulsed, and no activity is observed on the output. In part (b) the other input (pin 12) is pulsed, and again there is no activity on the output. This test indicates that the gate output is open. Notice that the unpulsed input should be connected to a HIGH.

Shorted input or output Although not as common as an open, an internal short to the dc supply voltage, the ground, another input, or an output can occur. When an input or output is shorted to the supply voltage, it will be stuck in the HIGH state (+5 V). If an input or output is shorted to the ground, it will be stuck in the LOW state (0 V).

FIGURE 3–55
The effect of an open output on a logic gate.

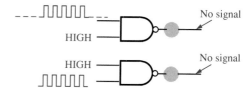

FIGURE 3–56
Troubleshooting an open gate output.

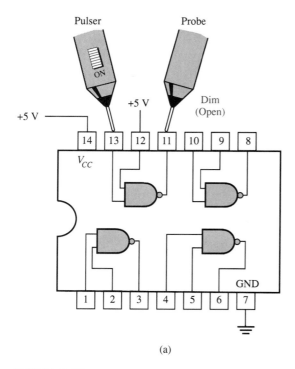

(a)

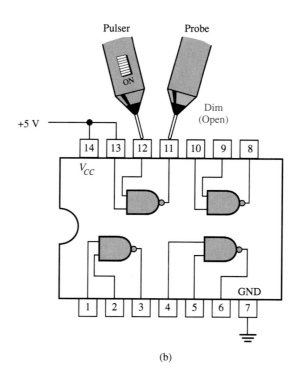

(b)

If two inputs or an input and an output are shorted together, they will always be at the same level.

We will reserve further discussion of the troubleshooting of shorts until Chapter 5, where the use of a current tracer to locate a short will be demonstrated in situations in which several gates are interconnected.

External Opens and Shorts

Many failures involving digital ICs are due to faults that are external to the IC package. These include bad solder connections, solder splashes, wire clippings, improperly etched printed circuit (PC) boards, and cracks or breaks in wires or printed circuit interconnections. These open or shorted conditions have the same effect on the logic gate as the internal faults, and troubleshooting is done in basically the same ways. A visual inspection of any circuit that is suspected of being faulty is the first thing a technician should do.

EXAMPLE 3–22

You are checking a 7410 triple 3-input NAND gate IC that is one of many ICs located on a printed circuit (PC) board. You have checked pins 1 and 2 with your logic probe, and they are both HIGH. Now your logic pulser is placed on pin 13, and your logic probe is placed first on pin 12 and then on the PC board connection, as indicated in Figure 3–57. Based on the responses of the probe, what is the most likely problem?

FIGURE 3–57

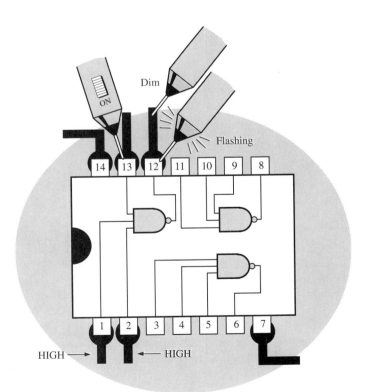

Solution   The flashing indicator on the probe shows that there is pulse activity on the gate output at pin 12 but no activity on the PC board connection. The gate is working properly, but the signal is not getting from pin 12 of the IC to the PC connection.

Most likely there is a bad solder connection between pin 12 of the IC and the PC board, which is creating an open. You should resolder that point and check it again.

In most cases you will be troubleshooting ICs that are mounted on printed circuit boards or prototype assemblies and interconnected with other ICs. As you progress through the book, you will learn how different types of digital ICs are used together to form system functions. At this point, however, we are concentrating on individual IC gates. This limitation does not prevent us from looking at the system concept at a very basic and simplified level, as we have already done several times.

To continue the emphasis on systems, the next two examples deal with troubleshooting the frequency counter system that was introduced in Section 3–8. In these examples, instead of logic probes, the oscilloscope is used to observe waveforms and levels.

EXAMPLE 3–23

After trying to operate the frequency counter, you find that it constantly reads out all 0s on its display, regardless of the input frequency. Determine the cause of this malfunction.

Solution   Here are several possible causes:
1. A constant active level on the counter reset input, which keeps the counter at zero.
2. No signal on the clock input to the counter because of an internal short in the counter. This problem would keep the counter from advancing after being reset to zero.
3. No signal on the clock input to the counter because of an open AND gate output or the absence of input signals, again keeping the counter from advancing from zero.

Figure 3–58 shows the oscilloscope displays for various points in the circuit. The scope shows that there is a LOW level at the reset to the counter (this is the nonactive level and is therefore not the problem). The scope also shows that both signals are present at the AND gate inputs. When the scope is connected to the output pin of the AND gate, an intermediate (floating) level is observed, neither HIGH nor LOW. This level indicates that the AND gate output is internally open. Replace the 7408 IC and check the operation again.

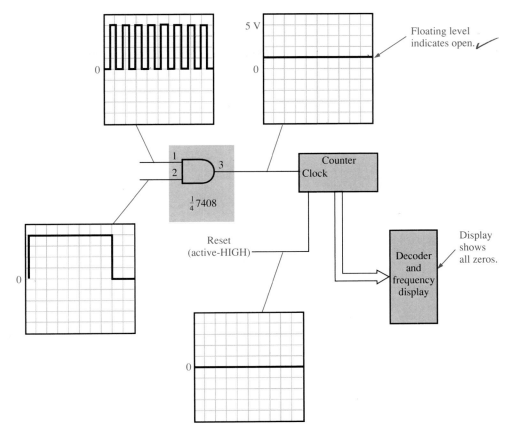

FIGURE 3–58

EXAMPLE 3–24

You find that the frequency counter shows a frequency that is much less than that of the signal you are feeding into the *A* input of the AND gate.

An oscilloscope check at several points gives you the indications shown in Figure 3–59 on page 130. Determine what is wrong.

Solution   Notice that the Enable signal, not the signal on input *A*, appears on the AND gate output. The counter is counting the frequency of the Enable signal, which is much lower than the value expected for input *A*.

The reason that the Enable signal appears on the gate output must be that the other input is open and acting as a HIGH. If the signal on input *A* appears directly on pin 1 of the IC, the open is internal. If the signal appears on the PC connection going to the pin but not on the pin itself, the open is external.

If the open is external, there is probably a bad solder joint at the pin. Resolder and check again. If the open is internal, replace the 7408 IC and check again.

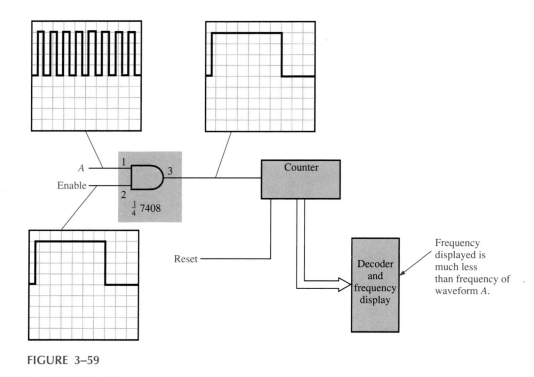

FIGURE 3-59

SECTION 3-9 REVIEW   .

Answers are found at the end of the chapter.
1. What are the most common types of failures in ICs?
2. If two input waveforms are applied to a two-input TTL NAND gate and the output waveform is just like one of the inputs, what is the most likely problem?
3. Name one advantage of the oscilloscope over the logic probe.
4. Name one advantage of the logic probe over the oscilloscope.

SUMMARY

☐ Distinctive shape symbols and truth tables for the various logic gates are shown in Figure 3-60.
☐ As a general rule, TTL is faster but consumes more power than CMOS.
☐ The specific devices presented in this chapter are as follows:

7400	Four 2-input NAND gates
7402	Four 2-input NOR gates
7404	Six inverters
7408	Four 2-input AND gates
7410	Three 3-input NAND gates
7411	Three 3-input AND gates
7420	Two 4-input NAND gates

7427 Three 3-input NOR gates
7430 One 8-input NAND gate
7432 Four 2-input OR gates
7486 Four 2-input XOR gates
74133 One 13-input NAND gate

☐ The meaning of prefixes in the 74 family of TTL and CMOS devices is as follows:

74 Standard TTL
74L Low-power TTL
74S Schottky TTL
74LS Low-power Schottky TTL
74ALS Advanced low-power Schottky TTL
74AS Advanced Schottky TTL
74C CMOS
74HC High-speed CMOS
74HCT High-speed CMOS with TTL compatibility

FIGURE 3–60

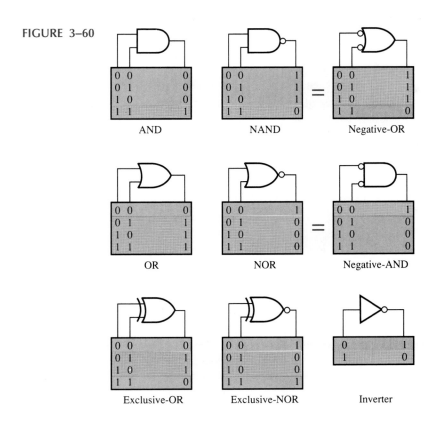

Note: Active states are shown in color.

Integrated Circuit Technologies (Chapter 13) may be covered after this chapter, covered after a later chapter, completely omitted, or partially omitted, depending on the needs and goals of your program. Coverage of Integrated Circuit Technologies is not prerequisite or corequisite to any other material in the book. Inclusion is purely optional.

GLOSSARY

ANSI   American National Standards Institute.

Bipolar   Having two opposite charge carriers within the transistor structure.

CMOS   Complementary metal oxide semiconductor. A type of transistor circuit which uses field-effect transistors.

Complementation   Inversion. LOW is the complement of HIGH, and 0 is the complement of 1.

Enable   To activate or put into an operational mode.

Exclusive-OR   A logic function in which a HIGH output occurs only when the two inputs are at opposite levels.

Fan-out   The number of equivalent gate inputs of the same family series that a logic gate can drive.

IEEE   Institute of Electrical and Electronic Engineers.

Inversion   Conversion of a HIGH level to a LOW level or vice versa.

NAND gate   A logic circuit in which a LOW output occurs if and only if all the inputs are HIGH.

NOR gate   A logic circuit in which the output is LOW when any or all of the inputs are HIGH.

Power dissipation   The product of the dc supply voltage and the dc supply current in an electronic circuit.

Propagation delay time   The time interval between the occurrence of an input transition and the occurrence of the corresponding output transition.

Speed-power product   A performance parameter that is the product of the propagation delay time and the power dissipation in a digital circuit.

Timing diagram   A diagram of input and output waveforms showing time relationships.

TTL   Transistor-transistor logic. A type of integrated circuit that uses bipolar junction transistors.

Unit load   A measure of fan-out. One gate input represents a unit load to the output of a gate within the same IC family.

SELF-TEST

Answers and solutions are found at the end of the book.

1. What is the output of an inverter for each of the following inputs? Which input states are equivalent?
   (a) HIGH   **(b)** LOW   **(c)** 1   **(d)** 0
2. Develop the truth table for a four-input AND gate.
3. Develop the truth table for a four-input OR gate.
4. A positive pulse is applied to each input of a two-input NAND gate. One pulse begins at $t = 0$ and ends at $t = 1$ ms. The other pulse begins at $t = 0.8$ ms and ends at $t = 3$ ms. Describe the output.

5. Answer question 4 for a two-input NOR gate.
6. Answer question 4 for an exclusive-OR gate.
7. A positive pulse is applied to an inverter. The time interval from the leading edge of the input to the leading edge of the output is 10 ns. Is this parameter t_{PHL} or t_{PLH}?
8. If $V_{CC} = 5$ V and $I_{CC} = 2$ mA, what is the power dissipation of the device?
9. A certain type of logic gate has a fan-out of 20. What does this mean?
10. Which TTL series has the faster switching speed, 74ALS or 74S? What do the prefixes ALS and S stand for?
11. In a certain application, propagation delay is not a major consideration, but power dissipation is critical. Which logic family would you use?
12. Define the type of logic gate for each of the following:
    (a) 7400   **(b)** 7404   **(c)** 7411   **(d)** 7420   **(e)** 7432   **(f)** 7427
13. Sketch the distinctive shape logic symbol for each of the gates: NOT, AND, OR, NAND, NOR, XOR, and XNOR.
14. Sketch the rectangular outline logic symbol for each of the gates in question 13.

PROBLEMS

Answers to odd-numbered problems are found at the end of the book.

Section 3–1 The Inverter

3–1 The input waveform shown in Figure 3–61 is applied to an inverter. Sketch the timing diagram of the output waveform in proper relation to the input.

3–2 A network of cascaded inverters is shown in Figure 3–62. If a HIGH is applied to point A, determine the logic levels at points B through F.

FIGURE 3–61

FIGURE 3–62

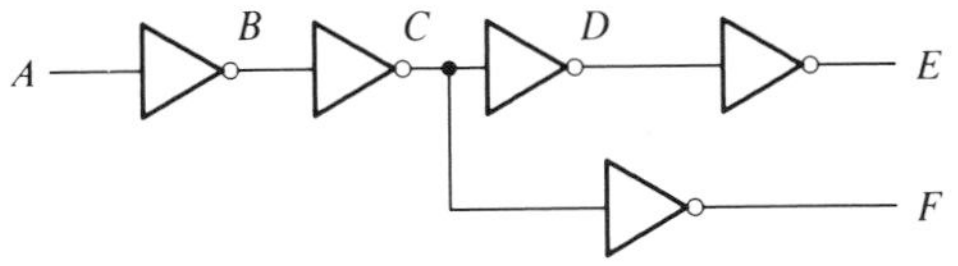

Section 3–2 The AND Gate

3–3 Determine the output, X, for a two-input AND gate with the input waveforms shown in Figure 3–63. Show the proper relationship of output to inputs.

3–4 Solve Problem 3–3 for the waveforms in Figure 3–64.

FIGURE 3–63

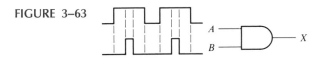

FIGURE 3–64

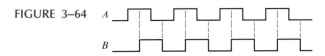

FIGURE 3–65

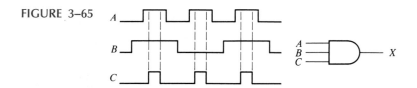

FIGURE 3–66

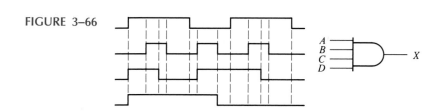

3–5 The input waveforms applied to a three-input AND gate are as indicated in Figure 3–65. Determine the output waveform in proper relation to the inputs.

3–6 The input waveforms applied to a four-input gate are as indicated in Figure 3–66. Determine the output waveform in proper relation to the inputs.

Section 3–3 The OR Gate

3–7 Determine the output for a two-input OR gate when the input waveforms are as in Figure 3–64.

3–8 Solve Problem 3–5 for a three-input OR gate.

3–9 Solve Problem 3–6 for a four-input OR gate.

3–10 For the five input waveforms in Figure 3–67, determine the output for a five-input AND gate and the output for a five-input OR gate.

FIGURE 3–67

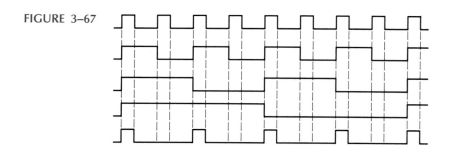

Section 3–4 The NAND gate

3–11 For the set of input waveforms in Figure 3–68, determine the output for the gate shown.

3–12 Determine the gate output for the input waveforms in Figure 3–69.

FIGURE 3–68

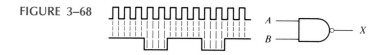

3–13 Determine the output waveform in Figure 3–70.

3–14 As you learned, the two logic symbols shown in Figure 3–71 are equivalent. The difference between the two is strictly from a *functional* viewpoint. For the NAND symbol, we are looking for two HIGHs on the inputs to give us a LOW output. For the negative-OR, we are looking for at least one LOW on the inputs to give us a HIGH on the output. Using these two functional viewpoints, show that each gate will produce the same output for the given inputs.

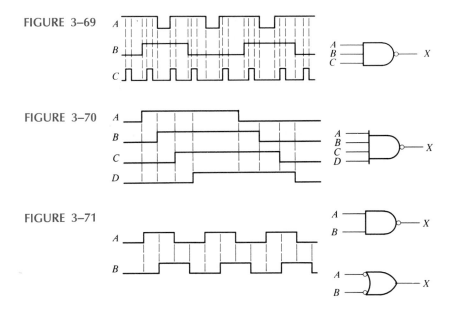

FIGURE 3–69

FIGURE 3–70

FIGURE 3–71

Section 3–5   The NOR Gate

3–15 Solve Problem 3–11 for a two-input NOR gate.

3–16 Determine the output waveform in Figure 3–72.

3–17 Solve Problem 3–13 for a four-input NOR gate.

3–18 For the NOR symbol, we are looking for at least one HIGH on the inputs to give us a LOW on the output. For the negative-AND, we are looking for two LOWs on the inputs to give us a HIGH output. Using these two functional points of view, show that both gates in Figure 3–73 will produce the same output for the given inputs.

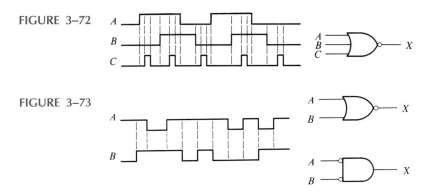

FIGURE 3–72

FIGURE 3–73

Section 3–6   The Exclusive-OR and Exclusive-NOR Gates

3–19   How does an exclusive-OR gate differ from an OR gate in its operation?

3–20   Solve Problem 3–11 for an Exclusive-OR gate.

3–21   Solve Problem 3–11 for an Exclusive-NOR gate.

3–22   Determine the output of an exclusive-OR gate for the inputs shown in Figure 3–64.

Section 3–7   Integrated Circuit Logic Families

3–23   In the comparison of certain logic devices, it is noted that the power dissipation for one particular type increases as the frequency increases. Is the device TTL or CMOS?

3–24   Using Table 3–10, determine which logic series offers the best performance at 100 kHz, considering both switching speed and power dissipation. Note: Find the speed-power product of each and compare the results.

3–25   Determine t_{PLH} and t_{PHL} from the oscilloscope display in Figure 3–74.

3–26   Gate A has $t_{PLH} = t_{PHL} = 6$ ns. Gate B has $t_{PLH} = t_{PHL} = 10$ ns. Which gate can be operated at a higher frequency than the other?

3–27   If a logic gate operates on a dc supply voltage of $+5$ V and draws an average current of 4 mA, what is its power dissipation?

3–28   The variable I_{CCH} represents the dc supply current from V_{CC} when all outputs of an IC are HIGH. The variable I_{CCL} represents the dc supply current when all outputs are LOW. For a 74ALS00A IC, determine the *typical* power dissipation when all four gate outputs are HIGH. (See data sheet in Appendix A.)

FIGURE 3–74

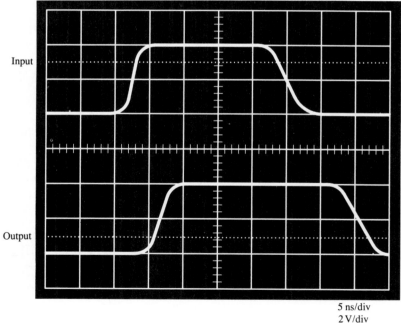

5 ns/div
2 V/div

Section 3–8   Applications

3–29   Sensors are used to monitor the pressure and the temperature of a chemical solution stored in a vat. The circuitry for each sensor produces a HIGH voltage when a specified maximum value is exceeded. An alarm requiring a LOW voltage input must be activated when either the pressure or the temperature is excessive. What type of logic gate is required for this application?

3–30 In a certain automated manufacturing process, electrical components are automatically inserted in a PC board. Before the insertion tool is activated, the PC board must be properly positioned, and the component must be in the chamber. Each of these prerequisite conditions is indicated by a HIGH voltage. The insertion tool requires a LOW voltage to activate it. Draw a diagram showing the logic gate required to implement this process, with its input and output connections to the system.

3–31 Modify the frequency counter in Figure 3–47 to operate with an Enable pulse that is LOW rather than HIGH during the 1-second interval.

3–32 Assume that the Enable signal in Figure 3–47 has the waveform shown in Figure 3–75. Assume that waveform *B* is also available. Devise a circuit that will produce an active-HIGH reset pulse to the counter only during the time that the Enable signal is LOW.

3–33 Design a circuit to fit in the colored block of Figure 3–76 that will cause the headlights of an automobile to be turned off automatically 15 seconds after the ignition switch is turned off, if the light switch is left on. Assume that a LOW is required to turn the lights off.

3–34 Expand the logic circuit for the intrusion alarm in Figure 3–50 so that two additional rooms, each with two windows and one door, can be protected.

FIGURE 3–75

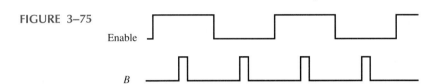

FIGURE 3–76

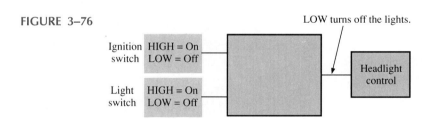

TBLS **Section 3–9 Troubleshooting**

3–35 Examine the conditions indicated in Figure 3–77, and identify the faulty gates.

3–36 Determine the faulty gates in Figure 3–78.

3–37 Using a logic probe and pulser, you make the observations indicated in Figure 3–79. For each observation determine the most likely gate failure.

FIGURE 3–77

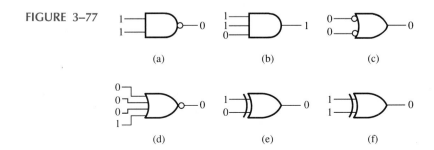

3–38 The seat belt alarm circuit in Figure 3–48 has malfunctioned. You find that when the ignition switch is turned on and the seat belt is unbuckled, the alarm comes on and will not go off. What is the most likely problem? How do you troubleshoot it?

3–39 Every time the ignition switch is turned on in the circuit of Figure 3–48, the alarm comes on for thirty seconds, even when the seat belt is buckled. What is the most probable cause of this malfunction?

FIGURE 3–78

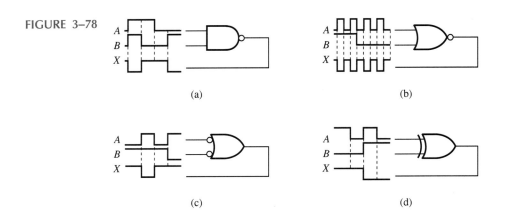

(a)

(b)

(c)

(d)

FIGURE 3–79

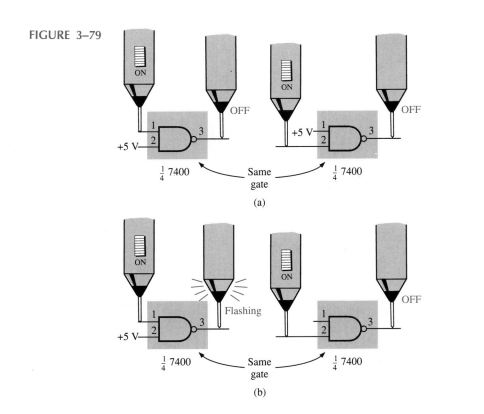

(a)

(b)

ANSWERS TO SECTION REVIEWS

Section 3–1   Review

1. 0   **2.** **(a)** the negation symbol ($\circ$) on the inverter output
(b) Output is LOW.

Section 3–2   Review

1. when all inputs are HIGH   **2.** when one or more inputs are LOW
3. $X = 1$ when $ABCD = 1111$, and $X = 0$ for all other combinations of $ABCD$.

Section 3–3   Review

1. when one or more inputs are HIGH   **2.** when all inputs are LOW
3. $X = 0$ when $ABC = 000$, and $X = 1$ for all other combinations of ABC.

Section 3–4   Review

1. when all inputs are HIGH   **2.** when one or more inputs are LOW
3. NAND: active-LOW output for all HIGH inputs; negative-OR: active-HIGH output for one or more LOW inputs; same truth tables.

Section 3–5   Review

1. when all inputs are LOW   **2.** when one or more inputs are HIGH
3. NOR: Active-LOW output for one or more HIGH inputs; negative-AND: active-HIGH output for all LOW inputs; same truth tables.

Section 3–6   Review

1. when the inputs are at opposite levels
2. when the inputs are at the same levels
3. Apply the bits to the inputs; when the output is HIGH, the bits are different.

Section 3–7   Review

1. $t_{PLH} = 8$ ns, $t_{PHL} = 10$ ns
2. I_{CCL}: supply current with gate output LOW; I_{CCH}: supply current with gate output HIGH
3. the maximum number of loads that a logic circuit can drive
4. **(a)** TTL, AS   **(b)** CMOS, 74HC (static), 4000 (100 kHz)   **(c)** TTL, AS
5. 12.75 pJ

Section 3–8   Review

1. 10^6 pulses
2. Use a five-input OR gate.

Section 3–9   Review

1. opens and shorts
2. a short between the input and the output or an open input
3. Waveforms can be observed with a scope.
4. easier to use

In 1854 George Boole published a book entitled *An Investigation of the Laws of Thought on Which Are Founded the Mathematical Theories of Logic and Probabilities*. It was in this publication that a "logical algebra," known today as Boolean algebra, was developed. Boolean algebra is a convenient and systematic way of expressing and analyzing the operation of digital circuits and systems.

The application of Boolean algebra to the analysis and design of digital logic circuits was first explored by Claude Shannon at MIT in a 1938 thesis entitled *A Symbolic Analysis of Relay and Switching Circuits*. This paper described a method by which any circuit consisting of combinations of switches and relays could be represented by mathematical expressions. Today semiconductor circuits have replaced mechanical switches and relays. However, the same logical analysis is still valid, and a basic knowledge of this area is essential to the study of digital electronics.

After completing this chapter, you will be able to

- □ express the operation of NOT, AND, OR, NAND, and NOR gates with Boolean algebra expressions.
- □ apply the basic laws and rules of Boolean algebra.
- □ apply DeMorgan's theorems to Boolean expressions.
- □ describe gate networks with Boolean expressions.
- □ evaluate Boolean expressions.
- □ simplify expressions by using the laws and rules of Boolean algebra.
- □ convert Boolean expressions of any form into a sum-of-products form.
- □ use a Karnaugh map to simplify Boolean expressions.
- □ use a Karnaugh map to simplify truth table functions.
- □ utilize "don't care" conditions to simplify logic functions.

4

Boolean Algebra

BOOLEAN OPERATIONS

4–1

Boolean algebra is the mathematics of digital systems. It is very important that you understand its principles thoroughly. A basic knowledge of Boolean algebra is indispensable to the study and analysis of digital logic circuits. In this section we will look at the basic Boolean operations and see how they apply to logic circuits.

Symbology

In the applications of Boolean algebra in this book, we will use capital letters to represent variables and functions of variables. Any single variable or a function of several variables can have either a 1 or a 0 value. In Boolean algebra the binary digits are utilized to represent the two levels that occur within digital circuits. A binary 1 will represent a HIGH level, and a binary 0 will represent a LOW level in Boolean equations. This is in keeping with our use in this text of positive logic, as explained in Chapter 1.

The *complement,* or inverse, of a variable is represented by a bar over the letter. For instance, for a variable represented by A, the complement of A is $\overline{A}$. So if $A = 1$, then $\overline{A} = 0$; or if $A = 0$, then $\overline{A} = 1$. The complement of a variable A is usually read "*A* bar" or "Not *A*." Sometimes a prime symbol rather than the bar symbol is used to denote the complement. For example, the complement of A can be written as A'.

The logical AND function of two variables is represented either by placing a dot between the two variables, as $A \cdot B$, or by simply writing the adjacent letters without the dot, as AB. We will normally use the latter notation because it is easier to write. The logical OR function of two variables is represented by a + between the two variables, as $A + B$.

Boolean Addition and Multiplication

Addition in Boolean algebra involves variables whose values are either binary 1 or binary 0. The basic rules for *Boolean addition* are as follows:

$$0 + 0 = 0$$
$$0 + 1 = 1$$
$$1 + 0 = 1$$
$$1 + 1 = 1$$

In the application of Boolean algebra to logic circuits, Boolean addition is the same as the OR function. Notice that it differs from binary addition in the case where two 1s are added.

Multiplication in Boolean algebra follows the basic rules governing binary multiplication, which were discussed in Chapter 2 and are as follows:

$$0 \cdot 0 = 0$$
$$0 \cdot 1 = 0$$
$$1 \cdot 0 = 0$$
$$1 \cdot 1 = 1$$

Boolean multiplication is the same as the AND function.

SECTION 4–1
REVIEW

Answers are found at the end of the chapter.
1. Addition in Boolean algebra is the same as the _____ function.
2. Multiplication in Boolean algebra is the same as the _____ function.
3. Represent each as a Boolean expression:
   (a) *A* AND *B*   **(b)** *A* OR *B*

BOOLEAN EXPRESSIONS

4–2

In this section you will learn how to express the operation of each of the basic logic gates in the terms of Boolean algebra. Expressions for NOT, AND, OR, NAND, and NOR are covered.

NOT

The operation of an inverter (NOT circuit) can be expressed with symbols as follows: If the input variable is called *A* and the output variable is called *X,* then $X = \overline{A}$. This expression states that the output is the complement of the input, so if $A = 0$, then $X = 1$, and if $A = 1$, then $X = 0$. Figure 4–1 illustrates this.

FIGURE 4–1
The inverter complements an input variable.

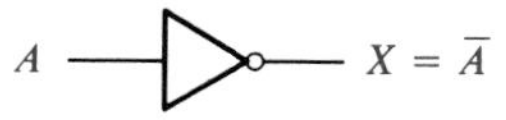

$$A \longrightarrow\!\!\!\!\triangleright\!\circ\!\longrightarrow X = \overline{A}$$

AND

The operation of a two-input AND gate can be expressed in equation form as follows: If one input variable is *A,* the other input variable is *B,* and the output variable is *X,* then the Boolean expression for this basic gate function is $X = AB$. Figure 4–2(a) shows the gate with the input and output variables indicated.

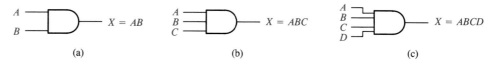

(a) (b) (c)

FIGURE 4–2
Boolean expressions for AND functions.

To extend the AND expression to more than two input variables, we simply use a new letter for each input variable. The function of a three-input AND gate, for example, can be expressed as $X = ABC$, where *A, B,* and *C* are the input variables. The expression for a four-input AND gate can be $X = ABCD$, and so on. Figures 4–2(b) and (c) show AND gates with three and four input variables, respectively.

An evaluation of AND gate operation can be made by using the Boolean expressions for the output. For example, each variable on the inputs can be either a 1 or a 0, so for the two-input AND gate, we can make substitutions in the equation for the output, $X = AB$, as shown in Table 4–1.

TABLE 4–1

A	B	$AB = X$
0	0	$0 \cdot 0 = 0$
0	1	$0 \cdot 1 = 0$
1	0	$1 \cdot 0 = 0$
1	1	$1 \cdot 1 = 1$

This evaluation simply tells us that the output X of an AND gate is a 1 (HIGH) only when both inputs are 1s (HIGHs). A similar analysis can be made for any number of input variables.

OR

The operation of a two-input OR gate can be expressed in equation form as follows: If one input is A, if the other input is B, and if the output is X, then the Boolean expression is $X = A + B$. Figure 4–3(a) shows the gate logic symbol, with input and output variables labeled.

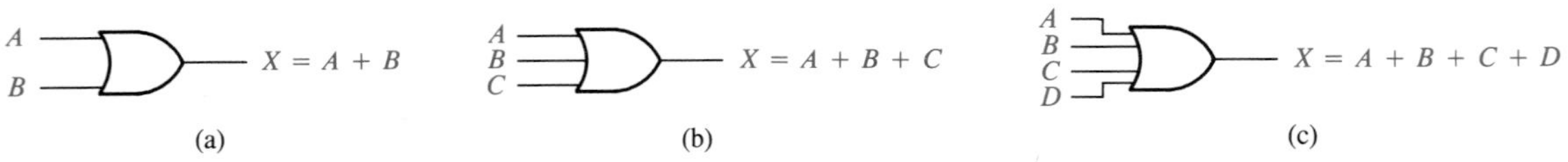

FIGURE 4–3
Boolean expressions for OR functions.

To extend the OR expression to more than two input variables, a new letter is used for each additional variable. For instance, the function of a three-input OR gate can be expressed as $X = A + B + C$. The expression for a four-input OR gate can be written as $X = A + B + C + D$, and so on. Figures 4–3(b) and (c) show OR gates with three and four input variables, respectively.

OR gate operation can be evaluated by using the Boolean expressions for the output X by substituting all possible combinations of 1 and 0 values for the input variables, as shown in Table 4–2 for a two-input OR gate.

TABLE 4–2

A	B	$A + B = X$
0	0	$0 + 0 = 0$
0	1	$0 + 1 = 1$
1	0	$1 + 0 = 1$
1	1	$1 + 1 = 1$

This evaluation shows that the output of an OR gate is a 1 (HIGH) when any one or more of the inputs are 1 (HIGH). A similar analysis can be extended to OR gates with any number of input variables.

NAND

The Boolean expression for a two-input NAND gate is $X = \overline{AB}$. This expression says that the two input variables, A and B, are first ANDed and then complemented, as indicated by the bar over the AND expression. This is a logical description in equation form of the operation of a NAND gate with two inputs. If we evaluate this expression for all possible values of the two input variables, the results are as in Table 4–3.

TABLE 4–3

A	B	$\overline{AB} = X$
0	0	$\overline{0 \cdot 0} = \overline{0} = 1$
0	1	$\overline{0 \cdot 1} = \overline{0} = 1$
1	0	$\overline{1 \cdot 0} = \overline{0} = 1$
1	1	$\overline{1 \cdot 1} = \overline{1} = 0$

Thus, once a Boolean expression is determined for a given logic function, that function can be evaluated for all possible values of the variables. The evaluation tells us exactly what the output of the logic circuit is for each of the input conditions, and it therefore gives us a complete description of the circuit's logical operation. The NAND expression can be extended to more than two input variables by including additional letters to represent all the variables.

NOR

The expression for a two-input NOR gate can be written as $X = \overline{A + B}$. This equation says that the two input variables are first ORed and then complemented, as indicated by the bar over the OR expression. Evaluating this expression, we get the results in Table 4–4.

TABLE 4–4

A	B	$\overline{A + B} = X$
0	0	$\overline{0 + 0} = \overline{0} = 1$
0	1	$\overline{0 + 1} = \overline{1} = 0$
1	0	$\overline{1 + 0} = \overline{1} = 0$
1	1	$\overline{1 + 1} = \overline{1} = 0$

SECTION 4–2 REVIEW

Answers are found at the end of the chapter.
1. Write the output expression for a five-input AND gate with input variables A, B, C, D, and E.
2. Write the output expression for a five-input OR gate with input variables F, G, H, I, and J.
3. (a) Answer question 1 for NAND gate.
   (b) Answer question 2 for a NOR gate.

RULES AND LAWS OF BOOLEAN ALGEBRA

4–3

As in other areas of mathematics, there are certain well-developed rules and laws that must be followed in order to properly apply Boolean algebra. The most important of these are presented in this section.

Three of the basic laws of Boolean algebra are the same as in ordinary algebra: the commutative laws, the associative laws, and the distributive law.

Commutative Laws

The *commutative law of addition* for two variables is written algebraically as

$$A + B = B + A \qquad (4\text{–}1)$$

This states that the order in which the variables are ORed makes no difference. Remember, in Boolean algebra terminology as applied to logic circuits, addition and the OR function are the same. Figure 4–4 illustrates the commutative law as applied to the OR gate.

The *commutative law of multiplication* for two variables is

$$AB = BA \qquad (4\text{–}2)$$

This states that the order in which the variables are ANDed makes no difference. Figure 4–5 illustrates this law as applied to the AND gate.

FIGURE 4–4
Application of commutative law of addition.

FIGURE 4–5
Application of commutative law of multiplication.

Associative Laws

The *associative law of addition* is stated as follows for three variables:

$$A + (B + C) = (A + B) + C \qquad (4\text{–}3)$$

This law states that in the ORing of several variables, the result is the same regardless of the grouping of the variables. Figure 4–6 illustrates this law as applied to OR gates.

The *associative law of multiplication* is stated as follows for three variables:

$$A(BC) = (AB)C \qquad (4\text{–}4)$$

This law tells us that it makes no difference in what order the variables are grouped when ANDing several variables. Figure 4–7 illustrates this law as applied to AND gates.

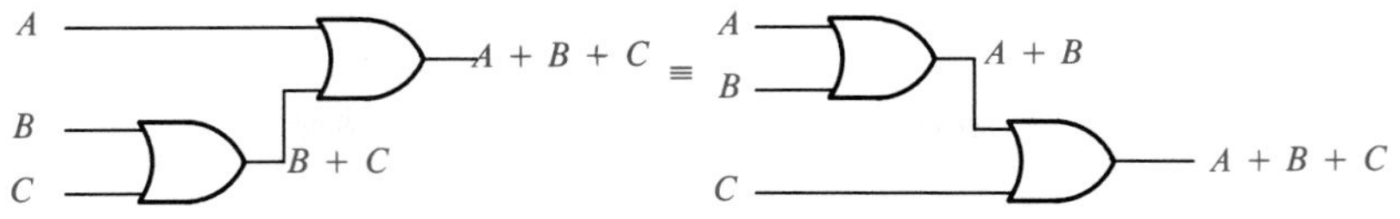

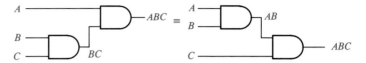

FIGURE 4–6
Application of associative law of addition.

FIGURE 4–7
Application of associative law of multiplication.

Distributive Law

The *distributive law* is written for three variables as follows:

$$A(B + C) = AB + AC \qquad (4\text{–}5)$$

This law states that ORing several variables and ANDing the result with a single variable is equivalent to ANDing the single variable with each of the several variables and then ORing the products. This law and the ones previously discussed should be familiar because they are the same as in ordinary algebra. Keep in mind that each of these laws can be extended to include any number of variables. Figure 4–8 illustrates this law in terms of gate implementation.

FIGURE 4–8
Application of distributive law.

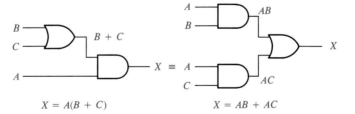

$$X = A(B + C) \qquad\qquad X = AB + AC$$

Rules for Boolean Algebra

Table 4–5 lists several basic rules that are useful in manipulating and simplifying Boolean algebra expressions.

We will now look at rules 1 through 9 of Table 4–5 in terms of their application to logic gates. Rules 10 through 12 will be derived in terms of the simpler rules and the laws previously discussed.

Rule 1 This rule can be understood by observing what happens when one input to an OR gate is always 0 and the other input, *A*, can take on either a 1 or a 0 value. If *A* is a 1, the output obviously is a 1, which is equal to *A*. If *A* is a 0, the output is a 0, which is also equal to *A*. Therefore, it follows that a variable ORed with a 0 is equal to

TABLE 4–5
Basic rules of Boolean algebra.

1. $A + 0 = A$
2. $A + 1 = 1$
3. $A \cdot 0 = 0$
4. $A \cdot 1 = A$
5. $A + A = A$
6. $A + \overline{A} = 1$
7. $A \cdot A = A$
8. $A \cdot \overline{A} = 0$
9. $\overline{\overline{A}} = A$
10. $A + AB = A$
11. $A + \overline{A}B = A + B$
12. $(A + B)(A + C) = A + BC$

NOTE: A, B, or C can represent a single variable or a combination of variables.

the value of the variable ($A + 0 = A$). This rule is further demonstrated in Figure 4–9, where the lower input is fixed at 0.

Rule 2   This rule is demonstrated when one input to an OR gate is always 1 and the other input, A, takes on either a 1 or a 0 value. A 1 on an input to an OR gate produces a 1 on the output, regardless of the value of the variable on the other input. Therefore, a variable ORed with a 1 is always equal to 1 ($A + 1 = 1$). This rule is illustrated in Figure 4–10, where the lower input is fixed at 1.

Rule 3   Rule 3 is demonstrated when a 0 is ANDed with a variable. Of course, any time one input to an AND gate is 0, the output is 0, regardless of the value of the variable on the other input. A variable ANDed with a 0 always produces a 0 ($A \cdot 0 = 0$). This rule is illustrated in Figure 4–11, where the lower input is fixed at 0.

Rule 4   To verify rule 4, AND a variable with a 1. If the variable A is a 0, the output of the AND gate is a 0. If the variable A is a 1, the output of the AND gate is a 1

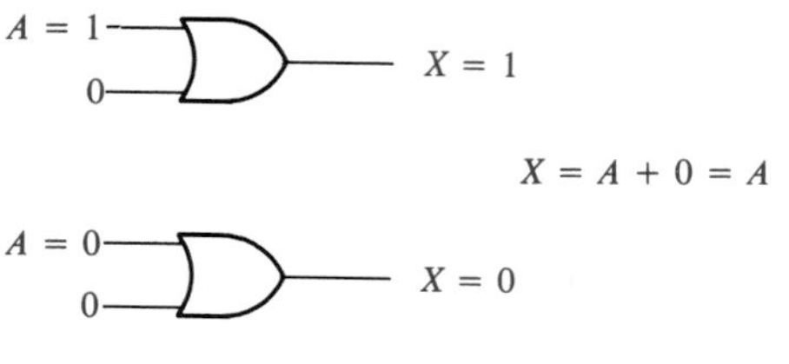

FIGURE 4–9
Illustration of rule 1.

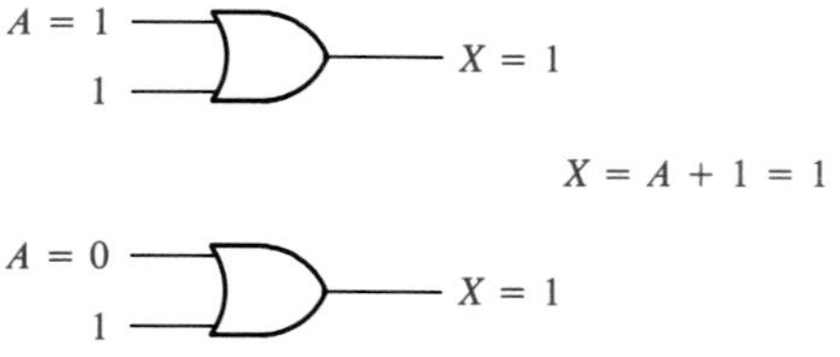

$X = A + 0 = A$

$X = A + 1 = 1$

FIGURE 4–10
Illustration of rule 2.

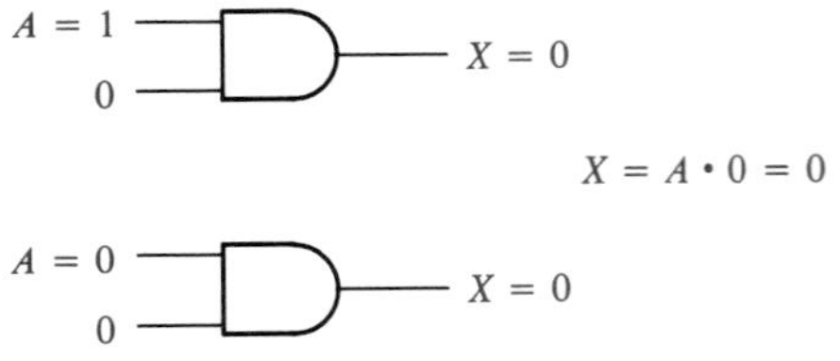

$X = A \cdot 0 = 0$

FIGURE 4–11
Illustration of rule 3.

FIGURE 4–12
Illustration of rule 4.

$$X = A \cdot 1 = A$$

FIGURE 4–13
Illustration of rule 5.

$$X = A + A = A$$

FIGURE 4–14
Illustration of rule 6.

$$X = A + \overline{A} = 1$$

because both inputs are now 1s. Therefore, the AND function of a variable and a 1 is equal to the value of the variable ($A \cdot 1 = A$). This is shown in Figure 4–12, where the lower input is fixed at 1.

Rule 5   This rule states that if a variable is ORed with itself, the output is equal to the variable. For instance, if A is a 0, then $0 + 0 = 0$, and if A is a 1, then $1 + 1 = 1$. This is shown in Figure 4–13, where both inputs are the same variable.

Rule 6   This rule can be explained as follows: If a variable and its complement are ORed, the result is always a 1. If A is a 0, then $0 + \overline{0} = 0 + 1 = 1$. If A is a 1, then $1 + \overline{1} = 1 + 0 = 1$. For further illustration of this rule, see Figure 4–14, where one input is the complement of the other.

Rule 7   Rule 7 states that if a variable is ANDed with itself, the result is equal to the variable. For example, if $A = 0$, then $0 \cdot 0 = 0$, and if $A = 1$, then $1 \cdot 1 = 1$. For either case, the output of an AND gate is equal to the value of the input variable A. Figure 4–15 illustrates this rule.

Rule 8   If a variable is ANDed with its complement, the result is 0. This is readily seen because either A or $\overline{A}$ will always be 0, and when a 0 is applied to the input of an AND gate, it ensures that the output will be 0 also. Figure 4–16 helps illustrate this rule.

$$X = AA = A$$

$$X = A\overline{A} = 0$$

FIGURE 4–15
Illustration of rule 7.

FIGURE 4–16
Illustration of rule 8.

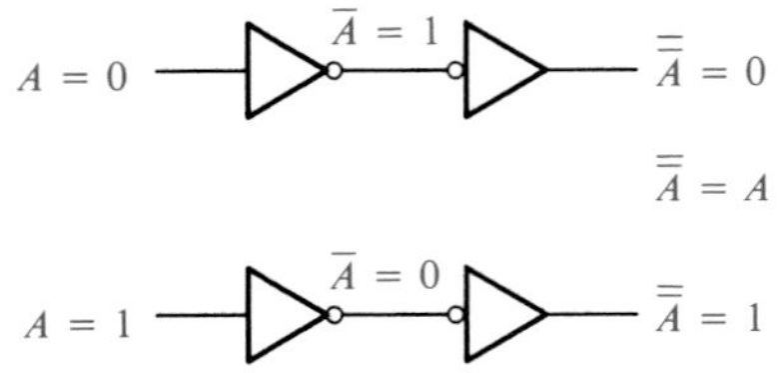

FIGURE 4–17
Illustration of rule 9.

Rule 9 This rule simply says that if a variable is complemented twice, the result is the variable itself. If we start with the variable A and complement (invert) it once, we get $\overline{A}$. If we then take $\overline{A}$ and complement (invert) it, we get A, which is the original variable. This is shown in Figure 4–17 with inverters.

Rule 10 This rule is proved by using the distributive law, rule 2, and rule 4 as follows:

$$
\begin{aligned}
A + AB &= A(1 + B) && \text{distributive law} \\
&= A \cdot 1 && \text{rule 2} \\
&= A && \text{rule 4}
\end{aligned}
$$

The proof can also be shown in tabular form, as in Table 4–6.

TABLE 4–6

A	B	AB	$A + AB$
0	0	0	0
0	1	0	0
1	0	0	1
1	1	1	1

equal

Rule 11 Rule 11 is proved as follows:

$$
\begin{aligned}
A + \overline{A}B &= (A + AB) + \overline{A}B && \text{rule 10} \\
&= (AA + AB) + \overline{A}B && \text{rule 7} \\
&= AA + AB + A\overline{A} + \overline{A}B && \text{rule 8 (adding } A\overline{A} = 0) \\
&= (A + \overline{A})(A + B) && \text{by factoring} \\
&= 1 \cdot (A + B) && \text{rule 6} \\
&= A + B && \text{rule 4}
\end{aligned}
$$

The proof can also be shown in tabular form as in Table 4–7.

TABLE 4–7

A	B	$\overline{A}B$	$A + \overline{A}B$	$A + B$
0	0	0	0	0
0	1	1	1	1
1	0	0	1	1
1	1	0	1	1

equal

Rule 12   This rule is proved as follows:

$$
\begin{aligned}
(A + B)(A + C) &= AA + AC + AB + BC && \text{distributive law} \\
&= A + AC + AB + BC && \text{rule 7} \\
&= A(1 + C) + AB + BC && \text{distributive law} \\
&= A \cdot 1 + AB + BC && \text{rule 2} \\
&= A(1 + B) + BC && \text{distributive law} \\
&= A \cdot 1 + BC && \text{rule 2} \\
&= A + BC && \text{rule 4}
\end{aligned}
$$

The proof can also be shown in tabular form as in Table 4–8.

TABLE 4–8

A	B	C	$A + B$	$A + C$	$(A + B)(A + C)$	BC	$A + BC$
0	0	0	0	0	0	0	0
0	0	1	0	1	0	0	0
0	1	0	1	0	0	0	0
0	1	1	1	1	1	1	1
1	0	0	1	1	1	0	1
1	0	1	1	1	1	0	1
1	1	0	1	1	1	0	1
1	1	1	1	1	1	1	1

⎣——— equal ———⎦

SECTION 4–3 REVIEW

Answers are found at the end of the chapter.
1. Apply the associative law of addition to the expression $A + (B + C + D)$.
2. Apply the distributive law to the expression $A(B + C + D)$.

DEMORGAN'S THEOREMS

4–4

DeMorgan, a logician and mathematician who was acquainted with Boole, proposed two theorems that are an important part of Boolean algebra. DeMorgan's theorems are actually a mathematical verification of the equivalency of the NAND gate and the negative-OR gate and of the NOR gate and the negative-AND gate, which were discussed in the last chapter.

DeMorgan's theorems are expressed in the following formulas for two variables:

$$\overline{XY} = \overline{X} + \overline{Y} \tag{4–6}$$

$$\overline{X + Y} = \overline{X}\,\overline{Y} \tag{4–7}$$

The theorem expressed in Equation (4–6) can be stated as follows:

The complement of a product is equal to the sum of the complements.

It really says that the complement of two or more variables ANDed is the same as the OR of the complements of the individual variables.

The theorem expressed in Equation (4–7) can be stated as follows:

The complement of a sum is equal to the product of the complements.

It says that the complement of two or more variables ORed is the same as the AND of the complements of the individual variables.

These theorems are illustrated by the gate equivalencies and truth tables in Figure 4–18.

As mentioned, DeMorgan's theorems also apply to expressions in which there are more than two variables. The following examples illustrate the application of DeMorgan's theorems to three- and four-variable expressions.

FIGURE 4–18

Gate equivalencies and corresponding truth tables illustrating DeMorgan's theorems. Notice the equality of the two output columns in each table.

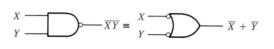

X	Y	$\overline{XY}$	$\overline{X} + \overline{Y}$
0	0	1	1
0	1	1	1
1	0	1	1
1	1	0	0

X	Y	$\overline{X + Y}$	$\overline{X}\,\overline{Y}$
0	0	1	1
0	1	0	0
1	0	0	0
1	1	0	0

EXAMPLE 4–1

Apply DeMorgan's theorem to the expressions $\overline{XYZ}$ and $\overline{X + Y + Z}$.

Solution

$$\overline{XYZ} = \overline{X} + \overline{Y} + \overline{Z}$$
$$\overline{X + Y + Z} = \overline{X}\,\overline{Y}\,\overline{Z}$$

EXAMPLE 4–2

Apply DeMorgan's theorems to the expressions $\overline{WXYZ}$ and $\overline{W + X + Y + Z}$.

Solution

$$\overline{WXYZ} = \overline{W} + \overline{X} + \overline{Y} + \overline{Z}$$
$$\overline{W + X + Y + Z} = \overline{W}\,\overline{X}\,\overline{Y}\,\overline{Z}$$

Each variable in DeMorgan's theorems as stated in Equations (4–6) and (4–7) can also represent a combination of other variables. For example, X can be equal to the term $AB + C$, and Y can be equal to the term $A + BC$. So if we apply DeMorgan's theorem to the expression $\overline{(AB + C)(A + BC)}$, we get the following result:

$$\overline{(AB + C)(A + BC)} = \overline{(AB + C)} + \overline{(A + BC)}$$

Now notice that in the preceding result we have two terms, $\overline{AB + C}$ and $\overline{A + BC}$, to which we can again apply DeMorgan's theorem individually, as follows:

$$\overline{(AB + C)} + \overline{(A + BC)} = \overline{(AB)}\overline{C} + \overline{A}\overline{(BC)}$$

Notice that we still have two terms to which DeMorgan's theorem can be applied. These terms are $\overline{AB}$ and $\overline{BC}$. A final application of DeMorgan's theorem gives the following result:

$$\overline{(AB)}\overline{C} + \overline{A}\overline{(BC)} = (\overline{A} + \overline{B})\overline{C} + \overline{A}(\overline{B} + \overline{C})$$

Although this result can be simplified by the use of Boolean rules and laws, DeMorgan's theorems cannot be used any further.

Applying DeMorgan's Theorems

The following procedure further illustrates the application of DeMorgan's theorems to a specific expression:

$$\overline{\overline{A + B\overline{C}} + D(\overline{E + \overline{F}})}$$

Step 1 Identify the terms to which you can apply DeMorgan's theorems, and think of each term as a single variable. Let $\overline{A + B\overline{C}} = X$ and $D(\overline{E + \overline{F}}) = Y$.

Step 2 Since $\overline{X + Y} = \overline{X}\overline{Y}$,

$$\overline{[\overline{A + B\overline{C}}] + [D(\overline{E + \overline{F}})]} = \overline{[\overline{A + B\overline{C}}]}\overline{[D(\overline{E + \overline{F}})]}$$

Step 3 Use rule 9 to cancel the double bars over the left term (this is not part of DeMorgan's theorem):

$$\overline{[\overline{A + B\overline{C}}]}\overline{[D(\overline{E + \overline{F}})]} = [A + B\overline{C}]\overline{[D(\overline{E + \overline{F}})]}$$

Step 4 In the right term, let $W = D$ and $Z = \overline{E + \overline{F}}$.

Step 5 Since $\overline{WZ} = \overline{W} + \overline{Z}$,

$$[A + B\overline{C}]\overline{[D(\overline{E + \overline{F}})]} = (A + B\overline{C})(\overline{D} + \overline{\overline{E + \overline{F}}})$$

Step 6 Use rule 9 to cancel the double bars over the $E + \overline{F}$ part of the term:

$$(A + B\overline{C})(\overline{D} + \overline{\overline{E + \overline{F}}}) = (A + B\overline{C})(\overline{D} + E + \overline{F})$$

The following examples will further illustrate how to use DeMorgan's theorem.

EXAMPLE 4–3

Apply DeMorgan's theorems to each of the following expressions.

(a) $\overline{(A + B + C)D}$ **(b)** $\overline{ABC + DEF}$ **(c)** $\overline{A\overline{B} + \overline{C}D + EF}$

Solutions

(a) Letting $A + B + C = X$ and $D = Y$, we have an expression of the form $\overline{XY} = \overline{X} + \overline{Y}$:

$$\overline{(A + B + C)D} = \overline{A + B + C} + \overline{D}$$

Next, we apply DeMorgan's theorem to the term $\overline{A + B + C}$:

$$\overline{A + B + C} + \overline{D} = \overline{A}\overline{B}\overline{C} + \overline{D}$$

(b) Letting $ABC = X$ and $DEF = Y$, we have an expression of the form $\overline{X + Y} = \overline{X}\overline{Y}$:

$$\overline{ABC + DEF} = \overline{(ABC)}\,\overline{(DEF)}$$

Next, we apply DeMorgan's theorem to each of the terms $\overline{ABC}$ and $\overline{DEF}$:

$$\overline{(ABC)}\overline{(DEF)} = (\overline{A} + \overline{B} + \overline{C})(\overline{D} + \overline{E} + \overline{F})$$

(c) Letting $A\overline{B} = X$, $\overline{C}D = Y$, and $EF = Z$, we have an expression of the form $\overline{X + Y + Z} = \overline{X}\overline{Y}\overline{Z}$;

$$\overline{A\overline{B} + \overline{C}D + EF} = \overline{(A\overline{B})}\,\overline{(\overline{C}D)}\,\overline{(EF)}$$

Next, we apply DeMorgan's theorem to each of the terms $\overline{A\overline{B}}$, $\overline{\overline{C}D}$, and $\overline{EF}$:

$$\overline{(A\overline{B})}\,\overline{(\overline{C}D)}\,\overline{(EF)} = (\overline{A} + B)(C + \overline{D})(\overline{E} + \overline{F})$$

EXAMPLE 4–4

Apply DeMorgan's theorems to each expression.

(a) $\overline{(A + B) + \overline{C}}$

(b) $\overline{(\overline{A} + B) + CD}$

(c) $\overline{(A + B)\overline{CD} + E + \overline{F}}$

Solutions

(a) $\overline{(A + B) + \overline{C}} = \overline{(A + B)} \cdot \overline{\overline{C}} = (A + B)C$

(b) $\overline{(\overline{A} + B) + CD} = \overline{(\overline{A} + B)}\,\overline{CD} = (\overline{\overline{A}}\overline{B})(\overline{C} + \overline{D})$
$$= A\overline{B}(\overline{C} + \overline{D})$$

(c) $\overline{(A + B)\overline{CD} + E + \overline{F}} = \overline{[(A + B)\overline{CD}]}\,\overline{(E + \overline{F})}$
$$= (\overline{AB} + C + D)\overline{E}F$$

EXAMPLE 4–5

The Boolean expression for an exclusive-OR gate is $A\overline{B} + \overline{A}B$. With this as a starting point, develop an expression for the exclusive-NOR gate, using DeMorgan's theorems and any other rules or laws that are applicable.

Solution   We start by complementing the exclusive-OR expression and then applying DeMorgan's theorems as follows:

$$\overline{A\overline{B} + \overline{A}B} = \overline{(A\overline{B})}\,\overline{(\overline{A}B)} = (\overline{A} + \overline{\overline{B}})(\overline{\overline{A}} + \overline{B}) = (\overline{A} + B)(A + \overline{B})$$

Next we apply the distributive law and rule 8:

$$(\overline{A} + B)(A + \overline{B}) = \overline{A}A + \overline{A}\overline{B} + AB + B\overline{B} = \overline{A}\overline{B} + AB$$

The final expression for the XNOR is $\overline{A}\overline{B} + AB$. You can see that this expression is a 1 any time both variables are 0s or both variables are 1s.

SECTION 4–4
REVIEW

Answers are found at the end of the chapter.

1. Apply DeMorgan's theorems to the following expressions:

(a) $X = \overline{ABC} + \overline{(\overline{D} + E)}$

(b) $X = \overline{(A + B)C}$

(c) $X = \overline{A + B + C} + \overline{\overline{DE}}$

BOOLEAN EXPRESSIONS FOR GATE NETWORKS

4–5

*When two or more logic gates are connected to perform a specified function, we have a gate **network**. Boolean algebra provides a concise way to express the operation of a gate network so that we can readily determine what the output will be for various combinations of input levels.*

The form of a given Boolean expression indicates the type of gate network it describes. For example, let us take the expression $A(B + CD)$ and determine what kind of logic circuit it represents. First, there are four variables: A, B, C, and D. Variable C is ANDed with D, giving CD; then CD is ORed with B, giving $(B + CD)$. Then this is ANDed with A to produce the final function. Figure 4–19 illustrates the gate network represented by this particular Boolean expression, $A(B + CD)$.

As you have seen, the form of the Boolean expression does determine how many logic gates are used, what type of gates are needed, and how they are connected together; this will be explored further in Chapter 5. The more complex an expression, the more complex the gate network will be. It is therefore an advantage to simplify an expression as much as possible in order to have the simplest gate network. We will cover simplification methods in a later section of this chapter.

There are two basic forms of Boolean expressions that are of primary importance because all other expressions can be converted to either of these two basic forms. These basic forms of Boolean expressions are the *sum-of-products (SOP)* and the *product-of-sums (POS)*. Of these two forms, the SOP will be emphasized because it can be converted easily to truth table format and because it lends itself well to a simplification technique that we will cover later in this chapter.

FIGURE 4–19
Logic gate implementation of the expression $A(B + CD)$.

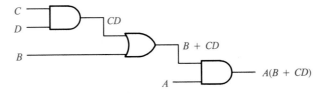

Sum-of-Products (SOP) Form

What does the sum-of-products form mean? First, let us review products in Boolean algebra. A product of two or more variables or their complements is simply the AND function of those variables. The product of two variables can be expressed as AB, the product of three variables as ABC, the product of four variables as $ABCD$, and so on. Recall that a sum in Boolean algebra is the same as the OR function, so a sum-of-products expression is two or more AND functions ORed together. For instance, $AB + CD$ is a sum-of-products expression. Several other examples of expressions in sum-of-products form follow:

☐ $AB + BCD$
☐ $ABC + DEF$
☐ $A\overline{B}C + D\overline{E}FG + AEG$
☐ $AB\overline{C} + \overline{A}B\overline{C} + ABC + A\overline{B}C$

An SOP expression can also contain a term with a single variable, as in $A + BCD + EFG$. In an SOP expression a single bar cannot extend over more than one variable in a term, although more than one variable in a term can have a bar over it. For example, we can have $\overline{A}\overline{B}C$ but we cannot have $\overline{ABC}$.

One reason the SOP is a useful form of Boolean expression is the straightforward manner in which it can be implemented with logic gates. We have AND functions that are ORed, as Example 4–6 illustrates.

EXAMPLE 4–6

Implement the expression $AB + BCD + EFGH$ with logic gates.

Solution

FIGURE 4–20

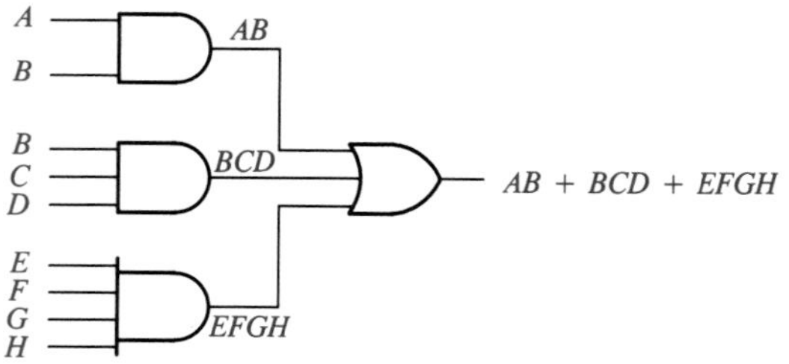

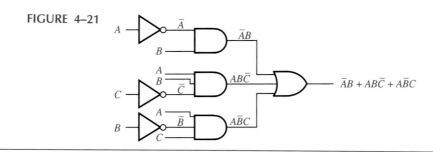

An important characteristic of the SOP form is that the corresponding implementation is always a two-level gate network; that is, the maximum number of gates through which a signal must pass in going from an input to the output is two, excluding input variable inversions. Inversions add another gate level, as the next example illustrates.

EXAMPLE 4–7

Implement the expression $\overline{A}B + AB\overline{C} + A\overline{B}C$ with logic gates.

Solution

FIGURE 4–21

Conversion to SOP Form

Any logic expression can be changed into SOP form by applying Boolean algebra techniques. For example, the expression $A(B + CD)$, which was implemented in Figure 4–19, can be converted to SOP form by applying the distributive law:

$$A(B + CD) = AB + ACD$$

When the SOP expression is implemented, as shown in Figure 4–22, the circuit is quite different from the one in Figure 4–19, although the two are equivalent; that is, they have the same output for each input combination.

FIGURE 4–22
Sum-of-products implementation of the circuit in Figure 4–19.

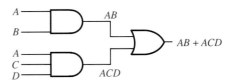

EXAMPLE 4–8

Convert each of the following expressions to SOP form.
(a) $AB + B(CD + EF)$
(b) $(A + B)(B + C + D)$
(c) $\overline{(A + B)} + C$

Solution
(a) $AB + B(CD + EF) = AB + BCD + BEF$
(b) $(A + B)(B + C + D) = AB + AC + AD + BB + BC + BD$
(c) $\overline{(A + B)} + C = \overline{(A + B)}\overline{C} = (A + B)\overline{C} = A\overline{C} + B\overline{C}$

Product-of-Sums (POS) Form

The product-of-sums form is, in terms of logic functions, the AND of two or more OR functions. For instance, $(A + B)(B + C)$ is a product-of-sums expression. Some other examples follow:

☐ $(A + B)(B + C + D)$
☐ $(A + B + C)(D + E + F)$
☐ $(A + \overline{B} + C)(D + E + \overline{F} + G)(A + \overline{E} + \overline{G})$

A POS expression can also contain a single variable term, as in

$$A(B + C + D)(E + F + G)$$

This form also lends itself to straightforward implementation with logic gates because it involves simply ANDing the outputs of two or more OR gates. A two-level gate network will always result, as the following example will show.

EXAMPLE 4–9

Construct the following POS function with logic gates:

$$(A + B)(C + D + E)(F + G + H + I)$$

Solution

FIGURE 4-23

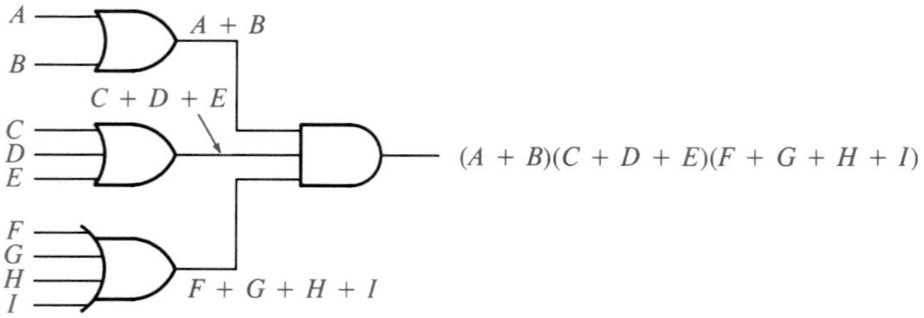

Answers are found at the end of the chapter.

1. Identify each expression as either a sum-of-products (SOP) or a product-of-sums (POS).
   (a) $X = AB + CD + EF$
   (b) $X = (A + B)(C + D)(E + F)$
   (c) $X = \overline{A}BC + A\overline{B}C + ABC + \overline{A}B\overline{C}$
2. Draw the logic gate diagram for each Boolean expression in question 1.

SIMPLIFICATION USING BOOLEAN ALGEBRA

4-6

Many times in the application of Boolean algebra, we have to reduce a particular expression to its simplest form or change its form to a more convenient one to implement the expression most efficiently. The approach taken in this section is to use the basic laws, rules, and theorems of Boolean algebra to manipulate and simplify an expression. This method depends on a thorough knowledge of Boolean algebra and considerable practice in its application.

The reason that we are interested in simplifying Boolean expressions is to use the fewest gates possible to implement a given expression. Several examples will illustrate Boolean simplification step by step.

EXAMPLE 4-10

Simplify the expression $AB + A(B + C) + B(B + C)$ using Boolean algebra techniques.

Solution   The following is not necessarily the only approach.

Step 1.   Apply the distributive law to the second and third terms in the expression, as follows:

$$AB + AB + AC + BB + BC$$

Step 2.   Apply rule 7 ($BB = B$):

$$AB + AB + AC + B + BC$$

Step 3. Apply rule 5 ($AB + AB = AB$):

$$AB + AC + B + BC$$

Step 4. Factor B out of the last two terms:

$$AB + AC + B(1 + C)$$

Step 5. Apply rule 2 ($1 + C = 1$):

$$AB + AC + B \cdot 1$$

Step 6. Apply rule 4 ($B \cdot 1 = B$):

$$AB + AC + B$$

Step 7. Factor B out of the first and third terms, as follows:

$$B(A + 1) + AC$$

Step 8. Apply rule 2 ($A + 1 = 1$):

$$B \cdot 1 + AC$$

Step 9. Apply rule 4 ($B \cdot 1 = B$):

$$B + AC$$

At this point we have simplified the expression as much as possible. Notice that it is in SOP form. It should be noted that once you gain experience in applying Boolean algebra, you can combine many individual steps. See if you can find an alternative approach.

Figure 4–24 shows that the simplification process in Example 4–10 has significantly reduced the number of logic gates required to implement the expression. Part (a) shows that five gates are required to implement the expression in its original form; only two gates are needed for the simplified expression, shown in part (b). It is important to realize that these two gate networks are equivalent. That is, for any combination of levels on the A, B, and C inputs, we get the same output from either circuit.

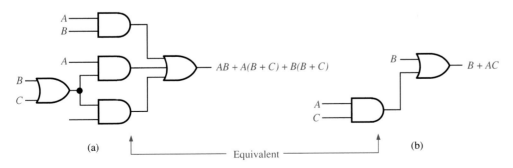

FIGURE 4–24
Gate networks for Example 4–10.

EXAMPLE 4–11   Simplify the expression $[A\overline{B}(C + BD) + \overline{A}\,\overline{B}]C$ as much as possible. Note that brackets and parentheses mean the same thing: simply that the expression inside is multiplied (ANDed) with the term outside.

Solution

Step 1.   Apply the distributive law to the terms within the brackets:

$$(A\overline{B}C + A\overline{B}BD + \overline{A}\,\overline{B})C$$

Step 2.   Apply rule 8 to the second term in the parentheses:

$$(A\overline{B}C + A \cdot 0 \cdot D + \overline{A}\,\overline{B})C$$

Step 3.   Apply rule 3 to the second term:

$$(A\overline{B}C + 0 + \overline{A}\,\overline{B})C$$

Step 4:   Apply rule 1 within the parentheses:

$$(A\overline{B}C + \overline{A}\,\overline{B})C$$

Step 5.   Apply the distributive law:

$$A\overline{B}CC + \overline{A}\,\overline{B}C$$

Step 6.   Apply rule 7 to the first term:

$$A\overline{B}C + \overline{A}\,\overline{B}C$$

Step 7.   Factor out $\overline{B}C$:

$$\overline{B}C(A + \overline{A})$$

Step 8.   Apply rule 6:

$$\overline{B}C \cdot 1$$

Step 9.   Apply rule 4:

$$\overline{B}C$$

EXAMPLE 4–12   Simplify the SOP expression:

$$\overline{A}BC + A\overline{B}\,\overline{C} + \overline{A}\,\overline{B}\,\overline{C} + A\overline{B}C + ABC$$

Solution

Step 1.   Factor BC out of the first and last terms:

$$BC(\overline{A} + A) + A\overline{B}\,\overline{C} + \overline{A}\,\overline{B}\,\overline{C} + A\overline{B}C$$

Step 2. Apply rule 6 to the first term, and factor $A\overline{B}$ from the second and last terms:

$$BC \cdot 1 + A\overline{B}(\overline{C} + C) + \overline{A}B\overline{C}$$

Step 3. Apply rule 4 to the first term and rule 6 to the second term:

$$BC + A\overline{B} \cdot 1 + \overline{A}B\overline{C}$$

Step 4. Apply rule 4 to the second term:

$$BC + A\overline{B} + \overline{A}B\overline{C}$$

Step 5. Factor $\overline{B}$ from the second and third terms:

$$BC + \overline{B}(A + \overline{A}\overline{C})$$

Step 6. Apply rule 11 to the second term:

$$BC + \overline{B}(A + \overline{C})$$

Step 7. Use the distributive and commutative laws to get the expression into SOP form:

$$BC + A\overline{B} + \overline{B}\overline{C}$$

EXAMPLE 4–13

Simplify the following expression:

$$\overline{AB + AC} + \overline{A}BC$$

Solution

Step 1. Apply DeMorgan's theorem to the first term:

$$\overline{(AB)}\,\overline{(AC)} + \overline{A}BC$$

Step 2. Apply DeMorgan's theorem to each term in parentheses:

$$(\overline{A} + \overline{B})(\overline{A} + \overline{C}) + \overline{A}BC$$

Step 3. Apply the distributive law to the two terms in parentheses:

$$\overline{A}\overline{A} + \overline{A}\overline{C} + \overline{A}\overline{B} + \overline{B}\overline{C} + \overline{A}BC$$

Step 4. Apply rule 7 to the first term, and apply rule 10 to the third and last terms:

$$\overline{A} + \overline{A}\overline{C} + \overline{A}\overline{B} + \overline{B}\overline{C}$$

Step 5. Apply rule 10 to the first and second terms:

$$\overline{A} + \overline{A}\overline{B} + \overline{B}\overline{C}$$

Step 6. Apply rule 10 to the first and second terms:

$$\overline{A} + \overline{B}\overline{C}$$

Answers are found at the end of the chapter.

1. Simplify the following expressions if possible:
   (a) $A + AB + A\overline{B}C$
   (b) $(\overline{A} + B)C + ABC$
   (c) $A\overline{B}C(BD + CDE) + A\overline{C}$

2. Implement each expression in question 1 as originally stated with the appropriate logic gates. Then implement the simplified expression, and compare the number of gates.

SIMPLIFICATION USING THE KARNAUGH MAP

4–7

*The **Karnaugh map** provides a systematic method for simplifying a Boolean expression or a truth table function and, if properly used, will produce the simplest sum-of-products expression possible.*

As you have seen, the effectiveness of algebraic simplification depends on your familiarity with all the laws, rules, and theorems of Boolean algebra and on your ability to apply them. Cleverness is often an important factor in algebraic simplification. The Karnaugh map (K map), on the other hand, provides basically a "cookbook" approach.

The K Map Format

The Karnaugh map is composed of an arrangement of adjacent *cells,* each representing one particular combination of variables in product form. Since the total number of combinations of n variables and their complements is 2^n, the Karnaugh map consists of 2^n cells. For example, there are four combinations of the products of two variables (A and B) and their complements: $\overline{A}\overline{B}$, $\overline{A}B$, $A\overline{B}$, and AB. Therefore, the Karnaugh map must have four cells, with each cell representing one of the variable combinations, as illustrated in Figure 4–25(a). The variable combinations are labeled in the cells only for purposes of illustration. In practice, the map is actually arranged with the variable labels outside the cells, as shown in Figure 4–25(b). The variable to the left of a row of cells applies to each cell in that row. The variable above a column of cells applies to each cell in that column.

Extensions of the Karnaugh map to three and four variables are shown in Figure 4–26. Notice that the cells are arranged so that there is only a single variable change between any adjacent cells (this is the characteristic that determines adjacency). For example, in Figure 4–26(a), the upper left cell is for $\overline{A}\overline{B}\overline{C}$, and the lower left cell is for $A\overline{B}\overline{C}$. Notice that only the variable A has changed between these two cells, thus making them adjacent. The other two cells that are adjacent to the upper left cell are

FIGURE 4–25
Format of a two-variable Karnaugh map.

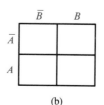

(a)   (b)

FIGURE 4–26

Formats for three- and four-variable Kar-naugh maps.

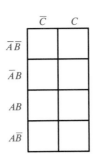

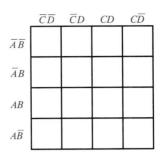

(a) Three-variable map ($2^3 = 8$ cells) (b) Four-variable map ($2^4 = 16$ cells)

the $\overline{A}B\overline{C}$ cell (upper right) and the $AB\overline{C}$ cell (immediately below the $\overline{A}B\overline{C}$ cell). In part (b) of the figure, the upper left cell is adjacent to the cells $\overline{A}B\overline{C}D$, $\overline{A}BCD$, $\overline{A}BC\overline{D}$, and $A\overline{B}\,\overline{C}\,\overline{D}$. Adjacency is determined for each of the other cells in a similar way.

Karnaugh maps can also be used for five, six, or more variables, but such maps are not covered in this book. Beyond six variables, the Karnaugh map is impractical except when implemented on a computer.

Plotting a Sum-of-Products Expression

Once a Boolean expression is in the sum-of-products form, you can plot it on the Karnaugh map by placing a 1 in each cell corresponding to a term in the sum-of-products expression. For example, the three-variable expression $\overline{A}BC + AB\overline{C} + ABC$ is plotted on the Karnaugh map in Figure 4–27(a), and the four-variable expression $\overline{A}B\overline{C}D + \overline{A}BCD + A\overline{B}\,\overline{C}D + A\overline{B}CD$ is plotted on the map in part (b). Zeros are placed in all the other cells.

FIGURE 4–27

Examples of plotting a Boolean expression on a Karnaugh map.

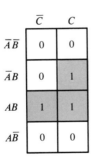

(a) $\overline{A}BC + AB\overline{C} + ABC$ (b) $\overline{A}B\overline{C}D + \overline{A}BCD + A\overline{B}\,\overline{C}D + A\overline{B}CD$

Sometimes an SOP expression does not have a complete set of variables in each of its terms. For example, the expression $\overline{A}B + AB\overline{C} + ABC$ contains a term that does not include the variable C or its complement. A missing variable in an SOP expression is often called a *suppressed variable*.

Before plotting an SOP expression such as the one mentioned above on the K map, we must expand the shortened term to include the missing variable and its complement. This expansion is based on rule 6 of Boolean algebra, which states that $C + \overline{C} = 1$. Multiplying the AB term by ($C + \overline{C}$), we get two terms as follows:

$$\overline{A}B = \overline{A}B \cdot 1 = \overline{A}B(C + \overline{C}) = \overline{A}BC + \overline{A}B\overline{C}$$

The important thing to recognize here is that we can replace the term $\overline{A}B$ in the original SOP expression with $\overline{A}BC + \overline{A}B\overline{C}$ because they are equivalent. The full SOP expression is written as

$$\overline{A}BC + \overline{A}B\overline{C} + AB\overline{C} + ABC$$

and we can now plot it on the K map.

Grouping Cells for Simplification

You can group 1s that are in adjacent cells according to the following rules, by drawing a loop around those cells:

1. Adjacent cells are cells that differ by only a single variable (for example, $ABCD$ and $AB\overline{C}D$ are adjacent).
2. The 1s in adjacent cells must be combined in groups of 2, 4, 8, 16, and so on.
3. Each group of 1s should be maximized to include the *largest* number of adjacent cells as possible in accordance with rule 2.
4. Every 1 on the map must be included in at least one group. There can be overlapping groups if they include noncommon 1s.

Grouping is illustrated by an example in Figure 4–28. Notice how the groups are overlapped to include all the 1s in the largest possible group.

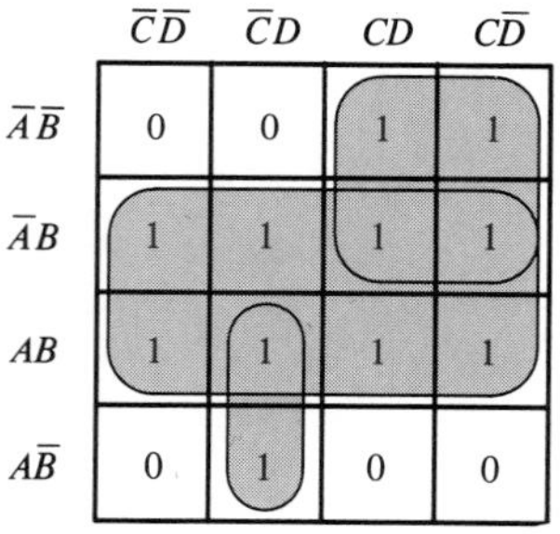

FIGURE 4–28
Example of the grouping of 1s on a four-variable Karnaugh map.

Simplifying the Expression

When all the 1s representing terms in the original Boolean expression are grouped, the mapped expression is ready for simplification. The following rules apply:

1. Each group of 1s creates a *product term* composed of all variables that appear in only one form (either uncomplemented or complemented) within the group. Variables that appear both uncomplemented *and* complemented within the group are eliminated. These are called *contradictory* variables.
2. The final simplified expression is formed by summing (ORing) the product terms of all the groups.

For example, in Figure 4–29 the product term for the eight-cell group is B because the cells within that group contain both A and $\overline{A}$, C and $\overline{C}$, and D and $\overline{D}$, so these

FIGURE 4–29

This plotted Boolean expression simplifies to
$B + \overline{A}C + A\overline{C}D$.

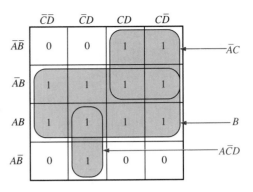

variables are eliminated. The four-cell group contains B, $\overline{B}$, D, and $\overline{D}$, leaving the product term $\overline{A}C$. The two-cell group contains B and $\overline{B}$, leaving $A\overline{C}D$ as the product term. The resulting Boolean expression is the sum of these product terms: $B + \overline{A}C + A\overline{C}D$.

The basis for eliminating contradictory variables within a K map grouping can be demonstrated by applying Boolean algebra to the four-cell group in Figure 4–28 as follows (B, $\overline{B}$, D, and $\overline{D}$ are the contradictory variables):

$$\overline{A}\overline{B}CD + \overline{A}\overline{B}C\overline{D} + \overline{A}BCD + \overline{A}BC\overline{D} = \overline{A}C(\overline{B}D + \overline{B}\overline{D} + BD + B\overline{D})$$
$$= \overline{A}C[\overline{B}(D + \overline{D}) + B(D + \overline{D})]$$
$$= \overline{A}C(\overline{B} \cdot 1 + B \cdot 1)$$
$$= \overline{A}C(\overline{B} + B)$$
$$= \overline{A}C$$

EXAMPLE 4–14

Minimize the expression $A\overline{B}C + \overline{A}BC + \overline{A}\overline{B}C + \overline{A}\overline{B}\overline{C} + A\overline{B}\overline{C}$.

FIGURE 4–30

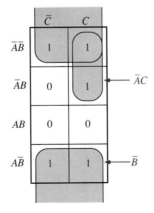

Solution Notice that this expression is already in a full sum-of-products form from which the 1s can be plotted very easily, as shown in Figure 4–30.

Four of the 1s appearing in adjacent cells can be grouped. The remaining 1 is absorbed in an overlapping group of two. The group of four 1s produces a single variable term, $\overline{B}$. This is determined by observing that within the group, $\overline{B}$ is the only variable that does not change from cell to cell. The group of two 1s produces a two-variable term, $\overline{A}C$. This is determined by observing that within the group, the variables $\overline{A}$ and C do not change from one cell to the next. To get the minimized function, the two terms that are produced are summed (ORed) as $\overline{B} + \overline{A}C$.

EXAMPLE 4–15

Reduce the following expression to its minimum SOP form:

$$\overline{A}\,\overline{B}\,\overline{C}\,\overline{D} + \overline{A}B\overline{C}\,\overline{D} + AB\overline{C}\,\overline{D} + A\overline{B}\,\overline{C}\,\overline{D} + \overline{A}\,\overline{B}CD + A\overline{B}CD + \overline{A}\,\overline{B}C\overline{D}$$
$$+ \overline{A}BC\overline{D} + ABC\overline{D} + A\overline{B}C\overline{D}$$

If all variables and their complements were available, this expression would take ten 4-input AND gates and one 10-input OR gate to implement.

FIGURE 4–31

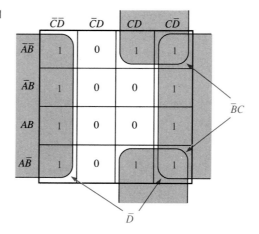

Solution   A group of eight 1s can be formed as shown in Figure 4–31 because the 1s in the outer columns are adjacent. A group of four 1s is formed by the wrap-around adjacency of the top and bottom cells to pick up the remaining two 1s. The minimum form of the original expression is $\overline{D} + \overline{B}C$.

The expression $\overline{D} + \overline{B}C$ takes one 2-input AND gate and one 2-input OR gate, excluding inverters. Compare this to the implementation of the original expression.

EXAMPLE 4–16

Reduce the following expression to its minimum sum-of-products form:

$$\overline{A}\,\overline{B}\,\overline{C}\,\overline{D} + \overline{A}\,\overline{B}C\overline{D} + A\overline{B}\,\overline{C}\,\overline{D} + \overline{A}CD + A\overline{B}C\overline{D}$$

FIGURE 4–32

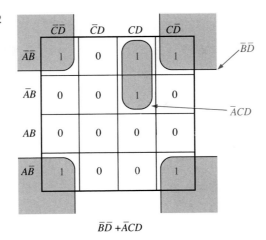

$$\overline{BD} + \overline{A}CD$$

Solution The function is plotted on the four-variable map and factored as indicated in Figure 4–32. Notice that the four corner cells are adjacent.

K Map Plotting from a Truth Table

A logic function can be stated either as a Boolean expression or in a truth table format. You have already seen how to plot a given function on a K map directly from a Boolean expression in SOP form. You will now learn how to plot a function on a K map when it is given in the form of a truth table.

Recall that a truth table gives the output of a specified logic function for all possible input combinations. An example is shown in Figure 4–33.

Notice in the truth table that the output X is a 1 for four different input combinations. For example, X is a 1 when all the inputs are 0s. This particular condition can be expressed in Boolean form as $\overline{A}\,\overline{B}\,\overline{C}$ because if $A = 0$, $B = 0$, and $C = 0$, then $\overline{A}\,\overline{B}\,\overline{C} = \overline{0} \cdot \overline{0} \cdot \overline{0} = 1 \cdot 1 \cdot 1 = X$. Similar reasoning can be applied to the other cases in which $X = 1$. The results are shown in Figure 4–34.

Once we have the Boolean terms for each combination of input variables for which $X = 1$, we can plot them directly on the K map, as illustrated in Figure 4–34,

FIGURE 4–33
Example of a logic function expressed in truth table format.

Inputs	Output
A B C	X
0 0 0	1
0 0 1	0
0 1 0	0
0 1 1	0
1 0 0	1
1 0 1	0
1 1 0	1
1 1 1	1

FIGURE 4–34

Example of plotting a truth table function on a Karnaugh map.

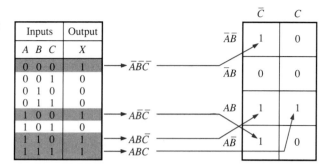

and simplify by the same procedures as before. When you become more familiar with translating truth tables, the intermediate step of writing down the Boolean terms can be omitted, and you can go directly from the truth table to the K map.

EXAMPLE 4–17

Use the simplest gate network possible to implement the logic function specified in the truth table of Figure 4–35.

Inputs			Output
A	B	C	X
0	0	0	0
0	0	1	1
0	1	0	1
0	1	1	1
1	0	0	1
1	0	1	1
1	1	0	0
1	1	1	0

FIGURE 4–35

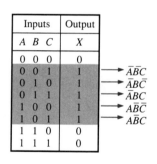

(a) Truth table

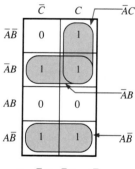

$X = \overline{A}C + \overline{A}B + A\overline{B}$

(b) K map

FIGURE 4–36

Solution

Step 1. Convert each combination of input variables for which $X = 1$ into a Boolean term, as shown in Figure 4–36(a).

Step 2. Plot a 1 in each cell of the K map that corresponds to a Boolean term derived from the truth table.

Step 3. Group the 1s, and write the reduced term for each group. Combine the reduced terms in SOP form. This step is shown in Figure 4–36(b).

Step 4. Using the minimized SOP expression derived from the K map, implement the logic function with gates, as shown in Figure 4–37.

FIGURE 4–37

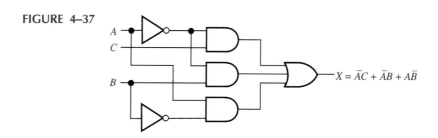

$$X = \overline{A}C + \overline{A}B + A\overline{B}$$

EXAMPLE 4–18

Reduce the function represented by the truth table in Figure 4–38 to a minimum SOP form.

Inputs				Output
A	B	C	D	X
0	0	0	0	0
0	0	0	1	1
0	0	1	0	0
0	0	1	1	0
0	1	0	0	1
0	1	0	1	0
0	1	1	0	0
0	1	1	1	0
1	0	0	0	1
1	0	0	1	1
1	0	1	0	0
1	0	1	1	1
1	1	0	0	1
1	1	0	1	1
1	1	1	0	0
1	1	1	1	1

FIGURE 4–38

Inputs				Output	
A	B	C	D	X	
0	0	0	0	0	
0	0	0	1	1	→ $\overline{A}\,\overline{B}\,\overline{C}D$
0	0	1	0	0	
0	0	1	1	0	
0	1	0	0	1	→ $\overline{A}B\overline{C}\,\overline{D}$
0	1	0	1	0	
0	1	1	0	0	
0	1	1	1	0	
1	0	0	0	1	→ $A\overline{B}\,\overline{C}\,\overline{D}$
1	0	0	1	1	→ $A\overline{B}\,\overline{C}D$
1	0	1	0	0	
1	0	1	1	1	→ $A\overline{B}CD$
1	1	0	0	1	→ $AB\overline{C}\,\overline{D}$
1	1	0	1	1	→ $AB\overline{C}D$
1	1	1	0	0	
1	1	1	1	1	→ $ABCD$

(a)

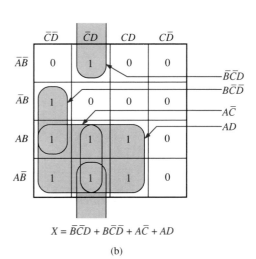

$$X = \overline{B}\,\overline{C}D + B\overline{C}\,\overline{D} + A\overline{C} + AD$$

(b)

FIGURE 4–39

Solution

Step 1. Convert each combination of input variables for which $X = 1$ on the truth table to a four-variable Boolean term, as shown in Figure 4–39(a).

Step 2. Plot the 1s on the K map, and group them as shown in Figure 4–39(b).

Step 3. Write the reduced term for each group, and combine the resulting terms in SOP form. The minimized SOP expression is $\overline{B}\,\overline{C}D + B\overline{C}\,\overline{D} + A\overline{C} + AD$. Notice how much the expression has been simplified from its full SOP form, taken directly from the truth table:

$$X = \overline{A}\,\overline{B}\,\overline{C}D + \overline{A}B\overline{C}\,\overline{D} + A\overline{B}\,\overline{C}\,\overline{D} + A\overline{B}\,\overline{C}D + A\overline{B}CD + AB\overline{C}\,\overline{D} + AB\overline{C}D + ABCD$$

"Don't Care" Conditions

Sometimes a situation arises in which some input variable combinations are not allowed. For example, recall that in the BCD code covered in Chapter 2, there are six invalid combinations: 1010, 1011, 1100, 1101, 1110, and 1111. Since these unallowed states will never occur in an application involving the BCD code, they can be treated as *"don't care"* terms with respect to their effect on the output. That is, for these "don't care" terms we may assign either a 1 or a 0 to the output because it really does not matter.

The "don't care" terms can be used to advantage when plotting and reducing a function on the K map. Figure 4–40 shows that for each "don't care" term, we place an *X* in the cell. When grouping the 1s, we can treat the *X*s as 1s to make a larger grouping, or we can treat them as 0s if they cannot be used to advantage. The larger we can make a group, the simpler the resulting term will be.

The truth table in Figure 4–40 describes a logic function that has a 1 output only when the BCD code for 7, 8, or 9 is present on the inputs. If we take advantage of the "don't cares" and use them as 1s, the resulting expression for the function is $A + BCD$, as indicated. If we do not use the "don't cares" as 1s, the resulting expression is $A\overline{B}\,\overline{C} + \overline{A}BCD$. So you can see the advantage of using "don't care" terms to get the simplest expression.

Inputs				Output
A	B	C	D	Y
0	0	0	0	0
0	0	0	1	0
0	0	1	0	0
0	0	1	1	0
0	1	0	0	0
0	1	0	1	0
0	1	1	0	0
0	1	1	1	1
1	0	0	0	1
1	0	0	1	1
1	0	1	0	X
1	0	1	1	X
1	1	0	0	X
1	1	0	1	X
1	1	1	0	X
1	1	1	1	X

Don't cares

(a)

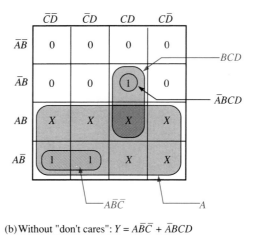

(b) Without "don't cares": $Y = A\overline{B}\,\overline{C} + \overline{A}BCD$
With "don't cares": $Y = A + BCD$

FIGURE 4–40
Example of the use of "don't care" conditions to simplify an expression.

SECTION 4–7
REVIEW

Answers are found at the end of the chapter.
1. Lay out a Karnaugh map for each of the following numbers of variables:
   (a) two   (b) three   (c) four
2. Group the 1s and write the simplified expression for each Karnaugh map in Figure 4–27.
3. Write the original sum-of-products expression that is plotted on the Karnaugh map in Figure 4–28.

SUMMARY

☐ Boolean expressions for an inverter and two-input gates:

FIGURE 4–41

FIGURE 4–41

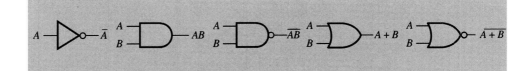

☐ Commutative laws:

$$A + B = B + A$$
$$AB = BA$$

☐ Associative laws:

$$A + (B + C) = (A + B) + C$$
$$A(BC) = (AB)C$$

☐ Distributive law:

$$A(B + C) = AB + AC$$

☐ Boolean rules:

1. $A + 0 = A$
2. $A + 1 = 1$
3. $A \cdot 0 = 0$
4. $A \cdot 1 = A$
5. $A + A = A$
6. $A + \overline{A} = 1$
7. $A \cdot A = A$
8. $A \cdot \overline{A} = 0$
9. $\overline{\overline{A}} = A$
10. $A + AB = A$
11. $A + \overline{A}B = A + B$
12. $(A + B)(A + C) = A + BC$

☐ DeMorgan's theorems:

1. The complement of a product is equal to the sum of the complements of the terms in the product:

$$\overline{XY} = \overline{X} + \overline{Y}$$

2. The complement of a sum is equal to the product of the complements of the terms in the sum:

$$\overline{X + Y} = \overline{X}\,\overline{Y}$$

Integrated Circuit Technologies (Chapter 13) may be covered after this chapter, covered after a later chapter, completely omitted, or partially omitted, depending on the needs and goals of your program. Coverage is not prerequisite or corequisite to any other material in the book. Inclusion is purely optional.

GLOSSARY

Boolean addition   In Boolean algebra, the OR operation.

Boolean multiplication   In Boolean algebra, the AND operation.

Cell   An area on a Karnaugh map that represents a unique combination of variables in product form.

Complement   In Boolean algebra, the inverse function. The complement of a 1 is a 0, and vice versa. Expressed with a bar over a variable.

"Don't care"   A combination of inputs that cannot occur.

Karnaugh map   An arrangement of cells representing the combinations of variables in a Boolean expression and used for a systematic simplification of the expression.

Network   An arrangement or interconnection of logic gates designed to perform a specified function.

Product-of-sums (POS)   A form of Boolean expression that is basically the ANDing of ORed terms.

Sum-of-products (SOP)   A form of Boolean expression that is basically the ORing of ANDed terms.

Suppressed variable   A variable that is not included in a given term in a sum-of-products expression but that is included in other terms in the same expression.

SELF-TEST

Answers and solutions are found at the end of the book.
1. (a) The expression AB means that the two variables are _____.
    (b) The expression $A + B$ means that the two variables are _____.
    (c) The expression $\overline{A}$ means that the variable is _____.
2. Determine the logic gate required to implement each of the following terms:
    (a) ABC (b) $A + B + C$ (c) $\overline{ABC}$ (d) $\overline{A + B + C}$
3. Apply the commutative law, the associative law, or the distributive law to each expression as appropriate.
    (a) $A + B$ (b) AB (c) CD (d) $(A + B) + C$
    (e) $A(B + C + D)$ (f) $A(BCD)$
4. Use the basic rules of Boolean algebra to simplify each of the following expressions:
    (a) $B + 1$ (b) $0 + B$ (c) $B \cdot 1$ (d) $B \cdot B$
    (e) $C + C$ (f) $D \cdot D$ (g) $\overline{\overline{C}}$ (h) $B + BC$
5. According to _____, $\overline{CD} = \overline{C} + \overline{D}$.
6. How many and what type of logic gates are required to implement the following expressions as they appear:
    (a) $AB + CD$ (b) $A + B + \overline{CD}$ (c) $A(B + C + DE)$
7. (a) The expression $ABC + \overline{A}BC + AB\overline{C}$ is an example of the _____ form.

    (b) The expression $(A + \overline{B} + C)(\overline{A} + B + C)(A + B + C)$ is an example of the
_____ form.

8. Determine whether or not the following equalities are correct:
    (a) $ABC + A\overline{BC} = A$
    (b) $A + BC + \overline{A}C = BC$
    (c) $A(\overline{A}BC + ABCD) = ABCD$

9. Convert the following expression to SOP form:

$$AB(\overline{A} + \overline{B}C) + (C + \overline{D})(A + A\overline{C}) + \overline{A}\overline{B}C\overline{D}$$

10. Simplify the expression in question 9 to its minimum SOP form.

PROBLEMS

Answers to odd-numbered problems are found at the end of the book.

Section 4–1 Boolean Operations

4–1   Using Boolean notation, express a function X that is a 1 whenever one or more of its variables (A, B, C, and D) are 1s.

4–2   Express a function Y that is a 1 only if all of its variables (A, B, C, D, and E) are 1s.

4–3   Express a function Z that is a 1 when one or more of its variables (A, B, and C) are 0s.

4–4   Evaluate the following:
    (a) $0 + 0 + 1$ **(b)** $1 + 1 + 1$
    (c) $1 \cdot 0 \cdot 0$ **(d)** $1 \cdot 1 \cdot 1$
    (e) $1 \cdot 0 \cdot 1$ **(f)** $1 \cdot 1 + 0 \cdot 1 \cdot 1$

Section 4–2 Boolean Expressions

4–5   Find the value of X for all possible values of the variables.
    (a) $X = AB$ **(b)** $X = ABC$ **(c)** $X = A + B$
    (d) $X = A + B + C$ **(e)** $X = AB + C$ **(f)** $X = \overline{A} + B$
    (g) $A = A\overline{B}C$ **(h)** $X = AB + \overline{A}C$ **(i)** $X = A(B + C)$
    (j) $X = \overline{A}(\overline{B} + \overline{C})$

4–6   Find the value of X for all possible values of the variables.
    (a) $X = (A + B)C + B$ **(b)** $X = (\overline{A + B})C$
    (c) $X = A\overline{B}C + AB$ **(d)** $X = (A + B)(\overline{A} + B)$
    (e) $X = (A + BC)(\overline{B} + \overline{C})$

4–7   Write the Boolean expression for each of the logic gates in Figure 4–42.

FIGURE 4–42

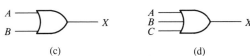

4–8 Write the Boolean expression for each of the logic gates in Figure 4–43.

FIGURE 4–43

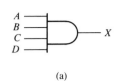

(a)

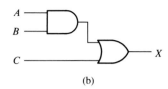

(b)

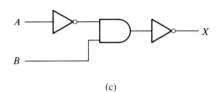

(c)

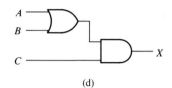

(d)

4–9 Construct a truth table for each of the following Boolean expressions:
(a) $A + B$ (b) AB (c) $AB + BC$
(d) $(A + B)C$ (e) $(A + B)(\overline{B} + C)$

Section 4–3   Rules and Laws of Boolean Algebra

4–10 Identify the law of Boolean algebra upon which each of the following equalities is based:
(a) $A\overline{B} + CD + A\overline{C}D + B = B + A\overline{B} + A\overline{C}D + CD$
(b) $ABCD + \overline{A}\overline{B}\overline{C} = DCBA + \overline{C}\overline{B}\overline{A}$
(c) $AB(CD + \overline{E}\overline{F} + GH) = ABCD + AB\overline{E}\overline{F} + ABGH$

4–11 Identify the Boolean rule(s) on which each of the following equalities is based.
(a) $\overline{\overline{AB} + \overline{CD} + \overline{EF}} = AB + CD + EF$
(b) $A\overline{A}B + AB\overline{C} + AB\overline{B} = AB\overline{C}$
(c) $A(\overline{B}C + BC) + AC = A(BC) + AC$
(d) $AB(C + \overline{C}) + AC = AB + AC$
(e) $A\overline{B} + A\overline{B}C = A\overline{B}$
(f) $ABC + \overline{A}\overline{B} + \overline{A}BCD = ABC + \overline{A}\overline{B} + D$

Section 4–4   De Morgan's Theorems

4–12 Apply DeMorgan's theorems to each expression.
(a) $\overline{A + \overline{B}}$ (b) $\overline{\overline{A}B}$ (c) $\overline{A + B + C}$
(d) $\overline{ABC}$ (e) $\overline{A(B + C)}$ (f) $\overline{AB + CD}$
(g) $\overline{AB + CD}$ (h) $\overline{(A + B)(\overline{C} + D)}$

4–13 Apply DeMorgan's theorems to each expression.
(a) $\overline{A\overline{B}(C + \overline{D})}$ (b) $\overline{AB(CD + EF)}$
(c) $\overline{(A + \overline{B} + C + \overline{D}) + ABC\overline{D}}$ (d) $\overline{(\overline{A} + B + C + D)(A\overline{B}\overline{C}D)}$
(e) $\overline{AB(CD + \overline{E}F)(AB + \overline{C}D)}$

4–14 Apply DeMorgan's theorems to the following:
(a) $\overline{\overline{(ABC)(EFG)} + \overline{(HIJ)(KLM)}}$ (b) $\overline{(A + \overline{B}\overline{C} + CD) + \overline{B}\overline{C}}$
(c) $\overline{\overline{(A + B)(C + D)(E + F)(G + H)}}$

Section 4–5   Boolean Expressions for Gate Networks

4–15   Convert the following expressions to sum-of-product forms:
(a) $(A + B)(C + D)$ (b) $(A + \overline{BC})D$
(c) $(A + C)(ABC + ACD)$

4–16   Convert the following expressions to sum-of-product forms:
(a) $AB + CD(A\overline{B} + CD)$ (b) $AB(\overline{BC} + BC)$
(c) $A + B[AC + (B + \overline{C})D]$

4–17   Write an expression for each of the gate networks in Figure 4–44, and identify the form of the expression as SOP or POS.

FIGURE 4–44

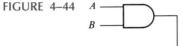
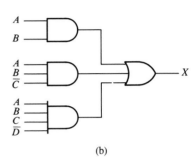
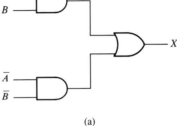
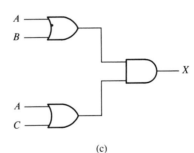
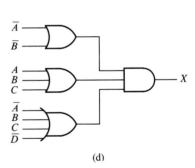

(a)   (b)

(c)   (d)

4–18   Write a Boolean expression for the following statement:
X is a 1 only if A is a 1 and B is a 1 or if A is a 0 and B is a 0.

4–19   Write a Boolean expression for the following statement:
X is a 1 only if A, B, and C are all 1s or if only one of the variables is a 0.

4–20   Write a Boolean expression for the following conditions:
X is a 0 if any two of the three variables A, B, and C are 1s. X is a 1 for all other conditions.

4–21   Draw the logic circuit represented by each of the following expressions:
(a) $A + B + C$ (b) ABC (c) $AB + C$
(d) $AB + CD$ (e) $\overline{AB}(C + \overline{D})$

4–22   Draw the logic circuit represented by each expression.
(a) $A\overline{B} + \overline{A}B$ (b) $AB + \overline{A}\overline{B} + \overline{A}BC$
(c) $A + B[C + D(B + \overline{C})]$

Section 4–6 Simplification Using Boolean Algebra

4–23 Using Boolean algebra techniques, simplify the following expressions as much as possible:

(a) $A(A + B)$ (b) $A(\overline{A} + AB)$ (c) $BC + \overline{B}C$

(d) $A(A + \overline{A}B)$ (e) $A\overline{B}C + A\overline{B}C + \overline{A}BC$

4–24 Using Boolean algebra, simplify the following expressions:

(a) $(A + \overline{B})(A + C)$ (b) $\overline{A}B + \overline{A}B\overline{C} + \overline{A}BCD + \overline{A}BC\overline{D}E$

(c) $AB + \overline{A}BC + A$ (d) $(A + \overline{A})(AB + AB\overline{C})$

(e) $AB + (\overline{A} + \overline{B})C + AB$

4–25 Using Boolean algebra, simplify each expression.

(a) $BD + B(D + E) + \overline{D}(D + F)$

(b) $\overline{A}\,\overline{B}C + \overline{(A + B + \overline{C})} + \overline{A}\overline{B}C\overline{D}$

(c) $(B + BC)(B + \overline{B}C)(B + D)$

(d) $ABCD + AB(\overline{CD}) + (\overline{AB})CD$

(e) $ABC[AB + \overline{C}(BC + AC)]$

4–26 Reduce the following expressions to minimum SOP forms:

(a) $A\overline{B}C[AC + B\overline{C}(\overline{A}C)] + (\overline{A} + \overline{C})(AC + \overline{B}\,\overline{C})$

(b) $(\overline{A}\overline{B}\,\overline{C} + \overline{D})(D + \overline{C}A + 1) + AB + \overline{A}B$

(c) $\overline{A}\overline{B}C\overline{D} + ABC\overline{D} + A\overline{B}\,\overline{C}D + \overline{A}\,\overline{B}\,\overline{C}\,\overline{D}$

4–27 Determine which of the gate networks in Figure 4–45 are equivalent.

FIGURE 4–45

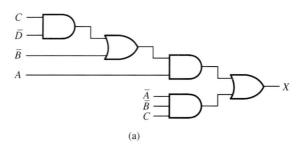

(a)

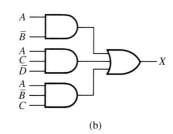

(b)

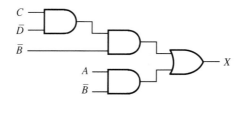

(c)

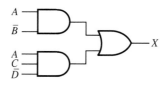

(d)

Section 4–7 Simplification Using the Karnaugh Map

4–28 Using the Karnaugh map method, simplify the following expressions to their minimum SOP forms:

(a) $X = \overline{A}\,\overline{B} + A\overline{B}$ (b) $X = \overline{A}B + \overline{A}B$

    (c) $X = A\overline{B} + AB$ **(d)** $X = \overline{A}\overline{B} + A\overline{B} + AB$

    (e) $X = A(\overline{B} + AB)$ **(f)** $X = \overline{A}\overline{B} + AB$

4–29   Use a Karnaugh map to find the minimum SOP form for each expression.

    (a) $X = \overline{A}\overline{B}\overline{C} + \overline{A}\overline{B}C + A\overline{B}C$

    (b) $X = AC(\overline{B} + C)$

    (c) $X = \overline{A}(BC + B\overline{C}) + A(BC + B\overline{C})$

    (d) $X = \overline{A}\overline{B}C + A\overline{B}C + \overline{A}BC + ABC$

4–30   Use a Karnaugh map to simplify each function to a minimum SOP form.

    (a) $X = \overline{A}\overline{B}\overline{C} + A\overline{B}C + \overline{A}BC + AB\overline{C}$

    (b) $X = AC[\overline{B} + B(B + \overline{C})]$

    (c) $X = D\overline{E}\overline{F} + \overline{D}E\overline{F} + \overline{D}\overline{E}F$

4–31   Expand each expression containing terms with suppressed variables to a full SOP form.

    (a) $X = AB + A\overline{B}C + ABC$

    (b) $X = A + BC$

    (c) $X = A\overline{B}\overline{C}D + AC\overline{D} + B\overline{C}D + \overline{A}BC\overline{D}$

    (d) $X = A\overline{B} + A\overline{B}\overline{C}D + CD + B\overline{C}D + ABCD$

4–32   Plot and simplify each expression in Problem 4–31 with a Karnaugh map.

4–33   Use a Karnaugh map to reduce each expression to a minimum SOP form.

    (a) $X = A + B\overline{C} + CD$

    (b) $X = \overline{A}\overline{B}\overline{C}D + \overline{A}BC\overline{D} + ABCD + AB\overline{C}\overline{D}$

    (c) $X = \overline{A}B(\overline{C}\overline{D} + \overline{C}D) + AB(\overline{C}\overline{D} + \overline{C}D) + A\overline{B}\overline{C}D$

    (d) $X = (\overline{A}\overline{B} + A\overline{B})(CD + C\overline{D})$

    (e) $X = \overline{A}\overline{B} + A\overline{B} + \overline{C}\overline{D} + C\overline{D}$

4–34   Convert to a Boolean term each of the conditions for which $X = 1$ in the truth table of Figure 4–46.

4–35   Reduce the function specified in Figure 4–46 to its minimum SOP form by using a K map.

4–36   Use the Karnaugh map method to implement in the simplest possible SOP form the logic function specified in the truth table in Figure 4–47.

4–37   Solve Problem 4–36 for a situation in which the last six combinations are not allowed.

Inputs			Output
A	B	C	X
0	0	0	1
0	0	1	1
0	1	0	0
0	1	1	1
1	0	0	1
1	0	1	1
1	1	0	0
1	1	1	1

FIGURE 4–46

A	B	C	D	X
0	0	0	0	0
0	0	0	1	1
0	0	1	0	1
0	0	1	1	0
0	1	0	0	0
0	1	0	1	0
0	1	1	0	1
0	1	1	1	1
1	0	0	0	1
1	0	0	1	0
1	0	1	0	1
1	0	1	1	0
1	1	0	0	1
1	1	0	1	1
1	1	1	0	0
1	1	1	1	1

FIGURE 4–47

ANSWERS TO SECTION REVIEWS

Section 4–1   Review

1. OR   2. AND   3. (a) $X = AB$ (b) $X = A + B$

Section 4–2   Review

1. $X = ABCDE$
2. $Y = F + G + H + I + J$
3. (a) $X = \overline{ABCDE}$ (b) $Y = \overline{F + G + H + I + J}$

Section 4–3   Review

1. $(A + B + C) + D$ 2. $AB + AC + AD$

Section 4–4   Review

1. (a) $\overline{A} + \overline{B} + \overline{C} + D\overline{E}$ (b) $\overline{A}\overline{B} + \overline{C}$ (c) $\overline{A}\overline{B}\overline{C} + D + \overline{E}$

Section 4–5   Review

1. (a) **sum-of-products**   (b) product-of-sums
   (c) sum-of-products
2. (a) three AND gates, one OR gate   (b) three OR gates, one AND gate
   (c) four AND gates, one OR gate

Section 4–6   Review

1. (a) A (b) $C(\overline{A} + B)$ (c) $A(\overline{C} + \overline{B}DE)$
2. (a) original: three gates; simplified: no gates (straight connection)
   (b) original: four gates; simplified: two gates
   (c) original: seven gates; simplified: three gates

Section 4–7   Review

1. (a) four cells   (b) eight cells   (c) sixteen cells
2. (a) $X = AB + BC$ (b) $X = \overline{A}BD + A\overline{B}D$
3. $X = \overline{A}\overline{B}CD + \overline{A}\overline{B}C\overline{D} + \overline{A}B\overline{C}D + \overline{A}BCD + \overline{A}BC\overline{D} + A\overline{B}C\overline{D} + AB\overline{C}D + AB\overline{C}\overline{D}$
   $+ ABCD + ABC\overline{D} + A\overline{B}\overline{C}D$

In Chapters 3 and 4, logic gates were studied on an individual basis and in simple combinations. When logic gates are connected together to produce a specified output for certain specified combinations of input variables, with no storage involved, the resulting network is called *combinational logic*. In combinational logic the output level is at all times dependent on the combination of input levels. In this chapter methods of analysis and design of combinational logic circuits are examined, and troubleshooting techniques are introduced.

Specific devices introduced in this chapter are 7451 AND-OR-INVERT logic and 7486 quad exclusive-OR gates.

After completing this chapter, you will be able to

- [ ] analyze basic combinational logic circuits, such as AND-OR, AND-OR-INVERT, exclusive-OR, exclusive-NOR, and other general combinational networks.
- [ ] use AND-OR and AND-OR-INVERT circuits to implement sum-of-products (SOP) and product-of-sums (POS) expressions.
- [ ] write the Boolean output expression for any combinational logic circuit.
- [ ] develop a truth table from the output expression for a combinational logic circuit.
- [ ] use the Karnaugh map to expand an output expression containing suppressed variable terms into a full SOP form.
- [ ] design a combinational logic circuit for a given Boolean output expression.
- [ ] design a combinational logic circuit for a given truth table.
- [ ] simplify a combinational logic circuit to its minimum SOP form.
- [ ] use NAND gates to implement any combinational logic function.
- [ ] use NOR gates to implement any combinational logic function.
- [ ] troubleshoot faulty nodes in logic networks.
- [ ] troubleshoot logic circuits by using signal tracing and waveform analysis.

5

Combinational Logic

ANALYSIS OF COMBINATIONAL LOGIC

5–1

In this section we will analyze several combinational logic circuits. We will examine several specific types of logic, including AND-OR, AND-OR-INVERT, exclusive-OR, exclusive-NOR, and a general combinational circuit. The importance of AND-OR and AND-OR-INVERT is that they directly implement sum-of-products (SOP) and product-of-sums (POS) expressions.

AND-OR Logic

Figure 5–1 shows a combinational logic circuit consisting of two AND gates and one OR gate. Each of the three gates has two input variables as indicated. Each of the input variables can be either a HIGH (1) or a LOW (0). Because there are four input variables, there are sixteen possible combinations of the input variables ($2^4 = 16$). To illustrate an analysis procedure, we will assign one of the sixteen possible input combinations and see what the corresponding output value is.

First, make each input variable a LOW, and examine the output of each gate in the network in order to arrive at the final output, Y. If the inputs to gate G_1 are both LOW, the output of gate G_1 is LOW. Also, the output of gate G_2 is LOW because its inputs are LOW. As a result of the LOWs on the outputs of gates G_1 and G_2, both inputs to gate G_3 are LOW, and therefore its output is LOW. We have determined that the output function of the logic circuit of Figure 5–1 is LOW when all of its inputs are LOW. This condition is illustrated in the first row of Table 5–1, which also contains the remaining fifteen input combinations. You should verify each of these conditions on the logic diagram.

Now, as an alternative method of analyzing the logical operation of the circuit of Figure 5–1, we can develop a logic equation for the output function and, using Boolean algebra, evaluate the equation for each of the sixteen combinations of input variables.

Since gate G_1 is an AND gate and its two inputs are A and B, its output can be expressed as AB. Gate G_2 is an AND gate and its two inputs are C and D, so its output can be expressed as CD. Gate G_3 is an OR gate and its two inputs are the outputs of gates G_1 and G_2, so its output can be expressed as $AB + CD$. The output of gate G_3 is

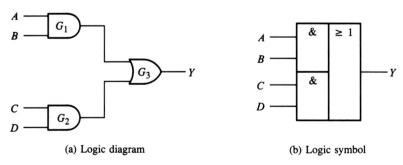

(a) Logic diagram   (b) Logic symbol

FIGURE 5–1
AND-OR logic (ANSI/IEEE Std. 91-1984).

TABLE 5–1
Truth table for Figure 5–1 (1 = HIGH, 0 = LOW).

Inputs				G_1 Output	G_2 Output	G_3 Output
A	B	C	D	AB	CD	Y
0	0	0	0	0	0	0
0	0	0	1	0	0	0
0	0	1	0	0	0	0
0	0	1	1	0	1	1
0	1	0	0	0	0	0
0	1	0	1	0	0	0
0	1	1	0	0	0	0
0	1	1	1	0	1	1
1	0	0	0	0	0	0
1	0	0	1	0	0	0
1	0	1	0	0	0	0
1	0	1	1	0	1	1
1	1	0	0	1	0	1
1	1	0	1	1	0	1
1	1	1	0	1	0	1
1	1	1	1	1	1	1

the output function of the logic network, so $Y = AB + CD$. Figure 5–2 shows the logic functions at all points in the circuit.

We can now evaluate the SOP output expression using Boolean algebra by substituting into it the various combinations of input variable values. For example, when $A = 1$, $B = 1$, $C = 1$, and $D = 0$, the output expression is evaluated as follows:

$$Y = AB + CD = 1 \cdot 1 + 1 \cdot 0 = 1 + 0 = 1$$

The same procedure can be used for any of the fifteen other input variable combinations.

FIGURE 5–2
AND-OR logic with Boolean expressions.

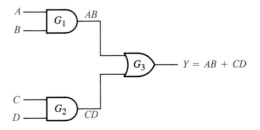

EXAMPLE 5–1

An AND-OR circuit is shown in Figure 5–3. Write the output expression, and determine the logic level at the output for the following input logic levels: *C, D, I,* and *G* are HIGH, and *A, B, E, F,* and *H* are LOW.

FIGURE 5–3

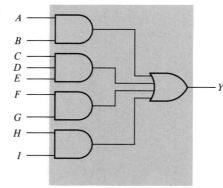

Solution The output expression is $Y = AB + CDE + FG + HI$. The output expression is evaluated for the given input levels as follows:

$$Y = AB + CDE + FG + HI = 0 \cdot 0 + 1 \cdot 1 \cdot 0 + 0 \cdot 1 + 0 \cdot 1$$
$$= 0 + 0 + 0 + 0 = 0$$

AND-OR-INVERT Logic

Figure 5–4(a) shows a combinational logic circuit consisting of two AND gates, one OR gate, and an inverter. As you can see, the operation is the same as that of the AND-OR circuit in Figure 5–1 except that the output is inverted. The output expression is $Y = \overline{AB + CD}$. An evaluation of this for the inputs $A = 1$, $B = 1$, $C = 1$, $D = 0$ is as follows:

$$Y = \overline{AB + CD} = \overline{1 \cdot 1 + 1 \cdot 0} = \overline{1 + 0} = \overline{1} = 0$$

Notice that DeMorgan's theorem can be applied to the AND-OR-INVERT output expression to get a product-of-sums (POS) form as follows:

$$Y = \overline{AB + CD} = (\overline{A} + \overline{B})(\overline{C} + \overline{D})$$

The resulting equivalent circuit is shown in Figure 5–4(b).

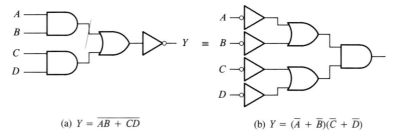

(a) $Y = \overline{AB + CD}$ (b) $Y = (\overline{A} + \overline{B})(\overline{C} + \overline{D})$

FIGURE 5–4
AND-OR-INVERT circuit and its POS equivalent.

EXAMPLE 5–2

The 7451 dual AND-OR-INVERT circuit is represented by a logic diagram and symbol in Figure 5–5. Notice that the inversion is indicated by a bubble on the output of the OR gate, effectively making it a NOR gate.

The following levels are applied to the inputs: pins 1, 2, 4, and 13 are HIGH, and pins 3, 5, 9, and 10 are LOW. Determine the output levels.

FIGURE 5–5
The 7451 dual AND-OR-INVERT.

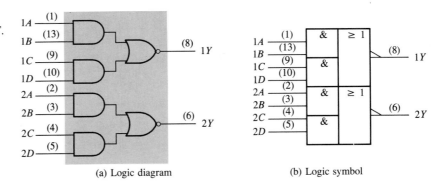

(a) Logic diagram   (b) Logic symbol

Solution   The $1Y$ output at pin 8 is LOW because both pins 1 and 13 are HIGH. The output $2Y$ at pin 6 is HIGH because neither AND gate in this circuit has both of its inputs HIGH.

Exclusive-OR Logic

The exclusive-OR gate was introduced in Chapter 3. Although, because of its importance, this circuit is considered a type of logic gate with its own unique symbol, it is actually a combination of AND gates, OR gates, and inverters as shown in Figure 5–6(a).

The output expression for this circuit is $Y = A\overline{B} + \overline{A}B$. Evaluation of this expression results in the truth table in Table 5–2. Notice that the output is HIGH only when the two inputs are at opposite levels. A special exclusive-OR operator, $\oplus$, is often used, so the expression $Y = A\overline{B} + \overline{A}B$ can also be written as $Y = A \oplus B$, stated as "Y is equal to A exclusive-OR B."

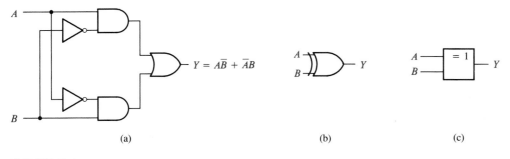

(a)   (b)   (c)

FIGURE 5–6
Exclusive-OR logic diagram and symbols.

TABLE 5–2
Truth table for an exclusive-OR.

A	B	Y
0	0	0
0	1	1
1	0	1
1	1	0

EXAMPLE 5–3

The 7486 quad exclusive-OR gate IC is shown in Figure 5–7. Certain input pins are connected *externally* to positive logic levels indicated by 1s and 0s, and others are connected to gate outputs as shown.

FIGURE 5–7
7486 quad exclusive-OR.

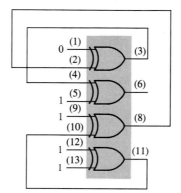

Solution   The final output on pin 6 is 0 (LOW). We determine this by tracing the logic with the use of the exclusive-OR truth table. The intermediate levels are as follows: pin 11 = 0, pin 8 = 1, and pin 3 = 1.

Exclusive-NOR Logic

As you know, the complement of the exclusive-OR function is the exclusive-NOR, which is derived as follows:

$$X = \overline{A\overline{B} + \overline{A}B} = (\overline{A\overline{B}})(\overline{\overline{A}B}) = (\overline{A} + B)(A + \overline{B}) = \overline{A}\,\overline{B} + AB$$

Notice that the output, *X,* is HIGH only when the two inputs, *A* and *B,* are at the same level.

The exclusive-NOR can be implemented by simply inverting the output of an exclusive-OR, as shown in Figure 5–8(a), or by directly implementing the expression $\overline{A}\,\overline{B} + AB$, as shown in part (b).

A General Combinational Logic Circuit

Figure 5–9 is the fourth combinational logic circuit that we will analyze. Unlike the previous circuits, it is not available as a specific IC device but is a general combination

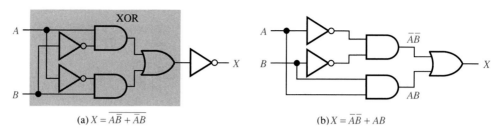

(a) $X = \overline{A\overline{B} + \overline{A}B}$ (b) $X = \overline{A}\,\overline{B} + AB$

FIGURE 5–8
Two equivalent ways of implementing the Exclusive-NOR.

FIGURE 5–9
A combinational logic circuit.

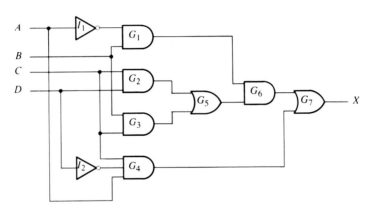

of gates. As before, the approach to analysis is to develop the logic expression for the circuit and evaluate that expression for all of the possible input conditions. In this case, there are four input variables with sixteen combinations of input states.

The logic expression for Figure 5–9 is developed step-by-step as follows:

$$\text{Output of } I_1 = \overline{A}$$
$$\text{Output of } G_1 = \overline{A}B$$
$$\text{Output of } G_2 = CD$$
$$\text{Output of } G_3 = BC$$
$$\text{Output of } I_2 = \overline{D}$$
$$\text{Output of } G_4 = AC\overline{D}$$
$$\text{Output of } G_5 = (\text{output of } G_2) + (\text{output of } G_3)$$
$$= CD + BC$$
$$\text{Output of } G_6 = (\text{output of } G_1) \cdot (\text{output of } G_5)$$
$$= \overline{A}B(CD + BC)$$
$$\text{Final output } X = \text{output of } G_7$$
$$= (\text{output of } G_4) + (\text{output of } G_6)$$
$$= AC\overline{D} + \overline{A}B(CD + BC)$$
$$= AC\overline{D} + \overline{A}BCD + \overline{A}BC$$

Figure 5–10 illustrates the logic expressions at each point in the circuit. Notice that the expression for the output X can be converted to a sum-of-products form as indicated.

We will now discuss how to develop from the output expression a truth table fully describing the circuit's operation.

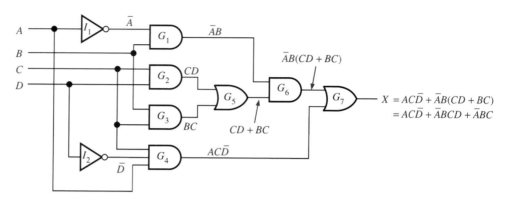

FIGURE 5–10

Expansion of the Output Expression Using Boolean Algebra

A sum-of-products expression with suppressed variables, in which some of the terms do not include all variables, can be expanded to include all variables in each term. The truth table can then be written directly from the expanded expression.

Notice that the expression $X = AC\overline{D} + \overline{A}BCD + \overline{A}BC$ in Figure 5–10 has two terms that do not include all four variables. Using Boolean algebra, expand each of the three-variable terms to include all four variables as follows:

$$AC\overline{D} = AC\overline{D}(B + \overline{B}) = ABC\overline{D} + A\overline{B}C\overline{D}$$
$$\overline{A}BC = \overline{A}BC(D + \overline{D}) = \overline{A}BCD + \overline{A}BC\overline{D}$$

The expanded output expression is

$$X = ABC\overline{D} + A\overline{B}C\overline{D} + \overline{A}BCD + \overline{A}BCD + \overline{A}BC\overline{D}$$

There are two $\overline{A}BCD$ terms, so one is dropped. The result is

$$X = ABC\overline{D} + A\overline{B}C\overline{D} + \overline{A}BCD + \overline{A}BC\overline{D}$$

Deriving the Truth Table from the Expanded Expression

The expanded output expression for a logic circuit shows all the combinations of the input variables that make the output a 1. An uncomplemented input variable stands for a 1, and a complemented input variable stands for a 0. For the expression $X = ABC\overline{D} + \overline{A}BC\overline{D} + \overline{A}BCD + \overline{A}BC\overline{D}$, $X = 1$ for four input variable combinations and $X = 0$ for all other input combinations, as shown in Table 5–3.

TABLE 5–3
Truth table for the circuit of Figure 5–10.

Inputs				Output
A	B	C	D	X
0	0	0	0	0
0	0	0	1	0
0	0	1	0	0
0	0	1	1	0
0	1	0	0	0
0	1	0	1	0
0	1	1	0	1
0	1	1	1	1
1	0	0	0	0
1	0	0	1	0
1	0	1	0	1
1	0	1	1	0
1	1	0	0	0
1	1	0	1	0
1	1	1	0	1
1	1	1	1	0

Expansion of the Output Expression Using a Karnaugh Map

The Karnaugh map can also be used to systematically expand a sum-of-products expression with suppressed variable terms, by the following steps:

Step 1. For each term in the expression, place a 1 in each cell where the variables in that term appear.

Step 2. Write the product term for each *individual* cell containing a 1 (do not group the cells).

Step 3. Write the sum of all the product terms.

To illustrate this procedure, the previous expression $X = AC\overline{D} + \overline{A}BCD + \overline{A}BC$ is expanded with the use of the sixteen-cell (four-variable) map in Figure 5–11. In part (a) of the figure, the three terms in the expression are expanded on the map. In part (b),

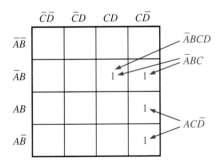

(a) $\overline{A}BC$, $\overline{A}BCD$, and $AC\overline{D}$ expanded onto map. For example, $\overline{A}BC$ can be expressed as $\overline{A}BCD + \overline{A}BC\overline{D}$.

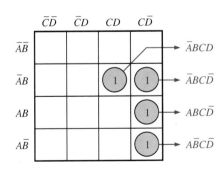

(b) Extracting the expanded expression
$X = \overline{A}BCD + \overline{A}BC\overline{D} + ABC\overline{D} + A\overline{B}C\overline{D}$

FIGURE 5–11
Sum-of-products expansion by Karnaugh map.

the four-variable terms are written for each 1, and the expanded expression is developed. Notice that it is the same as was obtained by Boolean expansion.

Answers are found at the end of the chapter.

1. Determine the output (1 or 0) of a four-variable AND-OR-INVERT circuit for each of the following input conditions:
   (a) $A = 1, B = 0, C = 1, D = 0$
   (b) $A = 1, B = 1, C = 0, D = 1$
   (c) $A = 0, B = 1, C = 1, D = 1$

2. Determine the output (1 or 0) of an exclusive-OR gate for each of the following input conditions:
   (a) $A = 1, B = 0$ **(b)** $A = 1, B = 1$
   (c) $A = 0, B = 1$ **(d)** $A = 0, B = 0$

3. Develop the truth table for a certain three-input logic circuit with the output expression $X = A\overline{B}C + \overline{A}BC + \overline{A}\overline{B}\overline{C} + AB\overline{C} + ABC$.

4. Draw the logic diagram for an exclusive-NOR circuit.

DESIGN OF COMBINATIONAL LOGIC

5–2

In this section we start with an equation or a truth table that describes a logic function and from it determine the circuit required to implement the function. As before, several examples are used to illustrate a general procedure. This section is optional, and its omission will not affect any other material.

Implementing a Circuit from the Boolean Expression

Equation (5–1) is the first in the series of examples that will be considered.

$$X = AB + CDE \tag{5–1}$$

A brief inspection reveals that this function is composed of two terms, AB and CDE, with a total of five variables. The first term is formed by ANDing A with B, and the second term is formed by ANDing C, D, and E. The two terms are then ORed to form the output X. These operations are indicated in the structure of the equation as follows:

$$X = AB + CDE$$

It should be noted that in this particular equation, the AND operations forming the two individual terms, AB and CDE, must be performed before the terms can be ORed.

To implement this logic function, a two-input AND gate is required to form the term AB, and a three-input AND gate is needed to form the term CDE. A two-input OR gate is then required to combine the two AND terms. The resulting logic circuit is shown in Figure 5–12.

As another example, we will first implement Equation (5–2) as expressed.

$$X = AB(C\overline{D} + EF) \tag{5–2}$$

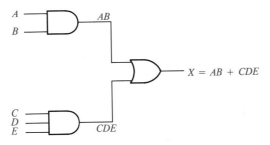

FIGURE 5–12
Logic circuit for Equation (5–1).

A breakdown of this equation shows that the terms A, B, and $C\overline{D} + EF$ are ANDed. The term $C\overline{D} + EF$ is formed by first ANDing C and $\overline{D}$ and ANDing E and F, and then ORing these two terms. This structure is indicated in relation to the equation as follows:

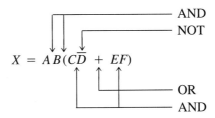

Before the output function X can be formed, we must have the term $C\overline{D} + EF$; but before we can get this term, we must have the terms $C\overline{D}$ and EF; but before we can get the term $C\overline{D}$, we must have $\overline{D}$. So, as you can see, there is a chain of logic operations that must be done in the proper order before the output function itself can be realized.

The logic gates required to implement Equation (5–2) are as follows:

1. one inverter to form $\overline{D}$
2. two 2-input AND gates to form $C\overline{D}$ and EF
3. one 2-input OR gate to form $C\overline{D} + EF$
4. one 3-input AND gate to form X

The logic circuit that produces this function is shown in Figure 5–13(a). Notice that there is a maximum of four gates between an input and output in this circuit (from input D to output). Often the total propagation delay time through a logic circuit is a major consideration. Propagation delays are additive, so the more gates between input and output, the greater the propagation delay time.

Unless an intermediate term, such as $C\overline{D} + EF$ in Figure 5–13(a), is required as an output for some other purpose, it is usually best to reduce a circuit to its minimum SOP form. Equation (5–2) is converted to SOP as follows, and the resulting circuit is shown in Figure 5–13(b).

$$AB(C\overline{D} + EF) = ABC\overline{D} + ABEF$$

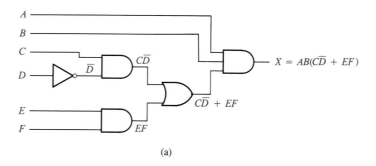

(a)

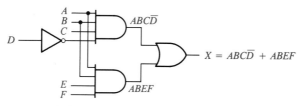

(b) Sum-of-products implementation
   of the circuit in part (a)

FIGURE 5–13
Logic circuits for Equation (5–2).

Now, as a third design example, we will begin with the truth table for a logic function, write the SOP Boolean expression from the truth table, and implement the logic circuit.

TABLE 5–4

Inputs			Output
A	B	C	X
0	0	0	0
0	0	1	0
0	1	0	0
0	1	1	1 $\longrightarrow \overline{A}BC$
1	0	0	1 $\longrightarrow A\overline{B}\,\overline{C}$
1	0	1	0
1	1	0	0
1	1	1	0

Table 5–4 specifies a logic function. The expression is obtained from the truth table by ORing the product terms for all occurrences of $X = 1$ (the unshaded rows). A 1 represents an uncomplemented variable, and a 0 represents a complemented variable. In this case, the expression is

$$X = \overline{A}BC + A\overline{B}\,\overline{C} \qquad \text{(5–3)}$$

The first term in the equation is formed by ANDing the three variables $\overline{A}$, B, and C. The second term is formed by ANDing the three variables A, $\overline{B}$, and $\overline{C}$.

The logic gates required to implement this function are as follows: three inverters to form the $\overline{A}$, $\overline{B}$, and $\overline{C}$ variables; two 3-input AND gates to form the terms $\overline{A}BC$ and $A\overline{B}\overline{C}$; and one 2-input OR gate to form the final output function, $\overline{A}BC + A\overline{B}\overline{C}$. Figure 5–14 illustrates the implementation of this logic function.

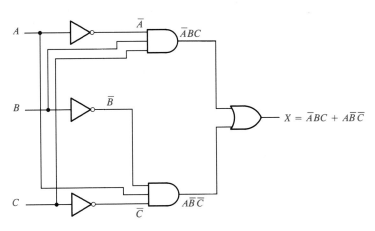

FIGURE 5–14
Logic circuit for Equation (5–3).

EXAMPLE 5–4

Design a logic circuit to implement the operation specified in the truth table of Table 5–5.

TABLE 5–5

Inputs			Output
A	B	C	X
0	0	0	0
0	0	1	0
0	1	0	0
0	1	1	1
1	0	0	0
1	0	1	1
1	1	0	1
1	1	1	0

Solution Notice that $X = 1$ for three and only three of the input conditions. Therefore, the logic expression is

$$X = \overline{A}BC + A\overline{B}C + AB\overline{C}$$

The logic circuit is shown in Figure 5–15.

FIGURE 5–15

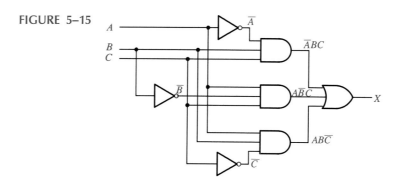

EXAMPLE 5–5

Develop a logic circuit with four input variables that will produce a 1 output when any three and only three input variables are 1s.

Solution Out of sixteen possible combinations of four variables, the combinations in which there are three 1s are listed in Table 5–6, along with the corresponding product term for each.

TABLE 5–6

A	B	C	D	
0	1	1	1	$\longrightarrow \overline{A}BCD$
1	0	1	1	$\longrightarrow A\overline{B}CD$
1	1	0	1	$\longrightarrow AB\overline{C}D$
1	1	1	0	$\longrightarrow ABC\overline{D}$

The product terms are ORed to get the logic expression

$$X = \overline{A}BCD + A\overline{B}CD + AB\overline{C}D + ABC\overline{D}$$

This expression is implemented in Figure 5–16 with AND-OR logic.

FIGURE 5–16

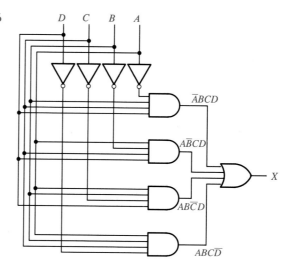

Gate Minimization

In many applications it is desirable to use the minimum number of gates in the simplest configuration possible to implement a given logic function. This simplification may be desirable for several reasons, such as economy, limitations of available power, minimization of delay times by reduction of logic levels, or—in the case of IC designs—maximum utilization of chip area.

Here we will use the two basic methods of gate minimization. First, the rules and laws of Boolean algebra will be applied to simplify the logic equation for the circuit in question and thereby reduce the number of gates required to implement the function. Second, the Karnaugh map method will be used to minimize the logic function. Both of these methods were studied in the previous chapter.

To begin, Boolean algebra will be applied to the equation for the circuit of Figure 5–9, which is redrawn for convenience in Figure 5–17. The output expression for this circuit was developed in Section 5–1 and is restated as Equation (5–4).

$$X = AC\overline{D} + \overline{A}B(CD + BC) \tag{5–4}$$

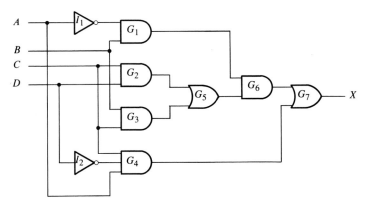

FIGURE 5–17

We know that five AND gates, two OR gates, and two inverters are needed to implement this function as it appears in Equation (5–4). We do not yet know whether this is the simplest form of the equation.

Inspection of Equation (5–4) shows that a possible first step in simplification is to apply the distributive law to the second term by multiplying the term $CD + BC$ by $\overline{A}B$. The result is

$$X = AC\overline{D} + \overline{A}BCD + \overline{A}BBC$$

The rule that $BB = B$ can be applied to the third term, yielding the sum-of-products form

$$X = AC\overline{D} + \overline{A}BCD + \overline{A}BC$$

Factoring $\overline{A}BC$ out of the second and third terms gives

$$X = AC\overline{D} + \overline{A}BC(D + 1)$$

Since $D + 1 = 1$,

$$X = AC\overline{D} + \overline{A}BC \qquad\qquad \textbf{(5–5)}$$

This appears to be the simplest SOP form.

This equation can be implemented with two 3-input AND gates, two inverters, and one 2-input OR gate, as shown in Figure 5–18.

Compare this circuit with the circuit in Figure 5–17, which has nine logic gates including inverters. The minimized circuit is equivalent but has only five gates including two inverters. At this point you should verify that the logic circuit of Figure 5–18 is indeed equivalent in logical operation to the logic circuit of Figure 5–17.

FIGURE 5–18
Logic circuit for Equation (5–5).

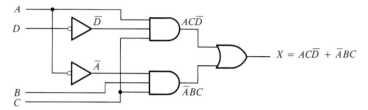

Karnaugh Map Approach

The function of Equation (5–4) is rewritten here for convenience:

$$X = AC\overline{D} + \overline{A}B(CD + BC)$$

To minimize this expression, convert it to the sum-of-products form:

$$X = AC\overline{D} + \overline{A}BCD + \overline{A}BC$$

This expression is expanded on the Karnaugh map in Figure 5–19(a) to include all variables in each term. As you can see, the fully expanded sum-of-products expression is

$$X = \overline{A}BCD + \overline{A}BC\overline{D} + ABC\overline{D} + A\overline{B}C\overline{D}$$

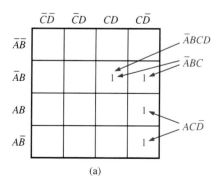

(a)

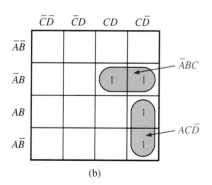

(b)

FIGURE 5–19
Karnaugh map for the function of Equation (5–4).

To simplify, the 1s are grouped as shown in Figure 5–19(b), and the resulting product terms are ORed to obtain the following minimum expression:

$$X = AC\overline{D} + \overline{A}BC$$

Notice that this is the same result that was achieved using algebraic simplification.

EXAMPLE 5–6

Reduce the combinational logic circuit in Figure 5–20 to a minimum form.

FIGURE 5–20

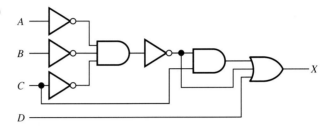

Solution   The expression for the output of the circuit is

$$X = (\overline{\overline{A}\,\overline{B}\,\overline{C}})C + \overline{\overline{A}\,\overline{B}\,\overline{C}} + D$$

Applying DeMorgan's theorem and Boolean algebra, we get the following reduction:

$$X = (\overline{\overline{A}} + \overline{\overline{B}} + \overline{\overline{C}})C + \overline{\overline{A}} + \overline{\overline{B}} + \overline{\overline{C}} + D$$
$$= AC + BC + CC + A + B + C + D$$
$$= AC + BC + C + A + B + \cancel{C} + D$$
$$= C(A + B + 1) + A + B + D$$
$$X = C + A + B + D$$

The simplified circuit is a four-input OR gate as shown in Figure 5–21.

FIGURE 5–21

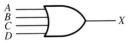

EXAMPLE 5–7

Minimize the combinational logic circuit in Figure 5–22.

FIGURE 5–22

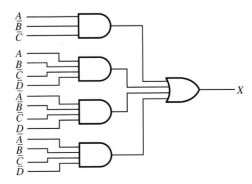

Solution   The output expression is

$$X = A\overline{B}\,\overline{C} + AB\overline{C}\overline{D} + \overline{A}\,\overline{B}CD + \overline{A}\,\overline{B}C\overline{D}$$

Expanding the first term to include the suppressed variables D and $\overline{D}$, we get

$$X = A\overline{B}\,\overline{C}\,(D + \overline{D}) + AB\overline{C}\overline{D} + \overline{A}\,\overline{B}CD + \overline{A}\,\overline{B}C\overline{D}$$
$$X = A\overline{B}\,\overline{C}D + A\overline{B}\,\overline{C}\,\overline{D} + AB\overline{C}\overline{D} + \overline{A}\,\overline{B}CD + \overline{A}\,\overline{B}C\overline{D}$$

This expanded SOP expression is plotted and simplified on the Karnaugh map in Figure 5–23(a). The simplified implementation is shown in part (b).

FIGURE 5–23

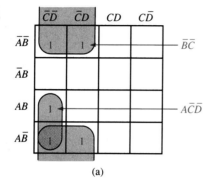

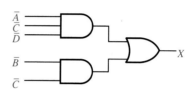

(a)   (b)

SECTION 5–2
REVIEW

Answers are found at the end of the chapter.
1. Implement the following logic expressions as they are written:
    (a) $X = ABC + AB + AC$
    (b) $X = AB(C + DE)$
2. Develop a logic circuit that will produce a 1 on its output only when all three inputs are 1s or when all three inputs are 0s.
3. Reduce the circuits in question 1 to minimum SOP form.

THE UNIVERSAL PROPERTY OF NAND AND NOR GATES

5–3

*Up to this point, combinational circuits using only AND and OR gates and inverters have been considered. In this section we will see that NAND gates and NOR gates can be used to produce **any** logic function. For this reason, they are referred to as universal gates.*

The Universal NAND Gate

The NAND gate can be used to generate the NOT function, the AND function, the OR function, and the NOR function. An inverter can be made from a NAND gate by connecting all of the inputs together and creating, in effect, a single input, as shown in Figure 5–24(a) for a two-input gate. An AND function can be generated by the use of NAND gates alone, as shown in part (b). Also, an OR function can be produced with

NAND gates, as illustrated in part (c). Finally, a NOR function is produced, as shown in part (d).

In Figure 5–24(b) a NAND gate is used to invert (complement) a NAND output to form the AND function, as indicated in the following equation:

$$X = \overline{\overline{AB}} = AB$$

In Figure 5–24(c) NAND gates G_1 and G_2 are used to invert the two input variables before they are applied to NAND gate G_3. The final OR output is derived as follows by application of DeMorgan's theorem:

$$X = \overline{\overline{A}\,\overline{B}} = A + B$$

In Figure 5–24(d) NAND gate G_4 is used as an inverter to produce the NOR function $\overline{A + B}$.

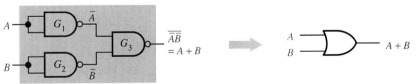

(a) A NAND gate used as an inverter

(b) Two NAND gates used as an AND gate

(c) Three NAND gates used as an OR gate

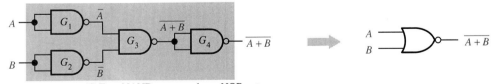

(d) Four NAND gates used as a NOR gate

FIGURE 5–24
Universal application of NAND gates.

The Universal NOR Gate

Like the NAND gate, the NOR gate can be used to generate the NOT, AND, OR, and NAND functions. A NOT circuit, or inverter, can be made from a NOR gate by connecting all of the inputs together to effectively create a single input, as shown in Figure 5–25(a) with two-input example. Also, an OR gate can be produced from NOR gates, as illustrated in Figure 5–25(b). An AND gate can be constructed by the use of NOR

(a) A NOR gate used as an inverter

(b) Two NOR gates used as an OR gate

(c) Three NOR gates used as an AND gate

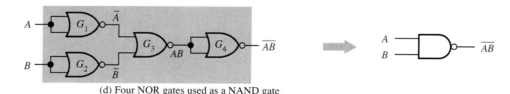

(d) Four NOR gates used as a NAND gate

FIGURE 5–25
Universal application of NOR gates.

gates, as shown in Figure 5–25(c). In this case the NOR gates G_1 and G_2 are used as inverters, and the final output is derived by the use of DeMorgan's theorem as follows:

$$X = \overline{\overline{A} + \overline{B}} = AB$$

Figure 5–25(d) shows NOR gates used to form a NAND function.

**SECTION 5–3
REVIEW**

Answers are found at the end of the chapter.
1. Use NAND gates to implement each expression.
    (a) $X = \overline{A} + B$ (b) $X = A\overline{B}$
2. Use NOR gates to implement each expression.
    (a) $X = \overline{A} + B$ (b) $X = A\overline{B}$

COMBINATIONAL LOGIC USING UNIVERSAL GATES

5–4

In this section you will see how combinations of NAND gates and combinations of NOR gates can be used to implement a logic function. We will make extensive use of the dual function of the NAND and NOR gates to make the analysis of a logic network easier and more straightforward.

NAND Logic

As you have already learned, a NAND gate can function as either a NAND or a negative-OR because, by DeMorgan's theorem,

$$\overline{AB} = \overline{A} + \overline{B}$$

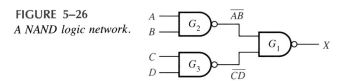

NAND ⌐ ⌐ negative-OR

Consider the NAND logic in Figure 5–26. The output expression is developed in the following steps:

$$
\begin{aligned}
X &= \overline{\overline{(AB)}\ \overline{(CD)}} \\
&= \overline{(\overline{A} + \overline{B})\ (\overline{C} + \overline{D})} \\
&= \overline{(\overline{A} + \overline{B})} + \overline{(\overline{C} + \overline{D})} \\
&= \overline{\overline{A}}\,\overline{\overline{B}} + \overline{\overline{C}}\,\overline{\overline{D}} \\
&= AB + CD
\end{aligned}
$$

Notice that in the last step of the development of the output expression, $AB + CD$ is in the form of two AND terms ORed together. This form of the expression shows that gates G_2 and G_3 act as AND gates and that gate G_1 acts as an OR gate, as illustrated in Figure 5–27(a). This circuit is redrawn in part (b) with NAND symbols for gates G_2 and G_3 and a negative-OR symbol for gate G_1.

FIGURE 5–26
A NAND logic network.

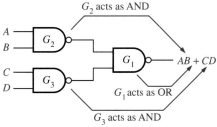

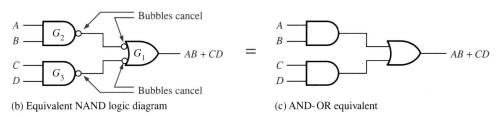

(a) NAND logic diagram

(b) Equivalent NAND logic diagram

(c) AND-OR equivalent

FIGURE 5–27
Why the circuit in Figure 5–26 is equivalent to an AND-OR circuit.

Notice the bubble-to-bubble connections between the outputs of gates G_2 and G_3 and the inputs of gate G_1. Since a bubble represents an inversion, two connected bubbles represent a double inversion and therefore cancel each other. This inversion cancellation can be seen in the previous development of the output expression and is indicated by the absence of barred terms in the output expression. Thus, the circuit in Figure 5–27(b) is effectively an AND-OR function as in part (c).

NAND Logic Diagrams

All logic diagrams using NAND gates should be drawn with each gate represented by either a NAND symbol or an equivalent negative-OR symbol to reflect the function of the gate within the logic network.

If we begin by representing the output gate with a negative-OR symbol, then we will use the NAND symbol for the level of gates right before that and will alternate the symbols for successive levels of gates as we move away from the output. Always use the gate symbols in such a way that every connection between a gate output and a gate input is either bubble-to-bubble or nonbubble-to-nonbubble. Never connect a bubble output to a nonbubble input or vice versa.

Figure 5–28 illustrates the procedure of alternating gate symbols for a NAND circuit with several gate levels. Although using all NAND symbols as in Figure 5–28(a) is correct and there is nothing wrong with it, the diagram in part (b) is much easier to read and is the preferred method. The shape of the gate indicates the way its inputs will appear in the output expression and thus shows how the gate functions within the logic

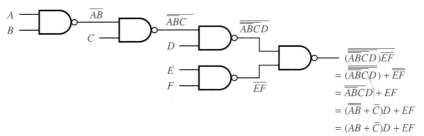

(a) Several Boolean steps are required to arrive at final output expression.

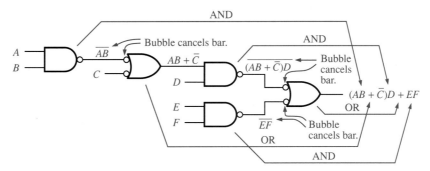

(b) Output expression is derived directly from diagram.

FIGURE 5–28
Example of the analysis of a NAND logic diagram.

network. For a NAND symbol the inputs appear ANDed in the output expression, and for a negative-OR symbol the inputs appear ORed in the output expression, as Figure 5–28(b) illustrates. You can see that the alternating-symbol diagram in part (b) makes it much easier to determine the output expression directly from the logic diagram, because each symbol itself indicates the relationship of its input variables as they appear in the output expression.

EXAMPLE 5–8

Redraw the logic diagram and develop the output expression for the circuit in Figure 5–29, using the appropriate alternating gate symbols.

FIGURE 5–29

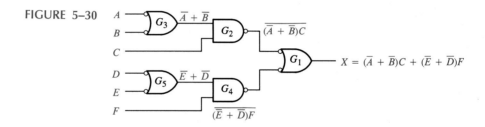

FIGURE 5–30

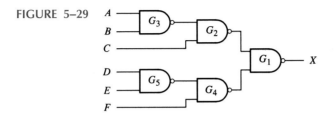

Solution Redrawing the network in Figure 5–29 with the use of equivalent negative-OR symbols as shown, we get Figure 5–30. Writing the expression for X directly from the indicated logic operation of each gate gives us

$$X = (\overline{A} + \overline{B})C + (\overline{D} + \overline{E})F$$

EXAMPLE 5–9

Implement each expression with NAND logic.
(a) $ABC + DE$ **(b)** $ABC + \overline{D} + \overline{E}$

Solution

FIGURE 5–31

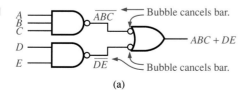

(a)

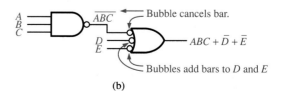

(b)

NOR Logic

A NOR gate can function as either a NOR or a negative-AND, as shown by De-Morgan's theorem:

$$\overline{A + B} = \overline{A}\,\overline{B}$$

NOR $\nearrow$ $\qquad$ $\nwarrow$ negative-AND

The output expression for the NOR logic in Figure 5–32 is developed as follows:

$$
\begin{aligned}
X &= \overline{\overline{A + B} + \overline{C + D}} \\
&= (\overline{\overline{A + B}})\,(\overline{\overline{C + D}}) \\
&= (A + B)\,(C + D)
\end{aligned}
$$

Notice that the expression $(A + B)(C + D)$ is two OR terms ANDed together. This shows that gates G_2 and G_3 act as OR gates and gate G_1 acts as an AND gate, as illustrated in Figure 5–33(a). This circuit is redrawn in part (b) with a negative-AND symbol for gate G_1.

As with NAND logic, the purpose for using the alternating symbols is to make the logic diagram easier to read and analyze, as illustrated in the NOR logic network in Figure 5–34. When the circuit in part (a) is redrawn with alternating symbols in part (b), notice that all output-to-input connections are bubble-to-bubble or nonbubble-to-

FIGURE 5–32
A NOR logic network.

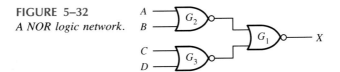

FIGURE 5–33

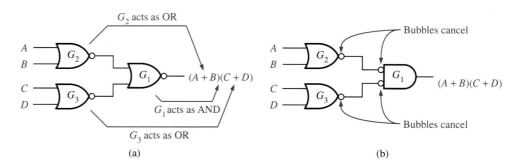

(a)

(b)

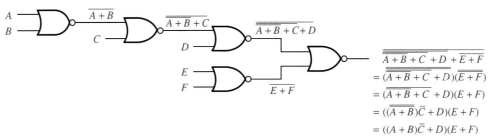

(a) Final output expression is obtained after several Boolean steps.

FIGURE 5–34
Example of analysis of a NOR logic diagram.

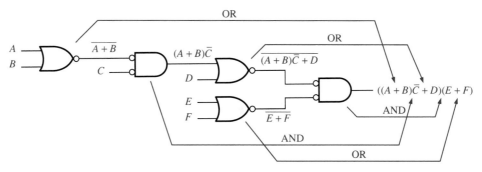

(b) Final output expression is obtained directly from logic diagram.

FIGURE 5–34
(Continued)

nonbubble. Again, you can see that the shape of each gate symbol indicates the type of term (AND or OR) that it produces in the output expression, thus making the output expression easier to determine.

EXAMPLE 5–10

Using alternating gate symbols, redraw the logic diagram and develop the output expression for the circuit in Figure 5–35.

FIGURE 5–35

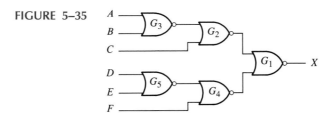

FIGURE 5–36

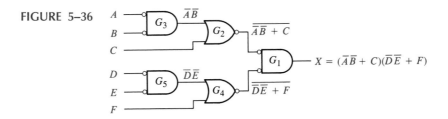

Solution Redraw the network with the equivalent negative-AND symbols as shown in Figure 5–36. Writing the expression for X directly from the indicated logic operation of each gate, we get

$$X = (\overline{A}\,\overline{B} + C)(\overline{D}\,\overline{E} + F)$$

**SECTION 5–4
REVIEW**

Answers are found at the end of the chapter.

1. Implement the expression $X = (\overline{\overline{A} + \overline{B} + \overline{C}})DE$ by using NAND logic.
2. Implement the expression $X = \overline{A}\,\overline{B}\,\overline{C} + (D + E)$ with NOR logic.

PULSED OPERATION

5–5

The operation of combinational logic circuits with pulsed inputs follows the same logical operation we have discussed in the preceding sections. The output of the logic network is dependent on the inputs at any given time; if the inputs are changing according to a time-varying pattern (waveform), the output depends on what the inputs happen to be at any instant.

The following is a summary of the operation of individual gates for use in analyzing combinational circuits with time-varying inputs:

1. The output of an AND gate is HIGH only when all inputs are HIGH at the same time.
2. The output of an OR gate is HIGH any time at least one of its inputs is HIGH. The output is LOW only when all inputs are LOW at the same time.
3. The output of a NAND gate is LOW only when all inputs are HIGH at the same time.
4. The output of a NOR gate is LOW any time at least one of its inputs is HIGH. The output is HIGH only when all inputs are LOW at the same time.

The logical operation of any gate is the same regardless of whether its inputs are pulsed or constant levels. The nature of the inputs (pulsed or constant levels) does not alter the truth table operation of a gate. The following examples illustrate the analysis of combinational logic circuits with pulsed inputs.

EXAMPLE 5–11

Determine the output waveform for the circuit in Figure 5–37(a), with the inputs as shown.

FIGURE 5–37

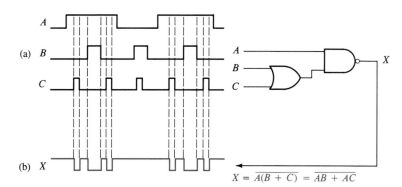

$$X = \overline{A(B + C)} = \overline{AB} + \overline{AC}$$

Solution The output expression, $\overline{AB + AC}$, indicates that the output X is LOW when both A and B are HIGH or when both A and C are HIGH. The output waveform X is shown in the proper time relationship to the inputs in Figure 5–37(b).

EXAMPLE 5–12

Determine the output waveform for the circuit in Figure 5–38(a) if the input waveforms, A and B, are as indicated in part (b).

FIGURE 5–38

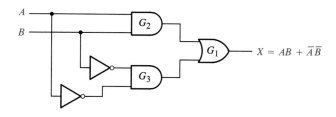

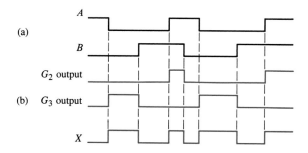

(a)

(b)

Solution When both inputs are HIGH or when both inputs are LOW, the output is HIGH. Notice that this is an *exclusive-NOR* circuit. The intermediate outputs of gates G_2 and G_3 are used to develop the final output. See Figure 5–38(c).

EXAMPLE 5–13

The pulse waveforms shown in Figure 5–39(a) are applied to the 7451 AND-OR-INVERT circuit in part (b). Sketch the output waveform at pin 6.

FIGURE 5–39

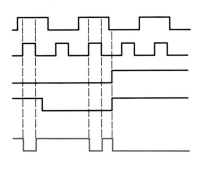

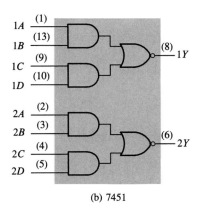

(b) 7451

Solution The output is LOW when 2A and 2B are HIGH or when 2C and 2D are HIGH. The output waveform is shown in part (c) in the proper time relationship to the inputs.

EXAMPLE 5–14

Determine the output waveform for the logic circuit in Figure 5–40(a) by first finding the intermediate waveform at each of points X_1, X_2, X_3, and X_4. The input waveforms are shown in Figure 5–40(b).

FIGURE 5–40

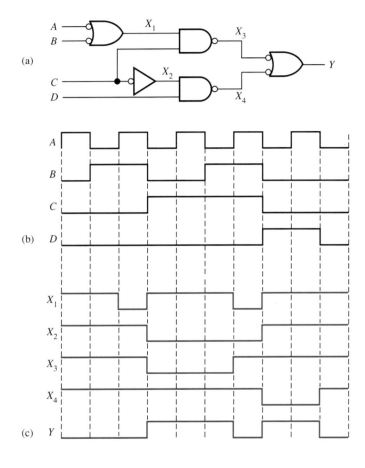

Solution   All the intermediate waveforms and the final output waveform are shown in Figure 5–40(c).

EXAMPLE 5–15

Determine the output waveform for the circuit in Example 5–14, Figure 5–40(a), directly from the output expression.

FIGURE 5–41

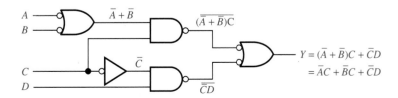

FIGURE 5–42

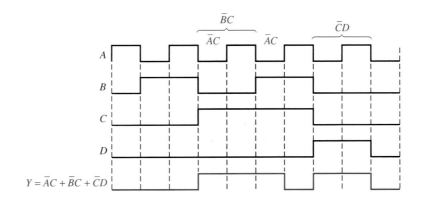

Solution   The output expression for the circuit is developed in Figure 5–41. The SOP form indicates that the output is HIGH when A is LOW and C is HIGH or when B is LOW and C is HIGH or when C is LOW and D is HIGH. The result is shown in Figure 5–42 and is the same as the one we got by the intermediate-waveform method in Example 5–14.

**SECTION 5–5
REVIEW**

Answers are found at the end of the chapter.

1. One pulse with $t_W = 50$ μs is applied to one of the inputs of an exclusive-OR circuit. A second positive pulse with $t_W = 10$ μs is applied to the other input beginning 15 μs after the leading edge of the first pulse. Sketch the output in relation to the inputs.

2. The pulse waveforms A and B in Figure 5–37(a) are applied to the exclusive-NOR circuit in Figure 5–38(a). Sketch a complete timing diagram, showing the output in relation to the inputs.

TROUBLESHOOTING GATE NETWORKS

5–6

The preceding sections have given you some insight into the operation of combinational logic networks and the relationships of inputs and outputs. This type of understanding is essential when you troubleshoot digital circuits because you must know what logic levels or waveforms to look for throughout the circuit for a given set of input conditions.

In this section you will learn how to use the logic pulser, the probe, and the current tracer to troubleshoot a logic network when a gate output is connected to several gate inputs. Also, you will see an example of signal tracing and waveform analysis as methods for locating a fault in a combinational logic circuit.

In a combinational logic network, the output of one gate may be connected to two or more gate inputs as shown in Figure 5–43. The interconnecting paths share a common electrical point known as a *node*.

Gate G_1 in the figure is *driving* the node, and the other gates represent *loads* connected to the node. Several types of failures are possible in this situation. Some of

FIGURE 5–43
Illustration of a node in a gate network.

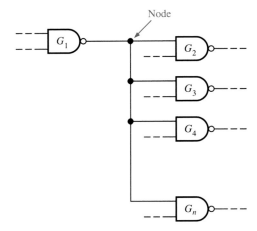

these failure modes are difficult to isolate to a bad gate because all the gates connected to the node are affected. The types of failures we will consider are as follows:

1. *Open output in driving gate.* This failure will cause a loss of signal to all load gates.
2. *Open input in a load gate.* This failure will not affect the operation of any of the other gates connected to the node, but it will result in loss of signal output from the faulty gate.
3. *Shorted output in driving gate.* This failure can cause the node to be stuck in the LOW state.
4. *Shorted input in a load gate.* This failure can also cause the node to be stuck in the LOW state.

We will now explore some approaches to troubleshooting each of the faults just listed.

Open Output in Driving Gate

In this situation there is no pulse activity on the node. With circuit power on, an open node will normally result in a "floating" level and will be indicated by a dim lamp on the logic probe, as illustrated in Figure 5–44.

FIGURE 5–44
Open output in driving gate.

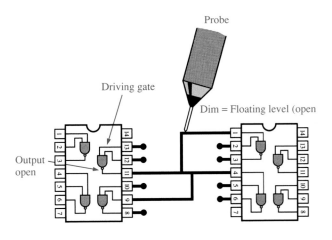

Open Input in a Load Gate

If the check for an open driver output is negative, then a check for an open input in a load gate should be performed. Apply the logic pulser tip to the node. Then check the output of each gate for pulse activity with the logic probe, as illustrated in Figure 5–45. If one of the inputs that is normally connected to the node is open, no pulses will be detected on that gate's output.

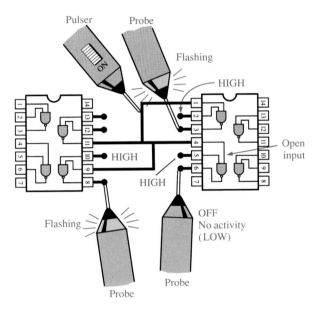

FIGURE 5–45
Open input in a load gate.

Shorted Output in Driving Gate

This fault can cause the node to be stuck LOW, as previously mentioned. A quick check with a pulser and a logic probe will indicate this, as shown in Figure 5–46(a). A short to ground in the driving gate's output or in any gate input will cause this symptom, and further checks must therefore be made to isolate the short to a particular gate.

If the driving gate's output is internally shorted to ground, then, essentially, no current activity will be present on any of the connections to the node. Thus, a current tracer will indicate no activity with circuit power on, as illustrated in Figure 5–46(b).

To further verify a shorted output, a pulser and a current tracer can be used with the circuit power off, as shown in Figure 5–46(c). When current pulses are applied to the node with the pulser, all of the current will flow into the shorted output, and none will flow through the circuit paths into the load gate inputs.

Shorted Input in a Load Gate

If one of the load gate inputs is internally shorted to ground, the node will be stuck in the LOW state. Again, as in the case of a shorted output, the logic pulser and current tracer can be used to isolate the faulty gate.

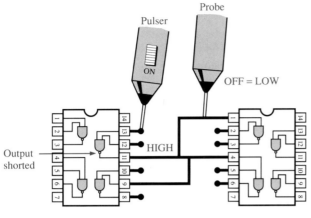

(a) Node is "stuck" LOW

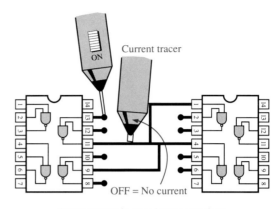

(b) No current in node interconnections

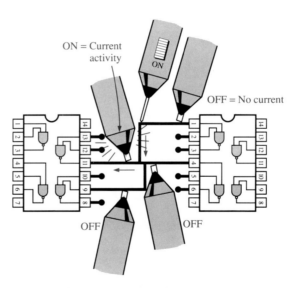

(c) Tracing current along path into shorted output

FIGURE 5–46
Shorted output in driving gate.

When the node is pulsed with circuit power off, essentially all the current will flow into the shorted input, and tracing its path with the current tracer will lead to the bad input, as illustrated in Figure 5–47.

Signal Tracing and Waveform Analysis

Although the methods of isolating an open or a short at a node point are very useful from time to time, a more general troubleshooting technique called *signal tracing* is of great value to the technician or technologist in just about every troubleshooting situa-

FIGURE 5–47
Shorted input in a load gate.

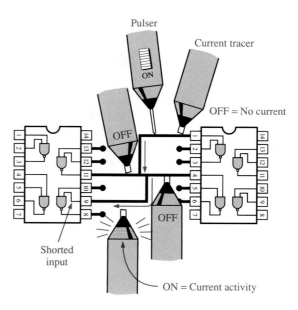

tion. Waveform (signal) measurement is accomplished with an oscilloscope or a logic analyzer.

Basically, the signal tracing method requires that you observe the waveforms and their time relationships at all accessible points in the logic circuit. You begin at the inputs and, from an analysis of the waveform timing diagram for each point, determine where an incorrect waveform first occurs. With this procedure you can usually isolate the fault to a specific gate. A procedure beginning at the output and working back toward the inputs can also be used.

The general procedure for starting at the inputs is outlined in the following steps, with a simple logic network used for illustration. The same approach can be applied to more complex circuits.

- ☐ Within a system, define the section of logic that is suspected of being faulty.
- ☐ Start at the inputs to the section of logic under examination. We assume, for this discussion, that the input waveforms coming from other sections of the system have been found to be correct.
- ☐ For each gate, beginning at the input and working toward the output of the logic network, observe the output waveform of the gate and compare it with the input waveforms by using the oscilloscope or the logic analyzer.
- ☐ Determine if the output waveform is correct, using your knowledge of the logical operation of the gate.
- ☐ If the output is incorrect, the gate under test may be faulty. Pull the IC containing the gate that is suspected of being faulty, and test it out-of-circuit. If the gate is found to be faulty, replace the IC. If it works correctly, the fault is in the external circuitry or in another IC to which the tested one is connected.
- ☐ If the output is correct, go to the next gate. Continue checking each gate until an incorrect waveform is observed.

Figure 5–48 illustrates this general procedure.

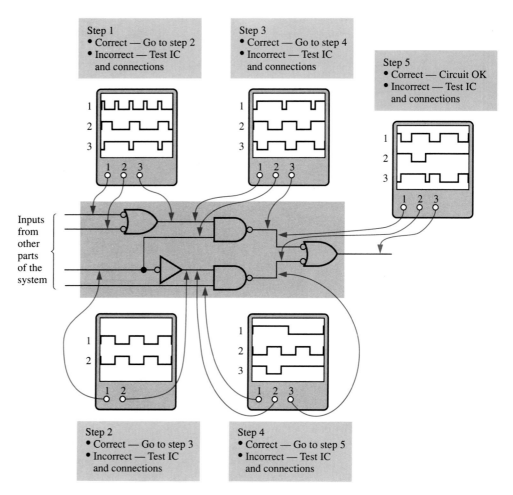

FIGURE 5–48

Example of signal tracing and waveform analysis in a portion of a digital system.

EXAMPLE 5–16

Determine the fault in the logic network of Figure 5–49(a) by using waveform analysis. Assume that you have observed the waveforms shown in black in Figure 5–49(b).

Solution

1. Determine what the correct waveform should be for each gate. The correct waveforms are shown in color, superimposed on the actual measured waveforms, in Figure 5–49(b).
2. Compare waveforms gate by gate until you find a measured waveform that does not match the correct waveform.

In this example everything tested is correct until gate G_4. The output of this gate is not correct, and an analysis of the waveforms indicates that if the D input to gate G_4

FIGURE 5–49

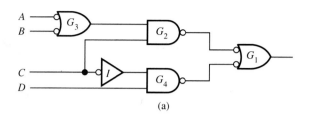

(a)

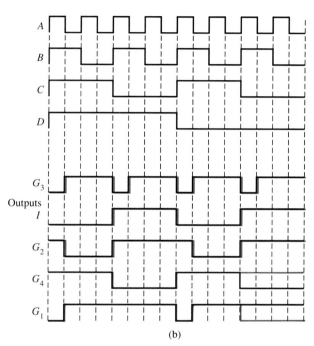

(b)

is open and acting as a HIGH, we will get the output waveform measured. Notice that the output of G_1 is correct for the inputs measured, although the input from G_4 is incorrect.

Replace the IC containing G_4, and check the circuit's operation again.

SECTION 5–6
REVIEW

Answers are found at the end of the chapter.
1. List four common internal failures in logic gates.
2. One input of a NOR gate is externally shorted to $+V_{CC}$. How does this condition affect the gate operation?
3. Determine the output of gate G_1 in Figure 5–49(a), with inputs as shown in part (b), for the following faults:
    (a) one input to G_1 shorted to ground
    (b) the inverter input shorted to ground
    (c) an open output in G_3

SUMMARY

☐ AND-OR logic produces an output expression in SOP form.

☐ AND-OR-INVERT logic produces a complemented SOP form, which is actually a POS form.

☐ The operational symbol for exclusive-OR is $\oplus$. An exclusive-OR expression can be stated in two equivalent ways:

$$A\overline{B} + \overline{A}B = A \oplus B$$

☐ To perform analysis of a logic circuit, start with the logic circuit, and develop the Boolean output expression or the truth table or both.

☐ Design of a logic circuit is the process in which we start with the Boolean output expression or the truth table and develop a logic circuit that produces that output function.

☐ All NAND or NOR logic diagrams should be drawn with alternating symbols so that bubble outputs are connected to bubble inputs and nonbubble outputs are connected to nonbubble inputs wherever possible.

☐ When two bubbles are connected, they effectively cancel each other.

Integrated Circuit Technologies (Chapter 13) may be covered after this chapter, covered after a later chapter, completely omitted, or partially omitted, depending on the needs and goals of your program. Coverage is not prerequisite or corequisite to any other material in the book. Inclusion is purely optional.

GLOSSARY

Alternate symbols The negative-AND is the alternate symbol for the NOR gate and the negative-OR is the alternate symbol for the NAND gate.

Combinational logic A combination of logic gates interconnected to produce a specified Boolean function with no storage or memory capability. Sometimes called *combinatorial logic*.

Node A common connection point in a circuit in which a gate output is connected to one or more gate inputs.

Signal tracing A troubleshooting technique in which waveforms are observed in a step-by-step manner beginning at the input and working toward the output. At each point the observed waveform is compared with the correct signal for that point.

Universal gate Either a NAND gate or a NOR gate. The term *universal* refers to the property of a gate that permits any logic function to be implemented by that gate or by a combination of gates of that kind.

SELF-TEST

Answers and solutions are found at the end of the book.

1. Write the output expression for an AND-OR logic circuit having one 4-input AND gate with inputs A, B, C, and D and one 2-input AND gate with inputs E and F.

2. Sketch the logic circuit described by each of the following output expressions:

   (a) $Y = \overline{A\overline{B}C} + AB\overline{C}$ **(b)** $Y = \overline{A}B + A\overline{B}$
   (c) $Y = A\overline{B}\,CD + ABC\overline{D} + \overline{A}BCD$

3. Write the truth table for a logic circuit that produces a 1 output only when its three input variables represent a two, five, or seven in binary and a 0 output for all other input states.

4. Draw the logic circuit for the truth table specified in Problem 3.

5. Which, if any, of the expressions in Problem 2 can be simplified?

6. Draw the logic diagram for a NOR circuit having the output expression

$$X = (A + B + C)(D + E)\overline{F}$$

7. Draw the logic diagram for a NAND circuit having the output expression

$$X = ABC + DE + \overline{F}$$

8. A certain application requires that the absence of input pulses (constant LOWs) on two input lines be indicated by a HIGH output. The condition in which both inputs have simultaneous positive pulses must also be indicated by a HIGH output. Draw the logic diagram for this circuit.

PROBLEMS

Answers to odd-numbered problems are found at the end of the book.

Section 5–1   Analysis of Combinational Logic

5–1 Determine the truth table for each of the circuits in Figure 5–50.

5–2 Determine the truth table for each circuit in Figure 5–51.

5–3 Determine the truth table for the circuit shown in Figure 5–52.

5–4 Write the logic expression for the output of each of the circuits in Figure 5–53.

5–5 Write the logic expression without simplification for the output of each of the circuits in Figure 5–54.

FIGURE 5–50

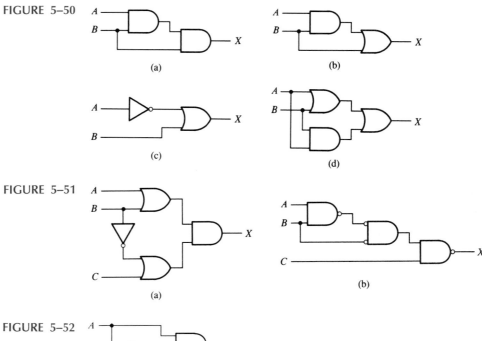

FIGURE 5–51

FIGURE 5–52

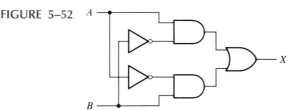

5–6 Develop the truth table for the circuit in Figure 5–55.

5–7 Repeat Problem 5–6 with each AND gate replaced by a NAND gate and each OR gate replaced by a NOR gate.

Section 5–2 Design of Combinational Logic

5–8 Use AND gates, OR gates, or combinations of both to implement the following logic expressions as stated:

(a) $X = AB$ **(b)** $X = A + B$

(c) $X = AB + C$ **(d)** $X = ABC + D$

(e) $X = A + B + C$ **(f)** $X = ABCD$

(g) $X = A(CD + B)$ **(h)** $X = AB(C + DEF) + CE(A + B + F)$

FIGURE 5–53

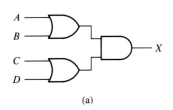

(a)

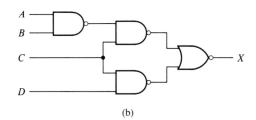

(b)

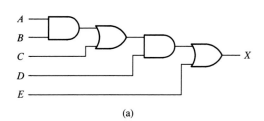

(a)

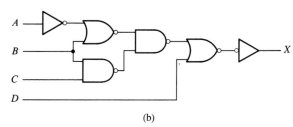

(b)

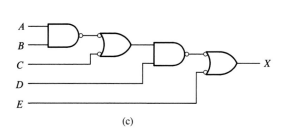

(c)

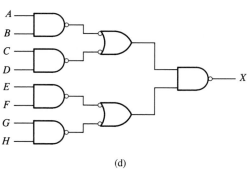

(d)

FIGURE 5–54

FIGURE 5–55

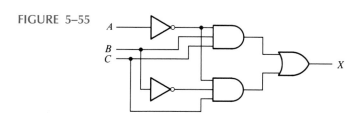

5–9 Use AND gates, OR gates, and inverters as needed to implement the following logic expressions as stated:

(a) $X = AB + \overline{B}C$ (b) $X = A(B + \overline{C})$

(c) $X = A\overline{B} + AB$ (d) $X = \overline{ABC} + B(EF + \overline{G})$

(e) $X = A[BC(A + B + C + D)]$ (f) $X = B(C\overline{D}E + \overline{E}FG)(\overline{AB} + C)$

5–10 Use NAND gates, NOR gates, or combinations of both to implement the following logic expressions as stated:

(a) $X = \overline{AB} + CD + \overline{(A + B)}(ACD + \overline{BE})$

(b) $X = \overline{ABCD} + \overline{DEF} + \overline{AF}$

(c) $X = \overline{A}[B + \overline{C}(D + E)]$

5–11 Design a logic circuit for the truth table in Table 5–7.

TABLE 5–7

Inputs			Output
A	B	C	X
0	0	0	1
0	0	1	0
0	1	0	1
0	1	1	0
1	0	0	1
1	0	1	0
1	1	0	1
1	1	1	1

5–12 Design a logic circuit for the truth table in Table 5–8.

TABLE 5–8

Inputs				Output
A	B	C	D	X
0	0	0	0	0
0	0	0	1	0
0	0	1	0	1
0	0	1	1	1
0	1	0	0	1
0	1	0	1	0
0	1	1	0	0
0	1	1	1	0
1	0	0	0	1
1	0	0	1	1
1	0	1	0	1
1	0	1	1	1
1	1	0	0	0
1	1	0	1	0
1	1	1	0	0
1	1	1	1	1

5–13 Develop the logic circuit necessary to meet the following requirements:

A lamp in a room is to be operated from two switches, one at the back door and one at the front door. The lamp is to be on if the front switch is on and the back switch is off, or if the front switch is off and the back switch is on. The lamp is to be off if both switches are off or if both switches are on. Let a HIGH output represent the on condition and a LOW output represent the off condition.

5–14 Design a logic circuit to produce a HIGH output if and only if the input, represented by a 4-bit binary number, is greater than twelve or less than three. First develop the truth table and then draw the logic diagram.

5–15 Simplify the circuit in Figure 5–56 as much as possible, and verify that the simplified circuit is equivalent to the original by showing that the truth tables are identical.

5–16 Repeat Problem 5–15 for the circuit in Figure 5–57.

5–17 Minimize the gates required to implement the functions in each part of Problem 5–9 in SOP form.

5–18 Minimize the gates required to implement the functions in each part of Problem 5–10 in SOP form.

5–19 Minimize the gates required to implement the function of the circuit in each part of Figure 5–54 in SOP form.

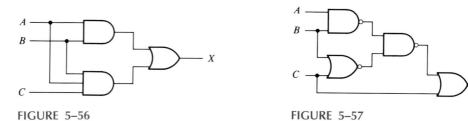

FIGURE 5–56   FIGURE 5–57

Section 5–3   The Universal Property of NAND and NOR Gates

5–20 Implement the logic circuit in Figure 5–55 using only NAND gates.

5–21 Implement the logic circuit in Figure 5–57 using only NAND gates.

5–22 Repeat Problem 5–20 using only NOR gates.

5–23 Repeat Problem 5–21 using only NOR gates.

Section 5–4   Combinational Logic Using Universal Gates

5–24 Show how the following expressions can be implemented as stated using only NOR gates:

(a) $X = ABC$ (b) $X = \overline{ABC}$

(c) $X = A + B$ (d) $X = A + B + \overline{C}$

(e) $X = \overline{AB} + \overline{CD}$ (f) $X = (A + B)(C + D)$

(g) $X = AB[C(\overline{DE} + \overline{AB}) + \overline{BCE}]$

5–25 Repeat Problem 5–24 using only NAND gates.

5–26 Implement each function as stated in Problem 5–8 by using only NAND gates.

5–27 Implement each function as stated in Problem 5–9 by using only NAND gates.

Section 5–5   Pulsed Operation

5–28 Given the logic circuit and the input waveforms in Figure 5–58, sketch the output waveform.

5–29 For the logic circuit in Figure 5–59, sketch the output waveform in proper relation to the inputs.

5–30 For the input waveforms in Figure 5–60, what logic circuit will generate the output waveform shown?

5–31 Repeat Problem 5–30 for the waveforms in Figure 5–61.

5–32 For the circuit in Figure 5–62, sketch the waveforms at the numbered points in the proper relationship to each other.

5–33 Assuming a propagation delay through each gate of 10 nanoseconds (ns), determine if the *desired* output waveform X in Figure 5–63 (a pulse with a minimum $t_W = 25$ ns positioned as shown) will be generated properly with the given inputs.

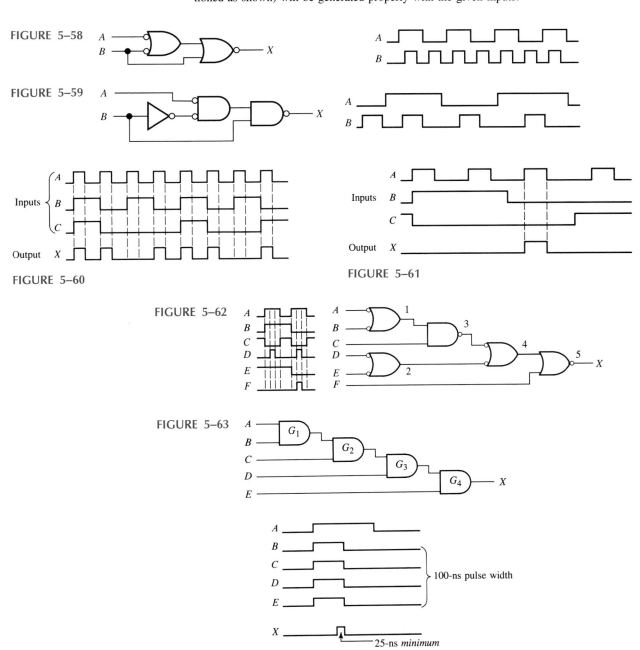

FIGURE 5–58

FIGURE 5–59

FIGURE 5–60

FIGURE 5–61

FIGURE 5–62

FIGURE 5–63

100-ns pulse width

25-ns *minimum*

Section 5–6   Troubleshooting Gate Networks

5–34   For the logic circuit and the input waveforms in Figure 5–64, the indicated output waveform is observed. Determine if this is the correct output waveform.

5–35   The output waveform in Figure 5–65 is incorrect for the inputs that are applied to the circuit. Assuming that one gate in the circuit has failed, with its output either an apparent constant HIGH or a constant LOW, determine the faulty gate and the type of failure (output open or shorted).

5–36   Repeat Problem 5–35 for the circuit in Figure 5–66, with input and output waveforms as shown.

5–37   The node in Figure 5–67 has been found to be stuck in the LOW state. A pulser and current tracer yield the indications shown in the diagram. From this information what would you conclude?

5–38   Figure 5–68(a) is a logic network under test. Figure 5–68(b) shows the waveforms as observed on a logic analyzer. The output waveform is incorrect for the inputs that are applied to the circuit. Assuming that one gate in the network has failed, with its output either an apparent constant HIGH or a constant LOW, determine the faulty gate and the type of failure.

5–39   The logic circuit in Figure 5–69 has the input waveforms shown.
(a)   Determine the correct output waveform in relation to the inputs.
(b)   Determine the output waveform if the output of gate G_3 is open.
(c)   Determine the output waveform if the upper input to gate G_5 is shorted to ground.

5–40   The logic circuit in Figure 5–70 has only one intermediate test point available besides the output, as indicated. For the inputs shown, you observe the indicated waveform at the test point. Is this waveform correct? If not, what are the possible faults that would cause it to appear as it does?

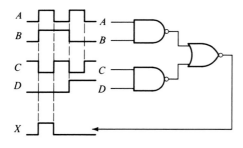

FIGURE 5–64

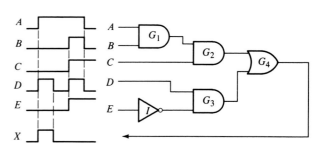

FIGURE 5–65

FIGURE 5–66

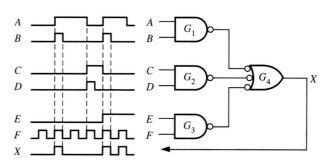

FIGURE 5–67

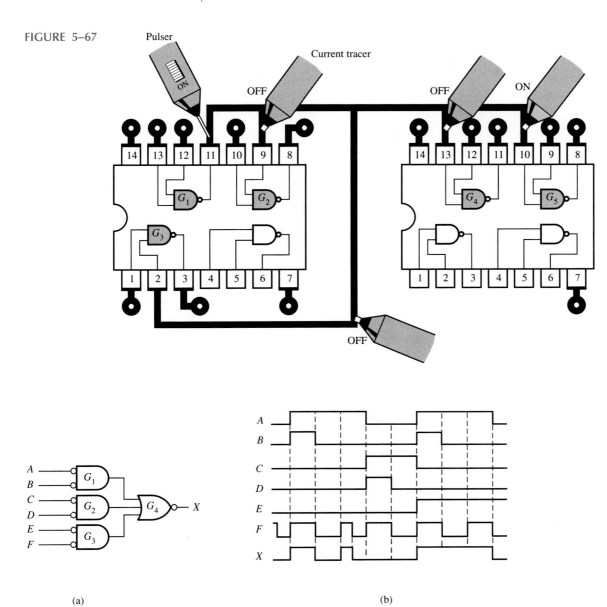

FIGURE 5–68

(a)

(b)

FIGURE 5–69

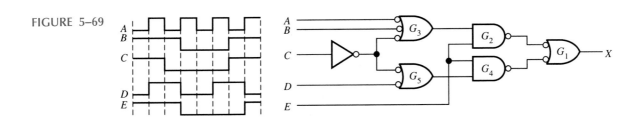

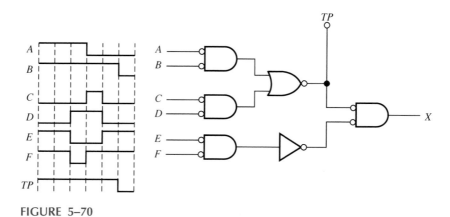

FIGURE 5–70

ANSWERS TO SECTION REVIEWS

Section 5–1 Review

1. (a) 1   (b) 0   (c) 0
2. (a) 1   (b) 0   (c) 1   (d) 0
3. $X = 1$ when $ABC = 000, 011, 101, 110$, and 111; $X = 0$ when $ABC = 001, 010$, and 100
4. $X = AB + \overline{AB}$; the circuit consists of two AND gates, one OR gate, and two inverters.

Section 5–2 Review

1. (a) three AND gates, one OR gate   (b) three AND gates, one OR gate
2. $X = ABC + \overline{ABC}$; two AND gates, one OR gate, and three inverters
3. (a) $X = AB + AC$ (b) $X = ABC + ABDE$

Section 5–3 Review

1. (a) a two-input NAND gate with A and $\overline{B}$ on its inputs
   (b) a two-input NAND with A and $\overline{B}$ on its inputs, followed by one NAND used as an inverter
2. (a) a two-input NOR with inputs $\overline{A}$ and B, followed by one NOR used as an inverter
   (b) a two-input NOR with $\overline{A}$ and B on its inputs

Section 5–4 Review

1. a three-input NAND with inputs A, B, and C, with its output connected to a second three-input NAND with two other inputs, D and E
2. a three-input NOR with inputs A, B, and C, with its output connected to a second three-input NOR with two other inputs, D and E

Section 5–5 Review

1. a 15-μs pulse followed by a 25-μs pulse, with a separation of 10 μs between the pulses
2. The output is HIGH when both inputs are HIGH or when both inputs are LOW.

Section 5–6 Review

1. input or output open; input or output shorted to ground
2. Output stuck LOW.
3. (a) constant HIGH
   (b) same as input D
   (c) Output is HIGH until rising edge of seventh pulse; then it goes LOW.

In this chapter several types of MSI combinational logic functions are studied, including adders, comparators, decoders, encoders, code converters, multiplexers (data selectors), demultiplexers, and parity generators/checkers.

Examples of systems applications of many of these devices are presented to provide a basic idea of how the devices can be used in practical situations. Also, a simple application of the logic analyzer and the oscilloscope in troubleshooting digital systems for glitches is introduced.

Specific devices introduced in this chapter are as follows:

1. 7482 two-bit binary adder
2. 7483A four-bit parallel binary adder
3. 7485 four-bit magnitude comparator
4. 74154 four-line-to-sixteen-line decoder
5. 7442A BCD-to-decimal decoder
6. 7449 BCD-to-seven-segment decoder/ driver
7. 74147 decimal-to-BCD priority encoder
8. 74150 sixteen-input data selector/multiplexer
9. 74151A eight-input data selector/multiplexer
10. 74157 quadruple two-input data selector/ multiplexer
11. 74138 three-line-to-eight-line decoder/demultiplexer
12. 74139 dual two-line-to-four-line decoder/ demultiplexer
13. 74180 nine-bit parity generator/checker

After completing this chapter, you will be able to
☐ recognize the difference between half adders and full adders.
☐ use full adders to implement multibit parallel binary adders.
☐ explain the differences between ripple carry and look-ahead carry parallel adders.
☐ use the magnitude comparator to determine the relationship between two binary numbers and use cascaded comparators to handle the comparison of larger numbers.
☐ implement a basic binary decoder.
☐ use BCD-to-seven-segment decoders and LEDs or LCDs in display systems.
☐ apply a decimal-to-BCD priority encoder in a simple keyboard application.
☐ convert from BCD to binary, binary to Gray code, and Gray code to binary by using logic devices.
☐ apply multiplexers in data selection, multiplexed displays, logic function generation, and simple communications systems.
☐ use decoders as demultiplexers.
☐ explain the meaning of *parity*.
☐ use parity generators and checkers to detect bit errors in digital systems.
☐ implement a simple data communications system.
☐ identify *glitches,* common bugs in digital systems.

6

Logic Functions

ADDERS

6–1

Adders are very important in many types of digital systems in which numerical data are processed. An understanding of their basic operation is fundamental to a thorough grasp of digital systems concepts.

*We will begin by developing a **half-adder** and then a **full-adder**. In Section 6–2 we will see how individual adders can be combined to form multiple-bit adders.*

Half-Adder

Recall the basic rules for binary addition as stated in Chapter 2.

$$
\begin{aligned}
0 + 0 &= 0 \\
0 + 1 &= 1 \\
1 + 0 &= 1 \\
1 + 1 &= 10
\end{aligned}
$$

These operations are performed by a logic circuit called a half-adder. The half-adder accepts two binary digits on its inputs and produces two binary digits on its outputs, a sum bit and a carry bit, as shown by the logic symbol in Figure 6–1.

From the logical operation of the half-adder as expressed in Table 6–1, expressions can be derived for the sum and carry outputs as functions of the inputs. Notice that the carry output (CO) is a 1 only when both P and Q are 1s; therefore, CO can be expressed as the AND of the input variables:

$$CO = PQ \tag{6–1}$$

Now observe that the sum output (Σ) is a 1 only if the input variables, P and Q, are not equal. The sum can therefore be expressed as the exclusive-OR of the input variables:

$$\Sigma = P \oplus Q \tag{6–2}$$

From these two expressions, the implementation required for the half-adder function is apparent. The carry output is produced with an AND gate with P and Q on the inputs, and the sum output is generated with an exclusive-OR gate, as shown in Figure 6–2.

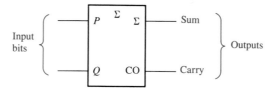

FIGURE 6–1
Logic symbol for a half-adder.

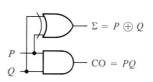

FIGURE 6–2
Half-adder logic diagram.

TABLE 6–1
Half-adder truth table.

P	Q	CO	Σ
0	0	0	0
0	1	0	1
1	0	0	1
1	1	1	0

Σ = sum
CO = carry output
P and Q = input variables (operands)

Full-Adder

The second basic category of adder is the full-adder. The full-adder accepts *three* inputs and generates a sum output and a carry output. So the basic difference between a full-adder and a half-adder is that the full-adder accepts an additional input, which allows it to handle input carries. A logic symbol for a full-adder is shown in Figure 6–3, and the truth table in Table 6–2 shows the operation of a full-adder.

FIGURE 6–3
Logic symbol for a full-adder.

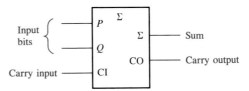

TABLE 6–2
Truth table for a full-adder.

P	Q	CI	CO	Σ
0	0	0	0	0
0	0	1	0	1
0	1	0	0	1
0	1	1	1	0
1	0	0	0	1
1	0	1	1	0
1	1	0	1	0
1	1	1	1	1

CI = carry input
CO = carry output
Σ = sum
P and Q = input variables (operands)

The full-adder must add the two input bits and the carry input bit. From the half-adder we know that the sum of the input bits P and Q is the exclusive-OR of those two variables, $P \oplus Q$. For the carry input (CI) to be added to the input bits, it must be exclusive-ORed with $P \oplus Q$, yielding the equation for the sum output of the full-adder:

$$\Sigma = (P \oplus Q) \oplus CI \qquad \qquad \text{(6–3)}$$

FIGURE 6–4
Full-adder logic.

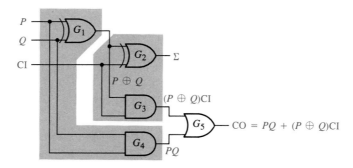

(a) Logic required to form the sum of three bits

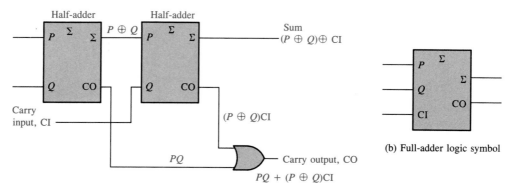

(b) Complete logic circuitry for a full-adder (each half-adder is shown in a shaded block).

FIGURE 6–5
Full-adder.

(a) Arrangement of two half-adders to form a full-adder

(b) Full-adder logic symbol

This means that to implement the full-adder sum function, two exclusive-OR gates can be used. The first must generate the term $P \oplus Q$, and the second has as its inputs the output of the first XOR gate and the carry input, as illustrated in Figure 6–4(a).

The carry output is a 1 when both inputs to the first XOR gate are 1s or when both inputs to the second XOR gate are 1s. You can verify this fact by studying Table 6–2. The carry output of the full-adder is therefore P ANDed with Q, $P \oplus Q$ ANDed with CI, and these two terms ORed, as expressed in Equation (6–4). This function is implemented and combined with the sum logic to form a complete full-adder circuit, as shown in Figure 6–4(b).

$$CO = PQ + (P \oplus Q)CI \qquad (6\text{–}4)$$

Note that in Figure 6–4(b) there are two half-adders, connected as shown in the block diagram of Figure 6–5(a), with their carry outputs ORed. The logic symbol shown in Figure 6–5(b) will normally be used to represent the full-adder.

EXAMPLE 6–1   Determine an alternative method for implementing the full-adder.

Solution   Going back to Table 6–2, we can write sum-of-products expressions for both
Σ and CO by observing the input conditions that make them 1s. The expressions are as
follows:

$$\Sigma = \overline{P}\overline{Q}\text{CI} + \overline{P}Q\overline{\text{CI}} + P\overline{Q}\overline{\text{CI}} + PQ\text{CI}$$
$$\text{CO} = \overline{P}Q\text{CI} + P\overline{Q}\text{CI} + PQ\overline{\text{CI}} + PQ\text{CI}$$

Plotting these two expressions on the Karnaugh maps in Figure 6–6, we find that the
sum (Σ) expression cannot be simplified. The expression for CO is reduced as indicated.
These two expressions are implemented with AND-OR logic as shown in Figure
6–7 to form a complete full-adder.

FIGURE 6–6

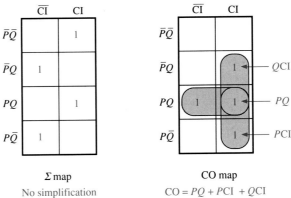

Σ map

No simplification

$\Sigma = \overline{P}\overline{Q}\text{CI} + \overline{P}Q\overline{\text{CI}} + P\overline{Q}\overline{\text{CI}} + PQ\text{CI}$

CO map

$\text{CO} = PQ + P\text{CI} + Q\text{CI}$

FIGURE 6–7

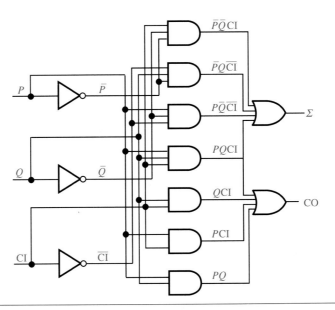

Answers are found at the end of the chapter.
1. Determine the sum (Σ) and the carry output (CO) of a half-adder for each set of input bits.
    (a) 01   **(b)** 00   **(c)** 10   **(d)** 11
2. A full-adder has CI = 1. What are the sum (Σ) and the carry output (CO) when $P = 1$ and $Q = 1$?

PARALLEL BINARY ADDERS

6–2

Most adders that are available in integrated circuit form are parallel adders. In this section you will learn the basics of this type of adder so that you will understand all the necessary input and output functions when working with these devices.

As you have seen, a single full-adder is capable of adding two one-bit numbers and an input carry. To add binary numbers with more than one bit, additional full-adders must be employed. When one binary number is added to another, each column generates a sum bit and a 1 or 0 carry bit to the next higher-order column, as illustrated here with two-bit numbers:

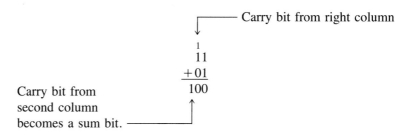

To implement the addition of two binary numbers, a full-adder is required for each bit in the numbers. So for two-bit numbers, two adders are needed; for three-bit numbers, three adders are used; and so on. The carry output of each adder is connected to the carry input of the next higher-order adder, as shown in Figure 6–8 for a two-bit

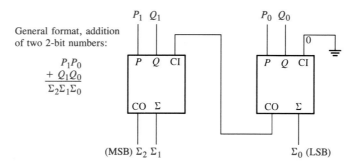

FIGURE 6–8
Block diagram of a basic two-bit parallel adder.

adder. It should be pointed out that either a half-adder can be used for the least significant position or the carry input of a full-adder can be made 0 (grounded), because there is no carry input to the least significant bit position.

In Figure 6–8 the least significant bits (LSB) of the two numbers are represented by P_0 and Q_0. The next higher-order bits are represented by P_1 and Q_1. The three sum bits are Σ_0, Σ_1, and Σ_2. Notice that the carry output from the left-most full-adder becomes the most significant bit (MSB) in the sum, Σ_2.

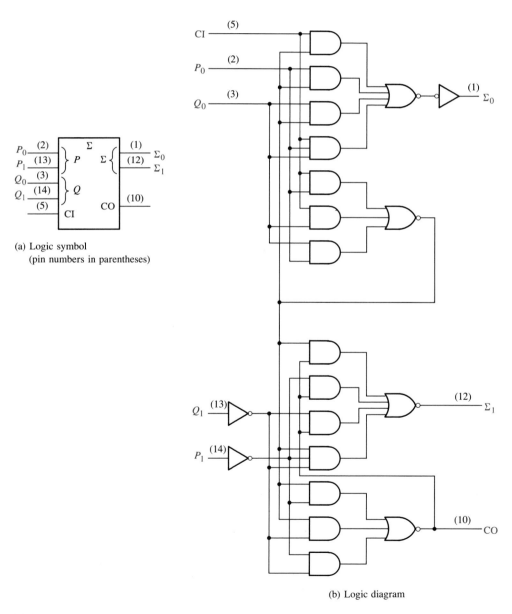

(a) Logic symbol
(pin numbers in parentheses)

(b) Logic diagram

FIGURE 6-9
7482 two-bit binary full-adder.

The 7482 Two-Bit Binary Full-Adder

The 7482 is an example of an integrated circuit two-bit adder. It is a TTL medium-scale integrated (MSI) circuit with two interconnected full-adders in one package. A logic symbol with pin numbers and a logic diagram are shown in Figure 6–9. Table 6–3 is the truth table for this device.

TABLE 6–3
Truth table for a 7482 two-bit binary full-adder.

	Inputs				Outputs				
				When CI = 0			When CI = 1		
P_0	Q_0	P_1	Q_1	Σ_0	Σ_1	CO	Σ_0	Σ_1	CO
0	0	0	0	0	0	0	1	0	0
1	0	0	0	1	0	0	0	1	0
0	1	0	0	1	0	0	0	1	0
1	1	0	0	0	1	0	1	1	0
0	0	1	0	0	1	0	1	1	0
1	0	1	0	1	1	0	0	0	1
0	1	1	0	1	1	0	0	0	1
1	1	1	0	0	0	1	1	0	1
0	0	0	1	0	1	0	1	1	0
1	0	0	1	1	1	0	0	0	1
0	1	0	1	1	1	0	0	0	1
1	1	0	1	0	0	1	1	0	1
0	0	1	1	0	0	1	1	0	1
1	0	1	1	1	0	1	0	1	1
0	1	1	1	1	0	1	0	1	1
1	1	1	1	0	1	1	1	1	·1

Notice the input and output labels on this device: P_0 and Q_0 are the LSB inputs, and P_1 and Q_1 are the MSB inputs. The input labeled CI is the carry input to the least significant bit adder; CO is the carry output of the most significant bit adder; and Σ_0 (LSB) and Σ_1 (MSB) are the sum outputs. (Labels may vary from one manufacturer to another.)

EXAMPLE 6–2

Verify that the two-bit parallel adder in Figure 6–10 properly performs the following addition:

$$\begin{array}{r} 11 \\ + 10 \\ \hline 101 \end{array}$$

Solution The logic level at each point in the circuit for the given input numbers is determined from the truth table of the relevant gate. By following these levels through

FIGURE 6–10

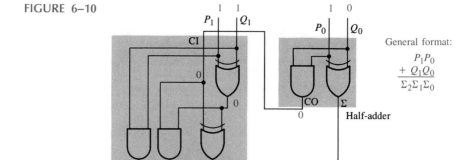

the circuit as indicated on the logic diagram, we find that the proper levels appear on the sum outputs.

Four-Bit Parallel Adders

A four-bit binary parallel adder is shown in Figure 6–11. Again, the LSBs in each number being added go into the right-most full-adder; the higher-order bits are applied

FIGURE 6–11

Four-bit parallel binary adder.

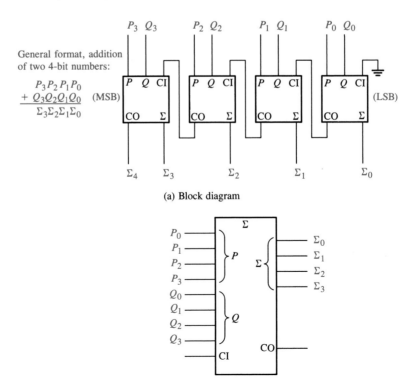

(a) Block diagram

(b) Logic symbol

as shown to the successively higher-order adders, with the MSBs in each number being applied to the left-most full-adder. The carry output of each adder is connected to the carry input of the next higher-order adder.

EXAMPLE 6–3

Show how to connect two 7482 adders to form a four-bit adder.

FIGURE 6–12
A four-bit adder using two 7482s.

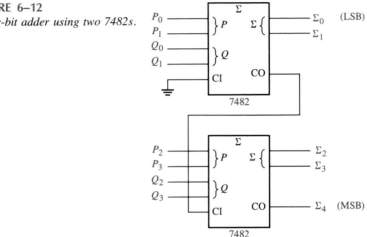

Solution   Figure 6–12 shows two 7482s used as four-bit adder. Additional bits can be added by connecting the CO of the most significant adder to the CI of the next adder, and so on. Note that the CI of the least significant adder is grounded (0) because there is no carry input to this stage.

EXAMPLE 6–4

Show how two four-bit parallel adders can be connected to form an eight-bit parallel adder. Show outputs for

$$P_7 P_6 P_5 P_4 P_3 P_2 P_1 P_0 \ = \ 10111001$$

and

$$Q_7 Q_6 Q_5 Q_4 Q_3 Q_2 Q_1 Q_0 \ = \ 10011110$$

Solution   Two 7483A four-bit parallel adders are used to implement the eight-bit adder. The only connection between the two 7483As is the carry output (CO) of the lower-order adder to the carry input (CI) of the higher-order adder, as shown in Figure 6–13. Pin 13 of the least significant 7483A is grounded (no carry input).

FIGURE 6–13
Two 7483A four-bit parallel binary adders connected to form an eight-bit parallel adder (pin numbers in parentheses).

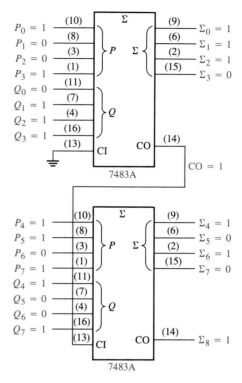

SECTION 6–2
REVIEW

Answers are found at the end of the chapter.

1. Two 2-bit binary numbers (11 and 10) are applied to a two-bit parallel adder. The carry input is 1. Determine the sum (Σ) and the carry output (CO).

2. Draw a diagram showing 7482s connected to form an eight-bit parallel adder.

RIPPLE CARRY VERSUS LOOK-AHEAD CARRY ADDERS

6–3

*Parallel adders can be placed into two categories based on the way in which internal carries from stage to stage are handled. Those categories are **ripple carry** and **look-ahead carry**. Externally, both types of adders are the same in terms of inputs and outputs. The difference is the speed at which they can add numbers. The look-ahead carry adder is much faster than the ripple carry adder. You will see why in this section. This material can be treated as optional.*

The Ripple Carry Adder

The 7482, covered in the last section, is an example of a two-bit *ripple carry* adder, in which the carry output of each full-adder stage is connected to the carry input of the

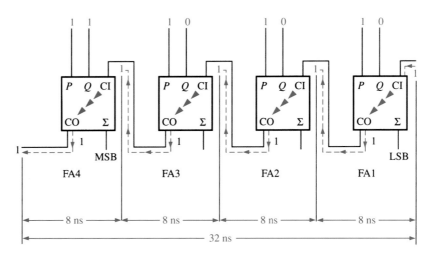

FIGURE 6–14
A four-bit parallel binary ripple carry adder showing "worst-case" carry propagation delays.

next higher-order stage. The sum and carry outputs of any stage cannot be produced until the carry input occurs; this leads to a time delay in the addition process, as illustrated in Figure 6–14. The carry propagation delay for each full-adder is the time from the application of the carry input until the carry output occurs, assuming that the P and Q inputs are present.

Full-adder 1 (FA1) cannot produce a potential carry output until a carry input is applied. Full-adder 2 (FA2) cannot produce a potential carry output until full-adder 1 produces a carry output. Full-adder 3 (FA3) cannot produce a potential carry output until a carry output is produced by full-adder 1 followed by a carry output from full-adder 2, and so on. As you can see, the carry input to the least significant stage has to *ripple* through all the adders before a final sum is produced. A cumulative delay through all the adder stages is a "worst-case" addition time. The total delay can vary, depending on the carry produced at each stage. If two numbers are added such that no carries occur between stages, the addition time is simply the propagation time through a single full-adder from the application of the data bits on the inputs to the occurrence of a sum output.

The Look-Ahead-Carry Adder

As you have seen, the speed with which an addition can be performed is limited by the time required for the carries to propagate, or ripple, through all the stages of the adder. One method of speeding up the process by eliminating this ripple carry delay is called *look-ahead carry* addition; this method is based on two functions of the full-adder, called the *carry generate* (CG) and the *carry propagate* (CP) functions.

The carry generate (CG) function indicates when an output carry is produced (generated) internally by the full-adder. A carry is *generated* only when both input bits are 1s. This condition is expressed as the AND function of the two input bits, P and Q:

$$CG = PQ \qquad\qquad \textbf{(6–5)}$$

A carry input may be *propagated* by the full-adder when either or both of the input bits are 1s. This condition is expressed as the OR function of the input bits:

$$CP = P + Q \qquad \text{(6-6)}$$

The carry generate and carry propagate conditions are illustrated in Figure 6–15.

How can the carry output of a full-adder be expressed in terms of the carry generate (CG) and the carry propagate (CP)? The carry output (CO) is a 1 if the carry generate is a 1 OR if the carry propagate is a 1 AND the carry input (CI) is a 1. In other words, we get a carry output of 1 if it is generated by the full-adder ($P = 1$ AND $Q = 1$) or if the adder can propagate the carry input ($P = 1$ OR $Q = 1$) AND CI = 1. This relationship is expressed as

$$CO = CG + CP \cdot CI \qquad \text{(6-7)}$$

Now we will see how this concept can be applied to a parallel adder, whose individual stages are shown in Figure 6–16 for a four-bit example. For each full-adder, the carry output is dependent on its carry generate (CG), its carry propagate (CP), and

FIGURE 6 –15
Illustration of carry generate and carry propagate conditions.

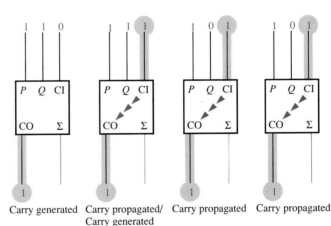

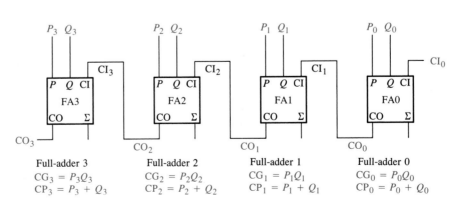

FIGURE 6–16
Carry generate and carry propagate functions in terms of the input bits to a four-bit adder.

its carry input (CI). The CG and CP functions for each stage are immediately available as soon as the input bits P and Q and the input carry to the LSB adder are applied, because they are dependent only on these bits. The carry input to each stage is the carry output of the previous stage. Based on this analysis, we will now develop expressions for the carry out, CO, of each full-adder stage for the four-bit example.

Full-adder 0:

$$CO_0 = CG_0 + CP_0CI_0 \tag{6-8}$$

Full-adder 1:

$$CI_1 = CO_0$$
$$CO_1 = CG_1 + CP_1CI_1$$
$$= CG_1 + CP_1CO_0$$
$$= CG_1 + CP_1(CG_0 + CP_0CI_0)$$
$$CO_1 = CG_1 + CP_1CG_0 + CP_1CP_0CI_0 \tag{6-9}$$

Full-adder 2:

$$CI_2 = CO_1$$
$$CO_2 = CG_2 + CP_2CO_1$$
$$= CG_2 + CP_2(CG_1 + CP_1G_0 + CP_1CP_0CI_0)$$
$$CO_2 = CG_2 + CP_2CG_1 + CP_2CP_1CG_0 + CP_2CP_1CP_0CI_0 \tag{6-10}$$

Full-adder 3:

$$CI_3 = CO_2$$
$$CO_3 = CG_3 + CP_3CI_3$$
$$= CG_3 + CP_3CO_2$$
$$= CG_3 + CP_3(CG_2 + CP_2CG_1 + CP_2CP_1CG_0 + CP_2CP_1CP_0CI_0)$$
$$CO_3 = CG_3 + CP_3CG_2 + CP_3CP_2CG_1 + CP_3CP_2CP_1CG_0 + CP_3CP_2CP_1CP_0CI_0$$
$$\tag{6-11}$$

Notice that in each of these expressions, the carry output for each full-adder stage is dependent only on the initial carry input (CI_0), the CG and CP functions of that stage, and the CG and CP functions of the preceding stages. Since each of the CG and CP functions can be expressed in terms of the P and Q inputs to the full-adders, all the output carries are immediately available (except for gate delays), and we do not have to wait for a carry to ripple through all the stages before a final result is achieved. Thus, the look-ahead carry technique speeds up the addition process.

Equations (6–8) through (6–11) can be implemented with logic gates and connected to the full-adders to create a look-ahead carry adder, as shown in Figure 6–17.

**SECTION 6–3
REVIEW**

Answers are found at the end of the chapter.
1. The input bits to a full-adder are $P = 1$ and $Q = 0$. Determine CG and CP.
2. Determine the carry output of a full-adder when $CI = 1$, $CG = 0$, and $CP = 1$.

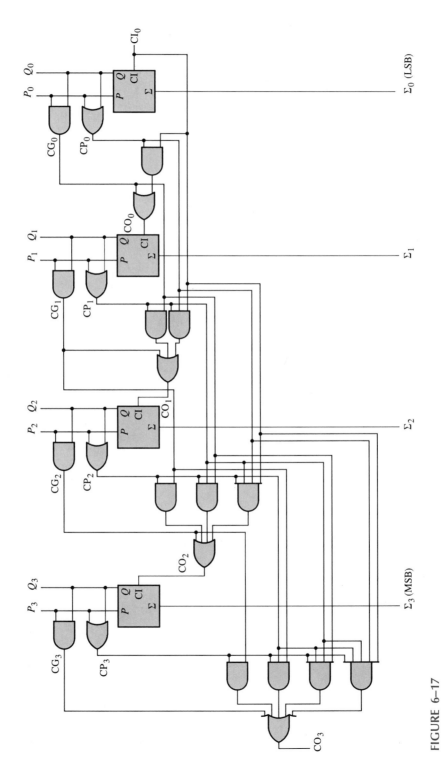

FIGURE 6-17
Logic diagram for a four-stage look-ahead carry adder.

COMPARATORS

6–4

The basic function of a comparator is to compare the magnitudes of two quantities to determine the relationship of those quantities. In its simplest form, a comparator circuit determines whether two numbers are equal.

The exclusive-OR gate can be used as a basic comparator because its output is a 1 if its two input bits are not equal and a 0 if the inputs are equal. Figure 6–18 shows the exclusive-OR as a two-bit comparator.

In order to compare binary numbers containing two bits each, an additional exclusive-OR gate is necessary. The two least significant bits (LSBs) of the two numbers are compared by gate G_1, and the two most significant bits (MSBs) are compared by gate G_2, as shown in Figure 6–19. If the two numbers are equal, their corresponding bits are the same, and the output of each exclusive-OR gate is a 0. If the corresponding sets of bits are not equal, a 1 occurs on that exclusive-OR gate output. In order to produce a *single* output indicating an equality or inequality of two numbers, two inverters and an AND gate can be used, as shown in Figure 6–19. The output of each exclusive-OR gate is inverted and applied to the AND gate input. When the two input bits for each exclusive-OR are equal, the corresponding bits of the numbers are equal, producing a 1 on both inputs to the AND gate and thus a 1 on the output. When the two numbers are not equal, one or both sets of corresponding bits are unequal, and a 0 appears on at least one input to the AND gate to produce a 0 on its output. Thus, the output of the AND gate indicates equality (1) or inequality (0) of the two numbers. Example 6–5 illustrates this operation for two specific cases. The exclusive-OR gate and inverter are replaced by an exclusive-NOR symbol.

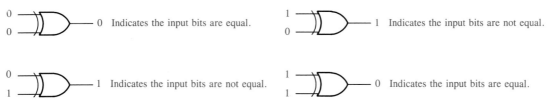

0
0 ── 0 Indicates the input bits are equal.

1
0 ── 1 Indicates the input bits are not equal.

0
1 ── 1 Indicates the input bits are not equal.

1
1 ── 0 Indicates the input bits are equal.

FIGURE 6–18
Basic comparator operation.

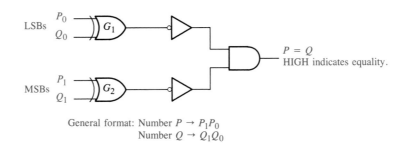

LSBs P_0 Q_0 G_1

MSBs P_1 Q_1 G_2

$P = Q$
HIGH indicates equality.

General format: Number $P \rightarrow P_1 P_0$
Number $Q \rightarrow Q_1 Q_0$

FIGURE 6–19
Logic diagram for comparison of two 2-bit binary numbers.

EXAMPLE 6–5

Apply each of the following sets of binary numbers to comparator inputs, and determine the output by following the logic levels through the circuit.

(a) 10 and 10   **(b)** 11 and 10

FIGURE 6–20

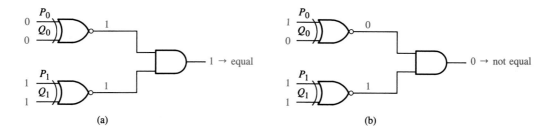

(a)   (b)

Solutions

(a) The output is 1 for inputs 10 and 10, as shown in Figure 6–20(a).

(b) The output is 0 for inputs 11 and 10, as shown in Figure 6–20(b).

The basic comparator circuit can be expanded to any number of bits, as illustrated in Figure 6–21 for two 4-bit numbers. The AND gate sets the condition that all corresponding bits of the two numbers must be equal if the two numbers themselves are equal.

MSI Comparators

Some integrated circuit comparators provide additional outputs that indicate which of the two numbers being compared is the larger. That is, there is an output that indicates when number P is greater than number Q $(P > Q)$ and an output that indicates when number P is less than number Q $(P < Q)$, as shown in the logic symbol for a four-bit comparator in Figure 6–22.

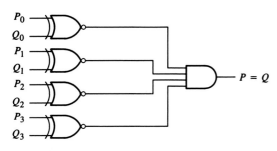

FIGURE 6–21

Logic diagram for the comparison of two 4-bit binary numbers, $P_3P_2P_1P_0$ and $Q_3Q_2Q_1Q_0$.

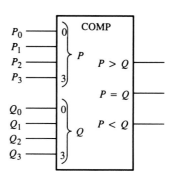

FIGURE 6–22

Logic symbol for a four-bit comparator with inequality indication.

A general method of implementing these two additional output functions is shown in Figure 6–23. In order to understand the logic circuitry required for the $P > Q$ and $P < Q$ functions, let us examine two binary numbers and determine what characterizes an inequality of the numbers.

For our purposes we will use two 4-bit binary numbers with the general format $P_3P_2P_1P_0$ for one number, which we will call number P, and $Q_3Q_2Q_1Q_0$ for the other number, which we will call number Q. To determine an inequality of numbers P and Q, we first examine the highest-order bit in each number. The following conditions are possible:

1. If $P_3 = 1$ and $Q_3 = 0$, number P is greater than number Q.
2. If $P_3 = 0$ and $Q_3 = 1$, number P is less than number Q.
3. If $P_3 = Q_3$, then we must examine the next lower bit position for an inequality.

The three observations are valid for each bit position in the numbers. The general procedure is to check for an inequality in a bit position, starting with the highest order. When such an inequality is found, the relationship of the two numbers is established, and any other inequalities in lower-order bit positions *must be ignored* because it is possible for an opposite indication to occur; the highest-order indication *must take precedence*. To illustrate, let us assume that number P is 0111 and number Q is 1000. Comparison of bits P_3 and Q_3 indicates that $P < Q$ because $P_3 = 0$ and $Q_3 = 1$. However, comparison of bits P_2 and Q_2 indicates $P > Q$ because $P_2 = 1$ and $Q_2 = 0$. The same is true for the remaining lower-order bits. In this case, priority must be given to P_3 and Q_3 because they determine the proper inequality condition.

Figure 6–23 shows a method of comparing two 4-bit numbers and generating a $P > Q$, or a $P < Q$, or a $P = Q$ output. The $P > Q$ condition is determined by gates G_6 through G_{10}. Gate G_6 checks for $P_3 = 1$ and $Q_3 = 0$, and its function is expressed as $P_3\overline{Q_3}$; gate G_7 checks for $P_2 = 1$ and $Q_2 = 0$ ($P_2\overline{Q_2}$); gate G_8 checks for $P_1 = 1$ and $Q_1 = 0$ ($P_1\overline{Q_1}$); gate G_9 checks for $P_0 = 1$ and $Q_0 = 0$ ($P_0\overline{Q_0}$). These conditions all indicate that number P is greater than number Q. The outputs of all these gates are ORed by gate G_{10} to produce the $P > Q$ output.

Notice that the output of gate G_1 is connected to inputs of gates G_7, G_8, and G_9. This provides a *priority inhibit* so that if the proper inequality occurs in bits P_3 and Q_3 ($P_3 < Q_3$), the lower-order bit checks will be inhibited. A priority inhibit is also provided by gate G_2 to gates G_8 and G_9, and by gate G_3 to G_9.

Gates G_{11} through G_{15} check for a $P < Q$ condition. Each AND gate checks a given bit position for the occurrence of a 0 in number P and a 1 in number Q. The AND gate outputs are ORed by gate G_{15} to provide the $P < Q$ output. Priority inhibiting is provided as previously discussed. When all four bits in P equal the bits in Q, the output of each exclusive-OR gate is 1. This enables G_5 to produce a 1 on the $P = Q$ output.

Example 6–6 shows the comparison of specific numbers and indicates the logic levels throughout the circuitry. You should also go through the analysis with numbers of your own choosing to verify the operation.

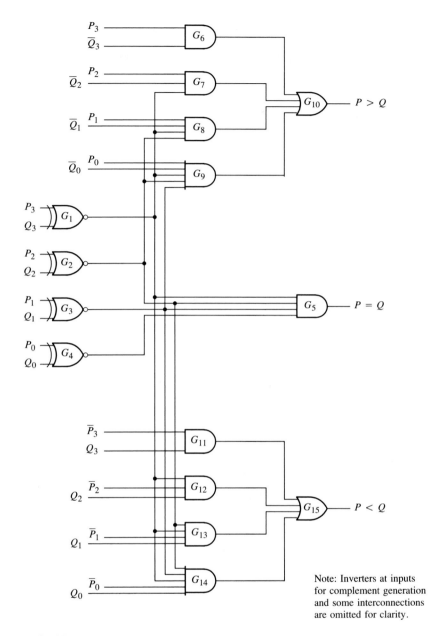

FIGURE 6–23
Logic diagram for a four-bit comparator.

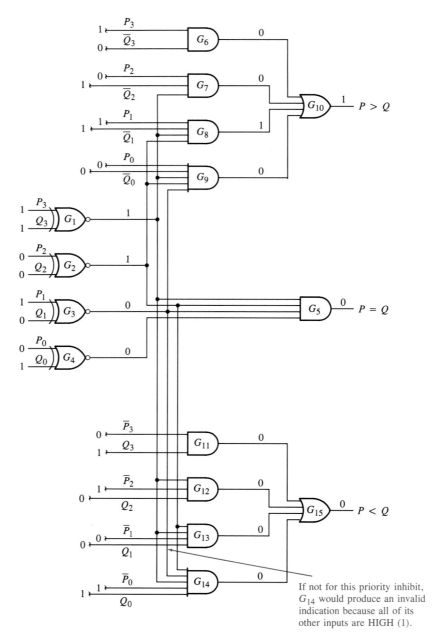

FIGURE 6–24

EXAMPLE 6–6

Analyze the comparator operation for the numbers $P_3P_2P_1P_0 = 1010$ and $Q_3Q_2Q_1Q_0 = 1001$.

Solution Figure 6–24 shows all logic levels within the comparator for the specified inputs.

The 7485 Four-Bit Magnitude Comparator

The 7485 is a representative integrated circuit comparator in the 54/74 family. The logic symbol is shown in Figure 6–25.

Notice that this device has all the inputs and outputs of the generalized comparator just discussed and, in addition, has three *cascading inputs*. ($<$, $=$, $>$). These inputs allow several comparators to be cascaded for comparison of any number of bits greater than four. To expand the comparator, the $P < Q$, $P = Q$, and $P > Q$ outputs of the less significant comparator are connected to the corresponding cascading inputs of the next higher comparator. The least significant comparator must have a HIGH on the $=$ input and LOWs on the $<$ and $>$ inputs.

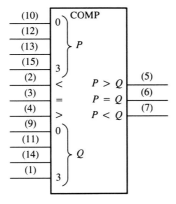

FIGURE 6–25
Logic symbol for the 7485 four-bit magnitude comparator (pin numbers are in parentheses).

EXAMPLE 6–7

Use 7485 comparators to compare the magnitudes of two 8-bit binary numbers. Show the comparators with proper interconnections.

Solution Two 7485s are required for eight bits. They are connected as shown in Figure 6–26, in a cascaded arrangement.

FIGURE 6–26

An eight-bit magnitude comparator using two 7485s.

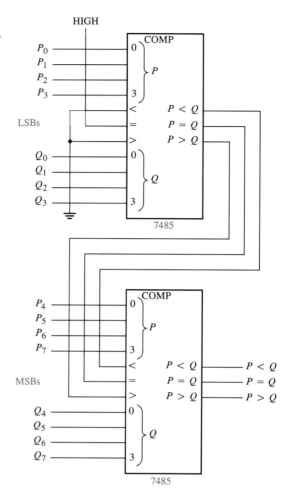

Answers are found at the end of the chapter.

1. The numbers $P = 1011$ and $Q = 1010$ are applied to the inputs of a 7485. Determine the outputs.

2. The numbers $P = 11001011$ and $Q = 11010100$ are applied to the eight-bit comparator in Figure 6–26. Determine the states of pins 5, 6, and 7 on each 7485.

DECODERS

6–5

The basic function of a decoder is to detect the presence of a specified combination of bits (code) on its inputs and to indicate that presence by a specified output level. In its general form a decoder has n input lines to handle n bits and from one to 2^n output lines to indicate the presence of one or more n-bit combinations.

The Basic Binary Decoder

Suppose we wish to determine when a binary 1001 occurs on the inputs of a digital circuit. An AND gate can be used as the basic decoding element because it produces a HIGH output only when all of its inputs are HIGH. Therefore, we must make sure that all of the inputs to the AND gate are HIGH when the binary number 1001 occurs; this can be done by inverting the two middle bits (the 0s), as shown in Figure 6–27.

The logic equation for the decoder of Figure 6–27(a) is developed as illustrated in Figure 6–27(b). You should verify that the output function is 0 except when $A_0 = 1$, $A_1 = 0$, $A_2 = 0$, and $A_3 = 1$ are applied to the inputs. A_0 is the LSB and A_3 is the MSB. In the representation of a binary number or other weighted code in this book, the LSB is always the right-most bit in a horizontal arrangement and the topmost bit in a vertical arrangement, unless specified otherwise.

If a NAND gate is used in place of the AND gate, as shown in Figure 6–28, a LOW output will indicate the presence of the proper binary code.

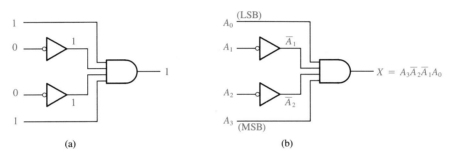

(a)   (b)

FIGURE 6–27
Decoding logic for 1001 with an active-HIGH output.

FIGURE 6–28
Decoding logic for 1001 with an active-LOW output.

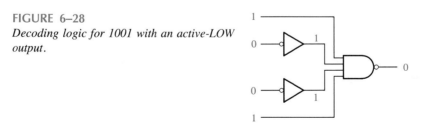

EXAMPLE 6–8

Determine the logic required to decode the binary number 1011 by producing a HIGH indication on the output.

Solution The decoding function can be formed by complementing only the variables that appear as 0 in the binary number, as follows:

$$X = A_3\overline{A}_2A_1A_0$$

FIGURE 6–29
Decoding logic for producing a HIGH output when 1011 is on the inputs.

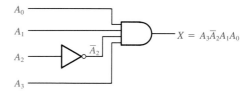

This function can be implemented by connecting the true (uncomplemented) variables A_0, A_1, and A_3 directly to the inputs of an AND gate, and inverting the variable A_2 before applying it to the AND gate input. The decoding logic is shown in Figure 6–29.

The Four-Bit Binary Decoder

In order to decode all possible combinations of four bits, sixteen decoding gates are required ($2^4 = 16$). This type of decoder is commonly called a *4-line-to-16-line* decoder because there are four inputs and sixteen outputs. A list of the sixteen binary code words and their corresponding decoding functions is given in Table 6–4.

If an active-LOW output is desired for each decoded number, the entire decoder can be implemented with NAND gates and inverters as follows:

First, since each variable and its complement are required in the decoder, as seen from Table 6–4, each complement can be generated once and then used for all decoding gates as required, rather than be generated by a separate inverter in each place that complement is used. This arrangement is shown in Figure 6–30.

TABLE 6–4
Decoding functions and truth table for a 4-line-to-16-line decoder.

Decimal Digit	Binary Inputs				Logic Function	Outputs															
	A_3	A_2	A_1	A_0		0	1	2	3	4	5	6	7	8	9	10	11	12	13	14	15
0	0	0	0	0	$\overline{A_3}\,\overline{A_2}\,\overline{A_1}\,\overline{A_0}$	0	1	1	1	1	1	1	1	1	1	1	1	1	1	1	1
1	0	0	0	1	$\overline{A_3}\,\overline{A_2}\,\overline{A_1}A_0$	1	0	1	1	1	1	1	1	1	1	1	1	1	1	1	1
2	0	0	1	0	$\overline{A_3}\,\overline{A_2}A_1\overline{A_0}$	1	1	0	1	1	1	1	1	1	1	1	1	1	1	1	1
3	0	0	1	1	$\overline{A_3}\,\overline{A_2}A_1A_0$	1	1	1	0	1	1	1	1	1	1	1	1	1	1	1	1
4	0	1	0	0	$\overline{A_3}A_2\overline{A_1}\,\overline{A_0}$	1	1	1	1	0	1	1	1	1	1	1	1	1	1	1	1
5	0	1	0	1	$\overline{A_3}A_2\overline{A_1}A_0$	1	1	1	1	1	0	1	1	1	1	1	1	1	1	1	1
6	0	1	1	0	$\overline{A_3}A_2A_1\overline{A_0}$	1	1	1	1	1	1	0	1	1	1	1	1	1	1	1	1
7	0	1	1	1	$\overline{A_3}A_2A_1A_0$	1	1	1	1	1	1	1	0	1	1	1	1	1	1	1	1
8	1	0	0	0	$A_3\overline{A_2}\,\overline{A_1}\,\overline{A_0}$	1	1	1	1	1	1	1	1	0	1	1	1	1	1	1	1
9	1	0	0	1	$A_3\overline{A_2}\,\overline{A_1}A_0$	1	1	1	1	1	1	1	1	1	0	1	1	1	1	1	1
10	1	0	1	0	$A_3\overline{A_2}A_1\overline{A_0}$	1	1	1	1	1	1	1	1	1	1	0	1	1	1	1	1
11	1	0	1	1	$A_3\overline{A_2}A_1A_0$	1	1	1	1	1	1	1	1	1	1	1	0	1	1	1	1
12	1	1	0	0	$A_3A_2\overline{A_1}\,\overline{A_0}$	1	1	1	1	1	1	1	1	1	1	1	1	0	1	1	1
13	1	1	0	1	$A_3A_2\overline{A_1}A_0$	1	1	1	1	1	1	1	1	1	1	1	1	1	0	1	1
14	1	1	1	0	$A_3A_2A_1\overline{A_0}$	1	1	1	1	1	1	1	1	1	1	1	1	1	1	0	1
15	1	1	1	1	$A_3A_2A_1A_0$	1	1	1	1	1	1	1	1	1	1	1	1	1	1	1	0

In order to decode each of the sixteen binary code words, sixteen NAND gates are required (AND gates can be used to produce active-HIGH outputs). The decoding gate arrangement is illustrated in Figure 6–30.

Rather than reproducing the complex logic diagram for the decoder each time it is required in a schematic, a simpler representation is normally used. A logic symbol for a 4-line-to-16-line decoder is shown in Figure 6–31. The BIN/DEC label indicates that a binary input makes the corresponding decimal output active. The input labels 1, 2, 4, and 8 represent the binary weights of the input bits.

FIGURE 6–30
Logic for a 4-line-to-16-line decoder.

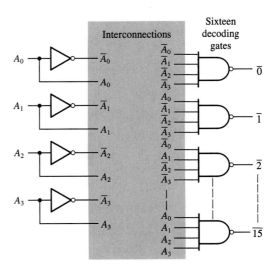

FIGURE 6–31
Logic symbol for a 4-line-to-16-line decoder.

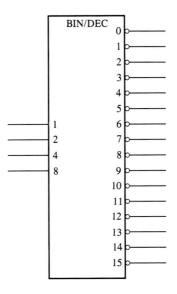

The 74154 4-Line-to-16-Line Decoder

The 74154 is a good example of a TTL MSI decoder. Its logic diagram is shown in Figure 6–32(a), and the logic symbol appears in Figure 6–32(b). The additional inverters on the inputs are required to prevent excessive loading of the driving source(s). Each input is connected to the input of only one inverter, rather than to the inputs of several NAND gates as in Figure 6–30. There is also an enable function provided on this particular device, which is implemented with a NOR gate used as a negative AND. A LOW level on each input, $\overline{G}_1$ and $\overline{G}_2$, is required in order to make the enable gate output (G) HIGH. The enable gate output is connected to an input of *each* NAND gate, so it must be HIGH for the gates to be enabled. If the enable gate is not activated (LOW on both inputs), then all sixteen decoder outputs will be HIGH regardless of the states of the four input variables, A_0, A_1, A_2, and A_3.

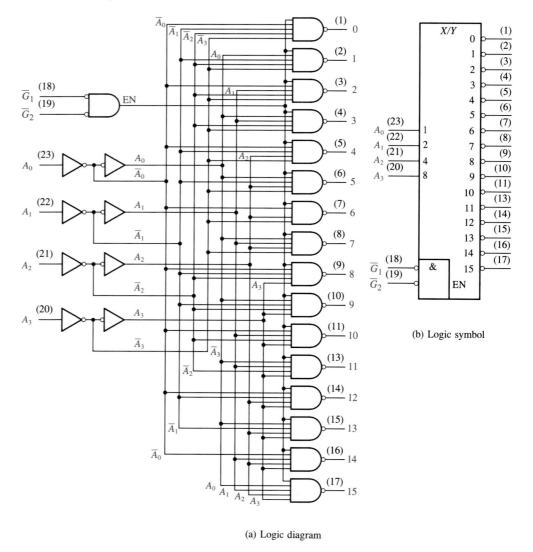

(a) Logic diagram

(b) Logic symbol

FIGURE 6–32
The 74154 4-line-to-16-line decoder.

EXAMPLE 6–9

A certain application requires that a five-bit binary number be decoded. Use 74154 decoders to implement the logic. The binary number is represented as $A_4A_3A_2A_1A_0$.

FIGURE 6–33

A five-bit binary decoder using 74154s.

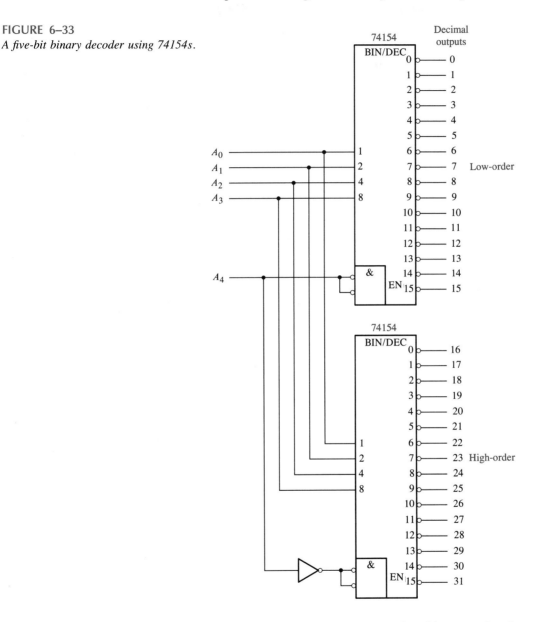

Solution Since the 74154 can handle only four bits, two decoders must be used to decode five bits. The fifth bit, A_4, is connected to the enable inputs, $\overline{G}_1$ and $\overline{G}_2$, of one decoder, and $\overline{A}_4$ is connected to the enable inputs of the other decoder, as shown in Figure 6–33. When the binary number is 15 or less, $A_4 = 0$, and the low-order decoder is enabled and the high-order decoder is disabled. When the binary number is greater than 15, $A_4 = 1$, and the high-order decoder is enabled and the low-order decoder is disabled.

The BCD-to-Decimal Decoder

The BCD-to-decimal decoder converts each BCD code word (8421 code) into one of ten possible decimal digit indications. It is frequently referred to as a *4-line-to-10-line* decoder, although other types of decoders also fall into this category (such as an Excess-3 decoder).

   The method of implementation is essentially the same as for the 4-line-to-16-line decoder previously discussed, except that only ten decoding gates are required because the BCD code represents only the ten decimal digits 0 through 9. A list of the ten BCD code words and their corresponding decoding functions is given in Table 6–5. Each of these decoding functions is implemented with NAND gates to provide active-LOW outputs. If an active-HIGH output is required, AND gates are used for decoding. The logic is identical to that of the first ten decoding gates in the 4-line-to-16-line decoder.

TABLE 6–5
BCD decoding functions.

Decimal Digit	BCD Code				Logic Function
	A_3	A_2	A_1	A_0	
0	0	0	0	0	$\overline{A_3}\,\overline{A_2}\,\overline{A_1}\,\overline{A_0}$
1	0	0	0	1	$\overline{A_3}\,\overline{A_2}\,\overline{A_1}\,A_0$
2	0	0	1	0	$\overline{A_3}\,\overline{A_2}\,A_1\,\overline{A_0}$
3	0	0	1	1	$\overline{A_3}\,\overline{A_2}\,A_1\,A_0$
4	0	1	0	0	$\overline{A_3}\,A_2\,\overline{A_1}\,\overline{A_0}$
5	0	1	0	1	$\overline{A_3}\,A_2\,\overline{A_1}\,A_0$
6	0	1	1	0	$\overline{A_3}\,A_2\,A_1\,\overline{A_0}$
7	0	1	1	1	$\overline{A_3}\,A_2\,A_1\,A_0$
8	1	0	0	0	$A_3\,\overline{A_2}\,\overline{A_1}\,\overline{A_0}$
9	1	0	0	1	$A_3\,\overline{A_2}\,\overline{A_1}\,A_0$

EXAMPLE 6–10

The 7442A is an integrated circuit BCD-to-decimal decoder. The logic symbol is shown in Figure 6–34. If the input waveforms in Figure 6–35 are applied to the inputs of the 7442A, sketch the output waveforms.

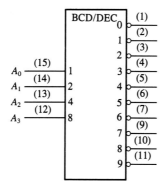

FIGURE 6–34
The 7442A BCD-to-decimal decoder.

FIGURE 6–35

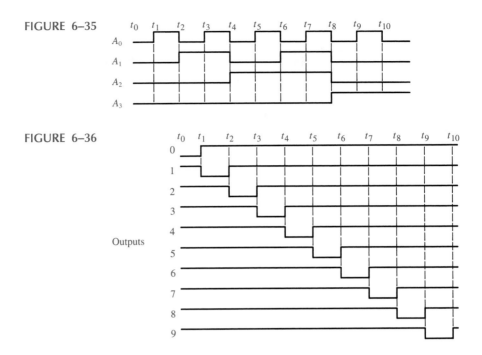

FIGURE 6–36

Outputs

Solution The output waveforms are shown in Figure 6–36. As you can see, the inputs are sequenced through the BCD for digits 0 through 9. The output waveforms indicate that sequence.

The BCD-to-Seven-Segment Decoder/Driver

This type of decoder accepts the BCD code on its inputs and provides outputs to energize seven-segment display devices to produce a decimal readout. Before proceeding with a discussion of this decoder, let us examine the basics of a seven-segment display device.

Figure 6–37 shows a common display format composed of seven elements or segments. Energizing certain combinations of these segments can cause each of the ten decimal digits to be produced. Figure 6–38 illustrates this method of digital display for each of the ten digits by using a darker segment to represent one that is energized. To produce a 1, segments *b* and *c* are energized; to produce a 2, segments *a, b, g, e,* and *d* are used; and so on.

FIGURE 6–37
Seven-segment display format showing arrangement of segments.

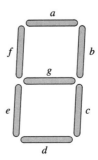

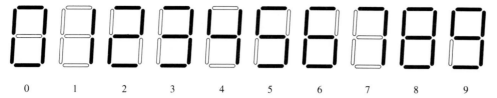

FIGURE 6–38
Display of decimal digits with a seven-segment device.

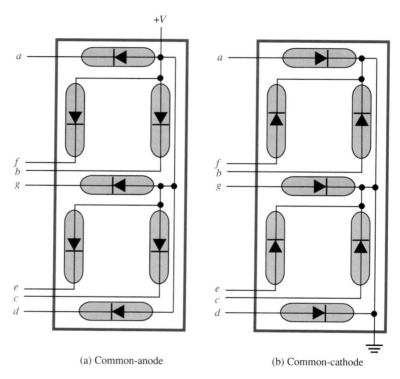

(a) Common-anode (b) Common-cathode

FIGURE 6–39
Arrangements of seven-segment LED displays.

LED displays One common type of seven-segment display consists of light-emitting diodes *(LEDS)* arranged as shown in Figure 6–39. Each segment is an LED that emits light when there is current through it. In Figure 6–39(a) the common-anode arrangement requires the driving circuit to provide a LOW level voltage in order to activate a given segment. When a LOW is applied to a segment input, the LED is turned on, and there is current through it. In Figure 6–39(b) the common-cathode arrangement requires the driver to provide a HIGH level voltage to activate a segment. When a HIGH is applied to a segment input, the LED is turned on and there is current through it.

The activated segments for each of the ten decimal digits are listed in Table 6–6. We will now examine the decoding logic required to produce the format for a seven-segment display with a BCD input. Notice that segment *a* is activated for digits 0, 2, 3, 5, 6, 7, 8, and 9; segment *b* is activated for digits 0, 1, 2, 3, 4, 7, 8, and 9; and so on. If we let the BCD inputs to the decoder be represented by the general form

$A_3A_2A_1A_0$, a Boolean expression can be found for each segment in the display, and this will tell us the logic circuitry required to drive or activate each segment. For example, the equation for segment a is as follows:

$$a = \overline{A_3}\overline{A_2}\overline{A_1}\overline{A_0} + \overline{A_3}\overline{A_2}A_1\overline{A_0} + \overline{A_3}\overline{A_2}A_1A_0 + \overline{A_3}A_2\overline{A_1}A_0$$
$$+ \overline{A_3}A_2A_1\overline{A_0} + \overline{A_3}A_2A_1A_0 + A_3\overline{A_2}\overline{A_1}\overline{A_0} + A_3\overline{A_2}\overline{A_1}A_0$$

This equation says that segment a is activated, or "true," if the BCD code is "0 OR 2 OR 3 OR 5 OR 6 OR 7 OR 8 OR 9." Table 6–7 lists the logic function for each of the seven segments.

TABLE 6–6
Seven-segment display format.

Digit	Segments Activated
0	a, b, c, d, e, f
1	b, c
2	a, b, g, e, d
3	a, b, c, d, g
4	b, c, f, g
5	a, c, d, f, g
6	a, c, d, e, f, g
7	a, b, c
8	a, b, c, d, e, f, g
9	a, b, c, d, f, g

TABLE 6–7
Seven-segment decoding functions.

Segment	Used in Digits	Logic Function
a	0, 2, 3, 5, 6, 7, 8, 9	$\overline{A_3}\overline{A_2}\overline{A_1}\overline{A_0} + \overline{A_3}\overline{A_2}A_1\overline{A_0} + \overline{A_3}\overline{A_2}A_1A_0 + \overline{A_3}A_2\overline{A_1}A_0 + \overline{A_3}A_2A_1\overline{A_0} + \overline{A_3}A_2A_1A_0 + A_3\overline{A_2}\overline{A_1}\overline{A_0} + A_3\overline{A_2}\overline{A_1}A_0$
b	0, 1, 2, 3, 4, 7, 8, 9	$\overline{A_3}\overline{A_2}\overline{A_1}\overline{A_0} + \overline{A_3}\overline{A_2}\overline{A_1}A_0 + \overline{A_3}\overline{A_2}A_1\overline{A_0} + \overline{A_3}\overline{A_2}A_1A_0 + \overline{A_3}A_2\overline{A_1}\overline{A_0} + \overline{A_3}A_2A_1A_0 + A_3\overline{A_2}\overline{A_1}\overline{A_0} + A_3\overline{A_2}\overline{A_1}A_0$
c	0, 1, 3, 4, 5, 6, 7, 8, 9	$\overline{A_3}\overline{A_2}\overline{A_1}\overline{A_0} + \overline{A_3}\overline{A_2}\overline{A_1}A_0 + \overline{A_3}\overline{A_2}A_1A_0 + \overline{A_3}A_2\overline{A_1}\overline{A_0} + \overline{A_3}A_2\overline{A_1}A_0 + \overline{A_3}A_2A_1\overline{A_0} + \overline{A_3}A_2A_1A_0 + A_3\overline{A_2}\overline{A_1}\overline{A_0} + A_3\overline{A_2}\overline{A_1}A_0$
d	0, 2, 3, 5, 6, 8, 9	$\overline{A_3}\overline{A_2}\overline{A_1}\overline{A_0} + \overline{A_3}\overline{A_2}A_1\overline{A_0} + \overline{A_3}\overline{A_2}A_1A_0 + \overline{A_3}A_2\overline{A_1}A_0 + \overline{A_3}A_2A_1\overline{A_0} + A_3\overline{A_2}\overline{A_1}\overline{A_0} + A_3\overline{A_2}\overline{A_1}A_0$
e	0, 2, 6, 8	$\overline{A_3}\overline{A_2}\overline{A_1}\overline{A_0} + \overline{A_3}\overline{A_2}A_1\overline{A_0} + \overline{A_3}A_2A_1\overline{A_0} + A_3\overline{A_2}\overline{A_1}\overline{A_0}$
f	0, 4, 5, 6, 8, 9	$\overline{A_3}\overline{A_2}\overline{A_1}\overline{A_0} + \overline{A_3}A_2\overline{A_1}\overline{A_0}, + \overline{A_3}A_2\overline{A_1}A_0 + \overline{A_3}A_2A_1\overline{A_0} + A_3\overline{A_2}\overline{A_1}\overline{A_0} + A_3\overline{A_2}\overline{A_1}A_0$
g	2, 3, 4, 5, 6, 8, 9	$\overline{A_3}\overline{A_2}A_1\overline{A_0} + \overline{A_3}\overline{A_2}A_1A_0 + \overline{A_3}A_2\overline{A_1}\overline{A_0} + \overline{A_3}A_2\overline{A_1}A_0 + \overline{A_3}A_2A_1\overline{A_0} + A_3\overline{A_2}\overline{A_1}\overline{A_0} + A_3\overline{A_2}\overline{A_1}A_0$

From the expressions in Table 6–7, the logic for the BCD-to-seven-segment decoder can be implemented. Each of the ten BCD codes is decoded, and then the decoding gates are ORed as dictated by the logic expression for each segment. For instance, segment a requires that the decoded BCD digits 0, 2, 3, 5, 6, 7, 8, and 9 be ORed.

The full SOP expression for segment *a* is first simplified by the use of a K map, as illustrated in Figure 6–40(a), and then implemented with the logic shown in part (b).

Segment *b* requires that the BCD digits 0, 1, 2, 3, 4, 7, 8, and 9 be ORed; segment *c* requires the ORing of 0, 1, 3, 4, 5, 6, 7, 8, and 9; and so on. A logic symbol for a complete seven-segment decoder is shown in Figure 6–41.

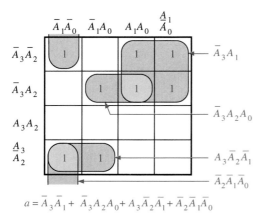

$$a = \bar{A}_3\bar{A}_1 + \bar{A}_3 A_2 A_0 + A_3 \bar{A}_2 \bar{A}_1 + \bar{A}_2 \bar{A}_1 \bar{A}_0$$

(a) K-map simplification of the full SOP expression for segment *a*.

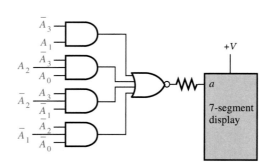

(b) Decoding logic for segment *a* with an active-LOW output for compatibility with the common-anode-type LED display.

FIGURE 6–40

*Implementation of the segment-*a* decoder.*

FIGURE 6–41

Logic symbol for a seven-segment decoder/ driver.

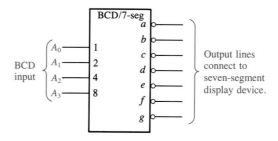

LCD displays Another common type of seven-segment display is the *liquid crystal display*. LCDs operate by polarizing light so that a nonactivated segment reflects incident light and thus appears invisible against its background. An activated segment does not reflect incident light and thus appears dark. LCDs consume much less power than LEDs but cannot be seen in the dark, while LEDs can.

LCDs operate from a low-frequency signal voltage (30 Hz to 60 Hz) applied between the segment and a common element called the *backplane* (bp). The basic operation is as follows: Figure 6–42 shows a square wave used as the source signal. Each segment in the display is driven by an exclusive-OR gate with one input connected to an output of the seven-segment decoder/driver and the other input connected to the signal source. When the decoder/driver output is HIGH (1), the exclusive-OR output is a square wave that is 180° out-of-phase with the source signal, as shown in Figure 6–42(a). You can verify this by reviewing the truth table operation of the exclusive-OR. The resulting voltage between the LCD segment and the backplane is also a square

FIGURE 6–42
Basic operation of an LCD.

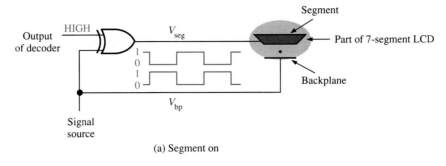

(a) Segment on

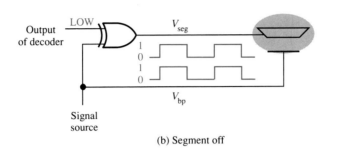

(b) Segment off

wave because when $V_{seg} = 1$, $V_{bp} = 0$, and vice versa. The voltage difference turns the segment on.

When the decoder/driver output is LOW (0), the exclusive-OR output is a square wave that is in-phase with the source signal, as shown in Figure 6–42(b). The resulting voltage difference between the segment and the backplane is 0 because $V_{seg} = V_{bp}$. This condition turns the segment off.

For driving LCDs, TTL is not recommended, because its LOW level voltage is typically a few tenths of a volt, thus creating a dc component across the LCD, which degrades its performance. Therefore, CMOS is used in LCD applications.

Zero suppression An additional feature found on many seven-segment decoders is the *zero suppression* logic. This extra function is useful in multidigit displays because it is used to blank out unnecessary zeros in the display. For example, the number 0006.400 would be displayed as 6.4, which is read more easily. Blanking of the zeros on the front of the number is called leading zero suppression, and blanking of the zeros after the number is called trailing zero suppression.

A block diagram of a four-digit display is shown in Figure 6–43. It will be used to illustrate the requirements for leading zero suppression. Notice that two additional functions have been added to each BCD-to-seven-segment decoder, a *ripple blanking input* (RBI) and a *ripple blanking output* (RBO). The highest-order digit position is always blanked if a zero code appears on its BCD inputs *and* the blanking input is HIGH. Each lower-order digit position is blanked if a zero code appears on its BCD inputs *and* the next higher-order digit is a zero as indicated by a HIGH on its blanking output. The ripple blanking output of any decoder indicates that it has a BCD zero on its inputs and that *all* higher-order digits are also zero. The blanking output of each stage is connected to the blanking input of the next lower-order stage, as shown in the

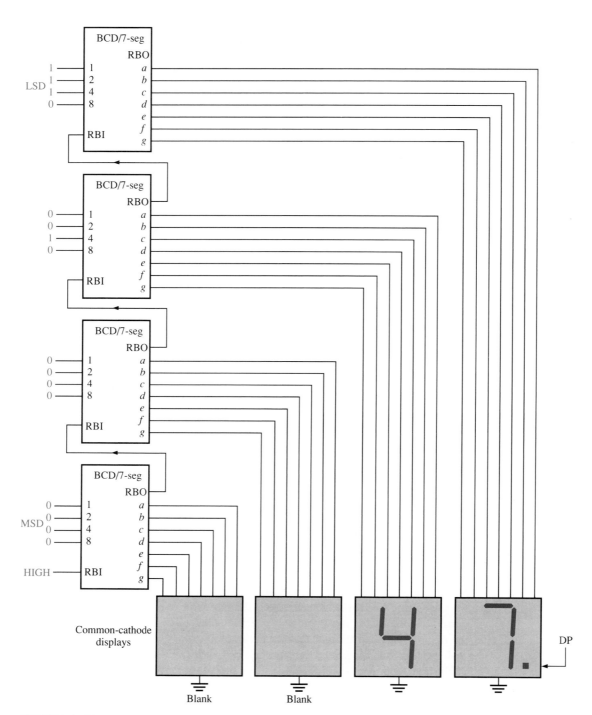

FIGURE 6–43
Illustration of leading zero suppression.

diagram. As an example, in Figure 6–43 the highest-order digit is a zero, which is therefore blanked. Also, the next digit is a zero, and because the highest-order digit is a zero, the second digit is also blanked. The remaining two digits are displayed.

For the fractional portion of a display (the digits to the right of the decimal point), trailing zero suppression is used. That is, the lowest-order digit is blanked if it is a zero, and each digit that is a zero *and* is followed by zeros in all the lower-order positions is also blanked. To illustrate, Figure 6–44 shows a block diagram in which there are three digits to the right of the decimal point. In this example, the lowest-order digit is blanked because it is a zero. The next digit is also a zero, and its blanking input is HIGH, so it is blanked. The highest-order digit is a 5, and it is displayed. Notice that the blanking output of each decoder stage is connected to the blanking input of the next higher-order stage.

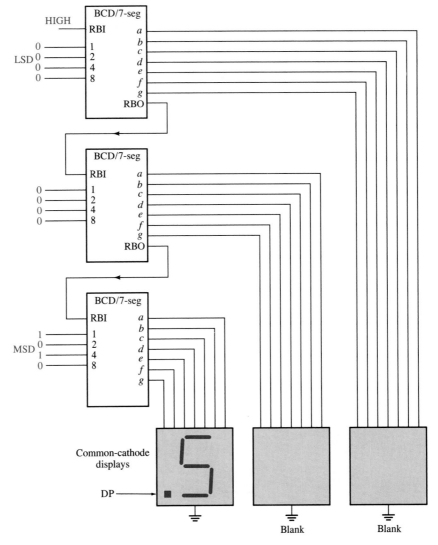

FIGURE 6–44
Illustration of trailing zero suppression.

In this section several decoders have been introduced. The basic principles can be extended to other types, such as the 2-line-to-4-line decoder, the 3-line-to-8-line decoder, and the Excess-3-to-decimal 4-line-to-10-line decoder, all of which are available in MSI.

**SECTION 6–5
REVIEW**

Answers are found at the end of the chapter.
1. A 3-line-to-8-line decoder can be used for octal-to-decimal decoding. When a binary 101 is on the inputs, which output line is activated?
2. How many 74154 4-line-to-16-line decoders are necessary to decode a six-bit binary number?
3. Would you select a decoder/driver with active-HIGH or active-LOW outputs to drive a common-anode seven-segment LED display?

ENCODERS

6–6

*An **encoder** is a combinational logic circuit that essentially performs a "reverse" decoder function. An encoder accepts an active level on one of its inputs representing a digit, such as a decimal or octal digit, and converts it to a coded output, such as binary or BCD. Encoders can also be devised to encode various symbols and alphabetic characters. The process of converting from familiar symbols or numbers to a coded format is called **encoding**.*

The Decimal-to-BCD Encoder

This type of encoder has ten inputs—one for each decimal digit—and four outputs corresponding to the BCD code, as shown in Figure 6–45. This is a basic 10-line-to-4-line encoder.

The BCD (8421) code is listed in Table 6–8, and from this table we can determine the relationship between each BCD bit and the decimal digits. For instance, the most

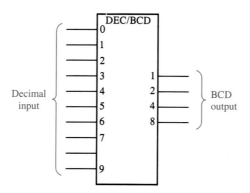

FIGURE 6–45
Logic symbol for a decimal-to-BCD encoder.

significant bit of the BCD code, A_3, is a 1 for decimal digit 8 or 9. The OR expression for bit A_3 in terms of the decimal digits can therefore be written

$$A_3 = 8 + 9$$

Bit A_2 is a 1 for decimal digit 4, 5, 6, or 7 and can be expressed as an OR function as follows:

$$A_2 = 4 + 5 + 6 + 7$$

Bit A_1 is a 1 for decimal digit 2, 3, 6, or 7 and can be expressed as

$$A_1 = 2 + 3 + 6 + 7$$

Finally, A_0 is a 1 for digit 1, 3, 5, 7, or 9. The expression for A_0 is

$$A_0 = 1 + 3 + 5 + 7 + 9$$

TABLE 6–8

Decimal Digit	BCD Code			
	A_3	A_2	A_1	A_0
0	0	0	0	0
1	0	0	0	1
2	0	0	1	0
3	0	0	1	1
4	0	1	0	0
5	0	1	0	1
6	0	1	1	0
7	0	1	1	1
8	1	0	0	0
9	1	0	0	1

Now we can implement the logic circuitry required for encoding each decimal digit to a BCD code by using the logic expressions just developed. It is simply a matter of ORing the appropriate decimal digit input lines to form each BCD output. The basic encoder logic resulting from these expressions is shown in Figure 6–46.

FIGURE 6–46

Basic logic diagram of a decimal-to-BCD encoder. A 0-digit input is not needed because the BCD outputs are all LOW when there are no HIGH inputs.

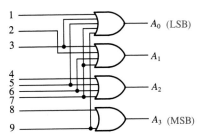

The basic operation is as follows: When a HIGH appears on one of the decimal digit input lines, the appropriate levels occur on the four BCD output lines. For instance, if input line 9 is HIGH (assuming all other input lines are LOW), this condition will produce a HIGH on outputs A_0 and A_3 and LOWs on outputs A_1 and A_2, which is the BCD code (1001) for decimal 9.

The Decimal-to-BCD Priority Encoder

The decimal-to-BCD *priority encoder* performs the same basic encoding function as that previously discussed. It also offers additional flexibility in that it can be used in applications requiring priority detection. The priority function means that the encoder will produce a BCD output corresponding to the highest-order decimal digit appearing on the inputs, and will ignore all others. For instance, if the 6 and the 3 inputs are both HIGH, the BCD output is 0110 (which represents decimal 6).

Now let us look at the requirements for the priority detection logic. The purpose of this logic circuitry is to prevent a lower-order digit input from disrupting the encoding of a higher-order digit. We will start by examining each BCD output (beginning with output A_0). Referring to Figure 6–46, note that A_0 is HIGH when 1, 3, 5, 7, or 9 is HIGH. Digit input 1 must be allowed to activate the A_0 output only if no higher-order digits *other than those that also activate A_0* are HIGH. This can be stated as follows:

1. A_0 is HIGH if 1 is HIGH and 2, 4, 6, and 8 are LOW.
2. A_0 is HIGH if 3 is HIGH and 4, 6, and 8 are LOW.
3. A_0 is HIGH if 5 is HIGH and 6 and 8 are LOW.
4. A_0 is HIGH if 7 is HIGH and 8 is LOW.
5. A_0 is HIGH if 9 is HIGH.

These five statements describe the priority of encoding for the BCD bit A_0. The A_0 output is HIGH if any of the conditions listed occur; that is, A_0 is true if statement 1, statement 2, statement 3, statement 4, or statement 5 is true. This can be expressed in the form of the following logic equation:

$$A_0 = 1 \cdot \overline{2} \cdot \overline{4} \cdot \overline{6} \cdot \overline{8} + 3 \cdot \overline{4} \cdot \overline{6} \cdot \overline{8} + 5 \cdot \overline{6} \cdot \overline{8} + 7 \cdot \overline{8} + 9$$

From this expression the logic circuitry required for the A_0 output with priority inhibits can be readily implemented, as shown in Figure 6–47.

The same reasoning process can be applied to output A_1, and the following logical statements can be made:

1. A_1 is HIGH if 2 is HIGH and 4, 5, 8, and 9 are LOW.
2. A_1 is HIGH if 3 is HIGH and 4, 5, 8, and 9 are LOW.
3. A_1 is HIGH if 6 is HIGH and 8 and 9 are LOW.
4. A_1 is HIGH if 7 is HIGH and 8 and 9 are LOW.

These statements are summarized in the following equation, and the logic implementation is shown in Figure 6–48.

$$A_1 = 2 \cdot \overline{4} \cdot \overline{5} \cdot \overline{8} \cdot \overline{9} + 3 \cdot \overline{4} \cdot \overline{5} \cdot \overline{8} \cdot \overline{9} + 6 \cdot \overline{8} \cdot \overline{9} + 7 \cdot \overline{8} \cdot \overline{9}$$

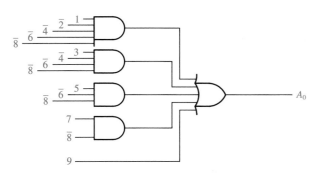

FIGURE 6–47
Logic for the A_0 output of a decimal-to-BCD priority encoder.

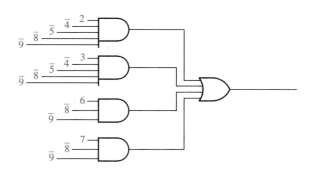

FIGURE 6–48
Logic for the A_1 output of a decimal-to-BCD priority encoder.

Output A_2 can be described as follows:

1. A_2 is HIGH if 4 is HIGH and 8 and 9 are LOW.
2. A_2 is HIGH if 5 is HIGH and 8 and 9 are LOW.
3. A_2 is HIGH if 6 is HIGH and 8 and 9 are LOW.
4. A_2 is HIGH if 7 is HIGH and 8 and 9 are LOW.

In equation form, output A_2 is

$$A_2 = 4 \cdot \overline{8} \cdot \overline{9} + 5 \cdot \overline{8} \cdot \overline{9} + 6 \cdot \overline{8} \cdot \overline{9} + 7 \cdot \overline{8} \cdot \overline{9}$$

The logic circuitry for the A_2 output appears in Figure 6–49.
Finally, for the A_3 output,

$$A_3 \text{ is HIGH if 8 is HIGH or if 9 is HIGH.}$$

This statement appears in equation form as follows:

$$A_3 = 8 + 9$$

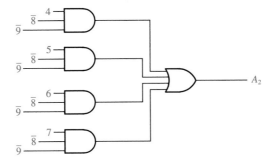

FIGURE 6–49
Logic for the A_2 output of a decimal-to-BCD priority encoder.

The logic for this output is shown in Figure 6–50. No inhibits are required for this one. We now have developed the basic logic for the decimal-to-BCD priority encoder. All the complements of the input digits are realized by inverting the inputs.

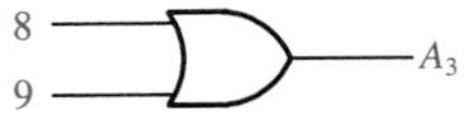

FIGURE 6–50

Logic for the A_3 output of a decimal-to-BCD priority encoder.

EXAMPLE 6–11

The 74147 is a decimal-to-BCD priority encoder. As indicated on the logic symbol in Figure 6–51, its inputs and outputs are all active-LOW.

If LOW levels appear on pins 1, 4, and 13, indicate the state of the four outputs. All other inputs are HIGH. Pin numbers are in parentheses.

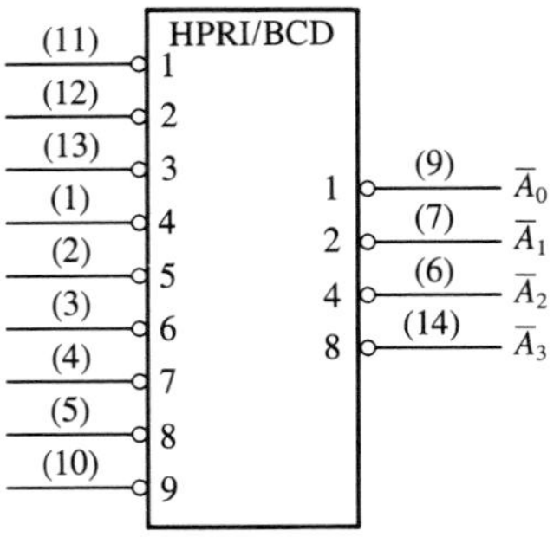

FIGURE 6–51

Logic symbol for the 74147 decimal-to-BCD priority encoder (HPRI means highest value input has priority).

Solution Pin 4 is the highest-order decimal input having a LOW level and represents decimal 7. Therefore, the output levels indicate the BCD code for decimal 7 where $\overline{A}_0$ is the LSB and $\overline{A}_3$ is the MSB. Output $\overline{A}_0$ is LOW, $\overline{A}_1$ is LOW, $\overline{A}_2$ is LOW, and $\overline{A}_3$ is HIGH.

An Encoder Application

A classic application example is a keyboard encoder. The ten decimal digits on the keyboard of a calculator, for example, must be encoded for processing by the logic circuitry. When one of the keypads is pressed, the decimal digit is encoded to the corresponding BCD code. Figure 6–52 shows a simple keyboard encoder arrangement using a 74147 priority encoder. Notice that there are ten push-button switches, each with a *pull-up* resistor to $+V$. The pull-up resistor ensures that the line is HIGH when a switch is not depressed. When a switch is depressed, the line is connected to ground, and a LOW is applied to the corresponding encoder input. The zero key is not connected because the BCD output represents zero when none of the other keys are depressed.

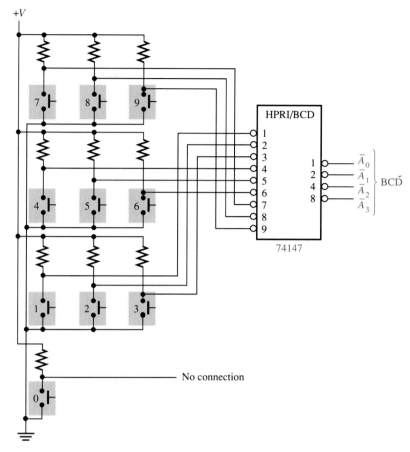

FIGURE 6–52
A simplified keyboard encoder.

The BCD output of the encoder goes into a storage device, and each successive BCD code is stored until the entire number has been entered. Methods of storing BCD numbers and binary data are covered in the following chapters.

SECTION 6–6
REVIEW

Answers are found at the end of the chapter.

1. Suppose that HIGH levels are applied to the 2 input and the 9 input of the circuit in Figure 6–46.
    (a) What are the states of the output lines?
    (b) Does this represent a valid BCD code?
    (c) What is the restriction on the encoder logic in Figure 6–46?
2. (a) What is the $\overline{A}_3\overline{A}_2\overline{A}_1\overline{A}_0$ output when LOWs are applied to pins 1 and 5 of the 74147 in Figure 6–51?
    (b) What does this output represent?

CODE CONVERTERS

6–7

In this section we will examine some methods of using combinational logic circuits to convert from one code to another.

BCD-to-Binary Conversion

One method of BCD-to-binary code conversion involves the use of adder circuits. The basic conversion process is as follows:

1. the value, or weight, of each bit in the BCD number is represented by a binary number.
2. All of the binary representations of bits that are 1s in the BCD number are added.
3. The result of this addition is the binary equivalent of the BCD number.

A more concise statement of this operation is that the binary numbers representing the weights of the BCD bits are summed to produce the total binary number.

We will examine an eight-bit BCD code (one that represents a two-digit decimal number) in order to understand the relationship between BCD and binary. For instance, we already know that the decimal number 87 can be expressed in BCD as

$$\underbrace{1000}_{8}\underbrace{0111}_{7}$$

The left-most four-bit group represents 80, and the right-most four-bit group represents 7. That is, the left-most group has a weight of 10, and the right-most has a weight of 1. Within each group, the binary weight of each bit is as follows:

	Tens Digit				*Units Digit*			
Weight:	80	40	20	10	8	4	2	1
Bit designation:	B_3	B_2	B_1	B_0	A_3	A_2	A_1	A_0

The binary equivalent of each BCD bit is a binary number representing the *weight* of that bit within the total BCD number. This representation is given in Table 6–9.

TABLE 6–9
Binary representations of BCD bit weights.

BCD Bit	BCD Weight	Binary Representation						
		64	32	16	8	4	2	1
A_0	1	0	0	0	0	0	0	1
A_1	2	0	0	0	0	0	1	0
A_2	4	0	0	0	0	1	0	0
A_3	8	0	0	0	1	0	0	0
B_0	10	0	0	0	1	0	1	0
B_1	20	0	0	1	0	1	0	0
B_2	40	0	1	0	1	0	0	0
B_3	80	1	0	1	0	0	0	0

If the binary representations for the weights of all the 1s in the BCD number are added, the result is the binary number corresponding to the BCD number. Example 6–12 illustrates this.

EXAMPLE 6–12

Convert the BCD numbers 00100111 (decimal 27) and 10011000 (decimal 98) to binary.

Solutions Write the binary representations of the weights of all 1s appearing in the numbers, and then add them together.

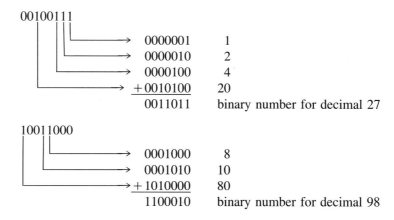

With this basic procedure in mind, let us determine how the process can be implemented with logic circuits. Once the binary representation for each 1 in the BCD number is determined, adder circuits can be used to add the 1s in each column of the binary representation. The 1s occur in a given column only when the corresponding BCD bit is a 1. The occurrence of a BCD 1 can therefore be used to generate the proper binary 1 in the appropriate column of the adder structure. To handle a two-decimal-digit (two-decade) BCD code, eight BCD input lines and seven binary outputs are required. (It takes seven binary bits to represent numbers through ninety-nine.)

Referring to Table 6–9, notice that the "1" (LSB) column of the binary representation has only a single 1 and no possibility of an input carry, so that a straight connection from the A_0 bit of the BCD input to the least significant binary output is sufficient. In the "2" column of the binary representation, the possible occurrence of the two 1s can be accommodated by adding the A_1 bit and the B_0 bit of the BCD number. In the "4" column of the binary representation, the possible occurrence of the two 1s is handled by adding the A_2 bit and the B_1 bit of the BCD number. In the "8" column of the binary representation, the possibility of the three 1s is handled by adding the A_3, B_0, and B_2 bits of the BCD number. In the "16" column, the B_1 and the B_3 bits are added. In the "32" column, only a single 1 is possible, so the B_2 bit is added to the carry from the "16" column. In the "64" column, only a single 1 can occur, so the B_3 bit is added only to the carry from the "32" column. A method of implementing these requirements with full-adders is shown in Figure 6–53.

Most MSI code converters are implemented with preprogrammed read only memories (ROMs). These are covered in a later chapter.

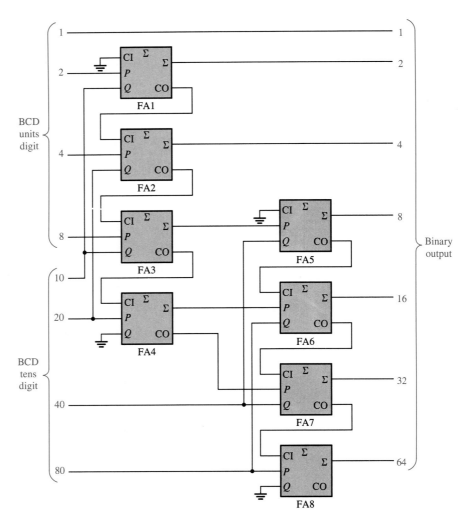

FIGURE 6–53
Two-digit BCD-to-binary converter using full-adders.

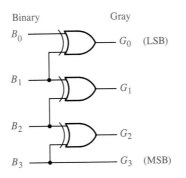

FIGURE 6–54
Four-bit binary-to-Gray conversion logic.

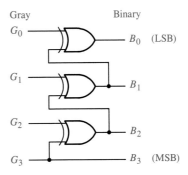

FIGURE 6–55
Four-bit Gray-to-binary conversion logic.

Binary-to-Gray and Gray-to-Binary Conversion

The basic process for Gray-binary conversions was covered in Chapter 2. We will now see how exclusive-OR gates can be used for these conversions. Figure 6–54 shows a four-bit binary-to-Gray code converter, and Figure 6–55 illustrates a four-bit Gray-to-binary converter.

EXAMPLE 6–13

Convert the following binary numbers to Gray code by using exclusive-OR gates:
(a) 0101   **(b)** 00111   **(c)** 101011

Solutions

(a) 0101_2 is 0111 Gray.
(b) 00111_2 is 00100 Gray.
(c) 101011_2 is 111110 Gray.
See Figure 6–56.

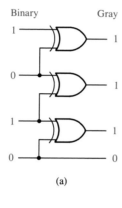

(a)

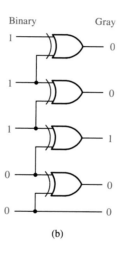

(b)

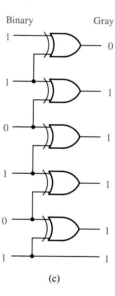

(c)

FIGURE 6–56

EXAMPLE 6–14

Convert the following Gray codes to binary by using exclusive-OR gates:
(a) 1011   **(b)** 11000   **(c)** 1001011

Solutions

(a) 1011 Gray is 1101_2.
(b) 11000 Gray is 10000_2.
(c) 1001011 Gray is 1110010_2.
See Figure 6–57.

FIGURE 6–57

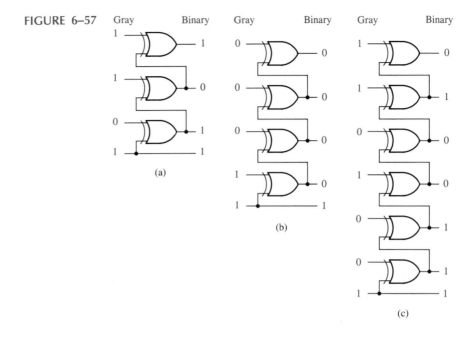

(a)

(b)

(c)

Answers are found at the end of the chapter.
1. Convert the BCD number 10000101 to binary.
2. Draw the logic diagram for converting an eight-bit binary number to Gray code.

MULTIPLEXERS (DATA SELECTORS)

6–8

A **multiplexer (MUX)** *is a device that allows digital information from several sources to be routed onto a single line for transmission over that line to a common destination. The basic multiplexer has several data-input lines and a single output line. It also has* **data-select** *inputs, which permit digital data on any one of the inputs to be switched to the output line.*

A logic symbol for a four-input multiplexer is shown in Figure 6–58. Notice that there are two data-select lines because with two select bits, any of the four data-input lines can be selected.

In reference to Figure 6–58, a two-bit binary code on the data-select inputs will allow the data on the selected data input to pass through to the data output. If a binary 0 ($S_1 = 0$ and $S_0 = 0$) is applied to the data-select lines, the data on input D_0 appear on the data-output line. If a binary 1 ($S_1 = 0$ and $S_0 = 1$) is applied to the data-select lines, the data on input D_1 appear on the data output. If a binary 2 ($S_1 = 1$ and $S_0 = 0$) is applied, the data on D_2 appear on the output. If a binary 3 ($S_1 = 1$ and $S_0 = 1$) is applied, the data on D_3 are switched to the output line. A summary of this operation is given in Table 6–10.

FIGURE 6–58
Logic symbol for a 1-of-4 data selector/multiplexer.

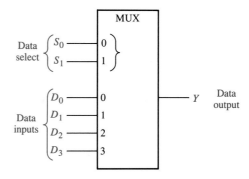

TABLE 6–10
Data selection for a four-input multiplexer.

Data-Select Inputs		
S_1	S_0	Input Selected
0	0	D_0
0	1	D_1
1	0	D_2
1	1	D_3

Now let us look at the logic circuitry required to perform this multiplexing operation. The data output is equal to the state of the *selected* data input. We should, therefore, be able to derive a logical expression for the output in terms of the data input and the select inputs. This can be done as follows:

The data output Y is equal to the data input D_0 if and only if $S_1 = 0$ and $S_0 = 0$:

$$Y = D_0 \overline{S}_1 \overline{S}_0$$

The data output is equal to D_1 if and only if $S_1 = 0$ and $S_0 = 1$:

$$Y = D_1 \overline{S}_1 S_0$$

The data output is equal to D_2 if and only if $S_1 = 1$ and $S_0 = 0$:

$$Y = D_2 S_1 \overline{S}_0$$

The data output is equal to D_3 if and only if $S_1 = 1$ and $S_0 = 1$:

$$Y = D_3 S_1 S_0$$

If these terms are ORed, the total expression for the data output is

$$Y = D_0 \overline{S}_1 \overline{S}_0 + D_1 \overline{S}_1 S_0 + D_2 S_1 \overline{S}_0 + D_3 S_1 S_0$$

The implementation of this equation requires four 3-input AND gates, a 4-input OR gate, and two inverters to generate the complements of S_1 and S_0, as shown in Figure 6–59. Because data can be *selected* from any of the input lines, this circuit is also referred to as a *data selector*.

FIGURE 6–59

Logic diagram for a four-input multiplexer.

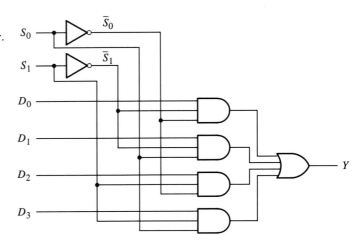

EXAMPLE 6–15

The data-input and data-select waveforms in Figure 6–60 are applied to the multiplexer in Figure 6–59. Determine the output waveform in relation to the inputs.

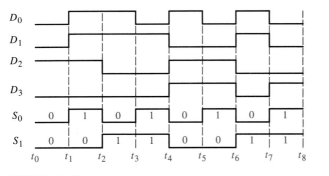

FIGURE 6–60

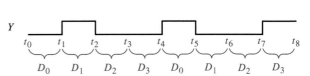

FIGURE 6–61

Solution The binary state of the data-select inputs during each interval determines which data input is selected. Notice that the data-select inputs go through a repetitive binary sequence 00, 01, 10, 11, 00, 01, 10, 11, and so on. The resulting output waveform is shown in Figure 6–61.

Three representative MSI multiplexers are now introduced:

1. The 74157 quadruple two-input data selector/multiplexer.
2. The 74151A eight-input data selector/multiplexer.
3. The 74150 sixteen-input data selector/multiplexer.

The 74157 Quadruple Two-Input Data Selector/Multiplexer

The 74157 consists of four separate two-input multiplexers on a single chip. Each of the four multiplexers shares a common data-select line and a common $\overline{\text{Enable}}$, as shown in Figure 6–62(a). Because there are only two inputs to be selected in each multiplexer, a single data-select input is sufficient ($2^1 = 2$).

As you can see in the logic diagram, the data-select input is ANDed with the B input of each two-input multiplexer, and the complement of data-select is ANDed with each A input.

A LOW on the $\overline{\text{Enable}}$ input allows the selected input data to pass through to the output. A HIGH on the $\overline{\text{Enable}}$ input prevents data from going through to the output; that is, it *disables* the multiplexers.

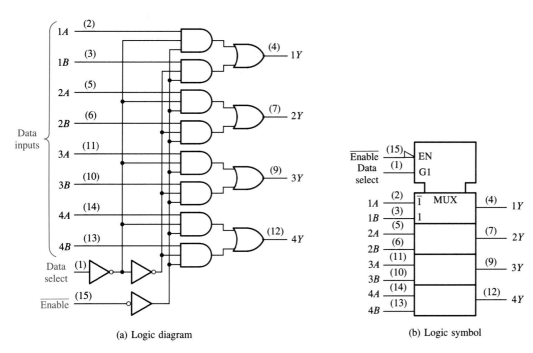

(a) Logic diagram

(b) Logic symbol

FIGURE 6–62
The 74157 quadruple two-input data selector/multiplexer.

The ANSI/IEEE Logic Symbol

The ANSI/IEEE logic symbol for this device is shown in Figure 6–62(b). Notice that the four multiplexers are indicated by the partitioned outline and that the inputs common to all four multiplexers are indicated as inputs to the notched block at the top, which is called the *common control block*. All labels within the upper MUX block apply to the other blocks below it.

Notice the 1 and $\overline{1}$ labels in the MUX blocks and the G1 label in the common control block. These labels are an example of the *dependency notation* system specified in the ANSI/IEEE Standard 91–1984. In this case G1 indicates an AND relationship between the data-select input and the data inputs with 1 or $\overline{1}$ labels. (The $\overline{1}$ means that the AND relationship applies to the complement of the G1 input.) In other words, when the data-select input is HIGH, the *B* inputs of the multiplexers are selected, and when the data-select input is LOW, the *A* inputs are selected. A G is always used to denote AND dependency.

Other aspects of dependency notation are introduced as appropriate throughout the book.

The 74151A Eight-Input Data Selector/Multiplexer

The 74151A has eight data inputs and, therefore, three data-select input lines. Three bits are required to select any one of the eight data inputs ($2^3 = 8$). A LOW on the $\overline{\text{Enable}}$ input allows the selected input data to pass through to the output. Notice that the data output and its complement are both available. The logic diagram is shown in Figure 6–63(a), and the ANSI/IEEE logic symbol is shown in part (b). In this case there is no need for a common control block on the logic symbol because there is only one multiplexer to be controlled, not four as in the 74157. The G_7^0 label within the logic symbol indicates the AND relationship between the data-select inputs and each of the data inputs 0 through 7.

The 74150 Sixteen-Input Data Selector/Multiplexer

The 74150 has sixteen data inputs and four data-select lines. In this case four bits are required to select any one of the sixteen data inputs ($2^4 = 16$). There is also an active-LOW $\overline{\text{Enable}}$ input. On this particular device, only the complement of the output is available. The logic symbol is shown in Figure 6–64.

EXAMPLE 6–16

Use 74150s and any other logic necessary to multiplex 32 data lines onto a single data-output line.

Solution An implementation of this system is shown in Figure 6–65. Five bits are required to select one of 32 data inputs ($2^5 = 32$). In this application the $\overline{\text{Enable}}$ input is used as the most significant data-select bit. When the MSB in the data-select code is LOW, the upper 74150 is enabled, and one of the data inputs (D_0 through D_{15}) is selected by the other four data-select bits. When the data-select MSB is HIGH, the lower 74150 is enabled, and one of the data inputs (D_{16} through D_{31}) is selected. The selected input data are then passed through to the negative-OR gate and onto the single output line.

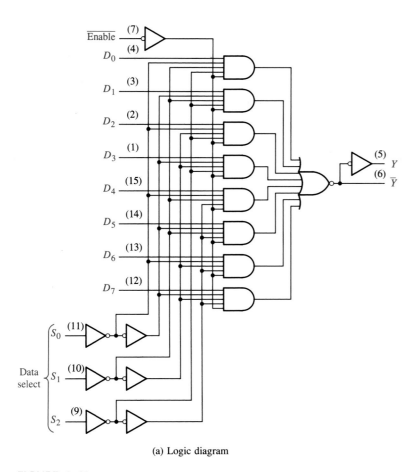

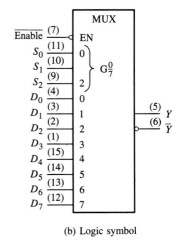

(a) Logic diagram

(b) Logic symbol

FIGURE 6–63
The 74151A eight-input data selector/multiplexer.

FIGURE 6–64
The 74150 sixteen-input data selector/
multiplexer.

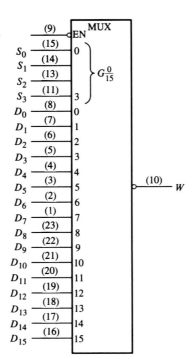

FIGURE 6–65
A 32-input multiplexer.

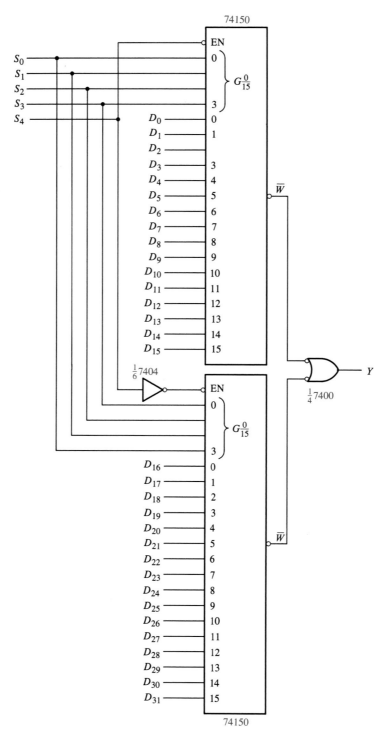

Data Selector/Multiplexer Applications

The seven-segment display multiplexer   Figure 6–66 shows a simplified method of multiplexing BCD numbers to a seven-segment display. In this example two-digit numbers are displayed on the seven-segment readout by the use of a single BCD-to-seven-segment decoder. This basic method of display multiplexing can be extended to displays with any number of digits. The basic operation is as follows:

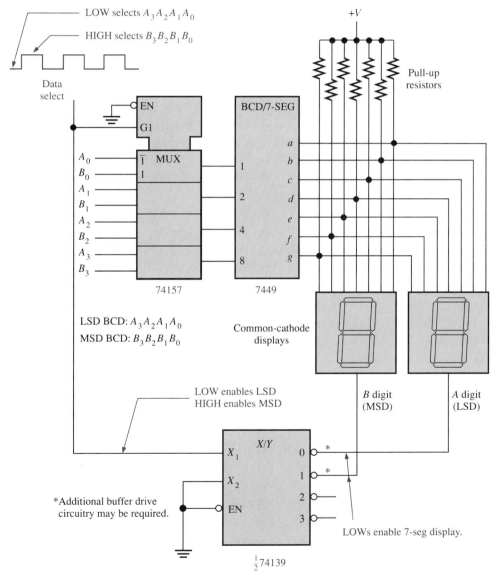

FIGURE 6–66
Simplified seven-segment display multiplexing logic.

Two BCD digits ($A_3A_2A_1A_0$ and $B_3B_2B_1B_0$) are applied to the multiplexer inputs. A square wave is applied to the *data-select* line, and when it is LOW, the A bits ($A_3A_2A_1A_0$) are passed through to the inputs of the 7449 BCD-to-seven-segment decoder. The LOW on the data-select also puts a LOW on the 1 input of the 74139 2-line-to-4-line decoder, thus activating its 0 output and enabling the A-digit display by effectively connecting its common terminal to ground. The A digit is now on and the B digit is off.

When the *data-select* line goes HIGH, the B bits ($B_3B_2B_1B_0$) are passed through to the inputs of the BCD-to-seven-segment decoder. Also, the 74139 decoder's 1 output is activated, thus enabling the B-digit display. The B digit is now on and the A digit is off. The cycle repeats at the frequency of the data-select square wave. This frequency must be high enough (about 30 Hz) to prevent visual flicker as the digit displays are multiplexed.

The logic function generator   A very useful application of the data selector/multiplexer is in the generation of combinational logic functions in sum-of-products form. When used in this way, the device can replace discrete gates, can often greatly reduce the number of ICs, and can make design changes much easier.

To illustrate, a 74151A eight-input data selector/multiplexer can be used to implement any specified three-variable logic function if the variables are connected to the data-select inputs and each data input is set to the logic level required in the truth table for that function. For example, if the function is a 1 when the variable combination is $\overline{A_2}A_1\overline{A_0}$, the 2 input (selected by 010) is connected to a HIGH. This HIGH is passed through to the output when this particular combination of variables occurs on the data-select lines. An example will help clarify this procedure.

EXAMPLE 6–17

Implement the logic function specified in Table 6–11 by using a 74151A eight-input data selector/multiplexer. Compare this method with a discrete logic gate implementation.

TABLE 6–11

Inputs			Output
A_2	A_1	A_0	Y
0	0	0	0
0	0	1	1
0	1	0	0
0	1	1	1
1	0	0	0
1	0	1	1
1	1	0	1
1	1	1	0

Solution   Notice from the truth table that Y is a 1 for the following input variable combinations: 001, 011, 101, and 110. For all other combinations, Y is 0. For this function to be implemented with the data selector, the data input selected by each of the above-mentioned combinations must be connected to a HIGH. All the other data inputs must be connected to a LOW, as shown in Figure 6–67.

The implementation of this function with logic gates would require as many as four 3-input AND gates, one 4-input OR gate, and three inverters.

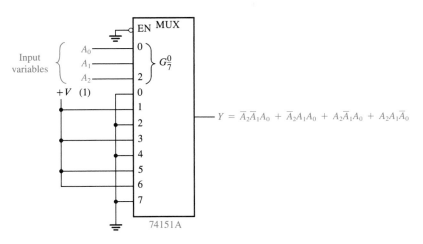

FIGURE 6–67
Data selector as a three-variable logic function generator.

Example 6–17 illustrated how the eight-input data selector can be used as a logic function generator for three variables. Actually, this device can be also used as a four-variable logic function generator by the utilization of one of the bits (A_0) in conjunction with the data inputs.

A four-variable truth table has sixteen combinations of input variables. When an eight-bit data selector is used, each input is selected twice: the first time when A_0 is 0 and the second time when A_0 is 1. With this in mind, the following rules can be applied (Y is the output, and A_0 is the least significant bit):

1. If $Y = 0$ both times a given data input is selected by $A_3A_2A_1$, connect that input to ground (0).
2. If $Y = 1$ both times a given data input is selected by $A_3A_2A_1$, connect that input to $+V$ (1).
3. If Y is different the two times a given data input is selected by $A_3A_2A_1$, and if $Y = A_0$, connect that input to A_0.
4. If Y is different the two times a given data input is selected by $A_3A_2A_1$, and if $Y = \overline{A_0}$, connect that input to A_0.

The following example illustrates this method.

EXAMPLE 6–18

Implement the logic function in Table 6–12 by using a 74151A eight-input data selector/multiplexer. Compare this method with a discrete logic gate implementation.

TABLE 6–12

	Inputs				Output
	A_3	A_2	A_1	A_0	Y
0	0	0	0	0	0
1	0	0	0	1	1
2	0	0	1	0	1
3	0	0	1	1	0
4	0	1	0	0	0
5	0	1	0	1	1
6	0	1	1	0	1
7	0	1	1	1	1
8	1	0	0	0	1
9	1	0	0	1	0
10	1	0	1	0	1
11	1	0	1	1	0
12	1	1	0	0	1
13	1	1	0	1	1
14	1	1	1	0	0
15	1	1	1	1	1

Solution The data-select inputs are $A_3A_2A_1$. In the first row of the table, $A_3A_2A_1 = 000$ and $Y = A_0$. In the second row, where $A_3A_2A_1$ again is 000, $Y = A_0$. Thus, A_0 is connected to the 0 input. In the third row of the table, $A_3A_2A_1 = 001$ and $Y = \overline{A_0}$. Also, in the fourth row, when $A_3A_2A_1$ again is 001, $Y = \overline{A_0}$. Thus, A_0 is inverted and connected to the 1 input. This analysis is continued until each input is properly connected according to the specified rules. The implementation is shown in Figure 6–68.

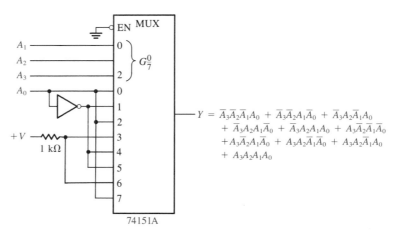

$$Y = \overline{A_3}\,\overline{A_2}\,\overline{A_1}A_0 + \overline{A_3}\,\overline{A_2}A_1\overline{A_0} + \overline{A_3}A_2\overline{A_1}A_0$$
$$+ \ \overline{A_3}A_2A_1\overline{A_0} + \overline{A_3}A_2A_1A_0 + A_3\overline{A_2}\,\overline{A_1}\,\overline{A_0}$$
$$+ A_3\overline{A_2}A_1\overline{A_0} + A_3A_2\overline{A_1}\,\overline{A_0} + A_3A_2\overline{A_1}A_0$$
$$+ \ A_3A_2A_1A_0$$

FIGURE 6–68
Data selector as a four-variable logic function generator.

If implemented with logic gates, the function would require as many as ten 4-input AND gates, one 10-input OR gate, and four inverters, although simplification would reduce this requirement.

SECTION 6–8
REVIEW

Answers are found at the end of the chapter.
1. In Figure 6–59, $D_0 = 1$, $D_1 = 0$, $D_2 = 1$, $D_3 = 0$, $S_0 = 1$, and $S_1 = 0$. What is the output?
2. Identify each MSI device.
   (a) 74157   **(b)** 74151A   **(c)** 74150
3. A 74151A has alternating LOW and HIGH levels on its data inputs beginning with $D_0 = 0$. The data-select lines are sequenced through a binary count (000, 001, 010, and so on) at a frequency of 1 kHz. The enable input is LOW. Draw the timing diagram for this operation.
4. Briefly describe the purpose of each of the following devices in Figure 6–66:
   (a) 74157   **(b)** 7449   **(c)** 74139

DEMULTIPLEXERS

6–9

*A **demultiplexer (DMUX)** basically reverses the multiplexing function. It takes data from one line and distributes them to a given number of output lines. For this reason, this device is also known as a **data distributor.***

Figure 6–69 shows a 1-line-to-4-line demultiplexer circuit. The data-input line goes to all of the AND gates. The two data-select lines enable only one gate at a time, and the data appearing on the data-input line will pass through the selected gate to the associated data-output line.

FIGURE 6–69
A 1-line-to-4-line demultiplexer.

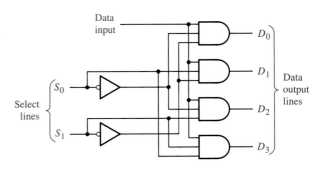

EXAMPLE 6–19

The serial data-input waveform and data-select inputs are shown in Figure 6–70. Determine the data-output waveforms for the demultiplexer in Figure 6–69.

FIGURE 6–70

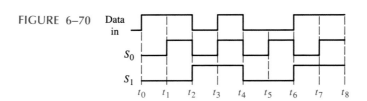

FIGURE 6–71

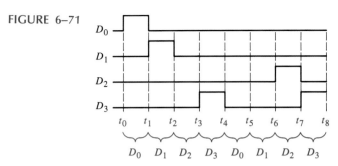

Solution Notice that the select lines go through a binary sequence so that each successive input bit is routed to D_0, D_1, D_2, and D_3 in sequence, as shown in Figure 6–71.

The 74154 as a Demultiplexer

We have already discussed the 74154 in its application as a 4-line-to-16-line decoder (Section 6–5). This device and other decoders can also be used in demultiplexing applications. The logic symbol for this device when used as a demultiplexer is shown in Figure 6–72.

In demultiplexer applications the input lines are used as the data-select lines. One of the Enable inputs is used as the data-input line, with the other Enable input held LOW to enable the internal negative-AND gate.

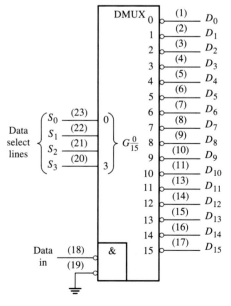

FIGURE 6–72
The 74154 used as a demultiplexer.

SECTION 6–9
REVIEW

Answers are found at the end of the chapter.
1. Generally, how can an MSI decoder be used as a demultiplexer?
2. The 74154 demultiplexer in Figure 6–72 has a binary code of 1010 on the data-select lines, and the data-input line is LOW. What are the states of the output lines?

PARITY GENERATORS/CHECKERS

6–10

Errors can occur as digital codes are being transferred from one point to another within a digital system or while codes are being transmitted from one system to another. The errors take the form of undesired changes in the bits that make up the coded information; that is, a 1 can change to a 0, or a 0 to a 1, because of component malfunctions or electrical noise. In most digital systems the probability that even a single bit error will occur is very small, and the likelihood that more than one will occur is even smaller. Nevertheless, when an error occurs undetected, it can cause serious problems in a digital system.

Parity

Many systems employ a *parity bit* as a means of detecting a bit error. Binary information is normally handled by a digital system in groups of bits called *words*. A word always contains either an even or an odd number of 1s. A parity bit is attached to the group of information bits to make the *total* number of 1s in a word always even or always odd. An *even parity* bit makes the total number of 1s even, and an *odd parity* bit makes the total odd.

A given system operates with even or odd parity, but not both. For instance, if a system operates with even parity, a check is made on each group of bits received to make sure the total number of 1s in that group is even. If there is an odd number of 1s, an error has occurred.

As an illustration of how parity bits are attached to a code word, Table 6–13 lists the parity bits for each 8421 BCD code number for both even and odd parity. The parity bit for each BCD number is in the *P* column.

The parity bit can be attached to the code group at either the beginning or the end, depending on system design. Notice that the total number of 1s, *including* the parity bit, is always even for even parity and always odd for odd parity.

TABLE 6–13
The 8421 BCD code with parity bits.

Even Parity		Odd Parity	
P	8421	*P*	8421
0	0000	1	0000
1	0001	0	0001
1	0010	0	0010
0	0011	1	0011
1	0100	0	0100
0	0101	1	0101
0	0110	1	0110
1	0111	0	0111
1	1000	0	1000
0	1001	1	1001

Detecting an Error

A parity bit provides for the detection of a *single* bit error (or any odd number of errors, which is very unlikely) but cannot check for two errors in one word. For instance, let us assume that we wish to transmit the BCD code 0101. (Parity can be used with any number of bits; we are using four for illustration.) The total code transmitted, including the even parity bit, is

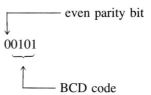

Now let us assume that an error occurs in the third bit from the left (the 1 becomes a 0, as follows:

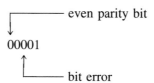

When this code is received, the parity check circuitry determines that there is only a single 1 (odd number), when there should be an even number of 1s. Because an even number of 1s does not appear in the code when it is received, an error is indicated.

Let us now consider what happens if two bit errors occur as follows:

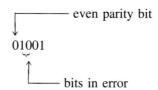

When a check is made, an even number of 1s appears, and although there are two errors, the parity check indicates a correct code.

An odd parity bit also provides in a similar manner for the detection of a single error in a given group of bits.

EXAMPLE 6–20

Assign the proper even parity bit to the following code words:
(a) 1010   (b) 111000   (c) 101101
(d) 100011100101 (e) 101101011111

Solutions   Make the parity bit either 1 or 0 to make the total number of 1s even. The parity bit will be the left-most bit (shaded).

(a) 0 1010
(b) 1 111000
(c) 0 101101
(d) 0 100011100101
(e) 1 101101011111

EXAMPLE 6–21

An odd parity system receives the following code words: 10110, 11010, 110011, 110101110100, and 1100010101010. Determine which ones, if any, are in error.

Solutions   Since odd parity is required, any code with an even number of 1s is incorrect. The following codes are in error: 110011 and 1100010101010.

Several specific codes also provide inherent error detection; a few of the most important, as well as an error-correcting code, are discussed in Appendix B.

Parity Logic

In order to check for or generate the proper parity in a given code word, a very basic principle can be used: *The sum* (disregarding carries) *of an even number of 1s is always 0, and the sum of an odd number of 1s is always 1*. Therefore, to determine if a given code word has even or odd parity, all the bits in that code word are summed. As we know, the sum of two bits can be generated by an exclusive-OR gate, as shown in Figure 6–73(a); the sum of three bits can be formed by two exclusive-OR gates connected as shown in Figure 6–73(b); and so on.

FIGURE 6–73

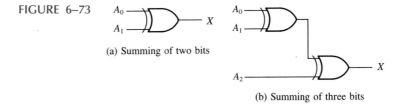

(a) Summing of two bits

(b) Summing of three bits

A typical five-bit generator/checker circuit is shown in Figure 6–74. It can be used for either odd or even parity. When used as an odd parity checker, as shown, the operation is as follows:

A five-bit code (four data bits and one parity bit) is applied to the inputs. The four data bits are on the exclusive-OR inputs, and the parity bit is applied to the ODD input line. When the number of 1s in the five-bit code is odd, the Σ ODD output is LOW, indicating proper parity. When there is an even number of 1s, the Σ ODD output is

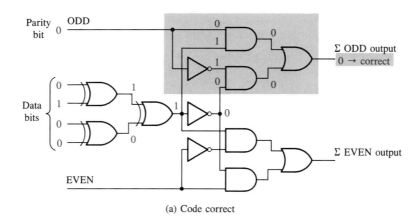

(a) Code correct

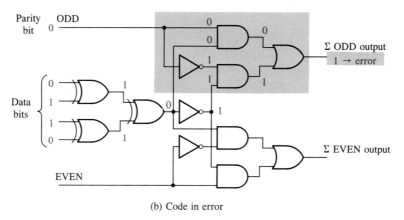

(b) Code in error

FIGURE 6–74
Examples of odd parity detection.

HIGH, indicating incorrect parity. Both of these conditions are illustrated in Figure 6–74. Similarly, even parity checks are illustrated for both nonerror and error conditions in Figure 6–75.

The same circuit can be used as a *parity generator,* as shown in Figure 6–76. The operation for odd parity generation is illustrated in part (a). A four-bit code is applied to the inputs, and the ODD line is held LOW (by grounding in this case). When the four-bit code has an even number of 1s, as shown, the Σ ODD output is HIGH (1). This 1 output is the odd parity bit and is combined with the four-bit code to form a five-bit odd parity code as shown. Similarly, a 0 parity bit is produced when there are an odd number of 1s in the input code.

Figure 6–76(b) shows an example of the same circuit used in an even parity system. Notice that the EVEN line is grounded in this case.

This basic logic can be expanded to accommodate any number of input bits by adding more exclusive-OR gates.

FIGURE 6–75
Examples of even parity detection.

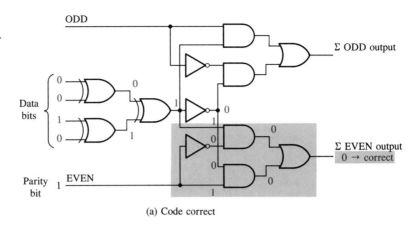

(a) Code correct

(b) Code in error

FIGURE 6–76
Examples of parity generation.

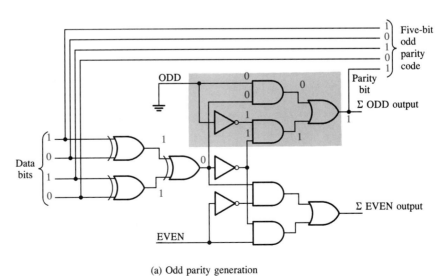

(a) Odd parity generation

FIGURE 6–76
(Continued)

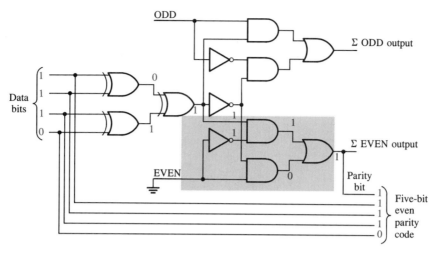

(b) Even parity generation

The 74180 Nine-Bit Parity Generator/Checker

The 74180 is represented in Figure 6–77. This particular MSI device can be used to check for odd or even parity on a nine-bit code (eight data bits and one parity bit), or it can be used to generate a nine-bit odd or even parity code. The truth table operation varies slightly from the simpler basic circuits just discussed, but the principle is the same.

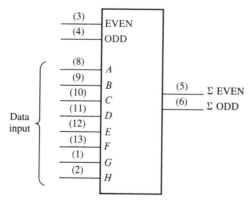

(a) Traditional logic symbol

Inputs			Outputs	
Σ of 1s at *A* through *H*	EVEN	ODD	Σ EVEN	Σ ODD
EVEN	1	0	1	0
ODD	1	0	0	1
EVEN	0	1	0	1
ODD	0	1	1	0
X	1	1	0	0
X	0	0	1	1

(b) Truth table ($X \equiv$ don't care)

FIGURE 6–77
The 74180 nine-bit parity generator/checker.

A Data Transmission System with Error Detection

A simplified data transmission system is shown in Figure 6–78 to illustrate an application of parity generators/checkers, as well as multiplexers and demultiplexers, and to illustrate the need for data storage in some applications.

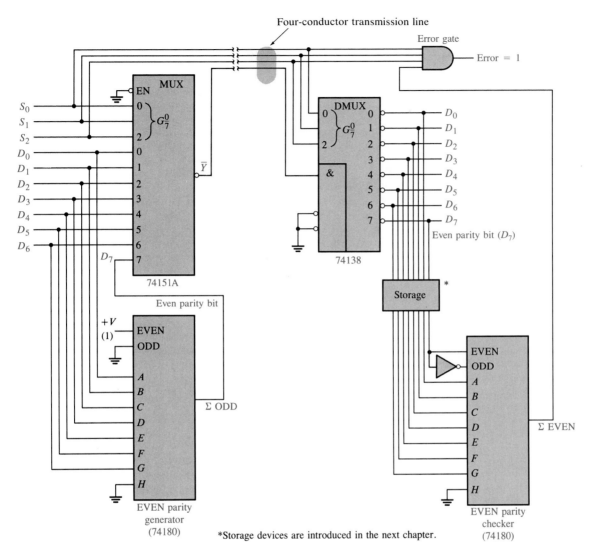

FIGURE 6–78

Simplified data transmission system with error detection.

In this application, digital data from seven sources are multiplexed onto a single line for transmission to a distant point. The seven data bits (D_0 through D_6) are applied to the multiplexer data inputs and, at the same time, to the even parity generator inputs. The Σ ODD output of the parity generator is used as the even parity bit. This bit is 0 if the number of 1s on the inputs A through H is even and is a 1 if the number of 1s on A through H is odd. This bit is D_7 of the transmitted code.

The data-select inputs are repeatedly cycled through a binary sequence, and each data bit, beginning with D_0, is serially passed through and onto the transmission line ($\overline{Y}$). In this example, the transmission line consists of four conductors: one carries the serial data and three carry the timing signals (data selects). There are more sophisticated ways of sending the timing information, but we are using this direct method to illustrate a basic application.

At the demultiplexer end of the system, the data-select signals and the serial data stream are applied to the demultiplexer. The data bits are distributed by the demultiplexer onto the output lines in the order in which they occurred on the multiplexer inputs. That is, D_0 comes out on the D_0 output, D_1 comes out on the D_1 output, and so on. The parity bit comes out on the D_7 output. These eight bits are temporarily stored and applied to the even parity checker. Not all of the bits are present on the parity checker inputs until the parity bit D_7 comes out and is stored. At this time, the error gate is enabled by the 111 data-select code. If the parity is correct, a 0 appears on the Σ EVEN output, keeping the ERROR output at 0. If the parity is incorrect, all 1s appear on the error gate inputs, and a 1 on the ERROR output results.

This particular application has demonstrated the need for data storage so that you will be better able to appreciate the usefulness of the storage devices introduced in the next chapter.

The timing diagram in Figure 6–79 illustrates a specific case in which two eight-bit words are transmitted, one with correct parity and one with an error.

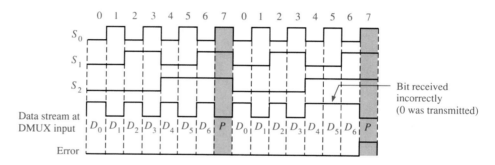

FIGURE 6–79
Example of data transmission with and without error for the system in Figure 6–78.

**SECTION 6–10
REVIEW**

Answers are found at the end of the chapter.
1. Add an even parity bit to each of the following codes:
   (a) 110100   **(b)** 01100011
2. Add an odd parity bit to each of the following codes:
   (a) 1010101   **(b)** 1000001
3. Check each of the even parity codes for an error.
   (a) 100010101   **(b)** 1110111001

TROUBLESHOOTING

6–11

*In this section, the problem of **decoder glitches** is introduced and examined from a troubleshooting standpoint. A **glitch** is any undesired voltage or current spike (pulse) of very short duration. A glitch can be interpreted as a valid signal by a logic circuit and may cause improper operation.*

A 3-line-to-8-line decoder (binary-to-octal) is used in Figure 6–80 to illustrate how glitches occur and how to identify their cause. The $A_2A_1A_0$ inputs of the decoder are sequenced through a binary count, and the resulting waveforms of the inputs and outputs can be displayed on the screen of a logic analyzer as shown in Figure 6–80.

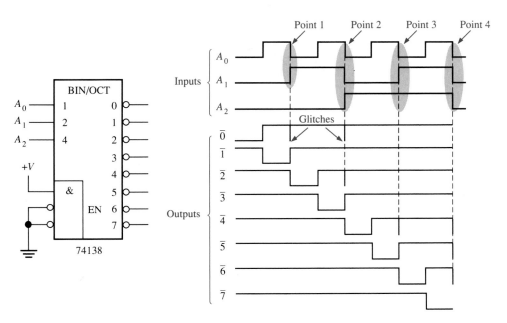

FIGURE 6–80
Decoder waveforms with output glitches.

The output waveforms are correct except for the glitches that occur on some of the output signals. An oscilloscope can be used to examine the critical timing details of the input waveforms in an effort to pinpoint the cause of the glitches. The oscilloscope is useful in this case because the waveforms can be "magnified" so that very small time differences between waveform transitions are observable.

The points of interest indicated by the shaded areas on the input waveforms in Figure 6–80 are displayed on an oscilloscope as shown in Figure 6–81. At point 1 there is a transitional state of 000 due to delay differences in the waveforms. This causes the first glitch on the $\overline{0}$ output of the decoder. At point 2 there are two transitional states, 010 and 000. These cause the glitch on the $\overline{2}$ output of the decoder and the second glitch on the $\overline{0}$ output, respectively. At point 3 the transitional state is 100, which causes the first glitch on the $\overline{4}$ output of the decoder. At point 4 the two transitional states, 110 and 100, result in the glitch on the $\overline{6}$ output and the second glitch on the $\overline{4}$ output, respectively.

One way to eliminate the glitch problem is a method called *strobing,* in which the decoder is enabled only during the times when the waveforms are not in transition. This method is discussed in a later chapter.

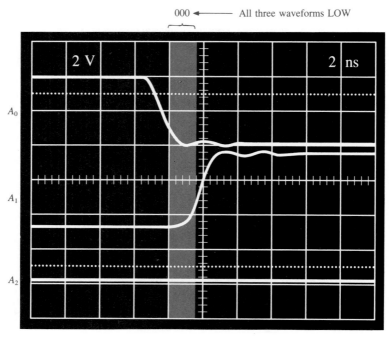

(a) Transitional state of 000 at point 1
of input waveforms in Figure 6–80.

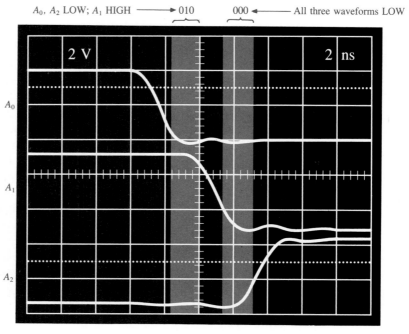

(b) Transitional states of 010 and 000
at point 2 of input waveforms in
Figure 6–80.

FIGURE 6–81
Oscilloscope displays of transitions of decoder input waveforms for glitch analysis.

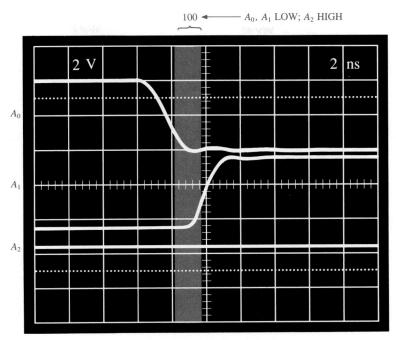

(c) Transitional state 100 at point 3
of input waveforms in Figure 6–80.

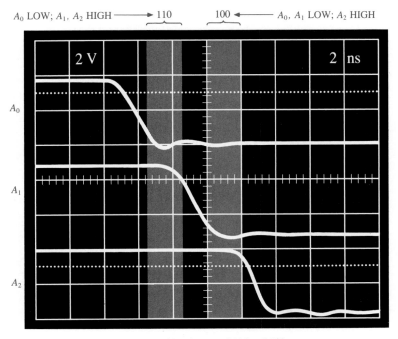

(d) Transitional states of 110 and 100
at point 4 of input waveforms
in Figure 6–80.

FIGURE 6–81
(Continued)

SECTION 6–11
REVIEW

Answers are found at the end of the chapter.
1. Define the term *glitch*.
2. Explain the basic cause of glitches in decoder logic.
3. Define the term *strobing*.

SUMMARY ☐

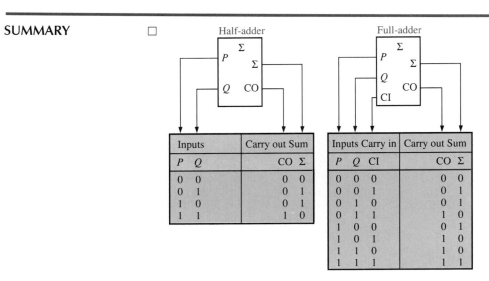

FIGURE 6–82

☐ Connection diagrams for the ICs used in this chapter are shown in Figure 6–83. Pin labeling may differ from some manufacturers' data sheets. The labeling is for consistency with text usage.

Integrated Circuit Technologies (Chapter 13) may be covered after this chapter, covered after a later chapter, completely omitted, or partially omitted, depending on the needs and goals of your program. Coverage is not prerequisite or corequisite to any other material in the book. Inclusion is purely optional.

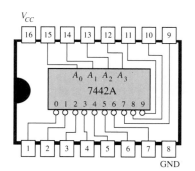

7442A BCD-to-decimal decoder

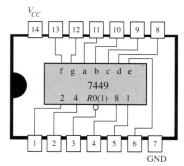

7449 BCD-to-7-segment decoder/driver

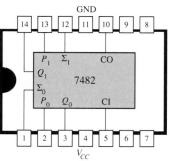

7482 2-bit binary full-adder

FIGURE 6–83

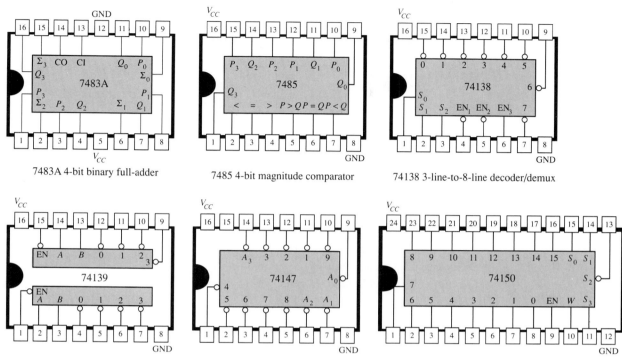

7483A 4-bit binary full-adder

7485 4-bit magnitude comparator

74138 3-line-to-8-line decoder/demux

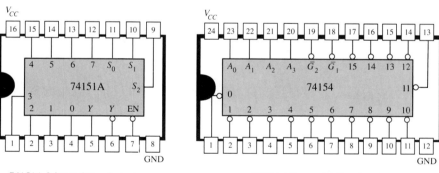

74139 Dual 2-line-to-4-line decoder/demux

74147 Decimal-to-BCD priority encoder

74150 16-input data selector/mux

74151A 8-input data selector/mux

74154 4-line-to-16-line decoder

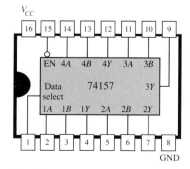

74157 Quad 2-input data selector/mux

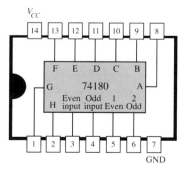

74180 9-bit parity generator/checker

FIGURE 6–83
(*Continued*)

GLOSSARY

Carry generate   The condition of a full-adder when both input bits are 1s.

Carry propagate   The condition of a full-adder when either input bit is a 1 and the carry input is a 1.

Cascading inputs   Inputs that can be connected to outputs of a similar device, allowing one device to drive another in order to expand the operational capability.

Code converter   A digital device that converts one type of coded information into another coded form.

Data selector   A circuit that selects data from several inputs in a time sequence and places them on the output. Also called a *multiplexer*.

Decoder   A digital device that converts coded information to a familiar form.

Demultiplexer (DMUX)   A digital device capable of distributing digital data from one source or line to several lines.

Dependency notation   A notational system for logic symbols that specifies input and output relationships, thus fully defining a given function. An integral part of ANSI/IEEE Std. 91–1984

Encoder   A digital device that converts information to coded form.

Even parity   The condition of having an even number of 1s in every group of bits.

Full-adder   A digital device that adds two binary digits and a carry input to produce a sum and a carry output.

Glitch   A voltage or current spike of short duration, usually unintentionally produced and unwanted.

Half-adder   A digital device that adds two bits and produces a sum and a carry output. It cannot handle input carries.

LCD   Liquid crystal display.

LED   Light-emitting diode.

Look-ahead carry   A method of binary addition whereby carries from preceding adder stages are anticipated, thus eliminating carry propagation delays.

Multiplexer (MUX)   A digital device capable of putting digital data from several sources onto a single line or transmission path.

Odd parity   The condition of having an odd number of 1s in every group of bits.

Parity bit   A bit attached to each group of information bits to make the total number of 1s odd or even for every group of bits.

Priority encoder   An encoder in which only the highest-value input digit is encoded and any other active input is ignored.

Pull-up resistor   A resistor used to keep a given point in a circuit HIGH when in the inactive state.

Strobing   A process of using a pulse to sample the occurrence of an event at a specified time in relation to the event.

Zero suppression   The process of blanking out leading or trailing zeros in a digital readout.

SELF-TEST

Answers and solutions are found at the end of the book.

1. Describe the difference between a half-adder and a full-adder.
2. The following input bits are applied to a half-adder. Determine the Σ and CO for each.
   (a) $P = 1, Q = 0$ **(b)** $P = 1, Q = 1$ **(c)** $P = 0, Q = 1$
   (d) $P = 0, Q = 0$
3. The following input bits are applied to a full-adder. Determine the Σ and CO for each.
   (a) $P = 1, Q = 0, CI = 0$ **(b)** $P = 1, Q = 0, CI = 1$
   (c) $P = 0, Q = 1, CI = 1$ **(d)** $P = 1, Q = 1, CI = 1$
4. A 7482 adder has the following inputs: $P_0 = 1, P_1 = 0, Q_0 = 1, Q_1 = 1$, and $CI = 0$. Determine Σ_0, Σ_1, and CO.

5. Show how to connect 7482 two-bit adders to form an eight-bit parallel adder.
6. Show how to connect 7483A four-bit adders to form a sixteen-bit adder.
7. Determine the carry generate (CG) and the carry propagate (CP) functions for a full-adder when the input bits are as follows:
   (a) $P = 1, Q = 0$ (b) $P = 0, Q = 0$ (c) $P = 1, Q = 1$
8. A 7485 four-bit magnitude comparator has $P = 1011$ and $Q = 1001$.
   (a) Determine the outputs.
   (b) Explain how to connect the $<$, $=$, and $>$ inputs if this is to be the least significant stage.
9. Use 7485 comparators to form a twelve-bit comparator. Show the diagram.
10. Determine which output is LOW for each set of inputs to a 74154 decoder.
    (a) $A_3A_2A_1A_0 = 1100$ (b) $A_3A_2A_1A_0 = 1000$ (c) $A_3A_2A_1A_0 = 0010$
11. Determine which output is LOW for each set of inputs to a 7442A decoder.
    (a) $A_3A_2A_1A_0 = 0101$ (b) $A_3A_2A_1A_0 = 1001$ (c) $A_3A_2A_1A_0 = 1100$
12. A BCD-to-seven-segment decoder/driver is connected to an LED display. Which segments are illuminated for each of the following input codes?
    (a) $A_3A_2A_1A_0 = 0001$ (b) $A_3A_2A_1A_0 = 0111$ (c) $A_3A_2A_1A_0 = 0011$
13. A 74147 decimal-to-BCD priority encoder has the following input pins LOW: 3, 4, and 12. The other inputs are HIGH. What is the BCD code on the outputs?
14. (a) Convert BCD 10010011 to binary.
    (b) Convert BCD 01100111 to binary.
15. Draw the logic diagram for a three-bit binary-to-Gray converter.
16. Draw the logic diagram for a nine-bit Gray-to-binary converter.
17. A 74157 quadruple two-input data selector/multiplexer (see Figure 6–62) has the following bits on its inputs:

$$1A = 1 \qquad 1B = 0$$
$$2A = 0 \qquad 2B = 0$$
$$3A = 1 \qquad 3B = 0$$
$$4A = 0 \qquad 4B = 1$$

The enable is LOW. Determine the outputs when (a) the data-select input is 0 and (b) the data-select input is 1.
18. The data-select inputs of a 74151A eight-input multiplexer (see Figure 6–63) are repetitively sequenced through a binary count from zero to seven. If there is a 1 on each odd-numbered data input and a 0 on all the others, determine the output waveform for one cycle through the binary count.
19. Show how to connect three 74151A multiplexers to form a 24-input multiplexer.
20. Implement the logic equation $Y = \overline{A_2}A_1\overline{A_0} + A_2\overline{A_1}A_0 + A_2A_1A_0$ with a 74151A data selector/multiplexer.
21. The 74154 is used as a demultiplexer. If the select lines are $S_3S_2S_1S_0 = 1011$, and a 1 is applied to the data input (pin 18), on which output line does the data bit appear?
22. Attach the proper even parity bit to the following codes:
    (a) 11010   (b) 1001   (c) 0111101
23. Answer question 22 for odd parity.

PROBLEMS

Answers to odd-numbered problems are found at the end of the book.

Section 6–1 Adders

6–1   For the full-adder of Figure 6–4, determine the logic state (1 or 0) at each gate output for the following inputs:
       (a) $P = 1, Q = 1, CI = 1$ (b) $P = 0, Q = 1, CI = 1$
       (c) $P = 0, Q = 1, CI = 0$

6–2   What are the full-adder inputs that will produce each of the following outputs?
   (a) $\Sigma = 0$, CO $= 0$ **(b)** $\Sigma = 1$, CO $= 0$
   (c) $\Sigma = 1$, CO $= 1$ **(d)** $\Sigma = 0$, CO $= 1$

6–3   Simplify, if possible, the full-adder circuit of Example 6–1 (Figure 6–7) by using the Karnaugh map method.

Section 6–2   Parallel Binary Adders

6–4   For the parallel adder in Figure 6–84, determine the sum by analysis of the logical operation of the circuit. Verify your result by longhand addition of the two input numbers.

6–5   Solve Problem 6–4 for the circuit and input conditions in Figure 6–85.

6–6   The input waveforms in Figure 6–86 are applied to a 7482 two-bit adder. Determine the waveforms for the sum and carry outputs in relation to the inputs.

FIGURE 6–84

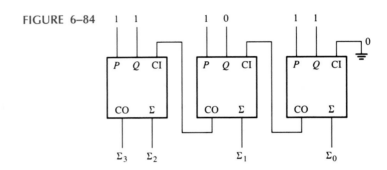

FIGURE 6–85

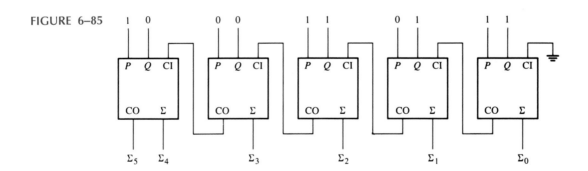

FIGURE 6–86

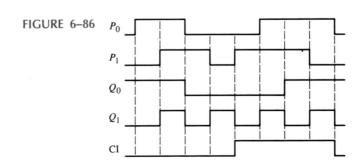

6–7 The following sequences of bits (right-most bit first) appear on the inputs to the adder developed in Example 6–3 (Figure 6–12). Determine the resulting sequence of bits on each sum output.

$$P_0 \quad 10010110$$
$$P_1 \quad 11101000$$
$$P_2 \quad 00001010$$
$$P_3 \quad 10111010$$
$$Q_0 \quad 11111000$$
$$Q_1 \quad 11001100$$
$$Q_2 \quad 10101010$$
$$Q_3 \quad 00100100$$

6–8 In the process of checking a 7483A four-bit full-adder, the following voltage levels are observed on its pins: 1-LOW, 2-HIGH, 3-HIGH, 4-HIGH, 6-HIGH, 7-HIGH, 8-LOW, 9-LOW, 10-LOW, 11-LOW, 13-LOW, 14-HIGH, 15-LOW, and 16-HIGH. Determine if the IC is functioning properly.

Section 6–3   Ripple Carry versus Look-Ahead Carry Adders

6–9 Each of the eight full-adders in an eight-bit parallel ripple carry adder exhibits the following propagation delays:

$$P \text{ to } \Sigma \text{ and CO:} \quad 40 \text{ ns}$$
$$Q \text{ to } \Sigma \text{ and CO:} \quad 40 \text{ ns}$$
$$\text{CI to } \Sigma: \quad 35 \text{ ns}$$
$$\text{CI to CO:} \quad 25 \text{ ns}$$

Determine the maximum total time for the addition of two eight-bit numbers.

6–10 Show the additional logic circuitry necessary to make the four-bit look-ahead carry adder in Figure 6–16 into a five-bit adder.

Section 6–4   Comparators

6–11 The waveforms in Figure 6–87 are applied to the comparator as shown. Determine the output $(P = Q)$ waveform.

6–12 For the four-bit comparator in Figure 6–88, plot each output waveform for the inputs shown. The outputs are active-HIGH.

6–13 For each set of binary numbers, determine the logic states at each point in the comparator circuit of Figure 6–23, and verify that the output indications are correct.

(a) $P_3P_2P_1P_0 = 1100$ (b) $P_3P_2P_1P_0 = 1000$ (c) $P_3P_2P_1P_0 = 0100$

$Q_3Q_2Q_1Q_0 = 1001$ $Q_3Q_2Q_1Q_0 = 1011$ $Q_3Q_2Q_1Q_0 = 0100$

FIGURE 6–87

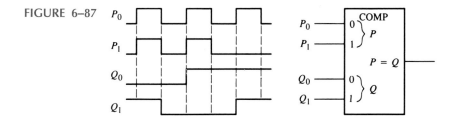

FIGURE 6–88

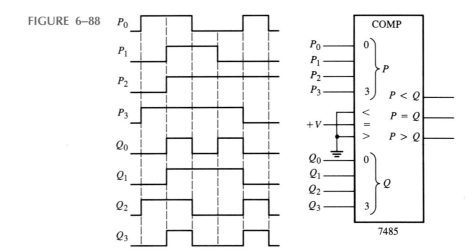

Section 6–5 Decoders

6–14 When a HIGH is on the output of each of the decoding gates in Figure 6–89, what is the binary code appearing on the inputs? The MSB is A_3.

6–15 Show the decoding logic for each of the following codes if an active-HIGH (1) indication is required:

(a) 1101 **(b)** 1000 **(c)** 11011 **(d)** 11100
(e) 101010 **(f)** 111110 **(g)** 000101 **(h)** 1110110

6–16 Solve Problem 6–15, given that an active-LOW (0) output is required.

6–17 We wish to detect only the presence of the codes 1010, 1100, 0001, and 1011. An active-HIGH output is required to indicate their presence. Develop the complete decoding logic with a single output that will tell us when any one of these codes is on the inputs. For any other code, the output must be LOW.

6–18 If the input waveforms are applied to the decoding logic as indicated in Figure 6–90, sketch the output waveform in proper relation to the inputs.

6–19 BCD numbers are applied sequentially to the BCD-to-decimal decoder in Figure 6–91. Draw the ten output waveforms, showing each in the proper relationship with the others and with the inputs.

6–20 A seven-segment decoder/driver drives the display in Figure 6–92. If the waveforms are applied as indicated, determine the sequence of digits that appears on the display.

FIGURE 6–89

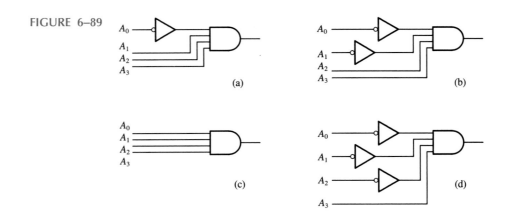

FIGURE 6–90

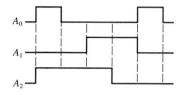 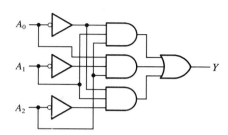

FIGURE 6–91

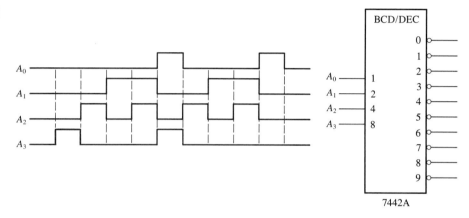

7442A

FIGURE 6–92

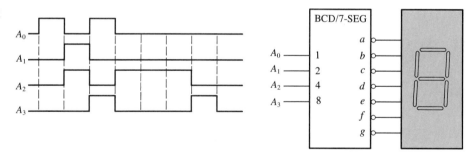

Section 6–6 Encoders

6–21 In the decimal-to-BCD encoder of Figure 6–46, assume that the 9 input and the 3 input are both HIGH. What is the output code? Is it a valid BCD (8421) code?

6–22 A 74147 encoder has LOW levels on pins 2, 5, and 12. What BCD code appears on the outputs if all the other inputs are HIGH?

Section 6–7 Code Converters

6–23 Show the logic required to convert a ten-bit binary number to Gray code, and use that logic to convert the following binary numbers to Gray code:
(a) 1010101010   **(b)** 1111100000
(c) 0000001110   **(d)** 1111111111

6–24 Show the logic required to convert a ten-bit Gray code to binary, and use that logic to convert the following Gray code words to binary:
(a) 1010000000   **(b)** 0011001100
(c) 1111000111   **(d)** 0000000001

6–25 Convert the following decimal numbers first to BCD. Then, using the logical operation of the BCD-to-binary converter of Figure 6–53, convert the BCD to binary. Verify the result in each case.

(a) 2   **(b)** 8   **(c)** 13   **(d)** 26   **(e)** 33

Section 6–8 Multiplexers (Data Selectors)

6–26 For the multiplexer in Figure 6–93, determine the output for the following input states: $D_0 = 0, D_1 = 1, D_2 = 1, D_3 = 0, S_0 = 1, S_1 = 0$.

6–27 If the data-select inputs to the multiplexer in Figure 6–93 are sequenced as shown by the waveforms in Figure 6–94, determine the output waveform with the data inputs specified in Problem 6–26.

6–28 The waveforms in Figure 6–95 are observed on the inputs of a 74151A eight-input multiplexer. Sketch the Y output waveform.

6–29 Modify the design of the seven-segment display multiplexing system in Figure 6–66 to accommodate two additional digits.

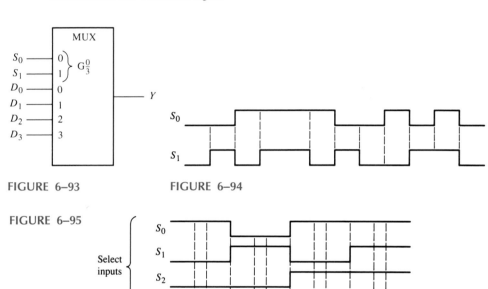

FIGURE 6–93   FIGURE 6–94

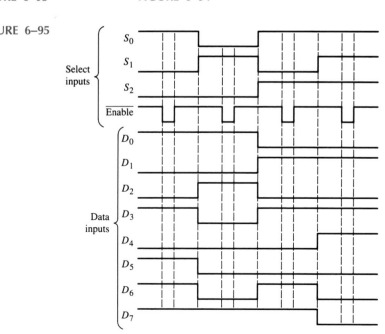

FIGURE 6–95

6–30 Implement the sum logic in Figure 6–77 by using a 74151A data selector. The three input variables are *P*, *Q*, and CI.

6–31 Implement the logic function specified in Table 6–14 by using a 74151A data selector.

TABLE 6–14

Inputs				Output
A_3	A_2	A_1	A_0	Y
0	0	0	0	0
0	0	0	1	0
0	0	1	0	1
0	0	1	1	1
0	1	0	0	0
0	1	0	1	0
0	1	1	0	1
0	1	1	1	1
1	0	0	0	1
1	0	0	1	0
1	0	1	0	1
1	0	1	1	1
1	1	0	0	0
1	1	0	1	1
1	1	1	0	0
1	1	1	1	1

Section 6–9   Demultiplexers

6–32 Develop the total timing diagram (inputs and outputs) for a 74154 used in a demultiplexing application in which the inputs are as follows: The data-select inputs are repetitively sequenced through a straight binary count beginning with 0000, and the data input is a serial data stream carrying BCD data representing the decimal number 2468. The least significant digit (8) is first in the sequence, with its LSB first, and it should appear in the first four-bit positions of the output.

Section 6–10   Parity Generators/Checkers

6–33 The waveforms in Figure 6–96 are applied to the four-bit parity logic. Determine the output waveform in proper relation to the inputs. How many times does even parity occur, and how is it indicated?

FIGURE 6–96

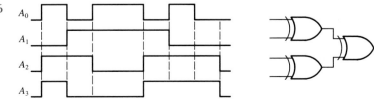

FIGURE 6–97

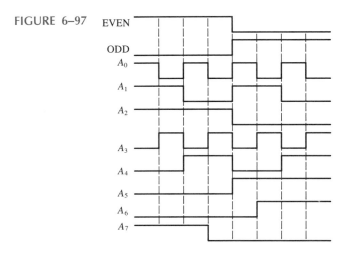

6–34 Determine the Σ EVEN and the Σ ODD outputs of a 74180 nine-bit parity generator/ checker for the inputs in Figure 6–97. Refer to the truth table in Figure 6–77.

Section 6–11 Troubleshooting

6–35 The full adder in Figure 6–98 is tested under all input conditions with the input wave-forms shown. From your observation of the Σ and CO waveforms, is it operating prop-erly, and if not, what is the most likely fault?

6–36 List the possible faults for each decoder/display in Figure 6–99.

FIGURE 6–98

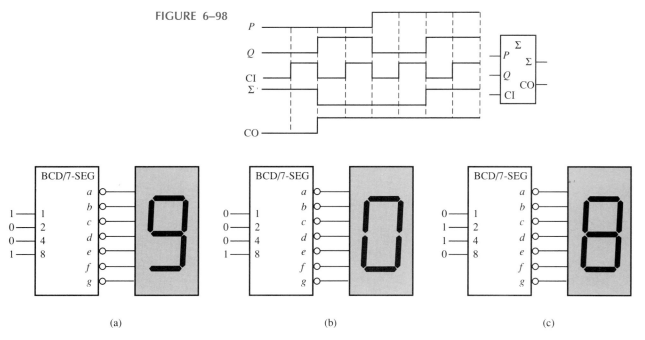

(a)   (b)   (c)

FIGURE 6–99

6–37 Develop a systematic test procedure to check out the complete operation of the keyboard encoder in Figure 6–52.

6–38 You are testing the BCD-to-binary converter in Figure 6–53. The procedure calls for applying BCD numbers in sequential order beginning with 0_{10} and checking for the correct binary output. What symptom or symptoms will appear on the binary outputs in the event of each of the following faults? For what BCD number is each fault *first* detected?
(a) The P input to FA1 is open.
(b) The CO of FA3 is open.
(c) The sum output of FA4 is shorted to ground.
(d) The CI of FA7 is shorted to ground.

6–39 For the display multiplexing system in Figure 6–66, determine the most likely cause or causes for each of the following symptoms:
(a) The B-digit (MSD) display does not turn on at all.
(b) Neither 7-segment display turns on.
(c) The f-segment of both displays appears to be on all the time.
(d) There is a visible flicker on the displays.

6–40 Develop a systematic procedure to fully test the 74151A data selector IC.

6–41 During the testing of the data transmission system in Figure 6–78, a code is applied to the D_0 through D_6 inputs that contains an odd number of 1s. A single bit error is deliberately introduced on the serial data transmission line between the MUX and the DMUX, but the system does not indicate an error (error output $= 0$). After some investigation, you check the inputs to the even parity checker and find that D_0 through D_6 contain an even number of 1s, as you would expect. Also, you find that the D_7 parity bit is a 1. What are the possible reasons for the system not indicating the error?

6–42 In general, describe how you would fully test the data transmission system in Figure 6–78, and specify a method for the introduction of parity errors.

ANSWERS TO SECTION REVIEWS

Section 6–1 Review
1. (a) $\Sigma = 1$, CO $= 0$ (b) $\Sigma = 0$, CO $= 0$ (c) $\Sigma = 1$, CO $= 0$
    (d) $\Sigma = 0$, CO $= 1$
2. $\Sigma = 1$, CO $= 1$

Section 6–2   Review
1. $CO\Sigma_1\Sigma_0 = 110$
2. Four 7482s are connected as indicated in Figure 6–12 for two 7482s.

Section 6–3   Review
1. CG $= 0$, CP $= 1$ 2. CO $= 1$

Section 6–4   Review
1. $P > Q = 1$, $P < Q = 0$, $P = Q = 0$
2. upper comparator: pin 7: $P < Q = 1$; pin 6: $P = Q = 0$; pin 5: $P > Q = 0$
    lower comparator: pin 7: $P < Q = 0$; pin 6: $P = Q = 0$; pin 5: $P > Q = 1$

Section 6–5   Review
1. the 5 input   2. four   3. active-LOW

Section 6–6 Review

1. **(a)** $\overline{A}_0 = 1, \overline{A}_1 = 1, \overline{A}_2 = 0, \overline{A}_3 = 1$ **(b)** no **(c)** Only one input can be active for valid output.
2. **(a)** $\overline{A}_3 = 0, \overline{A}_2 = 1, \overline{A}_1 = 1, \overline{A}_0 = 1$ **(b)** the complement of 0111 is 1000 (8)

Section 6–7 Review

1. 1010101_2
2. seven exclusive-OR gates in an arrangement like that in Figure 6–54.

Section 6–8 Review

1. 0
2. **(a)** quad two-input data selector **(b)** eight-input data selector **(c)** sixteen-input data selector
3. The data output alternates between LOW and HIGH as the data-select inputs sequence through the binary states.
4. **(a)** It multiplexes the two BCD codes to the seven-segment decoder.
   (b) It decodes the BCD to energize the display.
   (c) It enables the seven-segment displays alternately.

Section 6–9 Review

1. by using the input lines for data selection and an Enable line for data input
2. all HIGH except D_{10}, which is LOW

Section 6–10 Review

1. **(a)** 1110100 **(b)** 001100011
2. **(a)** 11010101 **(b)** 11000001
3. **(a)** Correct, four 1s **(b)** Error, seven 1s

Section 6–11 Review

1. A very short-duration voltage spike (usually unwanted).
2. Caused by transition states.
3. The enabling of a device for a specified period of time when the device is not in transition.

The flip-flop is a *bistable* logic circuit that is a type of *multivibrator*. There are three types of multivibrators: the *bistable,* the *monostable,* and the *astable*. In this chapter the bistable elements are emphasized, but coverage of the other two types is also provided.

There are four basic categories of bistable elements: the *latch,* the *edge-triggered flip-flop,* the *pulse-triggered (master-slave) flip-flop,* and the *data lock-out flip-flop*.

Bistable elements (latches and flip-flops) exhibit two stable states. They are capable of residing in either of these two states indefinitely. These two states are called SET and RESET. Because of their ability to retain a given state, bistable elements are useful as storage (memory) devices. The flip-flop is a basic building-block for registers, memories, counters, control logic, and other functions in digital systems.

The monostable multivibrator, commonly called a *one-shot,* has only one stable state. It produces a single pulse in response to a triggering input. The astable multivibrator has no stable states and is used primarily as an oscillator to generate periodic pulse waveforms for timing purposes.

Specific devices introduced in this chapter are as follows:

1. 7475 four-bit latch
2. 7474 dual edge-triggered D flip-flop
3. 74LS76A dual edge-triggered J-K flip-flop
4. 74L71 master-slave S-R flip-flop
5. 74107 dual master-slave J-K flip-flop
6. 74111 dual data lock-out J-K flip-flop
7. 74121 and 74122 one-shots
8. 555 timer

After completing this chapter, you will be able to

☐ use logic gates to construct basic latches.
☐ explain the difference between an S-R latch and a D latch.
☐ recognize the difference between a latch and a flip-flop.
☐ explain how S-R, D, and J-K flip-flops differ.
☐ explain how edge-triggered, master-slave, and data lock-out flip-flops differ.
☐ describe the advantages of a master-slave or data lock-out flip-flop over an edge-triggered type in certain situations.
☐ understand the significance of propagation delays, set-up time, hold time, maximum operating frequency, minimum clock pulse widths, and power dissipation in the application of flip-flops.
☐ apply flip-flops in basic applications.
☐ analyze circuits for race conditions and the occurrence of glitches.
☐ explain how retriggerable and nonretriggerable one-shots differ.
☐ connect a 555 timer to operate as either an astable multivibrator or a one-shot.
☐ approach the debugging of a new design.
☐ troubleshoot basic flip-flop and one-shot circuits.

7

Flip-Flops and Related Devices

LATCHES

7–1

*The latch is a type of bistable storage device that is normally placed in a category separate from that of flip-flops. Latches are basically similar to flip-flops because they are bistable devices that can reside in either of two states by virtue of a **feedback** arrangement, in which the outputs are connected back to the opposite inputs. The main difference between latches and flip-flops is in the method used for changing their state.*

The S-R Latch

An active-HIGH input S-R (SET-RESET) latch is formed with two cross-coupled NOR gates as shown in Figure 7–1(a); an active-LOW input S-R latch is formed with two cross-coupled NAND gates as shown in Figure 7–1(b). Notice that the output of each gate is connected to an input of the opposite gate. This produces the *regenerative feedback* that is characteristic of all multivibrators.

To understand the operation of the latch, we will use the NAND gate S-R latch in Figure 7–1(b). This latch is redrawn in Figure 7–2 with the negative-OR gate equivalents used for the NAND gates. This is done because LOWs on the $\overline{S}$ and $\overline{R}$ lines are the activating inputs.

The latch has two inputs, $\overline{S}$ and $\overline{R}$, and two outputs, Q and $\overline{Q}$. We will start by assuming that both inputs and the Q output are HIGH. Since the Q output is connected back to an input of gate G_2, and the $\overline{R}$ input is HIGH, the output of G_2 must be LOW. This LOW output is coupled back to an input of gate G_1, insuring that its output is HIGH.

When the Q output is HIGH, the latch is in the *SET* state. It will remain in this state indefinitely until a LOW is temporarily applied to the $\overline{R}$ input. With a LOW on the $\overline{R}$ input and a HIGH on $\overline{S}$, the output of gate G_2 is forced HIGH. This HIGH on the $\overline{Q}$ output is coupled back to an input of G_1, and since the $\overline{S}$ input is HIGH, the output of G_1 goes LOW. This LOW on the Q output is coupled back to an input of G_2, ensuring that the $\overline{Q}$ output remains HIGH even when the LOW on the $\overline{R}$ input is removed. When the Q output is LOW, the latch is in the *RESET* state. Now the latch remains indefinitely in the RESET state until a LOW is applied to the $\overline{S}$ input.

Notice that in the operation just described, the outputs are always complements of each other: when Q is HIGH, $\overline{Q}$ is LOW, and when Q is LOW, $\overline{Q}$ is HIGH.

A condition that is not allowed in the operation of an active-LOW input S-R latch occurs when LOWs are applied to both $\overline{S}$ and $\overline{R}$ at the same time. As long as the LOW

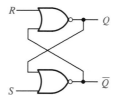

(a) Active-HIGH input S-R latch

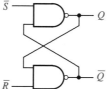

(b) Active-LOW input S-R latch

FIGURE 7–1
Two versions of SET-RESET latches.

FIGURE 7–2

levels are simultaneously held on the inputs, both the Q and $\overline{Q}$ outputs are forced HIGH, thus violating the basic complementary operation of the outputs. Also, if the LOWs are released simultaneously, both outputs will attempt to go LOW. Since there is always some small difference in the propagation delay of the gates, one of the gates will dominate in its transition to the LOW output state. This, in turn, forces the output of the slower gate to remain HIGH. In that situation we cannot reliably predict the next state of the latch.

Figure 7–3 illustrates the active-LOW input S-R latch operation for each of the four possible combinations of levels on the inputs. (The first three combinations are

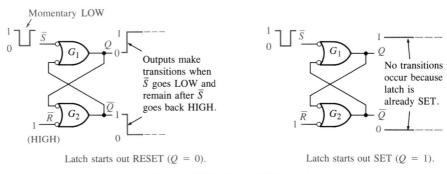

(a) Two possibilities for the SET operation

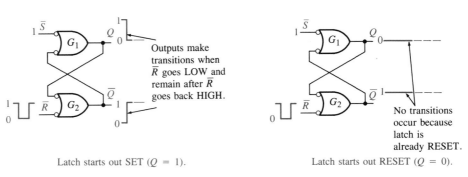

(b) Two possibilities for the RESET operation

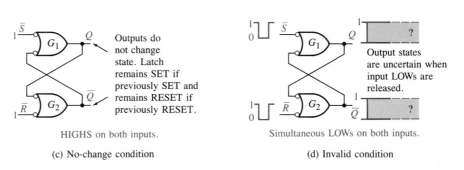

(c) No-change condition

(d) Invalid condition

FIGURE 7–3
The four modes of basic latch operation (SET, RESET, no-change, and invalid).

valid, but the last is not.) Table 7–1 summarizes the logical operation in truth table form. Operation of the active-HIGH input NOR gate latch in Figure 7–1(a) is similar but requires the use of opposite logic levels. You should analyze its operation as an exercise.

Logic symbols for the active-HIGH input and the active-LOW input S-R latches are shown in Figure 7–4.

The following example illustrates how an active-LOW input S-R latch responds to conditions on its inputs. We pulse LOW levels on each input in a certain sequence and observe the resulting Q output waveform. The $\overline{S} = 0, \overline{R} = 0$ condition is avoided because it results in an invalid mode of operation and is a major drawback of any SET-RESET type of latch.

TABLE 7–1
Truth table for an active-LOW input S-R latch.

Inputs		Outputs		
$\overline{S}$	$\overline{R}$	Q	$\overline{Q}$	Comments
1	1	NC	NC	No change. Latch remains in present state.
0	1	1	0	Latch SETS.
1	0	0	1	Latch RESETS.
0	0	1	1	Invalid condition

FIGURE 7–4
Logic symbols for the S-R latch.

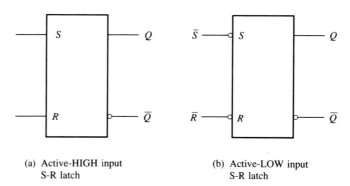

(a) Active-HIGH input S-R latch

(b) Active-LOW input S-R latch

EXAMPLE 7–1

If the $\overline{S}$ and $\overline{R}$ waveforms in Figure 7–5(a) are applied to the inputs of the latch of Figure 7–4(b), determine the waveform that will be observed on the Q output. Assume that Q is initially LOW.

FIGURE 7–5

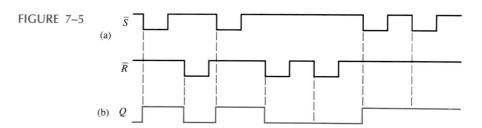

(a)

(b) Q

Solution See Figure 7–5(b).

The Latch as a Contact-Bounce Eliminator

A good example of an application of an S-R latch is in the elimination of mechanical switch contact "bounce." When the pole of a switch strikes the contact upon switch closure, it physically vibrates or bounces several times before finally making a solid contact. Although these bounces are minute, they produce voltage spikes that are often not acceptable in a digital system. This situation is illustrated in Figure 7–6(a).

An S-R latch can be used to eliminate the effects of switch bounce as shown in Figure 7–6(b). The switch is normally in position 1, keeping the $\overline{R}$ input LOW and the latch RESET. When the switch is thrown to position 2, $\overline{R}$ goes HIGH because of the pull-up resistor to V_{CC}, and $\overline{S}$ goes LOW on the first contact. Although $\overline{S}$ remains LOW for only a very short time before the switch bounces, this is sufficient to SET the latch. Any further voltage spikes on the $\overline{S}$ input due to switch bounce do not affect the latch, and it remains SET. Notice that the Q output of the latch provides a clean transition from LOW to HIGH, thus eliminating the voltage spikes caused by contact bounce. Similarly, a clean transition from HIGH to LOW is made when the switch is thrown back to position 1.

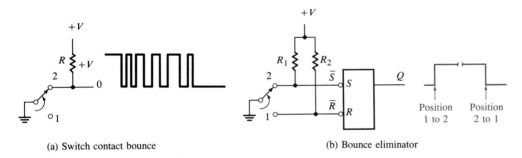

(a) Switch contact bounce (b) Bounce eliminator

FIGURE 7–6
The S-R latch used to eliminate switch contact bounce.

The Gated S-R Latch

A gated latch requires an Enable input, EN. The logic diagram and logic symbol for a gated S-R latch are shown in Figure 7–7. The S and R inputs control the state to which the latch will go upon application of a HIGH level on the EN input. The latch will not change until the EN input is HIGH, but as long as it remains HIGH, the output is

FIGURE 7–7
A gated S-R latch.

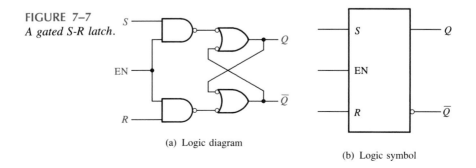

(a) Logic diagram

(b) Logic symbol

determined by the state of the *S* and *R* inputs. In this circuit the invalid state occurs when both *S* and *R* are simultaneously HIGH.

EXAMPLE 7–2

Determine the *Q* output waveform if the inputs shown in Figure 7–8(a) are applied to a gated S-R latch that is initially RESET.

FIGURE 7–8

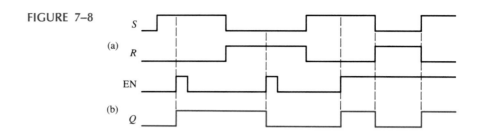

(a)

(b)

Solution The *Q* waveform is shown in Figure 7–8(b). Anytime *S* is HIGH and *R* is LOW, a HIGH on the EN input SETS the latch. Anytime *S* is LOW and *R* is HIGH, a HIGH on the EN input RESETS the latch.

The Gated D Latch

Another type of gated latch is called the D latch. It differs from the S-R latch in that it has only one input in addition to EN. This input is called the *D* input. Figure 7–9 contains a logic diagram and logic symbol of a D latch. When the *D* input is HIGH and the EN input is HIGH, the latch will SET. When the *D* input is LOW and the EN is HIGH, the latch will RESET. Stated another way, the output *Q* follows the input *D* when EN is HIGH.

FIGURE 7–9
A gated D latch.

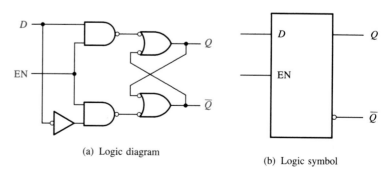

(a) Logic diagram

(b) Logic symbol

EXAMPLE 7–3

Determine the *Q* output waveform if the inputs shown in Figure 7–10(a) are applied to the gated D latch, which is initially RESET.

FIGURE 7–10

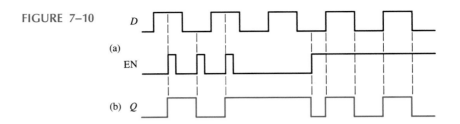

(a)

(b) Q

Solution   The Q waveform is shown in Figure 7–10(b). Whenever D is HIGH and the EN input is HIGH, Q goes HIGH. Whenever D is LOW and the EN input is HIGH, Q goes LOW. When the EN input is LOW, the state of the latch is not affected by the D input.

The 7475 Four-Bit Latches

An example of an IC D latch is the 7475 device represented by the logic symbol in Figure 7–11(a). This device has four latches on a single chip. Notice that each EN input is shared by two latches and is designated as a control input (C). The truth table for each latch is shown in Figure 7–11(b). The X in the truth table represents a "don't care" or irrelevant condition. In this case, when the EN input is LOW, it does not matter what the D input is because the outputs are unaffected and remain in their prior states.

FIGURE 7–11
The 7475 four-bit latches.

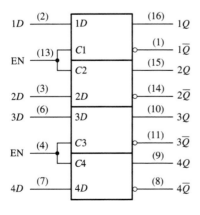

(a) Logic symbol

Inputs		Outputs		Comment
D	EN	Q	$\overline{Q}$	
0	1	0	1	RESET
1	1	1	0	SET
X	0	Q_0	$\overline{Q}_0$	No change

Q_0 is the prior output level.

(b) Truth table (each latch)

SECTION 7–1
REVIEW

Answers are found at the end of the chapter.
1. List three types of latches.
2. Develop the truth table for the active-HIGH input S-R latch in Figure 7–1(a).
3. What is the Q output of a D latch when EN = 1 and D = 1?

EDGE-TRIGGERED FLIP-FLOPS

7–2

Flip-flops *are* synchronous *bistable devices. In this case the term* synchronous *means that the output changes state* only *at a specified point on a triggering input called the* clock *(designated as a control input, C); that is, changes in the output occur in synchronization with the clock.*

An *edge-triggered flip-flop* changes state either at the positive edge (rising edge) or at the negative edge (falling edge) of the clock pulse and is sensitive to its inputs only at this transition of the clock. Three basic types of edge-triggered flip-flops are covered in this section: S-R, D, and J-K. The logic symbols for all of these are shown in Figure 7–12. Notice that each type can be either positive edge-triggered (no bubble at C input) or negative edge-triggered (bubble at C input). The key to identifying an edge-triggered flip-flop by its logic symbol is the small triangle inside the block at the clock (C) input. This triangle is called the *dynamic input indicator*.

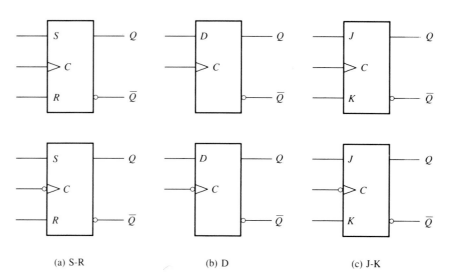

(a) S-R   (b) D   (c) J-K

FIGURE 7–12
Edge-triggered flip-flop logic symbols (top—positive edge-triggered; bottom—negative edge-triggered).

The Edge-Triggered S-R Flip-Flop

The S and R inputs of the S-R flip-flop are called the *synchronous inputs* because data on these inputs are transferred to the flip-flop's output only on the triggering edge of the clock pulse. When S is HIGH and R is LOW, the Q output goes HIGH on the triggering edge of the clock pulse, and the flip-flop is SET. When S is LOW and R is HIGH, the Q output goes LOW on the triggering edge of the clock pulse, and the flip-flop is RESET. When both S and R are LOW, the output does not change from its prior state. An *invalid* condition exists when both S and R are HIGH.

This basic operation is illustrated in Figure 7–13, and Table 7–2 is the truth table for this type of flip-flop. Remember, *the flip-flop cannot change state except on the triggering edge of a clock pulse*. The S and R inputs can be changed at any time when the clock input is LOW or HIGH (except for a very short interval around the triggering transition of the clock) without affecting the output.

The operation and truth table for a negative edge-triggered S-R flip-flop are the same as those for a positive except that the falling edge of the clock pulse is the triggering edge.

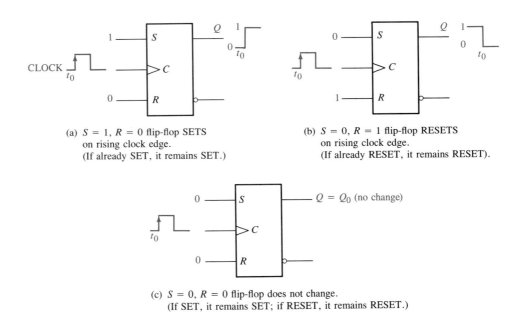

(a) $S = 1$, $R = 0$ flip-flop SETS
on rising clock edge.
(If already SET, it remains SET.)

(b) $S = 0$, $R = 1$ flip-flop RESETS
on rising clock edge.
(If already RESET, it remains RESET).

(c) $S = 0$, $R = 0$ flip-flop does not change.
(If SET, it remains SET; if RESET, it remains RESET.)

FIGURE 7–13
Operation of a positive edge-triggered S-R flip-flop.

TABLE 7–2
Truth table for a positive edge-triggered S-R flip-flop.

Inputs			Outputs		
S	R	C	Q	$\overline{Q}$	Comments
0	0	X	Q_0	$\overline{Q}_0$	No change
0	1	↑	0	1	RESET
1	0	↑	1	0	SET
1	1	↑	?	?	Invalid

↑ = clock transition LOW to HIGH

X = irrelevant ("don't care")

Q_0 = prior output level

EXAMPLE 7–4

Given the waveforms for the *S*, *R*, and *C* inputs in Figure 7–15(a) to the flip-flop of Figure 7–14, determine the *Q* and $\overline{Q}$ output waveforms. Assume that the positive edge-triggered flip-flop is initially RESET.

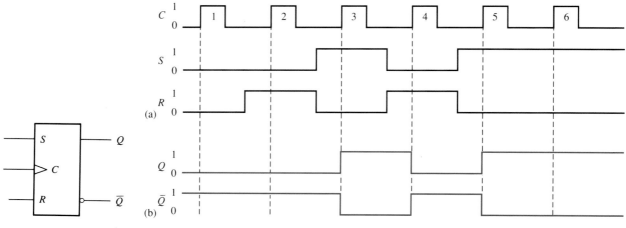

FIGURE 7–14

FIGURE 7–15

Solutions

1. At clock pulse 1, *S* is LOW and *R* is LOW, so *Q* does not change.
2. At clock pulse 2, *S* is LOW and *R* is HIGH, so *Q* remains LOW (RESET).
3. At clock pulse 3, *S* is HIGH and *R* is LOW, so *Q* goes HIGH (SET).
4. At clock pulse 4, *S* is LOW and *R* is HIGH, so *Q* goes LOW (RESET).
5. At clock pulse 5, *S* is HIGH and *R* is LOW, so *Q* goes HIGH (SET).
6. At clock pulse 6, *S* is HIGH and *R* is LOW, so *Q* stays HIGH.

Once *Q* is determined, $\overline{Q}$ is easily found since it is simply the complement of *Q*. The resulting waveforms for *Q* and $\overline{Q}$ are shown in Figure 7–15(b) for the input waveforms in (a).

A Method of Edge-Triggering

Although the internal circuitry of the flip-flops is not of primary concern to us, a simplified discussion of the basic operation of edge-triggering is presented here. A simplified implementation of an edge-triggered S-R flip-flop is illustrated in Figure 7–16(a) and is used to demonstrate the concept of edge-triggering. This coverage of the S-R flip-flop does not imply that it is the most important type. Actually, the D flip-flop and the J-K flip-flop are much more widely used and more available in IC form than is the S-R type. However, the S-R is a good base upon which to build because both the D and the J-K flip-flops are derived from the S-R.

Notice that the S-R flip-flop differs from the gated S-R latch only in that it has a pulse transition detector. The purpose of this circuit is to produce a very short-duration spike on the positive-going transition of the clock pulse. One basic type of pulse transition detector is shown in Figure 7–16(b). As you can see, there is a small delay on one input to the NAND gate so that the inverted clock pulse arrives at the gate input a few nanoseconds after the true clock pulse. This produces an output spike with a dura-

FIGURE 7–16
Edge triggering.

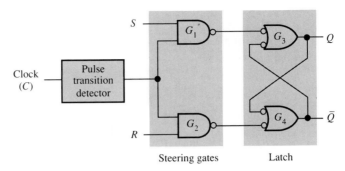

(a) A simplified logic diagram for a
positive edge-triggered S-R flip-flop.

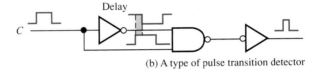

(b) A type of pulse transition detector

tion of only a few nanoseconds. A negative edge-triggered flip-flop inverts the clock pulse, thus producing a narrow spike on the negative-going edge.

Notice that the circuit in Figure 7–16 is partitioned into two sections, one labeled *steering gates,* and the other *latch.* The steering gates direct, or ''steer,'' the clock spike either to the input to gate G_3 or to the input to gate G_4, depending on the state of the S and R inputs. To understand the operation of this flip-flop, begin with the assumptions that it is in the RESET state ($Q = 0$) and that the S, R, and clock inputs are all LOW. For this condition the outputs of gate G_1 and gate G_2 are both HIGH. The LOW on the Q output is coupled back into one input of gate G_4, making the $\overline{Q}$ output HIGH. Because $\overline{Q}$ is HIGH, both inputs to gate G_3 are HIGH (remember, the output of gate G_1 is HIGH), holding the Q output LOW. If a pulse is applied to the clock input, the outputs of gates G_1 and G_2 remain HIGH because they are disabled by the LOWs on the S input and the R input; therefore, there is no change in the state of the flip-flop— it remains RESET.

We will now make S HIGH, leave R LOW, and apply a clock pulse. Because the S input to gate G_1 is now HIGH, the output of gate G_1 goes LOW for a very short time (spike) when the clock input goes HIGH, causing the Q output to go HIGH. Both inputs to gate G_4 are now HIGH (remember, gate G_2 output is HIGH because R is LOW), forcing the $\overline{Q}$ output LOW. This LOW on $\overline{Q}$ is coupled back into one input of gate G_3, ensuring that the Q output will remain HIGH. The flip-flop is now in the SET state. Figure 7–17 illustrates the logic level transitions that take place within the flip-flop for this condition.

Next, we will make S LOW and R HIGH, and we will apply a clock pulse. Because the R input is now HIGH, the positive-going edge of the clock produces a negative-going spike on the output of gate G_2, causing the $\overline{Q}$ output to go HIGH. Because of this HIGH on $\overline{Q}$, both inputs to gate G_3 are now HIGH (remember, the output of gate G_1 is HIGH because of the LOW on S), forcing the Q output to go LOW. This LOW on Q is coupled back into one input of gate G_4, ensuring that $\overline{Q}$ will remain HIGH. The flip-flop is now in the RESET state. Figure 7–18 illustrates the logic level transitions that occur within the flip-flop for this condition.

As with the gated latch, an invalid condition exists when both S and R are HIGH at the same time.

FIGURE 7–17

Flip-flop making transition from the RESET state to the SET state on the positive leading edge of the clock pulse.

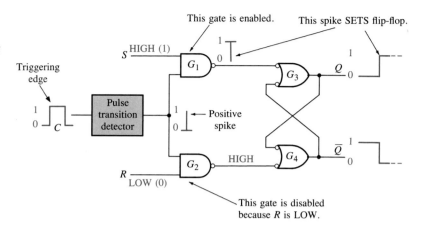

FIGURE 7–18

Flip-flop making transition from the SET state to the RESET state on the positive leading edge of the clock pulse.

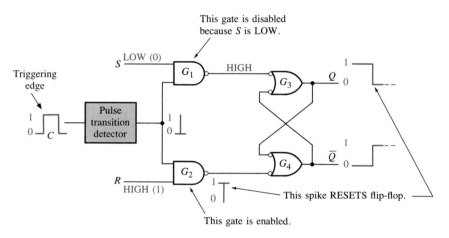

The Edge-Triggered D Flip-Flop

The D flip-flop is very useful when a single data bit (1 or 0) is to be stored. The simple addition of an inverter to an S-R flip-flop creates a basic D flip-flop, as in Figure 7–19, where a positive edge-triggered type is shown.

FIGURE 7–19

A positive edge-triggered D flip-flop formed with an S-R flip-flop and an inverter.

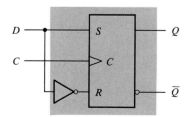

Notice that the flip-flop in Figure 7–19 has only one input in addition to the clock. This is called the D input. If there is a HIGH on the D input when a clock pulse is applied, the flip-flop will SET, and the HIGH on the D input will be "stored" by the flip-flop on the positive leading edge of the clock pulse. If there is a LOW on the D input when the clock pulse is applied, the flip-flop will RESET, and the LOW on the D input will thus be stored by the flip-flop on the leading edge of the clock pulse. In the SET state the flip-flop is storing a 1, and in the RESET state it is storing a 0.

The operation of the positive edge-triggered D flip-flop is summarized in Table 7–3. The operation of a negative edge-triggered device is, of course, the same, except that triggering occurs on the falling edge of the clock pulse. Remember, Q follows D at the clock edge.

TABLE 7–3
Truth table for a positive edge-triggered D flip-flop.

Inputs		Outputs		
D	C	Q	$\overline{Q}$	Comments
1	↑	1	0	SET (stores a 1)
0	↑	0	1	RESET (stores a 0)

↑ = clock transition LOW to HIGH

EXAMPLE 7–5

Given the waveforms in Figure 7–20(a) for the D input and the clock, determine the Q output waveform if the flip-flop starts out RESET.

FIGURE 7–20

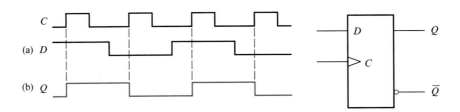

Solution The Q output assumes the state of the D input at the time of the positive leading clock edge. The resultant output is shown in Figure 7–20(b).

The Edge-Triggered J-K Flip-Flop

The J-K flip-flop is very versatile and is perhaps the most widely used type of flip-flop. The J and K designations for the inputs have no known significance except that they are adjacent letters in the alphabet.

The functioning of the J-K flip-flop is identical to that of the S-R flip-flop in the SET, RESET, and no-change conditions of operation. The difference is that the J-K flip-flop has no invalid state as does the S-R flip-flop. Therefore, the J-K flip-flop is a very versatile device that finds wide application in digital systems.

Figure 7–21 shows a basic positive edge-triggered J-K flip-flop. Notice that it differs from the S-R edge-triggered flip-flop in that the Q output is connected back to the input of gate G_2, and the $\overline{Q}$ output is connected back to the input of gate G_1. The

FIGURE 7–21

A simplified logic diagram for a positive edge-triggered J-K flip-flop.

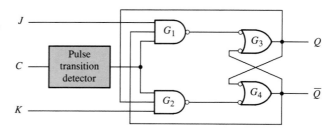

two control inputs are labeled J and K. A J-K flip-flop can also be of the negative edge-triggered type, in which case the clock input is inverted.

We will start by assuming that the flip-flop is RESET and that the J input is HIGH and the K input is LOW. When a clock pulse occurs, a leading-edge spike is passed through gate G_1 because $\overline{Q}$ is HIGH and J is HIGH. This will cause the latch portion of the flip-flop to change to the SET state.

The flip-flop is now SET. If we now make J LOW and K HIGH, the next clock spike will pass through gate G_2 because Q is HIGH and K is HIGH. This will cause the latch portion of the flip-flop to change to the RESET state.

Now if a LOW is applied to both the J and K inputs, the flip-flop will stay in its present state when a clock pulse occurs. So, a LOW on both J and K results in a *no-change* condition.

So far, the logical operation of the J-K flip-flop is the same as that of the S-R type in the SET, RESET, and no-change modes. The difference in operation occurs when both the J and K inputs are HIGH. To see this, assume that the flip-flop is RESET. The HIGH on the $\overline{Q}$ enables gate G_1, so the clock spike passes through to SET the flip-flop. Now there is a HIGH on Q, which allows the next clock spike to pass through gate G_2 and RESET the flip-flop. As you can see, on each successive clock spike, the flip-flop changes to the opposite state. This mode is called *toggle* operation. Figure 7–22 illustrates the transitions when the flip-flop is in the toggle mode.

Table 7–4 summarizes the operation of the edge-triggered J-K flip-flop in truth table form. Notice that there is no invalid state as there is with an S-R flip-flop. The truth table for a negative edge-triggered device is identical except that it is triggered on the falling edge of the clock pulse.

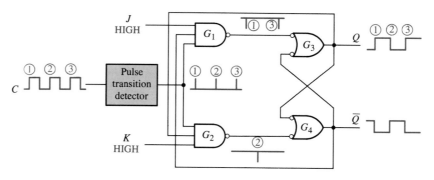

FIGURE 7–22

Transitions illustrating the toggle operation when $J = 1$ and $K = 1$.

TABLE 7–4

Truth table for a positive edge-triggered J-K flip-flop.

Inputs			Outputs		
J	K	C	Q	$\overline{Q}$	Comments
0	0	↑	Q_0	$\overline{Q_0}$	No change
0	1	↑	0	1	RESET
1	0	↑	1	0	SET
1	1	↑	$\overline{Q_0}$	Q_0	Toggle

↑ = clock transition LOW to HIGH

Q_0 = output level prior to clock transition

EXAMPLE 7–6

The waveforms in Figure 7–23(a) are applied to the J, K, and clock inputs as indicated. Determine the Q output, assuming that the flip-flop is initially RESET.

FIGURE 7–23

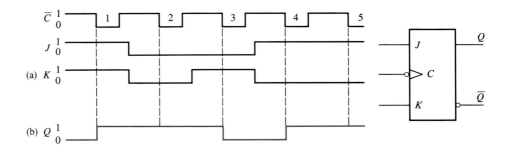

Solution

1. First, since this is a negative edge-triggered flip-flop, the Q output will change only on the negative-going edge of the clock pulse.
2. At the first clock pulse, both J and K are HIGH, and because this is a toggle condition, Q goes HIGH.
3. At clock pulse 2, a *no-change* condition exists on the inputs, keeping Q at a HIGH level.
4. When clock pulse 3 occurs, J is LOW and K is HIGH, resulting in a RESET condition; Q goes LOW.
5. At clock pulse 4, J is HIGH and K is LOW, resulting in a SET condition; Q goes HIGH.
6. A SET condition still exists on J and K when clock pulse 5 occurs, so Q will remain HIGH.

The resulting Q waveform is indicated in Figure 7–23(b).

Asynchronous Inputs

For the flip-flops just discussed, the *S-R, D,* and *J-K* inputs are called *synchronous inputs* because data on these inputs are transferred to the flip-flop's output only on the triggering edge of the clock pulse; that is, the data are transferred synchronously with the clock.

Most integrated circuit flip-flops also have *asynchronous inputs*. These are inputs that affect the state of the flip-flop independent of the clock. They are normally labeled *preset* (PRE) and *clear* (CLR), or *direct set* (S_D) and *direct reset* (R_D) by some manufacturers. An active level on the preset input will SET the flip-flop, and an active level on the clear input will RESET it. A logic symbol for a J-K flip-flop with preset and clear is shown in Figure 7–24. These inputs are active-LOW, as indicated by the bubbles. These preset and clear inputs must both be kept HIGH for synchronous operation.

Figure 7–25 shows the logic diagram for an edge-triggered J-K flip-flop with active-LOW *preset* ($\overline{PRE}$) and *clear* ($\overline{CLR}$) inputs. This figure illustrates basically how these inputs work. As you can see, they are connected directly to the latch portion of the flip-flop so that they override the effect of the synchronous inputs, *J, K,* and the clock.

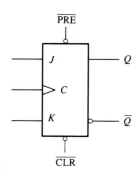

FIGURE 7–24
Logic symbol for a J-K flip-flop with active-LOW preset and clear inputs.

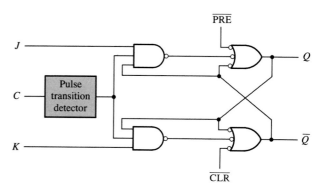

FIGURE 7–25
Logic diagram for a basic J-K flip-flop with active-LOW preset and clear.

EXAMPLE 7–7

For the positive edge-triggered J-K flip-flop with preset and clear inputs in Figure 7–26(a), determine the Q output for the inputs shown in the timing diagram if Q is initially LOW.

FIGURE 7–26

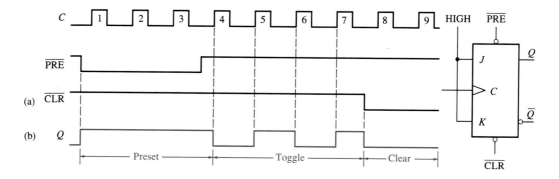

Solution

1. During clock pulses 1, 2, and 3, the preset ($\overline{PRE}$) is LOW, keeping the flip-flop SET regardless of the synchronous inputs.

2. For clock pulses 4, 5, 6, and 7, toggle operation occurs because J is HIGH, K is HIGH, and both $\overline{PRE}$ and $\overline{CLR}$ are HIGH.

3. For clock pulses 8 and 9, the clear ($\overline{CLR}$) input is LOW, keeping the flip-flop RESET regardless of the synchronous inputs. The resulting Q output is shown in Figure 7–26(b).

Specific Devices

Two specific IC edge-triggered flip-flops are now examined. They are representative of the various types available in IC form.

7474 dual D flip-flops This TTL device has two identical flip-flops on a single chip. The flip-flops are independent of each other except for sharing V_{CC} and ground. The flip-flops are positive edge-triggered and have active-LOW asynchronous preset and clear inputs. The logic symbols for the individual flip-flops within the package are shown in Figure 7–27(a), and an ANSI/IEEE standard single block symbol representing the entire device is shown in part (b) of the figure. The pin numbers are shown in parentheses.

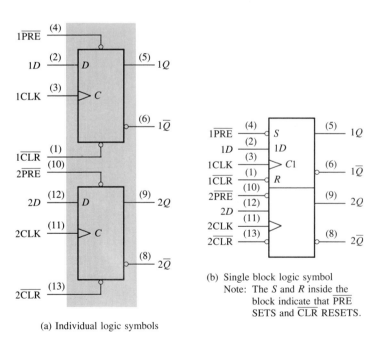

(a) Individual logic symbols

(b) Single block logic symbol
Note: The S and R inside the block indicate that $\overline{PRE}$ SETS and $\overline{CLR}$ RESETS.

FIGURE 7–27
Logic symbols for the 7474 dual positive edge-triggered D flip-flops.

74LS76A dual J-K flip-flops This TTL device also has two identical flip-flops that are negative edge-triggered and have asynchronous preset and clear inputs. The logic symbols are shown in Figure 7–28.

FIGURE 7–28

Logic symbols for the 74LS76A dual negative edge-triggered J-K flip-flops.

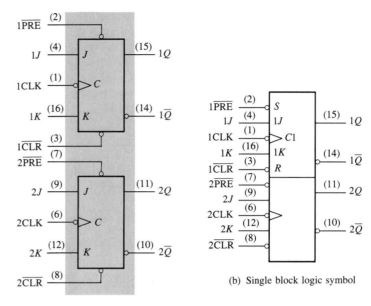

(a) Individual logic symbols

(b) Single block logic symbol

EXAMPLE 7–8

The *J*, *K*, *C*, $\overline{PRE}$, and $\overline{CLR}$ waveforms in Figure 7–29(a) are applied to one of the flip-flops in a 74LS76A package. Determine the *Q* output waveform.

FIGURE 7–29

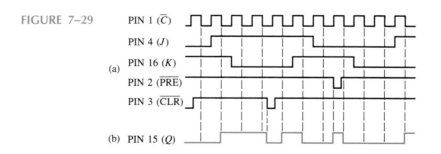

(a)

PIN 1 ($\overline{C}$)

PIN 4 (*J*)

PIN 16 (*K*)

PIN 2 ($\overline{PRE}$)

PIN 3 ($\overline{CLR}$)

(b) PIN 15 (*Q*)

Solution The resulting *Q* waveform is shown in Figure 7–29(b). Notice that each time a LOW is applied to the $\overline{PRE}$ or $\overline{CLR}$, the flip-flop is SET or RESET regardless of the states of the other inputs. Note that LOWs on both $\overline{PRE}$ and $\overline{CLR}$ at the same time cause both *Q* and $\overline{Q}$ to go HIGH, which is a logically invalid state.

SECTION 7–2
REVIEW

Answers are found at the end of the chapter.

1. Describe the main difference between a gated S-R latch and an edge-triggered S-R flip-flop.
2. How does a J-K flip-flop differ from an S-R flip-flop in its basic operation?
3. Assume that the flip-flop in Figure 7–20 is negative edge-triggered. Sketch the output waveform for the same *C* and *D* waveforms.

PULSE-TRIGGERED (MASTER-SLAVE) FLIP-FLOPS

7–3

The second class of flip-flop is the **pulse-triggered** *or* **master-slave.** *The term* **pulse-triggered** *means that data are entered into the flip-flop on the leading edge of the clock pulse but the output does not reflect the input state until the trailing edge of the clock pulse. The inputs must be set up prior to the clock pulse's leading edge, but the output is postponed until the trailing edge of the clock. A major restriction of the pulse-triggered flip-flop is that the data inputs must not change while the clock pulse is HIGH, because the flip-flop is sensitive to any change of input levels during this time.*

As with the edge-triggered flip-flops, there are three basic types of pulse-triggered flip-flops: S-R, D, and J-K. The J-K is by far the most commonly available in integrated circuit form. The logic symbols for all of these are shown in Figure 7–30. The key to identifying a pulse-triggered (master-slave) flip-flop by its logic symbol is the ANSI/IEEE *postponed output* symbol (⌐) at the outputs. This symbol means that the output does not change until the occurrence of the clock edge (either positive- or negative-going) following the triggering edge. Notice that there is no dynamic input indicator (▷) at the clock (*C*) input.

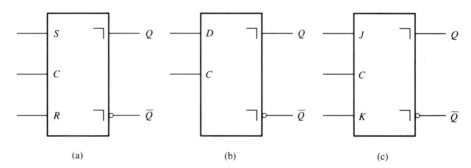

(a)   (b)   (c)

FIGURE 7–30
Pulse-triggered (master-slave) flip-flop logic symbols.

The Pulse-Triggered (Master-Slave) S-R Flip-Flop

A basic master-slave S-R flip-flop is shown in Figure 7–31. The truth table operation is the same as that for the edge-triggered S-R flip-flop except for the way it is clocked. Internally, though, it is quite different.

This type of flip-flop is composed of two sections, the *master* section and the *slave* section. The master section is basically a gated S-R latch, and the slave section is the same except that it is clocked on the inverted clock pulse and is controlled by the outputs of the master section rather than by the external *S-R* inputs.

The master section will assume the state determined by the *S* and *R* inputs at the leading edge (positive-going) of the clock pulse. The state of the master section is then transferred to the slave section on the trailing edge (negative-going) of the clock pulse, because the outputs of the master are applied to the inputs of the slave and the clock pulse to the slave is inverted. The state of the slave then immediately appears on the *Q* and $\overline{Q}$ outputs.

FIGURE 7–31

Logic diagram for a basic master-slave S-R flip-flop.

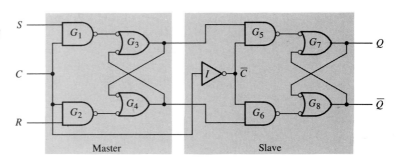

The timing diagram in Figure 7–32 illustrates the master-slave operation in reference to Figure 7–31. To begin, the flip-flop is RESET, S is HIGH, and R is LOW. The following statements describe what happens on the leading edge of the first clock pulse:

1. The output of gate G_1 goes from a HIGH to a LOW because both of its inputs are HIGH.
2. The output of gate G_3 goes from a LOW to a HIGH, and the output of gate G_4 goes from a HIGH to a LOW.
3. The inverted clock ($\overline{C}$) input to gates G_5 and G_6 goes LOW, thus disabling the G_5 and G_6 gates and forcing their outputs HIGH.

Let us review what has taken place on the leading edge of the clock pulse. The master section has been SET because the S input is HIGH and the R input is LOW. The

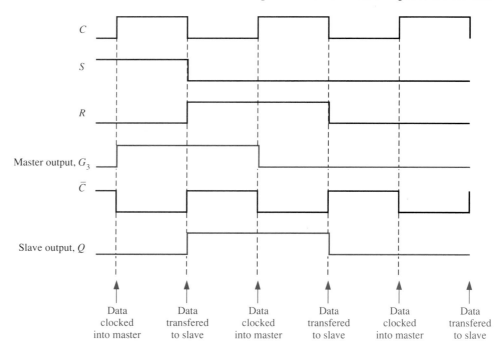

FIGURE 7–32

Timing diagram for master-slave flip-flop in Figure 7–31, showing SET, RESET, and no-change conditions.

slave section has not changed state, and therefore the Q and $\overline{Q}$ outputs remain in the initial RESET state.

Now let us look at what happens on the trailing edge of the first clock pulse:

1. The master section remains in the SET condition.
2. The output of gate G_5 goes from a HIGH to a LOW because both of its inputs are now HIGH.
3. The output of gate G_7 goes from a LOW to a HIGH, and the output of gate G_8 goes from a HIGH to a LOW.

The flip-flop is now in the SET state, and the Q output is therefore HIGH. It did not become SET until the trailing edge of the clock pulse, although the master section was SET on the leading edge. You should verify the operations for the RESET and no-change conditions on the second and third clock pulses, respectively.

The operation of the S-R master-slave flip-flop is summarized in Table 7–5. As mentioned, it is the same as for an edge-triggered S-R except for the manner in which it is triggered. This is indicated on the truth table by the pulse symbols in the C column, which mean that it takes the *complete pulse* to get data from the input to the output rather than just a single transition as with the edge-triggered type.

TABLE 7–5
Truth table for the S-R master-slave flip-flop.

Inputs			Outputs		
S	R	C	Q	$\overline{Q}$	Comments
0	0	⊓	Q_0	$\overline{Q}_0$	No change
0	1	⊓	0	1	RESET
1	0	⊓	1	0	SET
1	1	⊓	?	?	Invalid

⊓ = clock pulse
Q_0 = output level prior to clock pulse

The Pulse-Triggered (Master-Slave) D Flip-Flop

The truth table operation of the master-slave D flip-flop is shown in Table 7–6. Notice that it is the same as for the edge-triggered D flip-flop except for the way it is triggered. The basic method of master-slave triggering discussed in relation to the S-R flip-flop applies also to the D flip-flop, as well as to the J-K flip-flop.

TABLE 7–6
Truth table for the master-slave D flip-flop.

Inputs		Outputs		
D	C	Q	$\overline{Q}$	Comments
0	⊓	0	1	RESET
1	⊓	1	0	SET

EXAMPLE 7–9

Determine the Q output for the master-slave D flip-flop with the D and clock (C) inputs shown in Figure 7–33(a). The flip-flop is initially RESET.

FIGURE 7–33

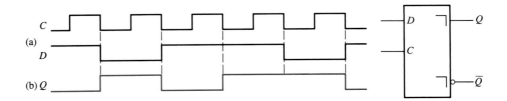

Solution The data are clocked into the master portion on the leading edges of the clock pulses, and the output changes accordingly on the trailing edges, as shown in Figure 7–33(b).

It was mentioned in the beginning of this section that a major restriction of the master-slave flip-flop is that the input data must be held constant while the clock is HIGH. If the D input changes while the clock pulse is HIGH, the new data bit is immediately stored by the master section; thus the data bit that was on the input at the start of the clock pulse is lost. That is, the flip-flop responds to any change in the input level when the clock is HIGH. The last input level seen while the clock pulse is HIGH is then triggered into the slave portion and appears on the output. This effect is illustrated in Figure 7–34.

Pulse-Triggered (Master-Slave) J-K Flip-Flop

Figure 7–35 is the basic logic diagram for a master-slave J-K flip-flop. A brief inspection of this diagram shows that this circuit is very similar to the master-slave S-R flip-flop. Notice the two main differences: First, the Q output is connected back into the input of gate G_2, and the $\overline{Q}$ output is connected back into the input of gate G_1. Second, one input is now designated J and the other input is now designated K. The logical operation is summarized in Table 7–7.

FIGURE 7–34
Example of the effect of a change in input level while clock is HIGH in a master-slave flip-flop.

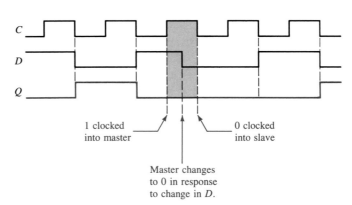

FIGURE 7–35

Logic diagram for a master-slave J-K flip-flop.

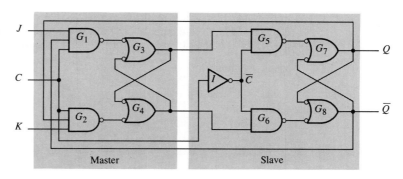

TABLE 7–7

Truth table for the master-slave J-K flip-flop.

Inputs			Outputs		
J	K	C	Q	$\overline{Q}$	Comments
0	0	⎍	Q_0	$\overline{Q}_0$	No change
0	1	⎍	0	1	RESET
1	0	⎍	1	0	SET
1	1	⎍	$\overline{Q}_0$	Q_0	Toggle

⎍ = clock pulse

Q_0 = output level before clock pulse

EXAMPLE 7–10

Determine the Q output of the master-slave J-K flip-flop for the input waveforms shown in Figure 7–36(a). The flip-flop starts out RESET. Notice that the clock is active-LOW.

FIGURE 7–36

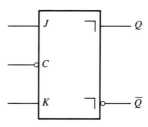

(a)

(b)

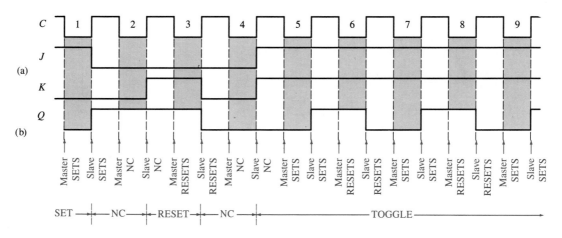

Solution The Q waveform is shown in Figure 7–36(b). The input states and events at each clock pulse are labeled to demonstrate the operation.

Specific Devices

Two specific TTL pulse-triggered flip-flops are now presented, and examples of their operation are given.

The 74L71 master-slave S-R flip-flop Two versions of the logic symbol for this flip-flop are shown in Figure 7–37. A more traditional symbol is in part (a), and the ANSI/IEEE Standard 91–1984 recommended symbol is shown in part (b). Notice that this particular device has AND gates for the S and R inputs. The three SET inputs, $S1$, $S2$, and $S3$, must all be HIGH to produce an $S = 1$ condition within the flip-flop. Likewise, $R1$, $R2$, and $R3$ must all be HIGH to produce an $R = 1$ condition. These AND-gated inputs simply provide additional flexibility in controlling the flip-flop. This device also has active-LOW preset ($\overline{PRE}$) and clear ($\overline{CLR}$) inputs.

In the symbol in part (b), the preset is labeled S inside the block because a LOW on the $\overline{PRE}$ line SETS the flip-flop. The clear is labeled R inside the block because a LOW on the $\overline{CLR}$ line RESETS the flip-flop. These two inputs are asynchronous, not dependent on the clock. The $1S$ and $1R$ labels inside the block are the synchronous SET and RESET functions. The 1 suffix after C and the 1 prefixes before S and R are examples of *control dependency notation*. The 1s indicate that $1S$ and $1R$ are dependent on the clock, $C1$.

FIGURE 7–37

Logic symbols for the 74L71 master-slave S-R flip-flop.

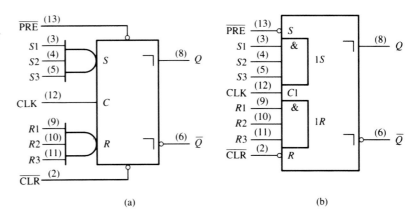

(a)   (b)

EXAMPLE 7–11

From the input waveforms in Figure 7–38(a), determine the Q and $\overline{Q}$ outputs for a 74L71 master-slave S-R flip-flop. Assume that $Q = 0$ initially.

Solution The resulting output waveforms are shown in Figure 7–38(b). The $\overline{PRE}$ and $\overline{CLR}$ inputs override synchronous operation.

FIGURE 7–38

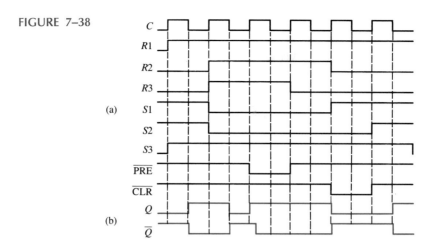

The 74107 dual master-slave J-K flip-flop Figure 7–39(a) shows the individual logic symbols for each flip-flop in the 74107 package. Part (b) represents the device with a single logic symbol, including dependency notation. Notice that this device has a $\overline{\text{CLR}}$ input but not a $\overline{\text{PRE}}$. This allows it to be asynchronously RESET for purposes of initialization. The labels inside the upper block in part (b) apply also to the lower block.

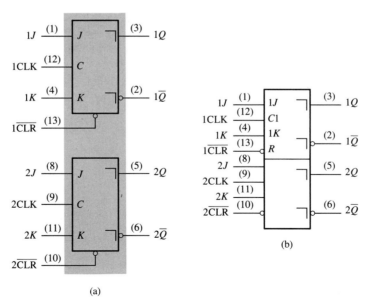

FIGURE 7–39
Logic symbols for the 74107 dual master-slave J-K flip-flops.

EXAMPLE 7–12

A clock pulse waveform with a frequency of 1 kHz is applied to a 74107 master-slave J-K flip-flop. The *J* and *K* inputs are both connected to V_{CC}. The $\overline{CLR}$ line is held LOW for the three initial clock pulses, and then it goes HIGH. Draw the timing diagram for ten clock pulses.

FIGURE 7–40

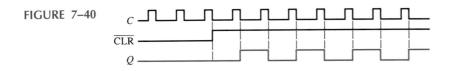

Solution The timing diagram is shown in Figure 7–40. Since *J* and *K* are connected to V_{CC}, the flip-flop is in the toggle condition ($J = 1$, $K = 1$). As long as the $\overline{CLR}$ is LOW, toggle operation is overridden, and the flip-flop is held in RESET. As soon as $\overline{CLR}$ is released from its LOW state, the flip-flop begins toggling on the fourth pulse.

SECTION 7–3 REVIEW

Answers are found at the end of the chapter.

1. Describe the basic difference between pulse-triggered and edge-triggered flip-flops.
2. Suppose that the *D* input of a flip-flop changes from LOW to HIGH in the middle of a clock pulse.
   (a) Describe what happens if the flip-flop is a positive edge-triggered type.
   (b) Describe what happens if the flip-flop is a pulse-triggered (master-slave) type.
3. What is the advantage of a J-K flip-flop over an S-R flip-flop?

DATA LOCK-OUT FLIP-FLOPS

7–4

*The data lock-out flip-flop is similar to the pulse-triggered (master-slave) flip-flop except that it has a **dynamic** clock input, which makes it sensitive to the data inputs only during a clock transition.*

In a data lock-out flip-flop, after the leading-edge clock transition, the data inputs are disabled (locked out) and do not have to be held constant while the clock pulse is HIGH. In essence, the master portion of this flip-flop is like an edge-triggered device, and the slave portion performs as in a pulse-triggered device to produce a postponed output on the trailing-edge transition of the clock.

A logic symbol for a data lock-out J-K flip-flop is shown in Figure 7–41. Notice that this symbol has both the dynamic input indicator for the clock and the postponed output indicators. This type of flip-flop is actually classified by most manufacturers as a master-slave with a special lock-out feature.

The 74111 is an example of this type of device. It is a dual J-K master-slave flip-flop with data lock-out. It has both preset and clear inputs. The logic symbols are shown in Figure 7–42. Again, part (a) is the more traditional symbol, and part (b) is a single block symbol as specified by ANSI/IEEE Standard 91–1984.

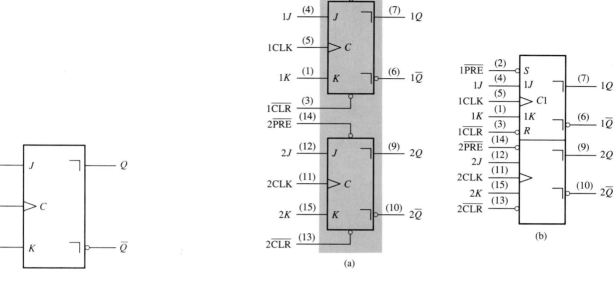

FIGURE 7–41
Logic symbol for a data lock-out J-K flip-flop.

FIGURE 7–42
Logic symbols for the 74111 dual J-K master-slave flip-flop with data lock-out.

EXAMPLE 7–13

The waveforms in Figure 7–43(a) are applied to the data lock-out flip-flop as shown. Determine the Q output, starting in the RESET state.

FIGURE 7–43

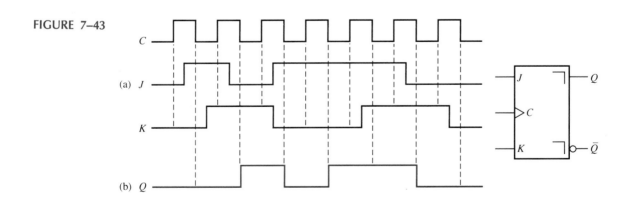

Solution The states of the J and K inputs are clocked into the master on the positive-going edge (triggering edge) of the clock pulse and into the slave (and thus onto the outputs) on the negative-going edge. A change in J or K after the triggering edge of the clock has no effect on the output, as shown in Figure 7–43, where J and K change several times while the clock is HIGH.

SECTION 7–4
REVIEW

Answers are found at the end of the chapter.

1. Describe how a data lock-out flip-flop differs from a pulse-triggered type.
2. What determines the synchronous output of a data lock-out flip-flop?

OPERATING CHARACTERISTICS

7–5

Several operating characteristics or parameters important in the application of flip-flops specify the performance, operating requirements, and operating limitations of the circuit. They are typically found in the data sheets for integrated circuits, and they are applicable to all flip-flops regardless of the particular form of the circuit.

Propagation Delay Time

A propagation delay is the interval of time required after an input signal has been applied for the resulting output change to occur. Several categories of propagation delay are important in the operation of a flip-flop:

1. Propagation delay t_{PLH} as measured from the triggering edge of the clock pulse to the LOW-to-HIGH transition of the output. This delay is illustrated in Figure 7–44(a).

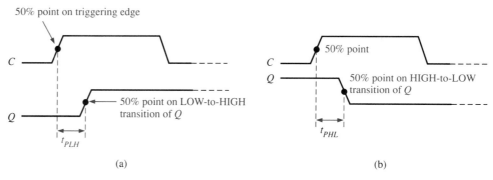

(a)   (b)

FIGURE 7–44
Propagation delays, clock to output.

2. Propagation delay t_{PHL} as measured from the triggering edge of the clock pulse to the HIGH-to-LOW transition of the output. This delay is illustrated in Figure 7–44(b).
3. Propagation delay t_{PLH} as measured from the preset input to the LOW-to-HIGH transition of the output. This delay is illustrated in Figure 7–45(a) for an active-LOW preset.
4. Propagation delay t_{PHL} as measured from the clear input to the HIGH-to-LOW transition of the output. This delay is illustrated in Figure 7–45(b) for an active-LOW clear.

FIGURE 7–45
Propagation delays, preset and clear to output.

Set-up Time

The *set-up time* (t_s) is the minimum interval required for the logic levels to be maintained constantly on the inputs (J and K, or S and R, or D) prior to the triggering edge of the clock pulse in order for the levels to be reliably clocked into the flip-flop. This interval is illustrated in Figure 7–46 for a D flip-flop.

FIGURE 7–46
Set-up time (t_s).

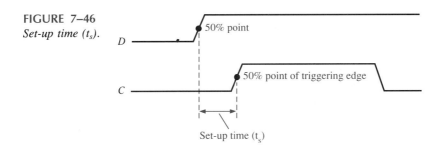

Hold Time

The *hold time* (t_h) is the minimum interval required for the logic levels to remain on the inputs after the triggering edge of the clock pulse in order for the levels to be reliably clocked into the flip-flop. This is illustrated in Figure 7–47 for a D flip-flop.

FIGURE 7–47
Hold time (t_h).

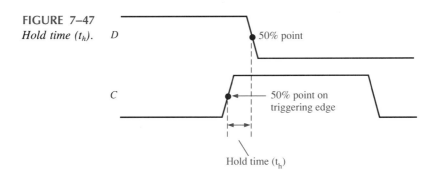

Maximum Clock Frequency

The maximum clock frequency (f_{max}) is the highest rate at which a flip-flop can be reliably triggered. At clock frequencies above the maximum, the flip-flop would be unable to respond quickly enough, and its operation would be impaired.

Pulse Widths

Minimum pulse widths (t_W) for reliable operation are usually specified by the manufacturer for the clock, preset, and clear inputs. Typically, the clock is specified by its minimum HIGH time and its minimum LOW time.

Power Dissipation

The power dissipation of any digital circuit is the total power consumption of the device. For example, if the flip-flop operates on a +5-V dc source and draws 50 mA of current, the power dissipation is

$$P = V_{CC} \times I_{CC} = 5 \text{ V} \times 50 \text{ mA} = 250 \text{ mW}$$

The power dissipation is very important in most applications in which the capacity of the dc supply is a concern. As an example, let us assume that we have a digital system that requires a total of ten flip-flops, and each flip-flop dissipates 250 mW of power. The total power requirement is

$$P_{\text{TOT}} = 10 \times 250 \text{ mW} = 2500 \text{ mW} = 2.5 \text{ W}$$

This tells us the output capacity required of the dc supply. If the flip-flops operate on +5 V dc, then the amount of current that the supply must provide is as follows:

$$I = \frac{2.5 \text{ W}}{5 \text{ V}} = 0.5 \text{ A}$$

We must use a +5-V dc supply that is capable of providing at least 0.5 A of current.

Comparison of Specific Flip-Flops

Table 7–8 provides a comparison, in terms of the operating parameters discussed in this section, of several TTL devices that have been studied in this chapter, as well as a CMOS device.

SECTION 7–5
REVIEW

Answers are found at the end of the chapter.
1. Define the following:
   (a) set-up time   (b) hold time
2. Which specific flip-flop can be operated at the highest frequency?

TABLE 7–8

Comparison of operating parameters of flip-flops.

Parameter (Times in ns)	TTL					CMOS
	7474	74LS76A	74L71	74107	74111	74HC112
t_{PHL} (CLK to Q)	40	20	150	40	30	31
t_{PLH} (CLK to Q)	25	20	75	25	17	31
t_{PHL} ($\overline{CLR}$ to Q)	40	20	200	40	30	41
t_{PLH} ($\overline{PRE}$ to Q)	25	20	75	25	18	41
t_s (set-up)	20	20	0	0	0	25
t_h (hold)	5	0	0	0	30	0
t_W (CLK HI)	30	20	200	20	25	25
t_W ($\overline{CLK\ LO}$)	37	—	200	47	25	25
t_W (CLR/PRE)	30	25	100	25	25	25
f_{max} (MHz)	25	45	3	20	25	20
Power (mW/F-F)	43	10	3.8	50	70	0.12

SOURCE: Compiled from Texas Instruments TTL data book, 1985.

NOTE: Values given are typical where available; otherwise, they may be maximum or minimum depending on availability.

BASIC FLIP-FLOP APPLICATIONS

7–6

In this section, several basic applications are presented to give you a better feel for how flip-flops can be applied in digital systems.

Parallel Data Storage

A common requirement in digital systems is to take several bits of data on parallel lines and store them simultaneously in a group of flip-flops. This operation is illustrated in Figure 7–48(a). Each of the four parallel data lines is connected to the D input of a flip-flop. The clock inputs of all the flip-flops are connected to a common clock input, so that each flip-flop is triggered by the same clock pulse. In this example, positive edge-triggered flip-flops are used, so the data on the D inputs are stored simultaneously by the flip-flops on the positive edge of the clock, as indicated in the timing diagram in Figure 7–48(b). Also, the asynchronous clear inputs are connected to a common $\overline{CLR}$ line, which resets all the flip-flops. In this application, resetting clears out all data bits previously stored.

Registers are groups of flip-flops used for data storage. In digital systems, data are normally stored in groups of bits that represent numbers, codes, or other information.

Data Transfer

It is often necessary to transfer a data bit from one flip-flop to another, as illustrated in Figure 7–49 with pulse-triggered (master-slave) flip-flops. A data bit is clocked into flip-flop A on the leading edge of the first clock pulse. (The data bit is a 1 in this case

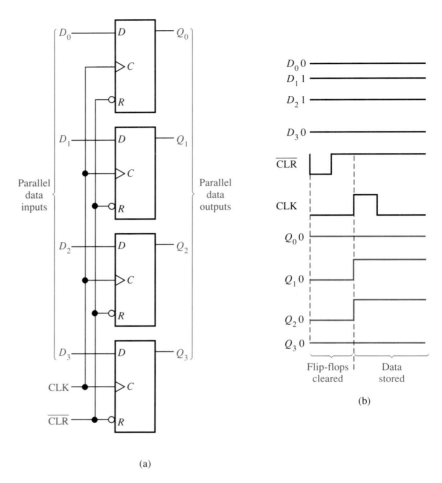

(a)

FIGURE 7–48
Example of flip-flops used for parallel data storage.

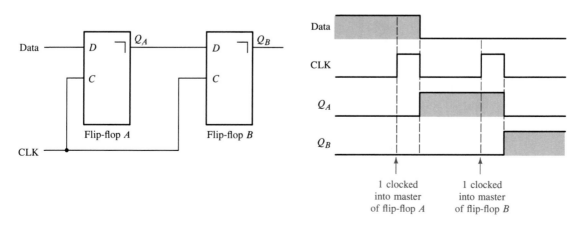

FIGURE 7–49
Data bit (indicated by shaded area) being transferred to flip-flop A and then to flip-flop B.

and is indicated by the shaded area in the timing diagram.) The Q_A output goes HIGH on the trailing edge of the clock pulse. Since the Q output of flip-flop A is connected to the D input of flip-flop B, the data bit is clocked into flip-flop B on the leading edge of the second clock pulse and appears on the Q_B output on the trailing edge. Thus, the transfer of the data bit from flip-flop A to flip-flop B has been achieved. This concept is important in certain types of registers that are covered in a later chapter.

This example is a good way to demonstrate the advantage of master-slave flip-flops over edge-triggered flip-flops in certain situations. Figure 7–50 shows the circuit of Figure 7–49 implemented with negative edge-triggered D flip-flops. The data bit is clocked into flip-flop A on the trailing edge of the first clock pulse, and Q_A goes HIGH at this time (actually t_{PLH} later). At the trailing edge of the second clock pulse, Q_A goes LOW a short time (t_{PHL}) after the data bit is clocked into flip-flop B, creating what is called a *race* condition. This produces a marginal condition because if Q_A (which is also the D input of flip-flop B) does not remain HIGH long enough after the triggering edge of the clock, the data bit may not be transferred. The Q_A output must remain HIGH for a time equal to the hold time (t_h) in order to be reliably clocked into flip-flop B. The data bit is properly transferred only if $t_h < t_{PHL}$ (CLK to Q). So a potentially unreliable data-transfer situation is created. The very low hold times on some newer edge-triggered flip-flops eliminate the need for master-slave flip-flops in many applications in which they previously had an advantage.

FIGURE 7–50

A case of data transfer using edge-triggered flip-flops. An unreliable transfer is possible because of a race condition.

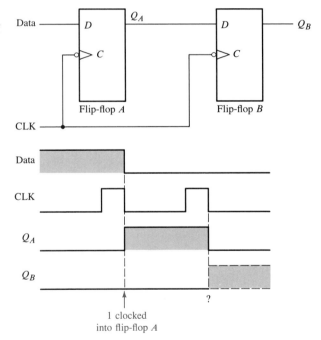

Frequency Division

Another basic application of a flip-flop is dividing (reducing) the frequency of a periodic waveform. When a pulse waveform is applied to the clock input of a J-K flip-flop that is connected to toggle, the Q output is a square wave with one-half the frequency of the clock input. Thus, a single flip-flop is a divide-by-2 device, as is illustrated in Figure

7–51. As you can see, the flip-flop changes state on each triggering clock edge (positive edge-triggered in this case). This results in an output that changes at half the frequency of the clock waveform.

Further division of a clock frequency can be achieved by using the output of one flip-flop as the clock input to a second flip-flop, as shown in Figure 7–52. The frequency of the Q_A output is divided by 2 by flip-flop B. The Q_B output is, therefore, one-fourth the frequency of the original clock input.

By connecting flip-flops in this way, a frequency division of 2^n is achieved, where n is the number of flip-flops. For example, three flip-flops divide the clock frequency by $2^3 = 8$; four flip-flops divide the clock frequency by $2^4 = 16$; and so on.

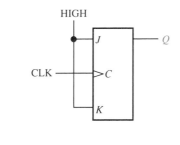

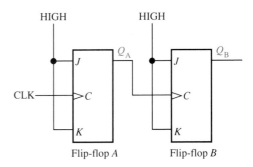

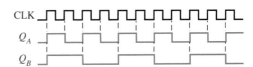

FIGURE 7–51
The J-K flip-flop as a divide-by-2 device.

FIGURE 7–52
Example of two J-K flip-flops used to divide the clock frequency by 4.

Counting

Another very important application of flip-flops is in digital counters, which are covered in detail in the next chapter. The concept is illustrated in Figure 7–53. The flip-flops are negative edge-triggered J-Ks. Both flip-flops are initially RESET. Flip-flop A toggles on the negative-going transition of each clock pulse. The Q output of flip-flop A clocks flip-flop B, so each time Q_A makes a HIGH-to-LOW transition, flip-flop B toggles. The resulting Q_A and Q_B waveforms are shown in the figure.

Observe the sequence of Q_A and Q_B. Prior to clock pulse 1, $Q_A = 0$ and $Q_B = 0$; after clock pulse 1, $Q_A = 1$ and $Q_B = 0$; after clock pulse 2, $Q_A = 0$ and $Q_B = 1$; and after clock pulse 3, $Q_A = 1$ and $Q_B = 1$. If we take Q_A as the least significant bit, a two-bit binary sequence is produced as the flip-flops are clocked. This binary sequence repeats every four clock pulses, as shown in the timing diagram of Figure 7–53. Thus, the flip-flops are counting in sequence from 0 to 3 (00, 01, 10, 11) and then recycling back to 0 to begin the sequence again.

FIGURE 7–53

Flip-flops used to generate a binary count sequence.

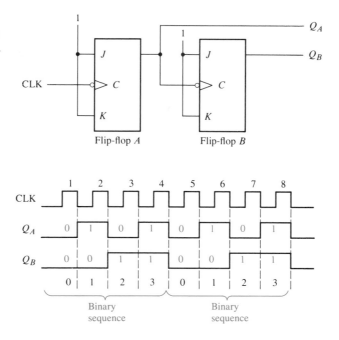

SECTION 7–6 REVIEW

Answers are found at the end of the chapter.
1. A group of flip-flops used for data storage is called a _____.
2. How must a J-K flip-flop be connected to function as a divide-by-2 element?
3. How many flip-flops are required to produce a divide-by-32 device?

ONE-SHOTS

7–7

*The **one-shot** is a **monostable** multivibrator. This device has only one stable state. When triggered, it changes from its stable to its unstable state and remains there for a specified length of time before returning automatically to its stable state.*

Figure 7–54 shows a basic one-shot circuit composed of a logic gate and an inverter. When a pulse is applied to the *trigger* input, the output of gate G_1 goes LOW. This HIGH-to-LOW transition is coupled through the capacitor to the input of inverter G_2. The apparent LOW on G_2 makes its output go HIGH. This HIGH is connected back into G_1, keeping its output LOW. Up to this point the trigger pulse has caused the output of the one-shot, Q, to go HIGH.

The capacitor immediately begins to charge through R toward the high voltage level. The rate at which it charges is determined by the RC time constant. When the capacitor charges to a certain level, which appears as a HIGH to G_2, the output goes back LOW.

To summarize, the output of inverter G_2 goes HIGH in response to the trigger input. It remains HIGH for a time set by the RC time constant. At the end of this time, it goes LOW. So a single narrow trigger pulse produces a single output pulse whose time duration is controlled by the *RC* time constant. This operation is illustrated in Figure 7–54. A typical one-shot logic symbol is shown in Figure 7–55(a), and the same symbol with an external *R* and *C* is shown in Figure 7–55(b).

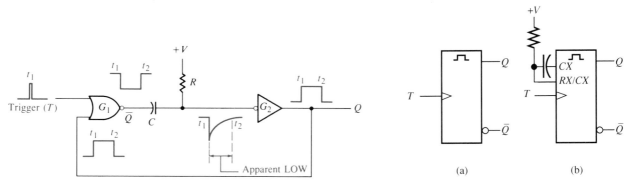

FIGURE 7–54
A simple one-shot circuit.

FIGURE 7–55
Basic one-shot logic symbols.

Integrated Circuit One-Shots

The two basic types of IC one-shots are *nonretriggerable* and *retriggerable*. A nonretriggerable type will not respond to any additional trigger pulses from the time it is triggered into its unstable state (fired) until it returns to its stable state (times out). In other words, the period of the trigger pulses must be greater than the time the one-shot remains in its unstable state. The time that the one-shot remains in its unstable state is the pulse width of the output.

Figure 7–56 shows the one-shot being triggered at intervals greater than its pulse width and at intervals less than the pulse width. Notice that in the second case, the additional pulses are ignored.

A retriggerable one-shot can be triggered before it times out. The result of retriggering is an extension of the pulse width as illustrated in Figure 7–57.

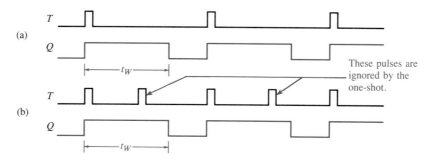

FIGURE 7–56
Nonretriggerable one-shot action.

FIGURE 7–57
Retriggerable one-shot action.

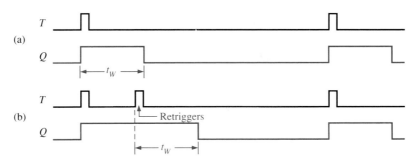

(a)

(b)

The 74121 Nonretriggerable One-Shot

The 74121 is an example of a nonretriggerable IC one-shot. It has provisions for external R and C, as shown in Figure 7–58. The inputs labeled $A1$, $A2$, and B are the gated trigger inputs. The R_{INT} label represents an internal timing resistor.

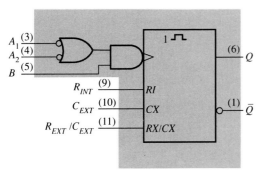

(a) Traditional logic symbol

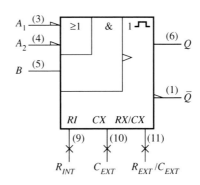

(a)*ANSI/IEEE* std. 91-1984 logic
symbol ($\times$ = nonlogic connection).
1 ⎍ is the qualifying symbol
for a nonretriggerable one-shot.

FIGURE 7–58
Logic symbols for the 74121 nonretriggerable one-shot (monostable multivibrator).

How to Set the Pulse Width

The output pulse width can be varied by selection of the external components. A minimum pulse width of about 30 ns is produced when no external timing components are used, as shown in Figure 7–59(a). The pulse width can be set anywhere between about 40 ns and 28 s by the external components.

Figure 7–59(b) shows connection of the internal resistor (2 kΩ) and an external capacitor. Part (c) illustrates connection of an external resistor and an external capacitor. The pulse width is set by the values of the resistor ($R_{INT} = 2$ kΩ, and R_{EXT} is selected) and the capacitor according to the following formula:

$$t_W = 0.7RC_{EXT}$$

(7–1)

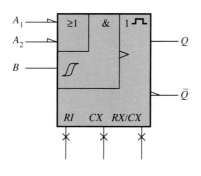

(a) No external components
($t_W \cong 30$ ns)

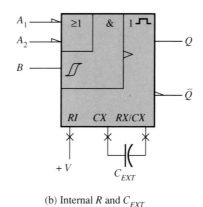

(b) Internal R and C_{EXT}

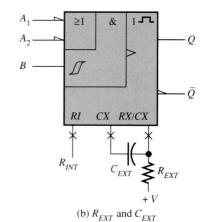

(b) R_{EXT} and C_{EXT}

FIGURE 7–59
Three ways to set the pulse width of a 74121.

EXAMPLE 7–14

A certain application requires a one-shot with a pulse width of approximately 1 s. Using a 74121, show the connections and the component values.

Solution Let us arbitrarily select $R_{EXT} = 10$ MΩ and calculate the capacitance that we need.

$$t_W = 0.7R_{EXT}C_{EXT}$$

$$C_{EXT} = \frac{t_W}{0.7R_{EXT}}$$

Since 10 MΩ = 10×10^6 Ω,

$$C_{EXT} = \frac{1\text{s}}{0.7(10 \times 10^6 \ \Omega)} = 0.143 \times 10^{-6}\text{F} = 0.143 \ \mu\text{F}$$

We can use a standard 0.15 μF capacitor. The proper connections are shown in Figure 7–60.

FIGURE 7–60

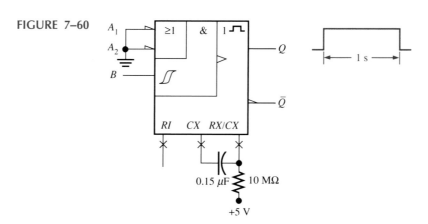

The Schmitt-Trigger Symbol

The symbol $\int\!\!\int$ indicates a Schmitt-trigger input. This type of input employs a special threshold circuit that produces *hysteresis,* a characteristic that prevents erratic switching between states when the input voltage hovers around the critical input level. This allows reliable triggering to occur even when the input is changing very slowly.

The 74122 Retriggerable One-Shot

The 74122 is an example of a retriggerable IC one-shot with a *clear* input. It also has provisions for external R and C, as shown in Figure 7–61. The inputs labeled $A1$, $A2$, $B1$, and $B2$ are the gated trigger inputs.

As with the 74121, the pulse width is set by the external resistor and capacitor connected to pins 10 and 11. A general formula for calculating the values of these components for a specified pulse width (t_W) is

$$t_W = KR_{\text{EXT}}C_{\text{EXT}}\left(1 + \frac{0.7}{R_{\text{EXT}}}\right) \qquad \textbf{(7–2)}$$

where K is a constant determined by the particular type of one-shot and is usually given on manufacturers' data sheets. For example, K is 0.32 for the 74122. (Notice the difference between this formula and that for the 74121, shown in Equation [7–1].)

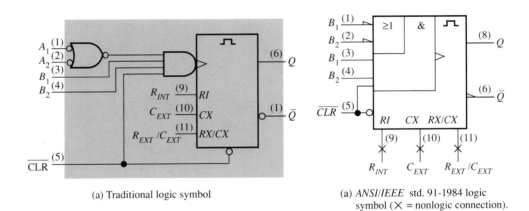

(a) Traditional logic symbol

(a) *ANSI/IEEE* std. 91-1984 logic symbol ($\times$ = nonlogic connection). $\sqcap$ is the qualifying symbol

FIGURE 7–61
Logic symbol for the 74122 retriggerable one-shot (monostable multivibrator).

EXAMPLE 7–15

Determine the values of R_{EXT} and C_{EXT} that will produce a pulse width of 1 μs when connected to a 74122.

Solution The manufacturer's data sheet gives $K = 0.32$ for this device.

$$t_W = KR_{\text{EXT}}C_{\text{EXT}}\left(1 + \frac{0.7}{R_{\text{EXT}}}\right)$$

Assume a value of $C_{EXT} = 560$ pF and then solve for R_{EXT}. t_w must be expressed in ns and C_{EXT} in pF.

$$t_W = KR_{EXT}C_{EXT} + 0.7\left(\frac{KR_{EXT}C_{EXT}}{R_{EXT}}\right)$$

$$= KR_{EXT}C_{EXT} + 0.7KC_{EXT}$$

$$R_{EXT} = \frac{t_W - 0.7KC_{EXT}}{KC_{EXT}} = \frac{t_W}{KC_{EXT}} - 0.7$$

$$= \frac{1000 \text{ ns}}{(0.32)560\text{pF}} - 0.7$$

$$= 4.88 \text{ k}\Omega$$

SECTION 7–7
REVIEW

Answers are found at the end of the chapter.
1. Describe the difference between a nonretriggerable and a retriggerable one-shot.
2. How is the output pulse width adjusted in most IC one-shots?

ASTABLE MULTIVIBRATORS AND TIMERS

7–8

*Another type of multivibrator is the **astable (free-running) type. This device has no stable states; it switches back and forth between two unstable states. The astable multivibrator is used as an oscillator to provide clock signals for timing purposes.***

Figure 7–62 shows a basic astable using two inverters. Notice the capacitive coupling from the output of each inverter to the input of the other. The operation is similar to that of the one-shot in its unstable state, but the two capacitive coupling networks prevent either inverter from having a stable state. If the circuit is designed properly, it will start oscillating on its own and requires no initial input trigger.

FIGURE 7–62
An example of an astable multivibrator.

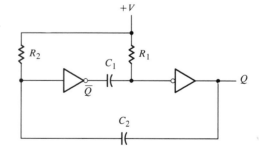

The 555 Timer

The 555 timer is an integrated circuit that can be used as an astable or monostable multivibrator and for many other applications. Figure 7–63 shows a functional diagram of a 555 timer.

The threshold and trigger levels are normally $\frac{2}{3}V_{CC}$ and $\frac{1}{3}V_{CC}$, respectively. These levels can be altered with external connections to the *control voltage* terminal. When

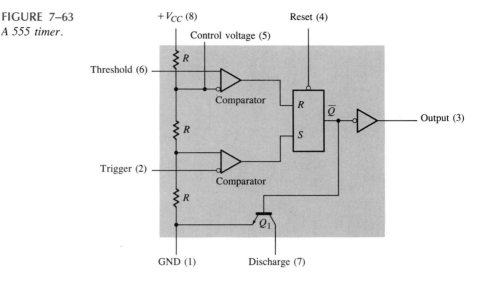

FIGURE 7–63
A 555 timer.

the trigger input goes below the trigger level, the internal latch is SET and the output goes HIGH. When the threshold input goes above the threshold level, the latch is RESET and the output goes LOW.

The RESET inputs can override the other inputs. Here is what happens when the RESET is LOW: The latch is RESET, causing the output to go LOW. This turns on transistor Q_1, providing a low-resistance path from the *discharge* terminal to ground.

Astable operation   A 555 timer is shown in Figure 7–64 connected as an astable multivibrator. Notice that the trigger and threshold inputs are connected together. The capacitor C charges through R_1 and R_2 and discharges through R_2. Therefore, the frequency and the duty cycle of the output waveform can be set by selecting proper values for these two resistors. Figure 7–65 shows typical waveforms produced by this device. It can be used as a source for pulse waveforms in various applications.

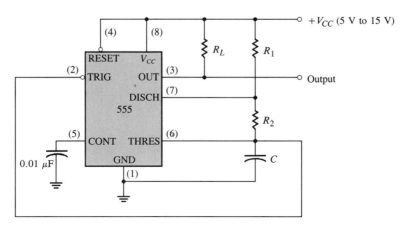

FIGURE 7–64
A 555 timer connected for astable operation.

FIGURE 7–65
Astable waveforms.

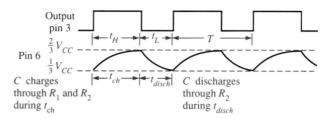

The following formulas can be used to set the frequency and the duty cycle of the 555 astable configuration.

$$f \cong \frac{1}{0.7(R_1 + 2R_2)C} \tag{7–3}$$

The HIGH time of the output waveform is

$$t_H \cong 0.7(R_1 + R_2)C \tag{7–4}$$

The LOW time of the output waveform is

$$t_L \cong 0.7R_2C \tag{7–5}$$

EXAMPLE 7–16

Determine the external component values for the 555 timer astable configuration in Figure 7–64 that will produce a frequency of 1 kHz and a duty cycle of 75%.

Solution   The period of the output waveform is

$$T = \frac{1}{f} = 1 \text{ ms}$$

For a duty cycle of 75%,

$$t_H = 0.75 \text{ ms}$$
$$t_L = 0.25 \text{ ms}$$

If we select a capacitor value of 0.1 μF, the resistance values are calculated as follows:
From Equation (7–4), we have

$$R_1 + R_2 \cong \frac{t_H}{0.7C} = \frac{0.75 \text{ ms}}{0.7(0.1 \text{ μF})} = 10{,}714 \text{ } \Omega$$

From Equation (7–5), we have

$$R_2 \cong \frac{t_L}{0.7C} = \frac{0.25 \text{ ms}}{0.7(0.1 \text{ μF})} = 3571 \text{ } \Omega$$

Then

$$R_1 = 10{,}714 \text{ } \Omega - R_2 = 10{,}714 \text{ } \Omega - 3571 \text{ } \Omega = 7143 \text{ } \Omega$$

Monostable operation . The 555 timer can also be set up to operate as a nonretriggerable one-shot. The connections for this configuration are shown in Figure 7–66. Compare this with the astable configuration.

A negative-going pulse on the trigger input initiates a positive-going output pulse with a duration determined by the following formula.

$$t_W = 1.1RC \qquad\qquad (7\text{–}6)$$

FIGURE 7–66

A 555 timer connected for monostable operation.

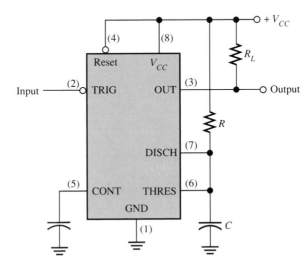

EXAMPLE 7–17

What is the output pulse width for a 555 monostable circuit with $R = 2.2$ kΩ and $C = 0.01$ µF?

Solution

$$t_W = 1.1RC = 1.1(2.2 \text{ k}\Omega)(0.01 \text{ µF}) = 24.2 \text{ µs}$$

SECTION 7–8 REVIEW

Answers are found at the end of the chapter.

1. Explain the difference in operation between an astable multivibrator and a monostable multivibrator.
2. For a certain astable multivibrator, $t_H = 15$ ms and $T = 20$ ms. What is the duty cycle of the output?

TROUBLESHOOTING

7–9

In industry, it is standard practice to test a new circuit design to be sure that it is operating as specified. New designs are usually "breadboarded" and tested before the design is finalized. The term **breadboard** *refers to a method of temporarily hooking up a circuit so that its operation can be verified and any faults (bugs) worked out before a prototype unit is built. In this section we will consider an example case.*

The purpose of the circuit in Figure 7–67(a) is to generate two clock waveforms (CLK A and CLK B) having an alternating occurrence of pulses. Each waveform is to be one-half the frequency of the original clock (CLK), as shown in the ideal timing diagram in part (b).

When the circuit is tested, the CLK A and CLK B waveforms appear on the oscilloscope as shown in Figure 7–68(a). Since glitches are observed on both waveforms, something is wrong with the circuit either in its basic design or in the way it is connected. Further investigation reveals that the glitches are caused by a race condition between the CLK signal and the Q and $\bar{Q}$ signals at the inputs of the AND gates. As displayed in Figure 7–68(b), the propagation delays between CLK and Q and $\bar{Q}$ create a short-duration coincidence of HIGH levels at the leading edges of alternate clock pulses. Thus, there is a basic design flaw.

The problem can be corrected by using a negative edge-triggered flip-flop in place of the positive edge-triggered device, as shown in Figure 7–69(a). Although the propagation delays between CLK and Q and $\bar{Q}$ still exist, they are initiated on the trailing edges of the clock (CLK), thus eliminating the glitches, as shown in the timing diagram of Figure 7–69(b).

SECTION 7–9 REVIEW

Answers are found at the end of the chapter.
1. Can a master-slave J-K flip-flop be used in the circuit of Figure 7–69 as effectively as the negative edge-triggered flip-flop?
2. What device can be used to provide the clock for the circuit in Figure 7–69?

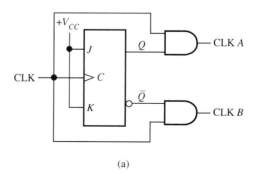

(a)

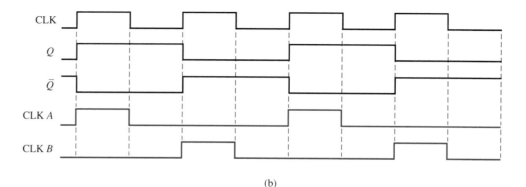

(b)

FIGURE 7–67
Two-phase clock generator with ideal waveforms.

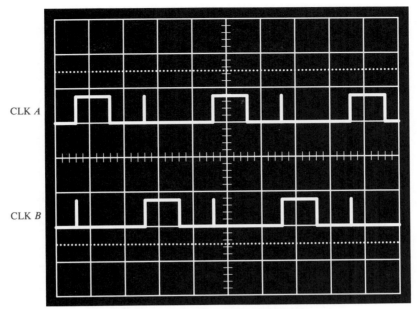

(a) Oscilloscope display of CLK A and CLK B waveforms

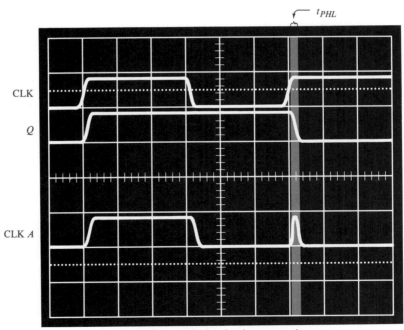

(b) Oscilloscope display showing propagation
delay that creates glitch on CLK A waveform

FIGURE 7–68
Oscilloscope displays for the circuit in Figure 7–67.

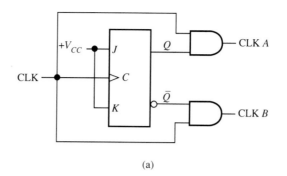

(a)

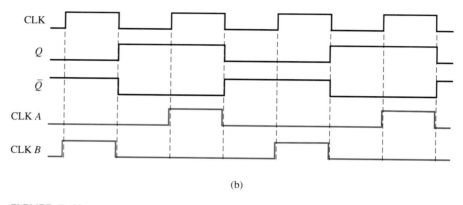

(b)

FIGURE 7–69

Two-phase clock generator using negative edge-triggered flip-flop to eliminate glitches.

SUMMARY

☐ The symbols for all devices in this summary conform with ANSI/IEEE Standard 91–1984. (Preset and clear inputs are not shown on devices.)

☐ Latches are bistable elements whose state normally depends on asynchronous inputs. Figure 7–70 shows symbols and truth tables for latches.

☐ Edge-triggered flip-flops are bistable elements with synchronous inputs whose state depends on the inputs only at the triggering transition of a clock pulse. Changes in the outputs occur at the triggering transition of the clock. Figure 7–71 shows symbols and truth tables for edge-triggered flip-flops.

☐ Pulse-triggered or master-slave flip-flops are bistable elements with synchronous inputs whose state depends on the inputs at the leading edge of the clock pulse, but whose output is postponed and does not reflect the internal state until the trailing clock edge. The synchronous inputs should not be allowed to change while the clock is HIGH. Figure 7–72 shows symbols and truth tables for master-slave flip-flops.

☐ Data lock-out flip-flops are similar to master-slave types in that they are sensitive to the inputs only at the leading clock edge but the output is postponed until the trailing clock edge. However, inputs can be changed while the clock is HIGH without affecting the operation. Figure 7–73 shows symbols and truth tables for data lock-out flip-flops.

FIGURE 7–70
Latches.

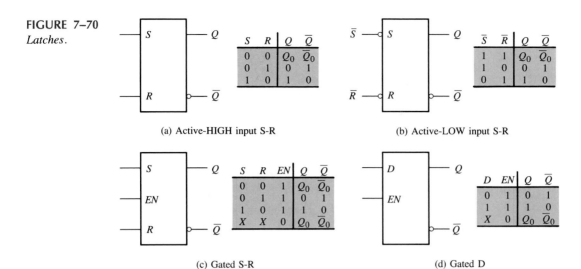

(a) Active-HIGH input S-R

(b) Active-LOW input S-R

(c) Gated S-R

(d) Gated D

Note: Q_0 is the initial state.

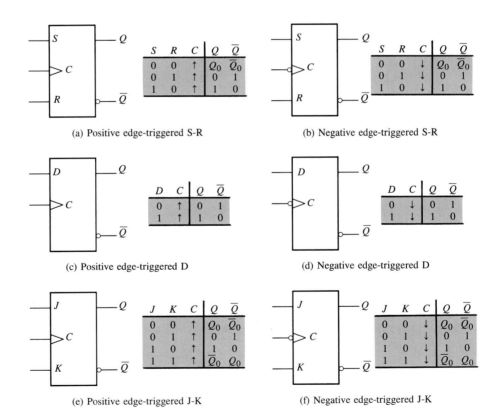

(a) Positive edge-triggered S-R

(b) Negative edge-triggered S-R

(c) Positive edge-triggered D

(d) Negative edge-triggered D

(e) Positive edge-triggered J-K

(f) Negative edge-triggered J-K

FIGURE 7–71
Edge-triggered flip-flops.

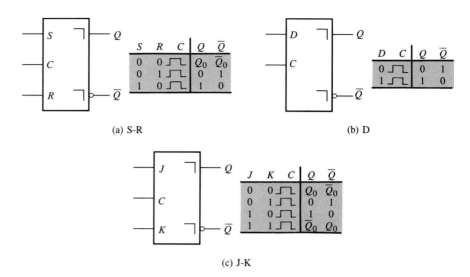

(a) S-R

(b) D

(c) J-K

FIGURE 7–72
Pulse-triggered (master-slave) flip-flops.

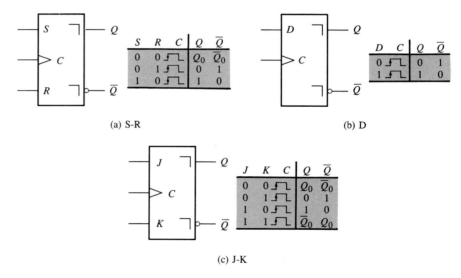

(a) S-R

(b) D

(c) J-K

FIGURE 7–73
Data lock-out flip-flops.

FIGURE 7–74
Monostable multivibrators (one-shots).

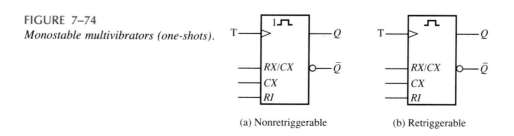

(a) Nonretriggerable

(b) Retriggerable

☐ Monostable multivibrators (one-shots) have one stable state. When the one-shot is triggered, the output goes to its unstable state for a time determined by an *RC* circuit. Figure 7–74 shows the symbols for nonretriggerable and retriggerable one-shots.

☐ Astable multivibrators have no stable states and are used as oscillators to generate timing waveforms in digital systems.

Integrated circuit technologies (Chapter 13) may be covered after this chapter, covered after a later chapter, completely omitted, or partially omitted, depending on the needs and goals of your program. Coverage is not prerequisite or corequisite to any other material in the book. Inclusion is purely optional.

GLOSSARY

Astable   Having no stable state. An astable multivibrator oscillates between two quasistable states.

Asynchronous   Having no fixed time relationship.

Bistable   Having two stable states. Flip-flops and latches are bistable multivibrators.

Clear   To reset, as in the case of a flip-flop.

D flip-flop   A type of bistable multivibrator in which the output follows the state of the D input.

Edge-triggered flip-flop   A type of flip-flop in which the data are entered and appear on the output on the same clock edge.

Flip-flop   A synchronous bistable device used for storing a bit of information.

Hold time   The time interval required for the control levels to remain on the inputs to a flip-flop after the triggering edge of the clock in order to reliably activate the device.

Hysteresis   A characteristic of a threshold-triggered circuit such as the Schmitt trigger.

J-K flip-flop   A type of flip-flop that can operate in the SET, RESET, no-change, and toggle modes.

Master-slave flip-flop   A type of flip-flop in which the input data are entered into the device on the leading edge of the clock pulse and appear at the output on the trailing edge. Also called *pulse-triggered flip-flop*.

Monostable   Having only one stable state. A monostable multivibrator, commonly called a *one-shot*, produces a single pulse in response to a triggering input.

Multivibrator   A class of digital circuits in which the output is connected back to the input (an arrangement called feedback) to produce either two stable states, one stable state, or no stable states, depending on the configuration.

One-shot   A monostable multivibrator.

Race   A condition in a logic network in which the differences in propagation times through two or more signal paths in the network can produce an erroneous output.

RESET   The state of a flip-flop or latch when the *Q* output is 0.

SET   The state of a flip-flop or latch when the *Q* output is 1.

Set-up time   The time interval required for the control levels to be on the inputs to a digital circuit, such as a flip-flop, prior to the triggering edge of the clock pulse.

S-R flip-flop   A SET-RESET flip-flop.

Synchronous   Having a fixed time relationship.

Toggle   The action of a flip-flop when it changes state on each clock pulse.

Trigger   A pulse used to initiate a change in the state of a logic circuit.

SELF-TEST

Answers and solutions are found at the end of the book.

1. An active-HIGH input S-R latch has a 1 on the S input and a 0 on the R input. What state is the latch in?

2. Why is the $S = 1$, $R = 1$ input state invalid in an S-R latch with active-HIGH inputs?

3. The following sequence of levels is applied to the inputs of a gated D latch: D goes HIGH, then EN goes HIGH, then D goes LOW, and then EN goes LOW and back HIGH. What is the sequence of levels on the Q output if Q is initially LOW?

4. List the three basic types of edge-triggered flip-flops classified by inputs.

5. Name the two types of edge-triggered flip-flops classified by the method of triggering, and explain the difference.

6. What advantage does a J-K flip-flop have over an S-R?

7. For a negative edge-triggered J-K flip-flop, $J = 0$, $K = 1$, and $Q = 0$. When the clock input goes HIGH, what happens to Q? When the clock input goes LOW, what happens to Q?

8. Describe basically how data are clocked into a pulse-triggered (master-slave) flip-flop.

9. What is the major restriction when operating a pulse-triggered flip-flop?

10. How can a data lock-out flip-flop be distinguished symbolically from a pulse-triggered flip-flop?

11. Typically, a manufacturer's data sheet specifies four different propagation delay times associated with a flip-flop. Name and describe each one.

12. The data sheet of a certain flip-flop specifies that the minimum HIGH time for the clock pulse is 30 ns and the minimum LOW time is 37 ns. What is the maximum operating frequency?

13. List four basic flip-flop applications.

14. Sketch the output of a one-shot with an output pulse width of 50 μs that is triggered with a 10-kHz signal.

15. How are astable multivibrators normally used in digital systems?

PROBLEMS

Answers to odd-numbered problems are found at the end of the book.

Section 7–1 Latches

7–1   If the waveforms in Figure 7–75 are applied to an active-LOW input S-R latch, sketch the resulting Q output waveform in relation to the inputs. Assume that Q starts LOW.

7–2   Solve Problem 7–1 for the input waveforms in Figure 7–76 applied to an active-HIGH S-R latch.

FIGURE 7–75

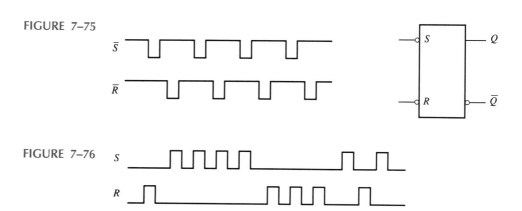

FIGURE 7–76

7–3   Solve Problem 7–1 for the input waveforms in Figure 7–77.

7–4   For a gated S-R latch, determine the Q and $\overline{Q}$ outputs for the given inputs in Figure 7–78. Show them in proper relation to the Enable. Assume that Q starts LOW.

7–5   Solve Problem 7–4 for the inputs in Figure 7–79.

7–6   Solve Problem 7–4 for the inputs in Figure 7–80.

7–7   For a gated D latch, the waveforms shown in Figure 7–81 are observed on its inputs. Sketch the output waveform you would expect to see at Q if the latch is initially RESET.

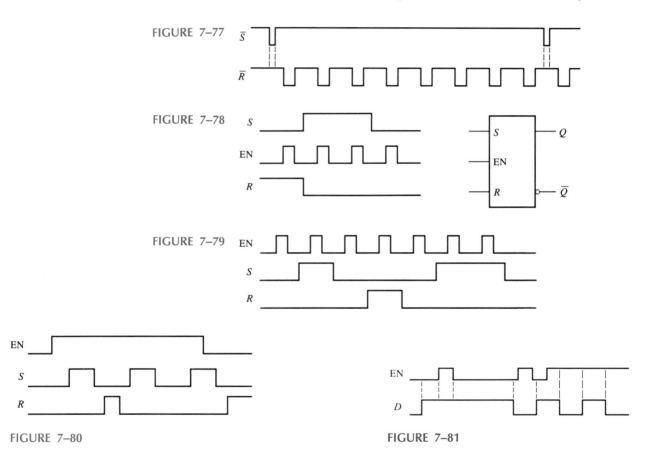

FIGURE 7–77

FIGURE 7–78

FIGURE 7–79

FIGURE 7–80

FIGURE 7–81

Section 7–2   Edge-Triggered Flip-Flops

7–8   Two edge-triggered S-R flip-flops are shown in Figure 7–82. If the inputs are as shown, sketch the Q output of each flip-flop, and explain the relative difference between the two. The flip-flops are initially RESET.

FIGURE 7–82

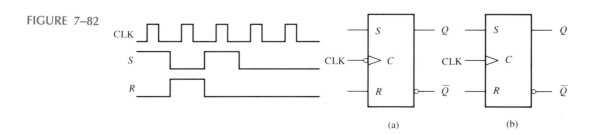

(a)   (b)

7–9 The Q output of the edge-triggered S-R flip-flop in Figure 7–83 is shown in relation to the clock signal. Determine the input waveforms on the S and R inputs that are required to produce this output if the flip-flop is a positive edge-triggered type.

7–10 Sketch the Q output for a D flip-flop with the inputs as shown in Figure 7–84. Assume positive edge-triggering and Q initially LOW.

7–11 Solve Problem 7–10 for the inputs in Figure 7–85.

7–12 For a positive edge-triggered J-K flip-flop with inputs as shown in Figure 7–86, determine the Q output. Assume that Q starts LOW.

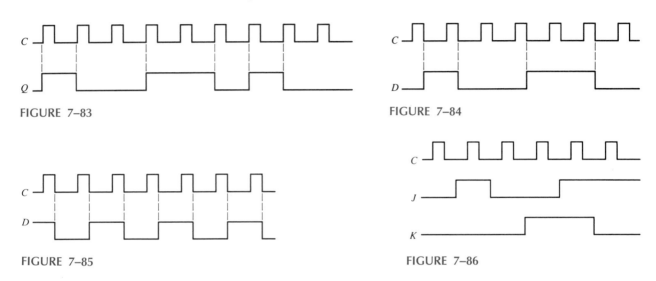

FIGURE 7–83

FIGURE 7–84

FIGURE 7–85

FIGURE 7–86

7–13 Solve Problem 7–12 for the inputs in Figure 7–87.

7–14 Determine the Q waveform if the signals shown in Figure 7–88 are applied to the inputs of the J-K flip-flop. Assume that Q is initially LOW.

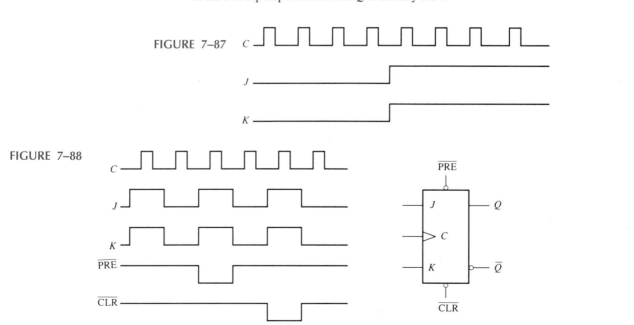

FIGURE 7–87

FIGURE 7–88

Section 7–3 Pulse-Triggered (Master-Slave) Flip-Flops

7–15 For a pulse-triggered (master-slave) J-K flip-flop with the inputs in Figure 7–89, sketch the Q output waveform. Assume that Q is initially LOW.

7–16 The following serial data are applied to the flip-flop indicated in Figure 7–90. Determine the resulting serial data that appear on the Q output. There is one clock pulse for each bit time. Assume that Q is initially 0. Right-most bits are applied first.

$$J_1: \quad 1010011$$
$$J_2: \quad 0111010$$
$$J_3: \quad 1111000$$
$$K_1: \quad 0001110$$
$$K_2: \quad 1101100$$
$$K_3: \quad 1010101$$

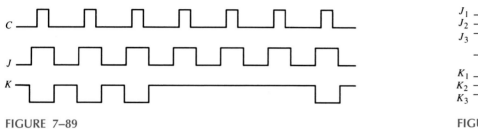

FIGURE 7–89

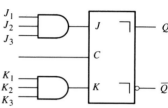

FIGURE 7–90

7–17 The 7472 is an AND-gated master-slave J-K flip-flop as shown in Figure 7–91(a). Complete the timing diagram in Figure 7–91(b) by sketching the Q output (which is initially LOW). Assume $\overline{PRE}$ and $\overline{CLR}$ remain HIGH.

7–18 Solve Problem 7–17 with the same J and K inputs but with the $\overline{PRE}$ and $\overline{CLR}$ inputs as shown in Figure 7–92 in relation to the clock.

FIGURE 7–91

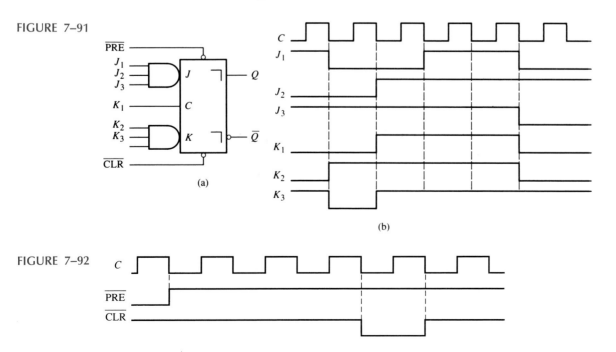

(a)

(b)

FIGURE 7–92

7–19 Determine the Q output in Figure 7–93. Assume that Q is initially LOW.

7–20 Sketch the Q output of flip-flop B in Figure 7–94 in proper relation to the clock. The flip-flops are initially RESET.

7–21 Sketch the Q output of flip-flop B in Figure 7–95 in proper relation to the clock. The flip-flops are initially RESET.

FIGURE 7–93

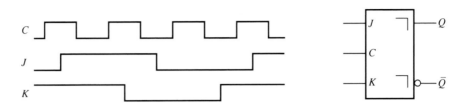

FIGURE 7–94

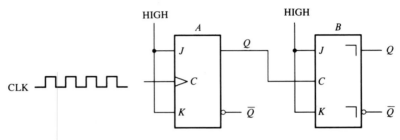

FIGURE 7–95

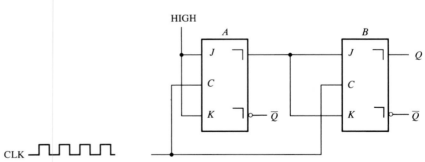

Section 7–4   Data Lock-Out Flip-Flops

7–22 For a data lock-out J-K flip-flop with the inputs in Figure 7–89, draw the timing diagram showing Q starting in the LOW state. Assume positive edge-triggering.

7–23 The D input and a single clock pulse are shown in Figure 7–96. Compare the resulting Q outputs for positive edge-triggered, negative edge-triggered, pulse-triggered, and data lock-out flip-flops. The flip-flops are initially RESET.

7–24 Solve Problem 7–19 for a data lock-out J-K flip-flop with positive edge-triggering.

7–25 Solve Problem 7–24 for a negative edge-triggered device.

FIGURE 7–96

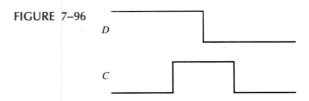

Section 7–5   Operating Characteristics

7–26   The flip-flop in Figure 7–97 is initially RESET. Show the relation between the Q output and the clock pulse if propagation delay t_{PLH} (clock to Q) is 8 ns.

7–27   The direct current required by a particular flip-flop that operates on a $+5$ V dc source is found to be 10 mA. A certain digital device uses 15 of these flip-flops. Determine the current capacity required for the $+5$ V dc supply and the total power dissipation of the system.

7–28   For the circuit in Figure 7–98, determine the maximum frequency of the clock signal for reliable operation if the set-up time for each flip-flop is 20 ns and the propagation delays (t_{PLH} and t_{PHL}) from clock to output are 50 ns for each flip-flop.

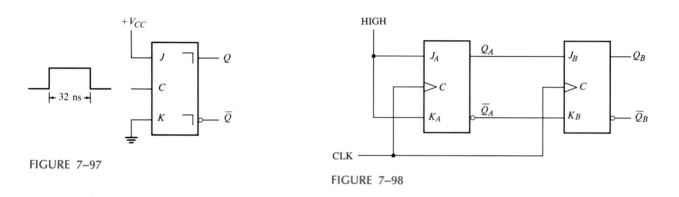

FIGURE 7–97

FIGURE 7–98

Section 7–6   Basic Flip-Flop Applications

7–29   An S-R flip-flop is connected as shown in Figure 7–99. Determine the Q output in relation to the clock. What specific function does this device perform?

7–30   For the circuit in Figure 7–98, develop a timing diagram for eight clock pulses, showing the Q_A and Q_B outputs in relation to the clock.

7–31   Show how to convert the circuit in Figure 7–98 to a two-bit serial-input shift register.

7–32   Draw the logic diagram for a circuit that will divide the input clock frequency by 16. Use negative edge-triggered J-K flip-flops.

7–33   Devise a basic counting circuit that produces a binary sequence from zero through seven by using negative edge-triggered J-K flip-flops.

7–34   In the shipping department of a softball factory, the balls roll down a conveyor and through a chute single file into boxes for shipment. Each ball passing through the chute activates a switch circuit that produces an electrical pulse. The capacity of each box is 32 balls. Design a logic circuit to indicate when a box is full so that an empty box can be moved into position.

FIGURE 7–99

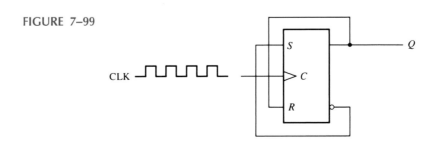

Section 7–7 One-Shots

7–35 Determine the pulse width of a 74121 one-shot if the external resistor is 3.3 kΩ and the external capacitor is 2000 pF.

7–36 An output pulse of 5 μs duration is to be generated by a 74122 one-shot. Using a capacitor of 10,000 pF and $K = 0.32$, determine the value of external resistance required.

Section 7–8 Astable Multivibrators and Timers

7–37 A 555 timer is configured to run as an astable multivibrator as shown in Figure 7–100. Determine its frequency.

7–38 Determine the values of the external resistors for a 555 timer used as an astable multivibrator with an output frequency of 20 kHz, if the external capacitor C is 0.002 μF and the duty cycle is to be approximately 75%.

7–39 Design a one-shot, using a 555 timer that will produce a 0.25 s pulse.

Section 7–9 Troubleshooting

7–40 The flip-flop in Figure 7–101 is tested under all input conditions as shown. Is it operating properly? If not, what is the most likely fault?

FIGURE 7–100

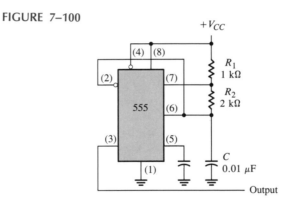

FIGURE 7–101

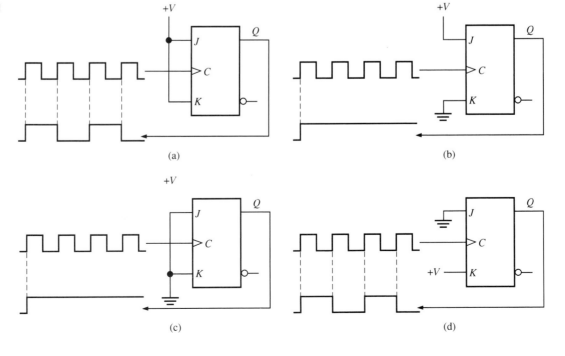

7–41 A 7400 quad NAND gate IC is used to construct a gated S-R latch on a protoboard in the lab as shown in Figure 7–102. The schematic in part (a) is used to connect the circuit in part (b). When you try to operate the latch, you find that the Q output stays HIGH no matter what the inputs are. Determine the problem.

7–42 Determine if the flip-flop in Figure 7–103 is operating properly, and if not, identify the most probable fault.

7–43 The parallel data storage circuit in Figure 7–48 does not operate properly. To check it out, you first make sure that V_{CC} and ground are connected, and then you apply LOW levels to all the D inputs and pulse the clock line. You check the Q outputs and find them all to be LOW; so far, so good. Next you apply HIGHs to all the D inputs and again pulse the clock line. When you check the Q outputs, they are still all LOW. What is the problem, and what procedure will you use to isolate the fault to a single device?

7–44 The flip-flop circuit in Figure 7–104(a) is used to generate a binary count sequence. The gate network is a decoder that is supposed to produce a HIGH when a binary zero or a binary three state occurs (00 or 11). When you check the Q_A and Q_B outputs on the scope, you get the display shown in part (b), which reveals glitches on the decoder output (X) in addition to the correct pulses. What is causing these glitches, and how can you get rid of them?

7–45 Determine the Q_A, Q_B, and X outputs over six clock pulses in Figure 7–104(a) for each of the following faults in the TTL circuits. Start with both Q_A and Q_B LOW.

(a) J_A input open   **(b)** K_B input open

(c) Q_B output open   **(d)** clock input to FFB shorted

(e) lower NAND gate output open

FIGURE 7–102

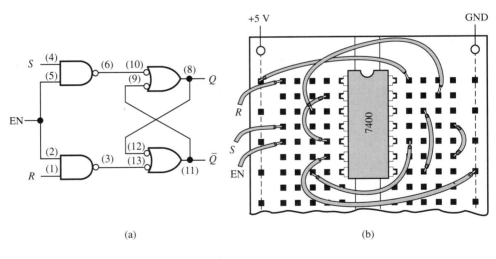

(a)   (b)

FIGURE 7–103

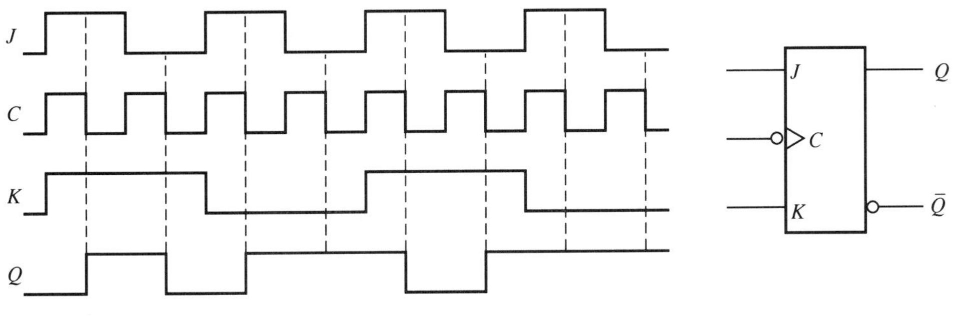

7-46 Two 74121 one-shots are connected on a circuit board as shown in Figure 7-105. After observing the oscilloscope display, do you conclude that the circuit is operating properly? If not, what is the most likely problem?

FIGURE 7-104

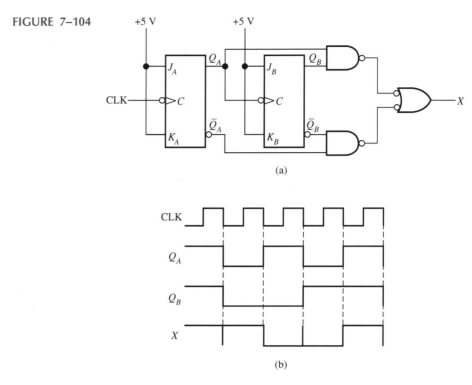

(a)

(b)

FIGURE 7-105

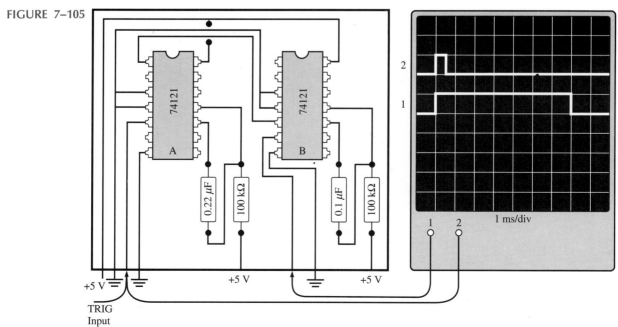

ANSWERS TO
SECTION REVIEWS

Section 7–1   Review

1. S-R, gated S-R, gated D
2. $SR = 00$, NC; $SR = 01$, $Q = 0$; $SR = 10$, $Q = 1$; $SR = 11$, invalid.
3. 1

Section 7–2   Review

1. The output of a gated S-R latch can change any time the gate Enable (EN) input is active. The output of an edge-triggered S-R flip-flop can change only on the triggering edge of a clock pulse.
2. The J-K flip-flop does not have an invalid state as does the S-R flip-flop.
3. Output Q goes HIGH on the trailing edge of the first clock pulse, LOW on the trailing edge of the second pulse, HIGH on the trailing edge of the third pulse, and LOW on the trailing edge of the fourth pulse.

Section 7–3   Review

1. In the pulse-triggered flip-flop (master-slave), a data bit goes into the master section on the leading edge of a clock pulse and is transferred to the output (slave) on the trailing edge. In the edge-triggered flip-flop, a data bit goes into the flip-flop and appears on the output on the same clock edge.
2. **(a)** Nothing happens.   **(b)** The output will change.
3. The J-K does not have an invalid state.

Section 7–4   Review

1. A data lock-out is sensitive to the input data only during a clock transition and not while the clock is in its active state.
2. The states of the inputs at the triggering edge of the clock.

Section 7–5   Review

1. **(a)** time required for input data to be present before triggering edge of clock
   (b) time required for data to remain on input after triggering edge of clock
2. 74LS76A, according to Table 7–8

Section 7–6   Review

1. register   **2.** toggle ($J = 1$, $K = 1$)   **3.** 5

Section 7–7   Review

1. Nonretriggerable times out before it can respond to another trigger input. Retriggerable responds to each trigger input.
2. with external R and C

Section 7–8   Review

1. Astable has no stable state. Monostable has one stable state.
2. 75%

Section 7–9   Review

1. yes
2. an astable multivibrator using a 555 timer

As you saw in the last chapter, flip-flops can be connected together to perform counting operations. Such a group of flip-flops is a counter. The number of flip-flops used and the way in which they are connected determine the number of states (called the *modulus*) and the sequence of states that the counter goes through in each complete cycle.

Counters are classified into two broad categories according to the way they are clocked: asynchronous and synchronous. In *asynchronous counters,* commonly called *ripple counters,* the first flip-flop is clocked by the external clock pulse, and then each successive flip-flop is clocked by the Q or $\overline{Q}$ output of the previous flip-flop. Therefore, in asynchronous counters, the flip-flops are not clocked simultaneously. In *synchronous counters,* the clock input is connected to all the flip-flops, and they are thus clocked simultaneously.

Within each of these two categories, counters are classified primarily by the sequence of states, by the number of states, or by the number of flip-flops (stages) within the counter.

Specific devices introduced in this chapter are the following:

1. 7493A four-bit asynchronous binary counter
2. 74LS160A synchronous decade counter
3. 74LS161A four-bit binary counter
4. 74LS163A four-bit synchronous binary counter
5. 74190 up/down decade counter

After completing this chapter, you will be able to

☐ describe the difference between an asynchronous and a synchronous counter.
☐ analyze counter timing diagrams.
☐ analyze counter circuits.
☐ explain how propagation delays affect the operation of a counter.
☐ determine the modulus of a counter.
☐ modify the modulus of a counter.
☐ recognize the difference between a four-bit binary counter and a decade counter.
☐ use an up/down counter to generate forward and reverse binary sequences.
☐ determine the sequence of a counter.
☐ use IC counters in various applications.
☐ design a counter that will have any specified sequence of states.
☐ cascade several counters to achieve a higher modulus.
☐ use logic gates to decode any given state of a counter.
☐ eliminate glitches in counter decoding.
☐ explain how a digital clock operates.
☐ troubleshoot counters for various types of faults.
☐ interpret counter logic symbols that use dependency notation.

8

Counters

ASYNCHRONOUS COUNTERS

8–1

*The term **asynchronous** refers to events that do not occur at the same time. With respect to counter operation, **asynchronous** means that the flip-flops within the counter are not made to change states at exactly the same time; the reason they are not is that the clock pulses are not connected directly to the **C** input of each flip-flop in the counter.*

Figure 8–1 shows a two-bit counter connected for asynchronous operation. Notice that the clock line (CLK) is connected to the clock input (C) of only the first stage, FF0. The second stage, FF1, is triggered by the $\overline{Q}_0$ output of FF0. FF0 changes state at the positive-going edge of each clock pulse, but FF1 changes only when triggered by a positive-going transition of the $\overline{Q}_0$ output of FF0. Because of the inherent propagation delay time through a flip-flop, a transition of the input clock pulse and a transition of the $\overline{Q}_0$ output of FF0 can never occur at exactly the same time. Therefore, the two flip-flops are never simultaneously triggered, so the counter operation is asynchronous.

Let us examine the basic operation of the counter of Figure 8–1 by applying four clock pulses to FF0 and observing the Q output of each flip-flop; Figure 8–2 illustrates the changes in the state of the flip-flop outputs in response to the clock pulses. Both flip-flops are connected for toggle operation ($J = 1$, $K = 1$) and are assumed to be initially RESET.

The positive-going edge of CLK_1 (clock pulse 1) causes the Q_0 output of FF0 to go HIGH. The $\overline{Q}_0$ output at the same time goes LOW, but it has no effect on FF1, because a *positive-going* transition must occur to trigger the flip-flop. After the leading edge of CLK_1, $Q_0 = 1$ and $Q_1 = 0$. The positive-going edge of CLK_2 causes Q_0 to go LOW. $\overline{Q}_0$ goes HIGH and triggers FF1, causing Q_1 to go HIGH. After the leading edge of CLK_2, $Q_0 = 0$ and $Q_1 = 1$. The positive-going edge of CLK_3 causes Q_0 to go HIGH again. Output $\overline{Q}_0$ goes LOW and has no effect on FF1. Thus, after the leading edge of

FIGURE 8–1

A two-bit asynchronous binary counter.

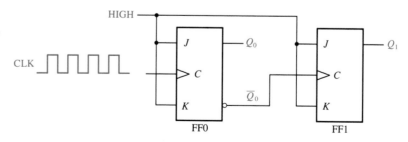

FIGURE 8–2

Timing diagram for the counter of Figure 8–1.

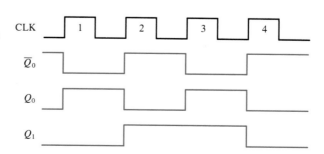

CLK_3, $Q_0 = 1$ and $Q_1 = 1$. The positive-going edge of CLK_4 causes Q_0 to go LOW, while $\overline{Q_0}$ goes HIGH and triggers FF1, causing Q_1 to go LOW. After the leading edge of CLK_4, $Q_0 = 0$ and $Q_1 = 0$. The counter has now recycled to its original state (both flip-flops are RESET).

The waveforms of the Q_0 and Q_1 outputs are shown relative to the clock pulses in Figure 8–2, which is a timing diagram. It should be pointed out that, for simplicity, the transitions of Q_0, Q_1, and the clock pulses are shown as simultaneous even though this is an asynchronous counter. There is, of course, some small delay between the CLK, Q_0, and Q_1 transitions.

Notice that the two-bit counter exhibits four different states, as you would expect with two flip-flops ($2^2 = 4$). Also, notice that if Q_0 represents the least significant bit (LSB) and Q_1 represents the most significant bit (MSB), the sequence of counter states is actually a sequence of binary numbers as shown in Table 8–1. In this book Q_0 is always the LSB unless otherwise specified.

Since it goes through a binary sequence, the counter in Figure 8–1 is a *binary counter*. It actually counts the number of clock pulses up to three, and on the fourth pulse it recycles to its original state ($Q_0 = 0$, $Q_1 = 0$). The term *recycle* is commonly applied to counter operation; it refers to the transition of the counter from its final state back to its original state.

TABLE 8–1

Clock Pulse	Q_1	Q_0
0	0	0
1	0	1
2	1	0
3	1	1

A Three-Bit Asynchronous Binary Counter

A three-bit asynchronous binary counter is shown in Figure 8–3(a). The basic operation, of course, is the same as that of the two-stage counter just discussed, except that it has eight states, due to its three flip-flops. A timing diagram appears in Figure 8–3(b) for eight clock pulses.

Notice that the counter progresses through a binary count of zero to seven and then recycles to the zero state. This counter sequence is presented in Table 8–2.

TABLE 8–2
State sequence for a three-stage binary counter.

Clock Pulse	Q_2	Q_1	Q_0
0	0	0	0
1	0	0	1
2	0	1	0
3	0	1	1
4	1	0	0
5	1	0	1
6	1	1	0
7	1	1	1

Asynchronous counters are commonly referred to as *ripple counters* for the following reason: The effect of the input clock pulse is first "felt" by FF0. This effect

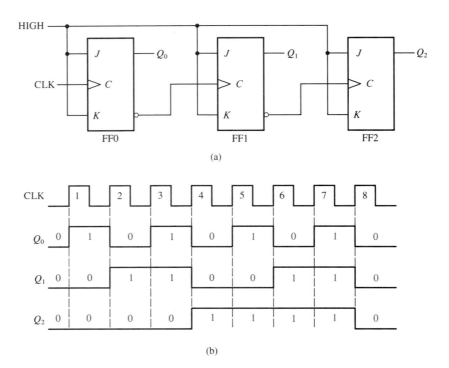

(a)

(b)

FIGURE 8–3

Three-bit asynchronous binary counter and its timing diagram for one cycle.

FIGURE 8–4

Propagation delays in a ripple-clocked binary counter.

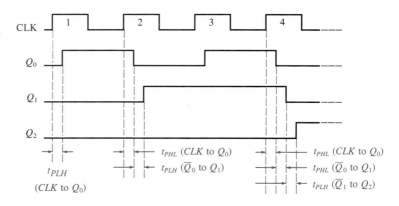

cannot get to FF1 immediately because of the propagation delay through FF0. Then there is the propagation delay through FF1 before FF2 can be triggered. Thus, the effect of an input clock pulse "ripples" through the counter, taking some time, due to propagation delays, to reach the last flip-flop. To illustrate, notice that all three flip-flops in the counter of Figure 8–3 change state as a result of CLK_4. The HIGH-to-LOW transition of Q_0 occurs one delay time after the positive-going transition of the clock pulse. The HIGH-to-LOW transition of Q_1 occurs one delay time after the positive-going transition of $\overline{Q}_0$. The LOW-to-HIGH transition of Q_2 occurs one delay time after the positive-going transition of $\overline{Q}_1$. As you can see, FF2 is not triggered until two delay times after the positive-going edge of the clock pulse, CLK_4. Thus, it takes three propagation delay times for the effect of the clock pulse (CLK_4) to ripple through the counter and change Q_2 from LOW to HIGH. This ripple clocking effect is illustrated in Figure 8–4 for the first four clock pulses, with the propagation delays shown.

This cumulative delay of an asynchronous counter is a major disadvantage in many applications because it limits the rate at which the counter can be clocked and creates decoding problems.

EXAMPLE 8–1

A four-stage asynchronous binary counter is shown in Figure 8–5(a). Each flip-flop is negative edge-triggered and has a propagation delay of 10 nanoseconds (ns). Draw a timing diagram showing the Q output of each stage, and determine the total propagation delay time from the triggering edge of a clock pulse until a corresponding change can occur in the state of Q_3.

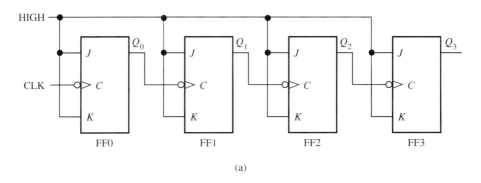

(a)

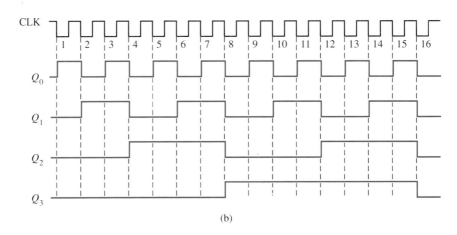

(b)

FIGURE 8–5
Four-bit asynchronous binary counter and its timing diagram.

Solutions The timing diagram with delays omitted is as shown in Figure 8–5(b). For the total delay time, the effect of CLK_8 or CLK_{16} must propagate through four flip-flops before Q_3 changes, so

$$t_p = 4 \times 10 \text{ ns} = 40 \text{ ns total delay}$$

Asynchronous Decade Counters

Regular binary counters, such as those previously introduced, have a maximum modulus; that is, they progress through all of their possible states. Recall that the maximum possible number of states (maximum *modulus*) of a counter is 2^n, where n is the number of flip-flops in the counter.

Counters can also be designed to have a number of states in their sequence that is less than 2^n. The resulting sequence is called a *truncated sequence*.

One common modulus for counters with truncated sequences is ten. Counters with ten states in their sequence are called *decade* counters. A decade counter with a count sequence of zero (0000) through nine (1001) is a *BCD decade counter* because its ten-state sequence is the BCD code. This type of counter is very useful in display applications in which BCD is required for conversion to a decimal readout.

To obtain a truncated sequence, it is necessary to force the counter to recycle before going through all of its normal states. For example, the BCD decade counter must recycle back to the 0000 state after the 1001 state.

A decade counter requires four flip-flops (three flip-flops are insufficient because $2^3 = 8$). We will now take a four-bit asynchronous counter such as the one in Figure 8–5(a) and modify its sequence in order to understand the principle of truncated counters. One method of achieving this recycling after the count of nine (1001) is to decode count ten (1010) with a NAND gate and connect the output of the NAND gate to the clear ($\overline{\text{CLR}}$) inputs of the flip-flops, as shown in Figure 8–6(a).

Notice that only Q_1 and Q_3 are connected to the NAND gate inputs. This arrangement is an example of *partial decoding,* in which the two unique states ($Q_1 = 1$ and $Q_3 = 1$) are sufficient to decode the count of ten, because none of the other states (zero through nine) have both Q_1 and Q_3 HIGH at the same time. When the counter goes into count ten (1010), the decoding gate output goes LOW and asynchronously RESETS all the flip-flops.

The resulting timing diagram is shown in Figure 8–6(b). Notice that there is a glitch on the Q_1 waveform. The reason for this glitch is that Q_1 must first go HIGH before the count of ten can be decoded. Not until several nanoseconds after the counter

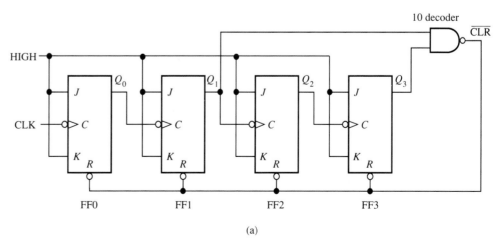

(a)

FIGURE 8–6

An asynchronously clocked decade counter with asynchronous recycling.

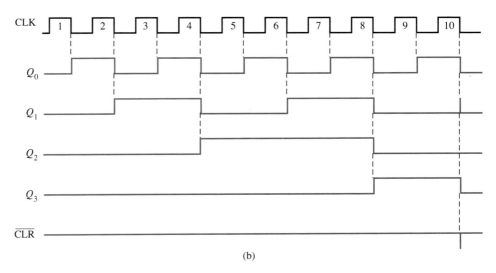

FIGURE 8–6
(Continued)

goes to the count of ten does the output of the decoding gate go LOW (both inputs are HIGH). Thus, the counter is in the 1010 state for a short time before it is RESET to 0000, thus producing the glitch on Q_1.

Other truncated sequences can be implemented in a similar way, as the next example shows.

EXAMPLE 8–2

Show how an asynchronous counter can be implemented having a modulus of twelve with a straight binary sequence from 0000 through 1011.

Solution Since three flip-flops can produce a maximum of eight states, four flip-flops are required to produce any modulus greater than eight but less than or equal to sixteen.

When the counter gets to its last state, 1011, it must recycle back to 0000 rather than going to its normal next state of 1100, as illustrated in the following sequence chart:

$$Q_3 \quad Q_2 \quad Q_1 \quad Q_0$$

Q_3	Q_2	Q_1	Q_0	
0	0	0	0	
.	.	.	.	Recycles
.	.	.	.	
1	0	1	1	
1	1	0	0	←—Normal next state

Observe that Q_0 and Q_1 both go to 0 anyway, but Q_2 and Q_3 must be forced to 0 on the twelfth clock pulse. Figure 8–7(a) shows the modulus-12 counter. The NAND gate partially decodes count twelve (1100) and RESETS flip-flop 2 and flip-flop 3. Thus, on the twelfth clock pulse, the counter is made to recycle from count eleven to count zero, as shown in the timing diagram of Figure 8–7(b). (It is in count twelve for only a few nanoseconds before it is RESET.)

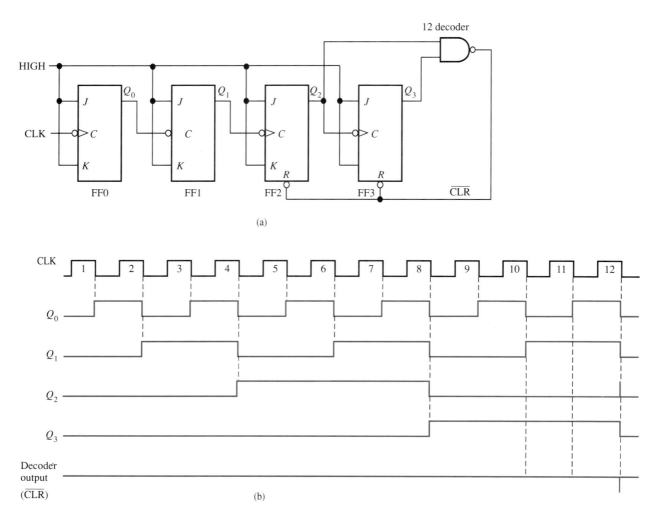

FIGURE 8–7
Asynchronously clocked modulus-12 counter with asynchronous recycling.

The 7493A Four-Bit Binary Counter

The 7493A is presented as an example of a specific integrated circuit asynchronous counter. As the logic diagram in Figure 8–8 shows, this device actually consists of a single flip-flop and a three-bit asynchronous counter. This arrangement is for flexibility. It can be used as a divide-by-2 device if only the single flip-flop is used, or it can be used as a modulus-8 counter if only the three-bit counter portion is used. This device also provides gated reset inputs, $R0(1)$ and $R0(2)$. When both of these inputs are HIGH, the counter is RESET to the 0000 state by $\overline{CLR}$.

Additionally, the 7493A can be used as a four-bit modulus-16 counter (counts 0 through 15) by connecting the Q_0 output to the $CLKB$ input as shown in Figure 8–9(a). It can also be configured as a decade counter (counts 0 through 9) with asynchronous recycling by using the gated reset inputs for partial decoding of count ten, as shown in Figure 8–9(b).

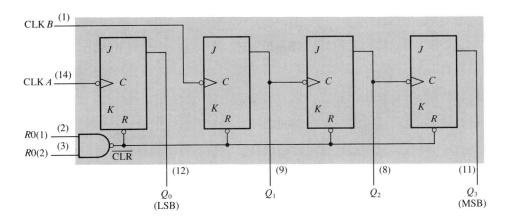

FIGURE 8–8
The 7493A four-bit binary counter logic diagram. (Pin numbers are in parentheses, and all J-K inputs are internally connected HIGH.)

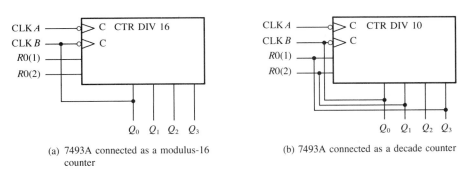

(a) 7493A connected as a modulus-16 counter

(b) 7493A connected as a decade counter

FIGURE 8–9
Two configurations of the 7493A asynchronous counter. (The qualifying label, CTR DIVn, indicates a counter with n states.)

EXAMPLE 8–3

Show how the 7493A can be used as a modulus-12 counter.

Solution Use the gated reset inputs, $R0(1)$ and $R0(2)$, to partially decode count 12 (remember, there is an internal NAND gate associated with these inputs). The count-12 decoding is accomplished by connecting Q_3 to $R0(1)$ and Q_2 to $R0(2)$, as shown in Figure 8–10. Output Q_0 is connected to CLKB to create a four-bit counter.

FIGURE 8–10
7493A connected as a modulus-12 counter.

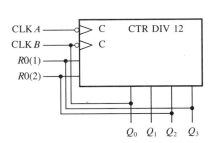

Immediately after the counter goes to count 12 (1100), it is RESET to 0000. The recycling, however, results in a glitch on Q_2 because the counter must reside in the 1100 state for several nanoseconds before recycling.

Answers are found at the end of the chapter.

1. What does the term *asynchronous* mean in relation to counters?
2. How many states does a modulus-14 counter have?

SYNCHRONOUS COUNTERS

8–2

The term **synchronous** *as applied to counter operation means that the counter is clocked in such a way that all flip-flops in the counter are triggered at the same time. This arrangement is accomplished by connecting the clock line to each stage of the counter.*

Figure 8–11 shows a two-stage counter. Notice that an arrangement different from that for the asynchronous counter must be used for the J and K inputs of FF1 in order to achieve a binary sequence.

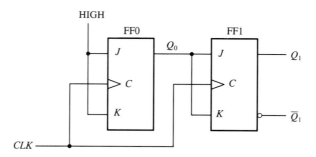

FIGURE 8–11
A two-bit synchronous binary counter.

The operation of this counter is as follows: First, we will assume that the counter is initially in the binary 0 state; that is, both flip-flops are RESET. When the positive edge of the first clock pulse is applied, FF0 will toggle, and Q_0 will therefore go HIGH. What happens to FF1 at the positive-going edge of CLK_1? To find out, let us look at the input conditions of FF1. Inputs J and K are both LOW because Q_0, to which they are connected, has not yet gone HIGH. Remember, there is a propagation delay from the triggering edge of the clock pulse until the Q output actually makes a transition. So, $J = 0$ and $K = 0$ when the leading edge of the first clock pulse is applied. This is a no-change condition, and therefore FF1 does not change state. A timing detail of this portion of the counter operation is given in Figure 8–12(a).

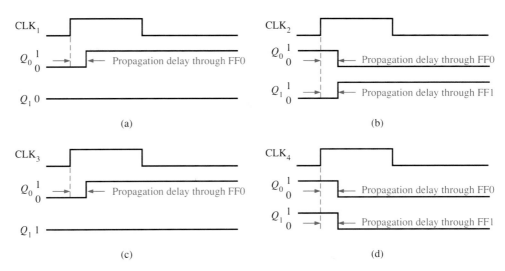

FIGURE 8–12
Timing details for the two-bit synchronous counter operation.

After CLK_1, $Q_0 = 1$ and $Q_1 = 0$ (which is the binary 1 state). At the leading edge of CLK_2, FF0 will toggle, and Q_0 will go LOW. Since FF1 "sees" a HIGH on its J and K inputs when the triggering edge of this clock pulse occurs, the flip-flop toggles, and Q_1 goes HIGH. Thus, after CLK_2, $Q_0 = 0$ and $Q_1 = 1$ (which is a binary 2 state). The timing detail for this condition is given in Figure 8–12(b). At the leading edge of CLK_3, FF0 again toggles to the SET state ($Q_0 = 1$), and FF1 remains SET ($Q_1 = 1$) because its J and K inputs are both LOW. After this triggering edge, $Q_0 = 1$ and $Q_1 = 1$ (which is a binary 3 state). The timing detail is shown in Figure 8–12(c).

Finally, at the leading edge of CLK_4, Q_0 and Q_1 go LOW because they both have a toggle condition on their J and K inputs. The timing detail is shown in Figure 8–12(d). The counter has now recycled to its original state, binary 0. The complete timing diagram is shown in Figure 8–13.

Notice that all the waveform transitions appear coincident; that is, the delays are not indicated. Although the delays are a very important factor in the counter operation—as we have seen in the preceding discussion—in an overall timing diagram they are normally omitted for simplicity. Major waveform relationships resulting from the logical operation of a circuit can be conveyed completely without showing small delay and timing differences.

FIGURE 8–13
Timing diagram for the counter of Figure 8–11.

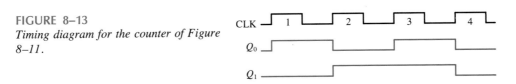

A Three-Bit Synchronous Binary Counter

A three-bit synchronous binary counter is shown in Figure 8–14, and its timing diagram in Figure 8–15. An understanding of this counter can be achieved by a careful examination of its sequence of states as shown in Table 8–3.

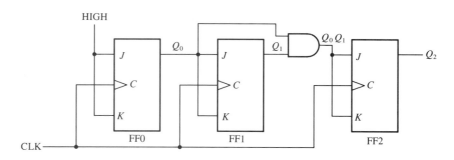

FIGURE 8–14
A three-bit synchronous binary counter.

FIGURE 8–15
Timing diagram for the counter of Figure 8–14.

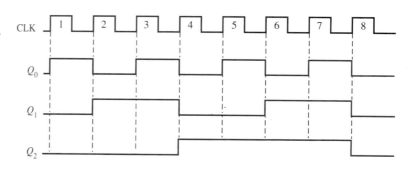

TABLE 8–3
State sequence for a three-stage binary counter.

Clock Pulse	Q_2	Q_1	Q_0
0	0	0	0
1	0	0	1
2	0	1	0
3	0	1	1
4	1	0	0
5	1	0	1
6	1	1	0
7	1	1	1

First, let us look at Q_0. Notice that Q_0 changes on each clock pulse as the counter progresses from its original state to its final state and then back to its original state. To produce this operation, FF0 must be held in the toggle mode by constant HIGHs on its J and K inputs. Now let us see what Q_1 does. Notice that it goes to the opposite state following each time Q_0 is a 1. This change occurs at CLK_2, CLK_4, CLK_6, and CLK_8. The CLK_8 pulse causes the counter to recycle. To produce this operation, Q_0 is connected to the J and K inputs of FF1. When Q_0 is a 1 and a clock pulse occurs, FF1 is in the toggle mode and therefore changes state. The other times, when Q_0 is a 0, FF1 is in the no-change mode and remains in its present state.

Next, let us see how FF2 is made to change at the proper times according to the binary sequence. Notice that both times Q_2 changes state, it is preceded by the unique condition in which both Q_0 and Q_1 are HIGH. This condition is detected by the AND gate and applied to the J and K inputs of FF2. Whenever both Q_0 and Q_1 are HIGH, the output of the AND gate makes the J and K inputs of FF2 HIGH, and FF2 toggles on the following clock pulse. At all other times, the J and K inputs of FF2 are held LOW by the AND gate output, and FF2 does not change state.

A Four-Bit Synchronous Binary Counter

Figure 8–16(a) shows a four-bit binary counter, and Figure 8–16(b) shows its timing diagram. This particular counter is implemented with negative edge-triggered flip-flops. The reasoning behind the J and K input control for the first three flip-flops is the same as presented previously for the three-stage counter. The fourth stage, FF3, changes only twice in the sequence. Notice that both of these transitions occur following the times

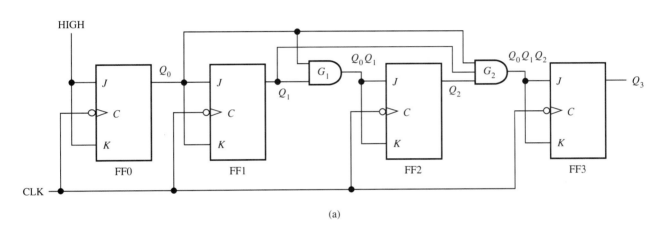

(a)

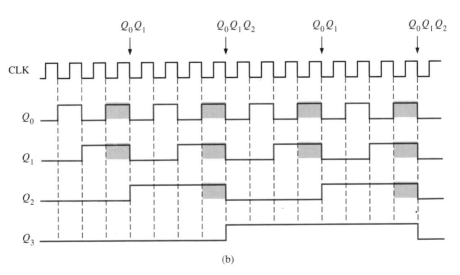

(b)

FIGURE 8–16
A four-bit synchronous binary counter and timing diagram.

Q_0, Q_1, and Q_2 are all HIGH. This condition is decoded by AND gate G_2 so that when a clock pulse occurs, FF3 will change state. For all other times the J and K inputs of FF3 are LOW, and it is in a no-change condition.

Synchronous Decade Counters

As you know, the BCD decade counter exhibits a truncated sequence and goes through a straight binary sequence through the 1001 state. Rather than going to the 1010 state, it recycles to the 0000 state. A synchronous BCD decade counter is shown in Figure 8–17.

The counter operation can be understood by examining the sequence of states in Table 8–4. First, notice that FF0 toggles on each clock pulse, so the logic equation for its J and K inputs is

$$J_0 = K_0 = 1$$

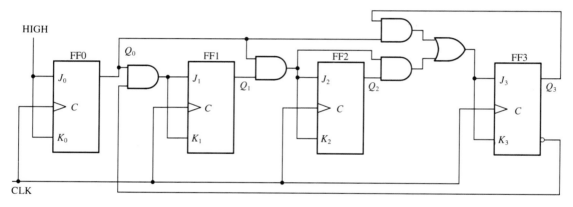

FIGURE 8–17
A synchronous BCD decade counter.

TABLE 8–4
States of a BCD decade counter.

Clock Pulse	Q_3	Q_2	Q_1	Q_0
0	0	0	0	0
1	0	0	0	1
2	0	0	1	0
3	0	0	1	1
4	0	1	0	0
5	0	1	0	1
6	0	1	1	0
7	0	1	1	1
8	1	0	0	0
9	1	0	0	1

This equation is implemented by connecting these inputs to a constant HIGH level. Next notice that FF1 changes on the next clock pulse each time $Q_0 = 1$ and $Q_3 = 0$, so the logic equation for its J and K inputs is

$$J_1 = K_1 = Q_0\overline{Q}_3$$

This equation is implemented by ANDing Q_0 and $\overline{Q}_3$ and connecting the gate output to the J and K inputs of FF1. Flip-flop 2 changes on the next clock pulse each time both $Q_0 = 1$ and $Q_1 = 1$. This requires an input logic equation as follows:

$$J_2 = K_2 = Q_0Q_1$$

This is implemented by ANDing Q_0 and Q_1 and connecting the gate output to the J and K inputs of FF2. Finally, FF3 changes to the opposite state on the next clock pulse each time $Q_0 = 1$, $Q_1 = 1$, and $Q_2 = 1$ (count 7), or when $Q_0 = 1$ and $Q_3 = 1$ (count 9). The equation for this is as follows:

$$J_3 = K_3 = Q_0Q_1Q_2 + Q_0Q_3$$

This function is implemented with the AND/OR logic connected to FF3 as shown in the logic diagram in Figure 8–17. Notice that the only difference between this decade counter and a modulus-16 binary counter is the Q_0Q_3 AND gate and the OR gate; this arrangement essentially detects the occurrence of the 1001 state and causes the counter to recycle properly on the next clock pulse. The timing diagram for the decade counter is given in Figure 8–18.

FIGURE 8–18
Timing diagram for the BCD decade counter (Q_0 is the LSB).

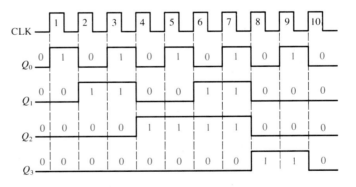

The 74LS163A Synchronous Four-Bit Binary Counter

The 74LS163A is an example of an integrated circuit synchronous binary counter. A logic symbol is shown in Figure 8–19. This counter has several features in addition to the basic functions previously discussed for the general synchronous binary counter.

First, the counter can be synchronously preset to any four-bit binary number by applying the proper levels to the data inputs. When a LOW is applied to the $\overline{\text{LOAD}}$ input, the counter will assume the state of the data inputs on the next clock pulse. Thus, the counter sequence can be started with any four-bit binary number.

FIGURE 8–19

The 74LS163A four-bit synchronous binary counter. (The qualifying label CTR DIV 16 indicates a counter with sixteen states.)

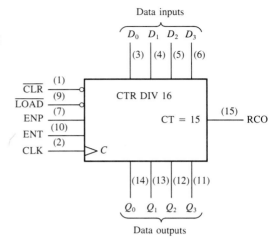

Also, there is an active-LOW clear input ($\overline{CLR}$), which synchronously RESETS all four flip-flops in the counter. There are two Enable inputs, ENP and ENT. These inputs must both be high for the counter to sequence through its binary states. When at least one is LOW, the counter is disabled. The ripple clock output (RCO) goes HIGH when the counter reaches the last state in the sequence, fifteen (1111). This output, in conjunction with the enable inputs, allows these counters to be cascaded for higher count sequences, as will be discussed later.

Figure 8–20 shows a timing diagram of this counter being preset to twelve (1100) and then counting up to its terminal count, fifteen (1111). Input D_0 is the least significant input bit, and Q_0 is the least significant output bit.

Let us examine this timing diagram in detail. This will aid you in interpreting timing diagrams found later in this chapter or in manufacturers' data sheets.

To begin, the LOW level pulse on the $\overline{CLR}$ input causes all the outputs (Q_0, Q_1, Q_2, and Q_3) to go LOW.

Next, the LOW level pulse on the $\overline{LOAD}$ input synchronously enters the data on the data inputs (D_0, D_1, D_2, and D_3) into the counter. These data appear on the Q outputs at the time of the first positive-going clock edge after $\overline{LOAD}$ goes LOW. This is the preset operation. In this particular example, Q_0 is LOW, Q_1 is LOW, Q_2 is HIGH, and Q_3 is HIGH. This, of course, is a binary 12 (Q_0 is the LSB).

The counter now advances through counts 13, 14, and 15 on the next three positive-going clock edges. It then recycles to 0, 1, 2 on the following clock pulses. Notice that both ENP and ENT inputs are HIGH during the count sequence. When ENP goes LOW, the count is inhibited, and the counter remains in the binary 2 state.

The 74LS160A Synchronous Decade Counter

This device has the same inputs and outputs as the 74LS163A binary counter previously discussed. It may be preset to any BCD count by the use of the data inputs and a LOW on the $\overline{LOAD}$ input. A LOW on the $\overline{CLR}$ will RESET the counter. The count Enable

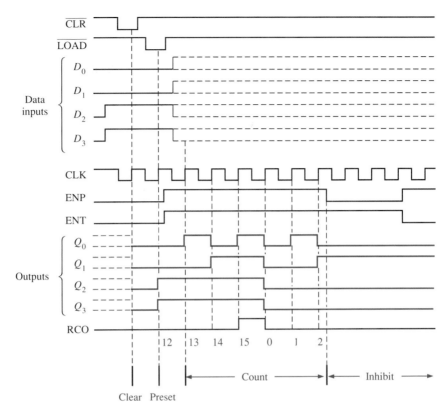

FIGURE 8–20
Timing example for a 74LS163A.

inputs ENP and ENT must both be high for the counter to advance through its sequence of states in response to a positive transition on the CLK input. As in the 74LS163A, the Enable inputs in conjunction with the ripple clock output (RCO; terminal count of 1001) provide for cascading several decade counters. Cascaded counters will be discussed later in this chapter.

Figure 8–21 shows a logic symbol, and Figure 8–22 is a timing diagram showing the counter being preset to count 7 (0111).

FIGURE 8–21
The 74LS160A synchronous decade (BCD) counter. (The qualifying label CTR DIV 10 indicates a counter with ten states.)

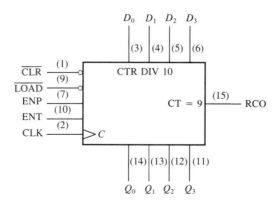

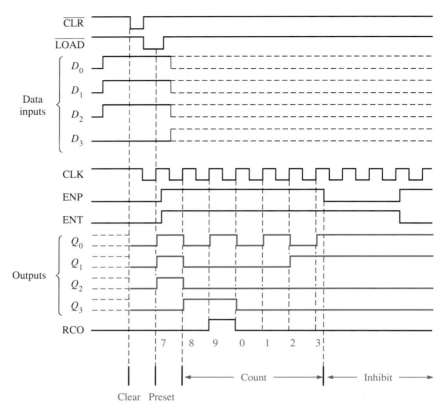

FIGURE 8–22
Timing example for a 74LS160A.

SECTION 8–2
REVIEW   Answers are found at the end of the chapter.
1.   How does a synchronous counter differ from an asynchronous counter?
2.   Explain the function of the presettable feature of counters such as the 74LS160A and the 74LS163A.
3.   Describe the purpose of the ENP and ENT inputs and the RCO output for the two specific counters introduced in this section.

UP/DOWN SYNCHRONOUS COUNTERS

8–3

*An **up/down counter** is one that is capable of progressing in **either direction** through a certain sequence. An up/down counter, sometimes called a **bidirectional** counter, can have any specified sequence of states. For example, a three-bit binary counter that advances upward through its sequence (0, 1, 2, 3, 4, 5, 6, 7) and then can be reversed so that it goes through the sequence in the opposite direction (7, 6, 5, 4, 3, 2, 1, 0) is an illustration of up/down sequential operation.*

In general, most up/down counters can be reversed at any point in their sequence. For instance, the three-bit binary counter mentioned can be made to go through the following sequence:

$$\overbrace{\text{up}} \qquad\qquad \overbrace{\text{up}}$$

0, 1, 2, 3, 4, 5, 4, 3, 2, 3, 4, 5, 6, 7, 6, 5, etc.

$$\underbrace{} \qquad\qquad \underbrace{}$$
$$\text{down} \qquad\qquad \text{down}$$

TABLE 8–5
Up/down sequence for a three-bit binary counter.

Clock Pulse	UP	Q_2	Q_1	Q_0	DOWN
0		0	0	0	
1		0	0	1	
2		0	1	0	
3		0	1	1	
4		1	0	0	
5		1	0	1	
6		1	1	0	
7		1	1	1	

Table 8–5 shows the complete up/down sequence for a three-bit binary counter. The arrows indicate the state-to-state movement of the counter for both its UP and its DOWN modes of operation. An examination of Q_0 for both the UP and DOWN sequences shows that FF0 toggles on each clock pulse. So the J and K inputs of FF0 are

$$J_0 = K_0 = 1$$

For the UP sequence, Q_1 changes state on the next clock pulse when $Q_0 = 1$. For the DOWN sequence, Q_1 changes on the next clock pulse when $Q_0 = 0$. Thus, the J and K inputs of FF1 must equal 1 under the conditions expressed by the following equation:

$$J_1 = K_1 = Q_0 \cdot \text{UP} + \overline{Q}_0 \cdot \text{DOWN}$$

For the UP sequence, Q_2 changes state on the next clock pulse when $Q_0 = Q_1 = 1$. For the DOWN sequence, Q_2 changes on the next clock pulse when $Q_0 = Q_1 = 0$. Thus, the J and K inputs of FF2 must equal 1 under the conditions expressed by the following equation:

$$J_2 = K_2 = Q_0 \cdot Q_1 \cdot \text{UP} + \overline{Q}_0 \cdot \overline{Q}_1 \cdot \text{DOWN}$$

Each of the conditions for the J and K inputs of each flip-flop produces a toggle at the appropriate point in the counter sequence.

Figure 8–23 shows a basic implementation of a three-bit up/down binary counter using the logic equations just developed for the J and K inputs of each flip-flop. Notice that the UP/$\overline{\text{DOWN}}$ control input is HIGH for UP and LOW for DOWN.

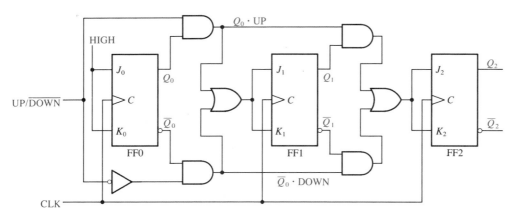

FIGURE 8–23
A basic three-bit up/down synchronous counter.

EXAMPLE 8–4

Determine the sequence of a synchronous four-bit binary up/down counter if the clock and up/down control inputs have waveforms as shown in Figure 8–24(a). The counter starts in the all 0s state and is positive edge-triggered.

Solution From the waveforms in Figure 8–24(b), the counter sequence is as follows:

Q_3	Q_2	Q_1	Q_0	
0	0	0	0	
0	0	0	1	
0	0	1	0	up
0	0	1	1	
0	1	0	0	
0	0	1	1	
0	0	1	0	
0	0	0	1	down
0	0	0	0	
1	1	1	1	
0	0	0	0	
0	0	0	1	up
0	0	1	0	
0	0	0	1	down
0	0	0	0	

FIGURE 8–24

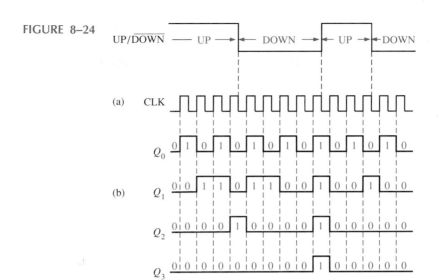

The 74190 Up/Down Decade Counter

Figure 8–25 shows a logic diagram for the 74190, a good example of an integrated circuit up/down counter. The direction of the count is determined by the level of the *up/down input* ($D/\overline{U}$). When this input is HIGH, the counter counts down; when it is LOW, the counter counts up. Also, this device can be preset to any desired BCD digit as determined by the states of the data inputs when the LOAD input is LOW.

The *MAX-MIN output* produces a HIGH pulse when the terminal count nine (1001) is reached in the up mode or when the terminal count zero (0000) is reached in the down mode. This MAX-MIN output, along with the *ripple clock output* ($\overline{RCO}$) and the *count Enable input* ($\overline{CTEN}$), is used when cascading counters. (The topic of cascading counters is discussed later in this chapter.)

Figure 8–26 is an example timing diagram showing the 74190 counter preset to seven (0111) and then going through a count-up sequence followed by a count-down sequence. The MAX/MIN output is HIGH when the counter is in either the all-0s state (MIN) or the 1001 state (MAX).

FIGURE 8–25
The 74190 up/down synchronous decade counter.

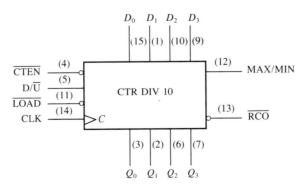

FIGURE 8–26
Timing example for a 74190.

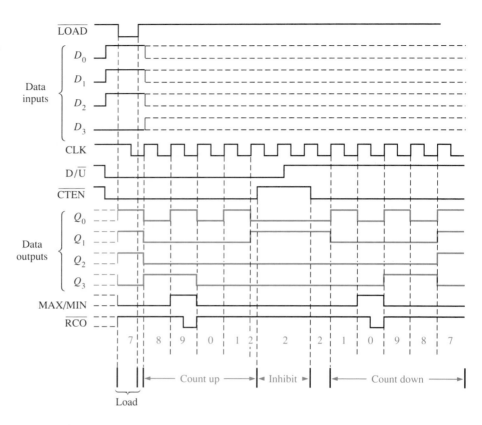

SECTION 8–3
REVIEW

Answers are found at the end of the chapter.
1. A four-bit up/down binary counter is in the down mode and in the 1010 state. On the next clock pulse, to what state does the counter go?
2. What is the terminal count of a four-bit binary counter in the up mode? In the down mode? What is the next state after the terminal count in the down mode?

DESIGN OF SEQUENTIAL CIRCUITS

8–4

This section is optional and may be omitted without affecting the flow of material in the remainder of the book. Inclusion of this material is recommended for those who need an introduction to counter design or to state machine design in general.

Before proceeding with a specific design technique, let us begin with a general definition of a *sequential circuit (state machine):* A general sequential circuit consists of a combinational logic section and a memory section (flip-flops), as shown in Figure 8–27. In a clocked sequential circuit, there is a clock input to the memory section as indicated.

The information stored in the memory section, as well as the inputs to the combinational logic ($I_0, I_1, \ldots, I_m$), is required for proper operation of the circuit. At any given time, the memory is in a state called the *present state* and will advance to a *next state* on a clock pulse as determined by conditions on the excitation lines ($Y_0, Y_1, \ldots, Y_p$). The present state of the memory is represented by the *state variables* ($Q_0, Q_1,$

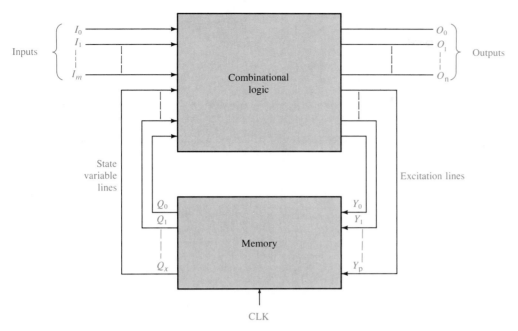

FIGURE 8–27
General clocked sequential circuit.

. . . , Q_x). These state variables, along with the inputs ($I_0, I_1, . . . , I_m$), determine the system outputs ($O_0, O_1, . . . , O_n$).

Not all sequential circuits have input and output variables as in the general model just discussed. However, all have excitation variables and state variables. Counters are a special case of clocked sequential circuits. In this section a general design procedure for sequential circuits as applied to counters is presented.

State Diagram

In this procedure, a clocked sequential circuit is first described by a *state diagram*, which shows the progression of states and input and output conditions, if any. Figure 8–28 is an example of a state diagram for a three-bit Gray code counter. The circuit

FIGURE 8–28
State diagram for a three-bit Gray code counter.

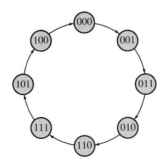

has no inputs other than the clock and no outputs other than its internal state (outputs are taken off each flip-flop in the counter).

Next-State Table

Once the sequential circuit is defined by a state diagram, the second step is to derive a *next-state table* which lists each state of the counter (present state) along with the corresponding next state. The next state is the state that the counter goes to from its present state upon application of a clock pulse. The next-state table is derived from the state diagram and is shown in Table 8–6 for the three-bit Gray code counter example.

TABLE 8–6

Next-state table for three-bit Gray code counter.

Present State			Next State		
Q_2	Q_1	Q_0	Q_2	Q_1	Q_0
0	0	0	0	0	1
0	0	1	0	1	1
0	1	1	0	1	0
0	1	0	1	1	0
1	1	0	1	1	1
1	1	1	1	0	1
1	0	1	1	0	0
1	0	0	0	0	0

Transition Table

The third step is to develop a *transition table* from the next-state table. The transition table shows the flip-flop inputs required to make the counter go from a present state to the proper next state.

If a flip-flop is RESET ($Q = 0$) in its present state and it must remain RESET in its next state (indicated by $0 \rightarrow 0$), it can be in either a no-change ($J = 0$, $K = 0$) condition or a RESET ($J = 0$, $K = 1$) condition when the clock pulse occurs. As you can see, K can be either a 1 or a 0; we do not care. Therefore, the flip-flop inputs are $J = 0$ and $K = X$ for a $Q = 0$ to $Q = 0$ transition. The X means "don't care."

If a flip-flop is SET ($Q = 1$) and it is to remain SET in its next state (indicated by $1 \rightarrow 1$), either a no-change ($J = 0$, $K = 0$) or a SET ($J = 1$, $K = 0$) condition is required. Therefore, $J = X$ and $K = 0$ for a $Q = 1$ to $Q = 1$ transition.

If a flip-flop is RESET ($Q = 0$) and must go to the SET ($Q = 1$) state (indicated by $0 \rightarrow 1$), either a SET ($J = 1$, $K = 0$) or a toggle ($J = 1$, $K = 1$) condition is required. Therefore, $J = 1$ and $K = X$ for a $Q = 0$ to $Q = 1$ transition.

Finally, if a flip-flop is SET ($Q = 1$) and must go to the RESET ($Q = 0$) state (indicated by $1 \rightarrow 0$), either a RESET ($J = 0$, $K = 1$) or a toggle ($J = 1$, $K = 1$) condition is required. Therefore, $J = X$ and $K = 1$ for a $Q = 1$ to $Q = 0$ transition.

The transition table for the three-bit Gray code counter using J-K flip-flops is shown in Table 8–7.

TABLE 8–7

Transition table for three-bit Gray code counter.

	Output State Transitions			Flip-Flop Inputs (present state)		
	Q_2	Q_1	Q_0			
	Present → Next	Present → Next	Present → Next	$J_2 K_2$	$J_1 K_1$	$J_0 K_0$
	0 → 0	0 → 0	0 → 1	0 X	0 X	1 X
	0 → 0	0 → 1	1 → 1	0 X	1 X	X 0
	0 → 0	1 → 1	1 → 0	0 X	X 0	X 1
	0 → 1	1 → 1	0 → 0	1 X	X 0	0 X
	1 → 1	1 → 1	0 → 1	X 0	X 0	1 X
	1 → 1	1 → 0	1 → 1	X 0	X 1	X 0
	1 → 1	0 → 0	1 → 0	X 0	0 X	X 1
	1 → 0	0 → 0	0 → 0	X 1	0 X	0 X

Karnaugh Maps

The fourth step is to transfer the *JK* states from the transition table to Karnaugh maps to derive a simplified Boolean expression for each flip-flop input. This procedure is shown in Figure 8–29, which shows a Karnaugh map for the *J* and *K* inputs of each flip-flop.

Input Logic

In the fifth step, the 1s in the Karnaugh maps of Figure 8–29 are grouped with "don't cares," and the following expressions for the *J* and *K* inputs of each flip-flop are obtained:

$$J_0 = Q_2 Q_1 + \overline{Q}_2 \overline{Q}_1$$
$$K_0 = Q_2 \overline{Q}_1 + \overline{Q}_2 Q_1$$
$$J_1 = \overline{Q}_2 Q_0$$
$$K_1 = Q_2 Q_0$$
$$J_2 = Q_1 \overline{Q}_0$$
$$K_2 = \overline{Q}_1 \overline{Q}_0$$

Implementation

The final step is to implement the combinational logic from the equations and connect the flip-flops to form the sequential circuit. The complete three-bit Gray code counter is shown in Figure 8–30.

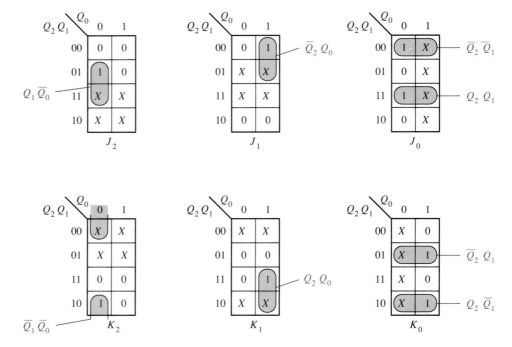

FIGURE 8–29

Karnaugh maps for present-state J and K inputs in Table 8–7.

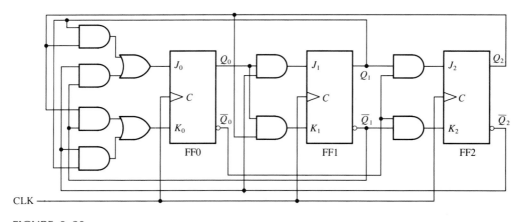

FIGURE 8–30

Three-bit Gray code counter.

A summary of steps used in the design of this counter follows. In general, these steps can be applied to any sequential circuit.

1. Specify the sequential circuit and draw a state diagram.
2. Derive a next-state table from the state diagram.
3. Develop a transition table showing the flip-flop inputs required to make each flip-flop go to its proper next state from each of its present states.
4. Transfer the *JK* states from the transition table to Karnaugh maps. There is a Karnaugh map for each input of each flip-flop.

5. Factor the maps to generate an expression for each flip-flop input.
6. Implement the expressions with combinational logic, and combine with the flip-flops.

This procedure is now applied to the design of another counter.

EXAMPLE 8–5

Develop a synchronous three-bit up/down counter with a Gray code sequence. The counter should count up when an up/down control input is 1 and count down when the control input is 0.

Solution

Step 1 The state diagram is shown in Figure 8–31. The 1 or 0 beside each arrow indicates the state of the up/down control input, Y.

FIGURE 8–31
State diagram for a three-bit up/down Gray code counter.

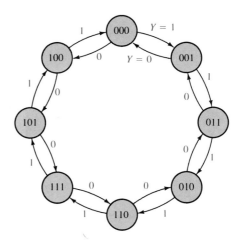

Step 2 The next-state table is derived from the state diagram and is shown in Table 8–8.

TABLE 8–8
Next-state table for three-bit up/down Gray code counter.

Present State			Next State					
			$Y = 0$ (down)			$Y = 1$ (up)		
Q_2	Q_1	Q_0	Q_2	Q_1	Q_0	Q_2	Q_1	Q_0
0	0	0	1	0	0	0	0	1
0	0	1	0	0	0	0	1	1
0	1	1	0	0	1	0	1	0
0	1	0	0	1	1	1	1	0
1	1	0	0	1	0	1	1	1
1	1	1	1	1	0	1	0	1
1	0	1	1	1	1	1	0	0
1	0	0	1	0	1	0	0	0

$Y = $ up/down control input.

Step 3 The transition table for the flip-flops is derived from the next-state table and is shown in Table 8–9. The control input, Y, is treated as a fourth variable along with the three present-state variables, Q_0, Q_1, and Q_2.

TABLE 8–9
Transition table for three-bit up/down Gray code counter.

Output State Transitions			Y	Flip-Flop Input (present state)		
Q_2	Q_1	Q_0	Down = 0 Up = 1			
Present → Next	Present → Next	Present → Next		J_2K_2	J_1K_1	J_0K_0
0 → 1	0 → 0	0 → 0	0	1 X	0 X	0 X
0 → 0	0 → 0	0 → 1	1	0 X	0 X	1 X
0 → 0	0 → 0	1 → 0	0	0 X	0 X	X 1
0 → 0	0 → 1	1 → 1	1	0 X	1 X	X 0
0 → 0	1 → 0	1 → 1	0	0 X	X 1	X 0
0 → 0	1 → 1	1 → 0	1	0 X	X 0	X 1
0 → 0	1 → 1	0 → 1	0	0 X	X 0	1 X
0 → 1	1 → 1	0 → 0	1	1 X	X 0	0 X
1 → 0	1 → 1	0 → 0	0	X 1	X 0	0 X
1 → 1	1 → 1	0 → 1	1	X 0	X 0	1 X
1 → 1	1 → 1	1 → 0	0	X 0	X 0	X 1
1 → 1	1 → 0	1 → 1	1	X 0	X 1	X 0
1 → 1	0 → 1	1 → 1	0	X 0	1 X	X 0
1 → 1	0 → 0	1 → 0	1	X 0	0 X	X 1
1 → 1	0 → 0	0 → 1	0	X 0	0 X	1 X
1 → 0	0 → 0	0 → 0	1	X 1	0 X	0 X

Step 4 The Karnaugh maps for the J and K inputs of the flip-flops are shown in Figure 8–32. As mentioned before, the control input Y is considered one of the state variables along with Q_0, Q_1, and Q_2. The information in the "Flip-Flop Input"

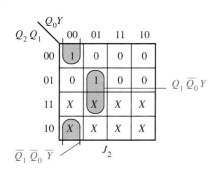

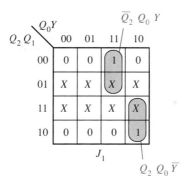

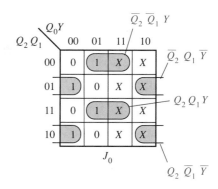

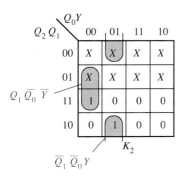

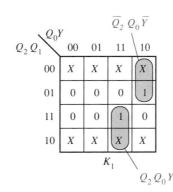

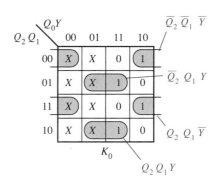

FIGURE 8–32
J and K maps for Table 8–9.

columns of Table 8–9 is transferred onto the maps as indicated for each present state of the counter.

Step 5   The 1s are combined in the largest possible groupings, with ''don't cares'' (X's) used where possible. The groups are factored, and the expressions for the J and K inputs are as follows:

$$J_0 = Q_2Q_1Y + Q_2\overline{Q}_1\overline{Y} + \overline{Q}_2Q_1\overline{Y} + \overline{Q}_2\overline{Q}_1Y$$
$$K_0 = \overline{Q}_2\overline{Q}_1\overline{Y} + \overline{Q}_2Q_1Y + Q_2Q_1\overline{Y} + Q_2Q_0Y$$
$$J_1 = \overline{Q}_2Q_0Y + Q_2Q_0\overline{Y}$$
$$K_1 = \overline{Q}_2Q_0\overline{Y} + Q_2Q_0Y$$
$$J_2 = Q_1\overline{Q}_0Y + \overline{Q}_1\overline{Q}_0\overline{Y}$$
$$K_2 = Q_1\overline{Q}_0\overline{Y} + \overline{Q}_1\overline{Q}_0Y$$

Step 6   The J and K equations are implemented with combinational logic, and the complete counter is shown in Figure 8–33.

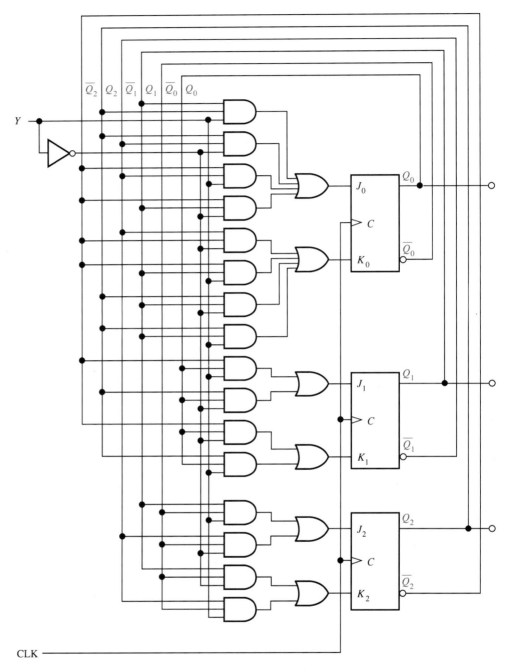

FIGURE 8–33
Three-bit up/down Gray code counter.

EXAMPLE 8–6

Design a counter with the irregular binary count sequence shown in the state diagram of Figure 8–34.

FIGURE 8–34

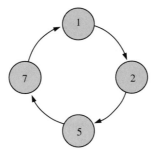

Solution   A three-bit counter is required to implement this sequence. Notice that the required sequence does not include all the possible binary states. The invalid states are 0, 3, 4, and 6. These invalid states can be treated as ''don't cares'' in the design. However, if the counter should erroneously get into an invalid state, we must make sure that it goes back to a valid state.

The next-state table is developed from the state diagram and is given in Table 8–10.

TABLE 8–10
Next-state table.

Present State			Next State		
Q_2	Q_1	Q_0	Q_2	Q_1	Q_0
0	0	1	0	1	0
0	1	0	1	0	1
1	0	1	1	1	1
1	1	1	0	0	1

Next, the transition table in Table 8–11 is developed from the next-state table.

TABLE 8–11
Transition table.

Output State Transitions			Flip-Flop Inputs		
Q_2	Q_1	Q_0	(present state)		
Present → Next	Present → Next	Present → Next	J_2K_2	J_1K_1	J_0K_0
$0 \rightarrow 0$	$0 \rightarrow 1$	$1 \rightarrow 0$	0 X	1 X	X 1
$0 \rightarrow 1$	$1 \rightarrow 0$	$0 \rightarrow 1$	1 X	X 1	1 X
$1 \rightarrow 1$	$0 \rightarrow 1$	$1 \rightarrow 1$	X 0	1 X	X 0
$1 \rightarrow 0$	$1 \rightarrow 0$	$1 \rightarrow 1$	X 1	X 1	X 0

Next, the J and K inputs are plotted on the present-state Karnaugh maps in Figure 8–35. Also "don't cares" can be placed in the cells corresponding to the invalid states of 000, 011, 100, and 110, as indicated by the colored Xs.

Group the 1s, taking advantage of as many of the "don't care" states as possible for maximum simplification, as shown in Figure 8–35. Notice that when *all* cells in a map are grouped, the expression is simply equal to 1. The expression for each J and K input taken from the maps is as follows:

$$J_0 = 1, K_0 = \overline{Q}_2$$
$$J_1 = K_1 = 1$$
$$J_2 = K_2 = Q_1$$

The implementation of the counter is shown in Figure 8–36.

FIGURE 8–35

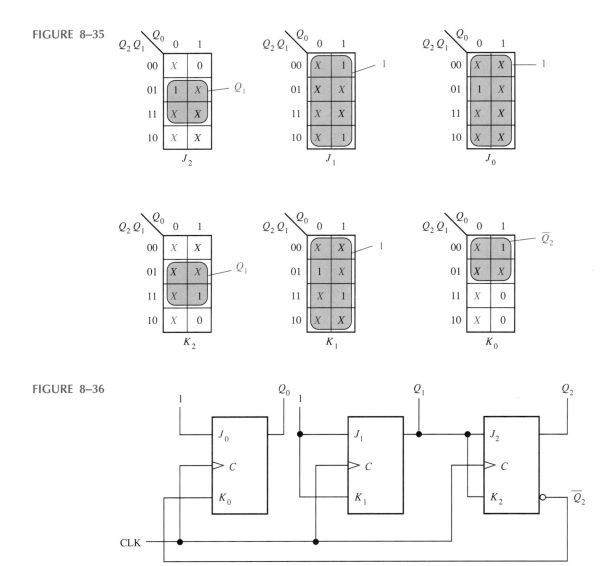

FIGURE 8–36

An analysis shows that if the counter, by accident, gets into one of the invalid states, it will always return to a valid state according to the following sequences: $0 \rightarrow 3 \rightarrow 4 \rightarrow 7$, and $6 \rightarrow 1$.

You have seen how sequential circuit design techniques can be applied specifically to counter design. In general, sequential circuits can be classified into two types: (1) those in which the output or outputs depend only on the present internal state (called *Moore* circuits) and (2) those in which the output or outputs depend on both the present state and the input or inputs (called *Mealy* circuits).

SECTION 8–4
REVIEW

Answers are found at the end of the chapter.
1. A flip-flop is presently in the RESET state and must go to the SET state on the next clock pulse. What must J and K be?
2. A flip-flop is presently in the SET state and must remain SET on the next clock pulse. What must J and K be?
3. A binary counter is in the 1010 state.
   (a) What is its next state?
   (b) What condition must exist on each flip-flop input to ensure that it goes to the proper next state on the clock pulse?

CASCADED COUNTERS

8–5

Counters can be connected in cascade to achieve higher-modulus operation. In essence, cascading means that the last-stage output of one counter drives the input of the next counter.

An example of two counters connected in cascade is shown for a two-bit and a three-bit ripple counter in Figure 8–37. The timing diagram is in Figure 8–38.

Notice that in the timing diagram of Figure 8–38, the final output of the modulus-8 counter, Q_4, occurs once for every 32 input clock pulses. The overall modulus of the cascaded counters is 32; that is, they act as a divide-by-32 counter.

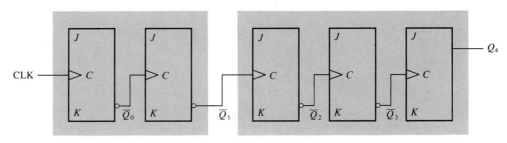

Modulus-4 counter   Modulus-8 counter

FIGURE 8–37
Two cascaded counters (all J, K inputs are HIGH).

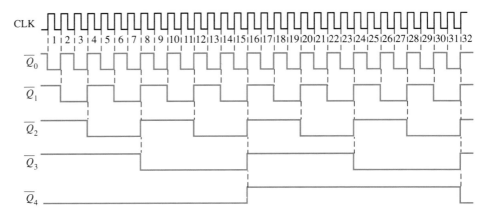

FIGURE 8–38

Timing diagram for the counter configuration of Figure 8–37.

In general, the overall modulus of cascaded counters is equal to the product of the individual moduli. For instance, for the counter in Figure 8–37, $4 \times 8 = 32$.

When operating synchronous counters in a cascaded configuration, it is necessary to use the count Enable and the terminal count functions to achieve higher-modulus operation. On some devices the count Enable is labeled simply CTEN or some other similar designation, and terminal count (TC) is analogous to *ripple clock output* (RCO) on some IC counters.

Figure 8–39 shows two decade counters connected in cascade. The terminal count (TC) output of counter 1 is connected to the count Enable (CTEN) input of counter 2. Counter 2 is inhibited by the LOW on its CTEN input until counter 1 reaches its last, or terminal, state and its terminal count output goes HIGH. This HIGH now enables counter 2, so that when the first clock pulse after counter 1 reaches its terminal count (CLK_{10}), counter 2 goes from its initial state to its second state. Upon completion of the entire second cycle of counter 1 (when counter 1 reaches terminal count the second time), counter 2 is again enabled and advances to its next state. This sequence continues. Since these are decade counters, counter 1 must go through ten complete cycles before counter 2 completes its first cycle. In other words, for every ten cycles of counter 1, counter 2 goes through one cycle. Thus, counter 2 will complete one cycle after one hundred clock pulses. The overall modulus of these two cascaded counters is $10 \times 10 = 100$.

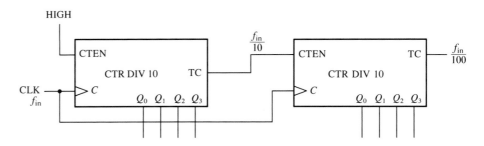

FIGURE 8–39

A modulus-100 cascaded counter using two decade counters.

When viewed as a frequency divider, the circuit of Figure 8–39 divides the input clock frequency by 100. Cascaded counters are often used to divide a high-frequency clock signal to obtain highly accurate pulse frequencies. Cascaded counter configurations used for such purposes are sometimes called *countdown chains*.

For example, suppose that we have a basic clock frequency of 1 MHz and we wish to obtain 100 kHz, 10 kHz, and 1 kHz; a series of cascaded counters can be used. If the 1-MHz signal is divided by 10, we get 100 kHz. Then if the 100-kHz signal is divided by 10, we get 10 kHz. Another division by 10 yields the 1-kHz frequency. The general implementation of this countdown chain is shown in Figure 8–40.

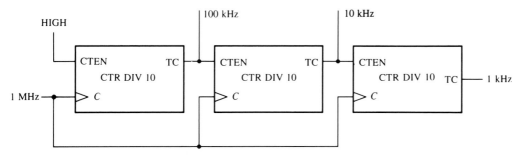

FIGURE 8–40
Three cascaded decade counters forming a divide-by-1000.

EXAMPLE 8–7

Determine the overall modulus of the two cascaded counter configurations in Figure 8–41.

FIGURE 8–41

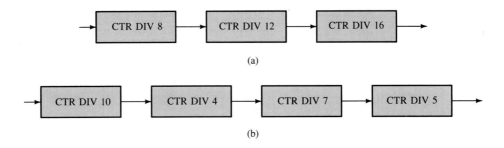

(a)

(b)

Solutions

(a) The overall modulus for the three-counter configuration is

$$8 \times 12 \times 16 = 1536$$

(b) The modulus for the four counters is

$$10 \times 4 \times 7 \times 5 = 1400$$

EXAMPLE 8–8

Use 74LS160A counters to obtain a 10-kHz waveform from a 1-MHz clock. Show the logic diagram.

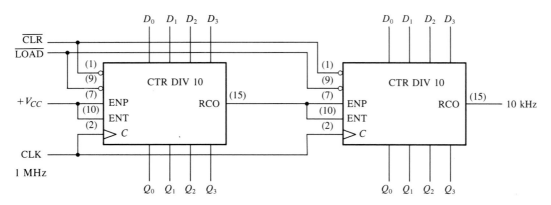

FIGURE 8–42

A divide-by-100 counter using two 74LS160A decade counters.

Solution   To obtain 10 kHz from a 1-MHz clock requires a division factor of 100. Two 74LS160A counters must be cascaded as shown in Figure 8–42. The left counter produces an RCO pulse for every 10 clock pulses. The right counter produces an RCO pulse for every 100 clock pulses.

Cascaded IC Counters with Truncated Sequences

The preceding material has shown how to achieve an overall modulus (*divide-by* factor) that is the product of the individual moduli of all the cascaded counters. This can be considered *full-modulus cascading*.

Often an application requires an overall modulus that is *less* than that achieved by full-modulus cascading. That is, a truncated sequence must be implemented with cascaded counters.

To illustrate this method, the cascaded counter configuration in Figure 8–43 is used as an example. This particular circuit uses four 74LS161A four-bit binary counters. If these four counters (sixteen bits total) were cascaded in a full-modulus arrangement, the modulus would be

$$2^{16} = 65,536$$

Let us assume that a certain application requires a divide-by-40,000 counter (modulus 40,000). The difference between 65,536 and 40,000 is 25,536, which is the number of states that must be deleted from the full-modulus sequence. The technique used in the circuit of Figure 8–43 is to preset the cascaded counter to 25,536 (63C0 in hexadecimal) each time it recycles, so that it will count from 25,536 up to 65,536 on each full cycle. Therefore, each full cycle of the counter consists of 40,000 states.

Notice that the RCO output of the right-most counter is inverted and applied to the $\overline{\text{LOAD}}$ input of each four-bit counter. Each time the count reaches its terminal value

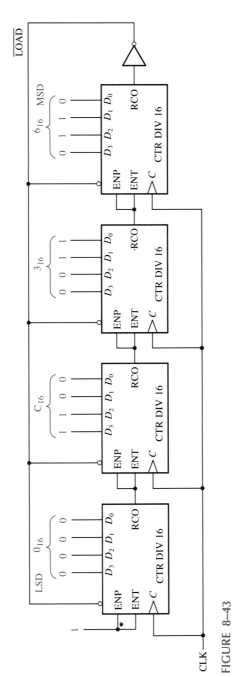

FIGURE 8—43

A divide-by-40,000 counter using 74LS161A four-bit binary counters. Note that the parallel data inputs are shown in binary order (the right-most bit is the LSB).

of 65,536, RCO goes HIGH and causes the number on the data inputs ($63C0_{16}$) to be preset into the counter. Thus, there is one RCO pulse from the right-most four-bit counter for every 40,000 clock pulses.

   With this technique any modulus can be achieved by the presetting of the counter to the appropriate initial state on each cycle.

SECTION 8–5
REVIEW

Answers are found at the end of the chapter.

1. How many decade counters are necessary to implement a divide-by-1000 (modulus-1000) counter? A divide-by-10,000?
2. Show with general block diagrams how to achieve each of the following, using a flip-flop, a decade counter, and a four-bit binary counter, or any combination of these:
   (a) Divide-by-20 counter   (b) Divide-by-32 counter
   (c) Divide-by-160 counter   (d) Divide-by-320 counter

COUNTER DECODING

8–6

In many digital applications it is necessary that some or all of the counter states be decoded. The decoding of a counter involves using decoders or logic gates to determine when the counter is in a certain state or states in its sequence. For instance, the terminal count function previously discussed is a single state (the last state) in the counter sequence.

For example, suppose that we wish to decode count 6 (110) of a three-bit binary counter. This can be done as shown in Figure 8–44. When $Q_2 = 1$, $Q_1 = 1$, and $Q_0 = 0$, a HIGH appears on the output of the decoding gate, indicating that the counter is at count 6. This is called *active-HIGH decoding*. Replacing the AND gate with a NAND gate provides *active-LOW decoding*.

FIGURE 8–44
Decoding of count 6.

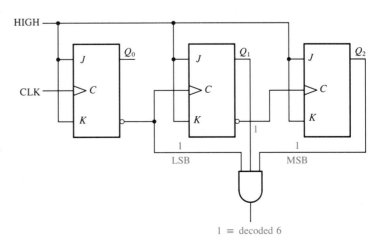

$1 \equiv$ decoded 6

EXAMPLE 8–9

Implement the decoding of state 2 and state 7 of a three-bit synchronous counter. Show the entire counter timing diagram and the output waveforms of the decoding gates. Binary $2 = \overline{Q_2}Q_1\overline{Q_0}$, and binary $7 = Q_2Q_1Q_0$.

Solution See Figure 8–45.

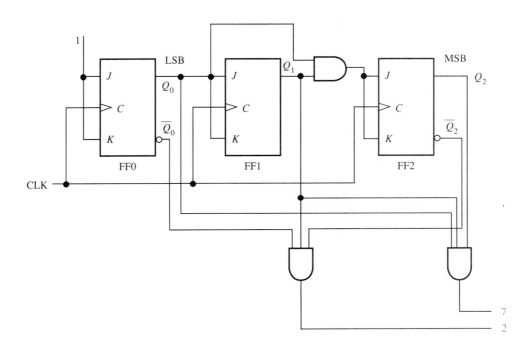

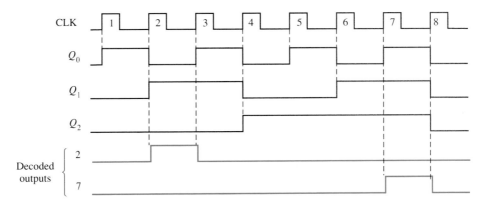

FIGURE 8–45

A divide-by-8 counter with active-HIGH decoding of count 2 and count 7.

Decoding Glitches

The problem of glitches produced by the decoding process was introduced in Chapter 6. As you have learned, the propagation delays due to the ripple effect in asynchronous counters create transitional states in which the counter outputs are changing at slightly different times. These transitional states produce undesired voltage spikes of short duration (glitches) on the outputs of a decoder. The glitch problem can also occur to some degree with synchronous counters because the propagation delays from clock to Q outputs of each flip-flop in a counter can vary slightly. As an example of the glitch problem in counter decoding, Figure 8–46 shows a basic asynchronous BCD counter connected to a BCD-to-decimal decoder.

To see what happens in this case, we will look at a timing diagram in which the propagation delays are taken into account, as in Figure 8–47. Notice that these delays cause false states of short duration. The decimal value of the false binary state at each critical transition is indicated on the diagram. The resulting glitches can be seen on the decoder outputs.

One way to eliminate the glitches is to enable the decoded outputs at a time *after* the glitches have had time to disappear. This method is known as *strobing* and can be accomplished in this case by using the LOW level of the clock to enable the decoder, as shown in Figure 8–48. The resulting improved timing diagram is shown in Figure 8–49.

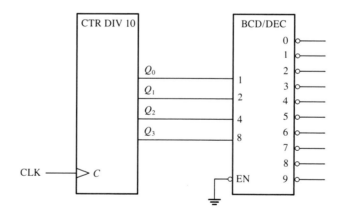

FIGURE 8–46
A basic decade (BCD) counter and decoder.

SECTION 8–6
REVIEW

Answers are found at the end of the chapter.
1. What transitional states are possible when an asynchronous four-bit binary counter changes from
   (a) count 2 to count 3
   (b) count 3 to count 4
   (c) count 10_{10} to count 11_{10}
   (d) count 15 to count 0

FIGURE 8–47
Outputs with glitches from the decoder in Figure 8–46. Glitch widths are exaggerated for illustration.

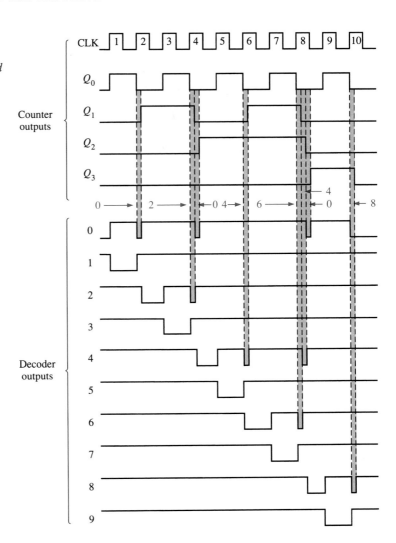

FIGURE 8–48
The basic decade counter and decoder with strobing to eliminate glitches.

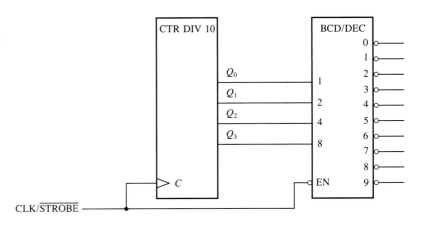

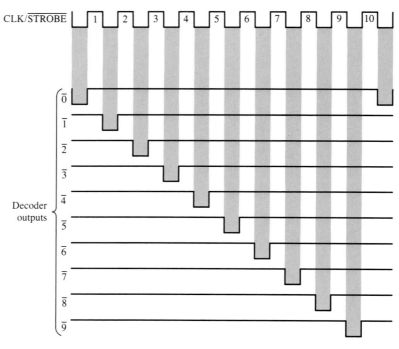

FIGURE 8–49
Strobed decoder outputs for the circuit of Figure 8–48.

COUNTER APPLICATIONS

8–7
The digital counter is a very useful and versatile device that is found in many applications. In this section, some representative counter applications are presented.

The Digital Clock

A very common example of a counter application is in timekeeping systems. Figure 8–50 is a simplified logic diagram of a digital clock that displays seconds, minutes, and hours.

First, a 60-Hz sinusoidal ac voltage is converted to a 60-Hz pulse waveform and divided down to a 1-Hz pulse waveform by a divide-by-60 counter formed by a divide-by-10 counter followed by a divide-by-6 counter. Both the *seconds* and *minutes* counts are also produced by divide-by-60 counters, the details of which are shown in Figure 8–51. These counters count from 0 to 59 and then recycle to 0; 74LS160A synchronous decade counters are used in this particular implementation. Notice that the divide-by-6 portion is formed with a decade counter with a truncated sequence achieved by using the decoded count 6 to asynchronously clear the counter. The terminal count, 5, is also decoded to enable the next counter in the chain.

The *hours* counter is implemented with a decade counter and a flip-flop as shown in Figure 8–52. Consider that initially both the decade counter and the flip-flop are RESET, and the decode-12 gate output is HIGH. The decade counter advances from

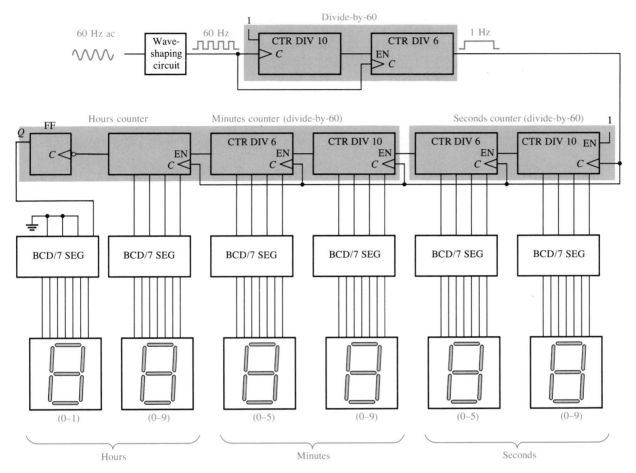

FIGURE 8–50
Simplified logic diagram for a 12-hour digital clock. Logic details using specific devices are shown in the following diagrams.

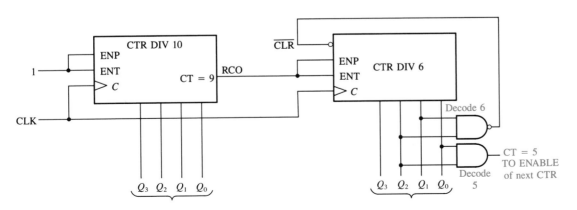

FIGURE 8–51
Logic diagram of typical divide-by-60 counter using 74LS160A synchronous decade counters. Note that the outputs are in binary order (the right-most bit is the LSB).

zero to nine, and as it recycles from nine back to zero, the flip-flop is toggled to the SET state by the HIGH-to-LOW transition of Q_3. This illuminates a 1 on the tens-of-hours display. The total count is now ten (the decade counter is in the zero state and the flip-flop is SET).

FIGURE 8–52
Logic diagram for hours counter and decoders. Note that on the counter inputs and outputs, the right-most bit is the LSB.

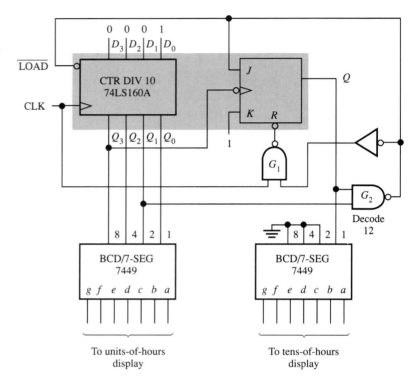

Next, the total count advances to eleven and then to twelve. In state 12 the Q_1 output of the decade counter is HIGH, the flip-flop is still SET, and thus the decode gate output is LOW. This activates the LOAD input of the decade counter. On the next clock pulse, the decade counter is preset to state 1 by the data inputs, and the flip-flop is RESET ($J = 0$, $K = 1$). As you can see, this logic always causes the counter to recycle from twelve back to one rather than back to zero.

Auto Parking Control

Now a simple application example illustrates the use of an up/down counter to solve an everyday problem. The problem is to devise a means of monitoring available spaces in a one-hundred-space parking garage and provide for an indication of a full condition by illuminating a display sign and lowering a gate bar at the entrance.

A system that solves this problem consists of (1) optoelectronic sensors at the entrance and exit of the garage, (2) an up/down counter and associated circuitry, and (3) an interface circuit that takes the counter output to turn the FULL sign on or off as required and lower or raise the gate bar at the entrance. A general block diagram of this system is shown in Figure 8–53.

A logic diagram of the up/down counter is shown in Figure 8–54. It consists of two cascaded 74190 up/down decade counters. The operation is as described in the following paragraphs.

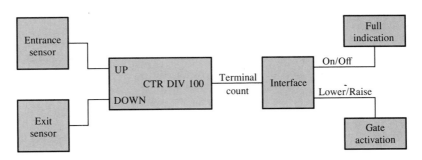

FIGURE 8–53
Functional block diagram for parking garage control.

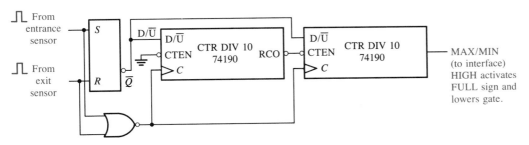

FIGURE 8–54
Logic diagram for modulus-100 up/down counter for auto parking control.

The counter is initially preset to 0. Each automobile entering the garage breaks a light beam, activating a sensor that produces an electrical pulse. This positive pulse SETS the S-R latch on its leading edge. The LOW on the $\overline{Q}$ output of the latch puts the counter in the UP mode. Also, the sensor pulse goes through the NOR gate and clocks the counter on the LOW-to-HIGH transition of its trailing edge. Each time an automobile enters the garage, the counter is advanced by one *(incremented)*. When the one-hundredth automobile enters, the counter goes to its last state (100_{10}). The MAX/MIN output goes HIGH and activates the interface circuit (no detail), which lights the FULL sign and lowers the gate bar to prevent further entry.

When an automobile exits, an optoelectronic sensor produces a positive pulse, which RESETS the S-R latch and puts the counter in the DOWN mode. The trailing edge of the clock decreases the count by one *(decrements)*.

If the garage is full and an automobile leaves, the MAX/MIN output of the counter goes LOW, turning off the FULL sign and raising the gate.

Parallel-to-Serial Data Conversion (Multiplexing)

A simplified example of data transmission using multiplexing and demultiplexing techniques was introduced in Chapter 6. Essentially, the parallel data bits on the multiplexer inputs were converted to serial data bits on the single transmission line. A group of bits appearing simultaneously on parallel lines is called *parallel data*. A group of bits appearing on a single line in a time sequence is called *serial data*.

Parallel-to-serial conversion is normally accomplished by the use of a counter to provide a binary sequence for the data-select inputs of a data selector/multiplexer, as illustrated in Figure 8–55. The Q outputs of the modulus-8 counter are connected to the data-select inputs of an eight-bit multiplexer. Figure 8–56 is a timing diagram illustrating the operation of this circuit.

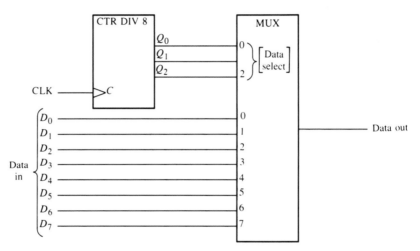

FIGURE 8–55
Parallel-to-serial data conversion logic.

The first byte (eight-bit group) of parallel data is applied to the multiplexer inputs. As the counter goes through a binary sequence from zero to seven, each bit, beginning with D_0, is sequentially selected and passed through the multiplexer to the output line. After eight clock pulses the data byte has been converted to a serial format and sent out on the transmission line. When the counter recycles back to 0, the next byte is applied to the data inputs and is sequentially converted to serial form as the counter cycles through its eight states. This process continues repeatedly as each parallel byte is converted to a serial byte. An end-of-chapter problem requires that this circuit be implemented with specific devices.

**SECTION 8–7
REVIEW**

Answers are found at the end of the chapter.
1. Explain the purpose of each NAND gate in Figure 8–52.
2. Identify the two recycle conditions for the hours counter in Figure 8–50, and explain the reason for each.

FIGURE 8–56

Example of parallel-to-serial conversion timing.

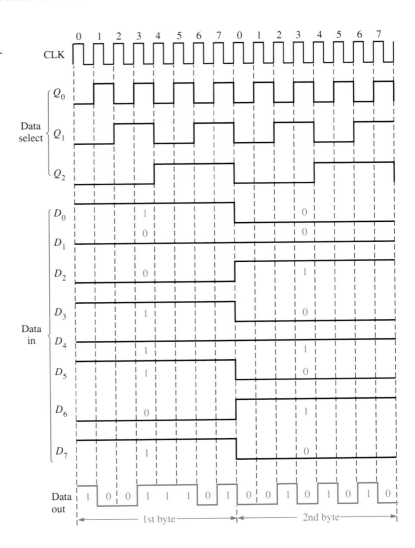

TROUBLESHOOTING

8–8

The troubleshooting of counters can be very simple or quite involved, depending on the type of counter and the type of fault. This section will give you some good practical insight into how to approach the troubleshooting of sequential circuits.

IC Counters

For an IC counter with a straightforward sequence that is not controlled by external logic, about the only thing to check (other than V_{CC} and ground) is the possibility of open or shorted inputs or outputs. An IC counter almost never alters its sequence of states because of an internal fault, so you need only check for pulse activity on the Q outputs to detect the existence of an open or a short, as shown in Figure 8–57(a). The absence of pulse activity on one of the Q outputs indicates an internal short or open. Absence of pulse activity on all the Q outputs indicates that the clock input is faulty or the clear input is stuck in its active state.

To check the clear input, apply a constant active level while the counter is clocked. You will observe a LOW on each of the Q outputs if it is functioning properly, as shown in part (b).

The parallel load feature on a counter can be checked by activating the $\overline{\text{LOAD}}$ input and exercising each stage as follows: Apply LOWs to the parallel data inputs, pulse the clock input, and check for LOWs on all the Q outputs. Next, apply HIGHs to all the parallel data inputs, pulse the clock input, and check for HIGHs on all the Q outputs. This is illustrated in Figure 8–57(c).

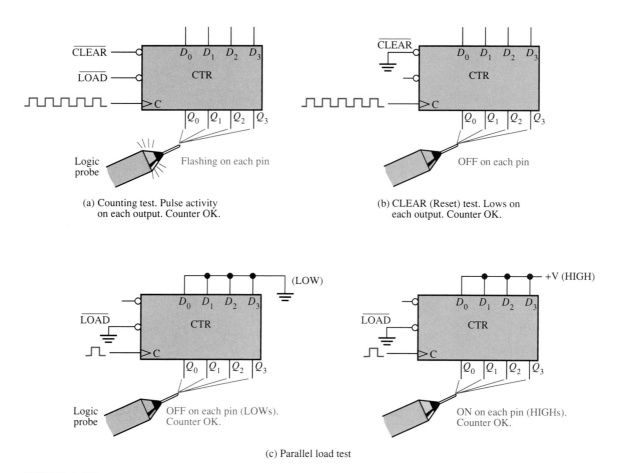

(a) Counting test. Pulse activity on each output. Counter OK.

(b) CLEAR (Reset) test. Lows on each output. Counter OK.

(c) Parallel load test

FIGURE 8–57
Procedures for testing an IC counter.

Cascaded IC Counters

A failure in one of the counters in a chain of cascaded counters can affect all the counters that follow it. For example, if a count Enable input opens, it effectively acts as a HIGH (TTL), and the counter is always enabled. This type of failure in one of the counters will cause that counter to run at the full clock rate and will also cause all the succeeding counters to run at higher than normal rates. This is illustrated in Figure 8–58 for a divide-by-1000 cascaded counter arrangement. Other faults that can affect "downstream" counter stages are open or shorted clock inputs or terminal count outputs. In some of these situations, pulse activity may be observed, but it may be at the wrong frequency. Exact frequency measurements must be made.

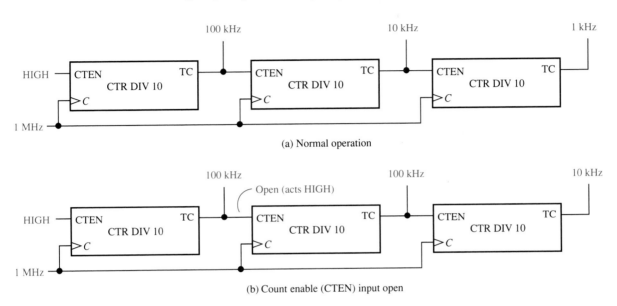

(a) Normal operation

(b) Count enable (CTEN) input open

FIGURE 8–58
Example of a failure that affects following counters in a cascaded arrangement.

Cascaded Counters with Truncated Sequences

The count sequence of a cascaded counter with a truncated sequence, such as that in Figure 8–59, can be affected by other types of faults in addition to those mentioned for full-modulus cascaded counters. For example, a failure in one of the parallel data inputs, the $\overline{\text{LOAD}}$ input, or the inverter can alter the preset count and thus change the modulus of the counter.

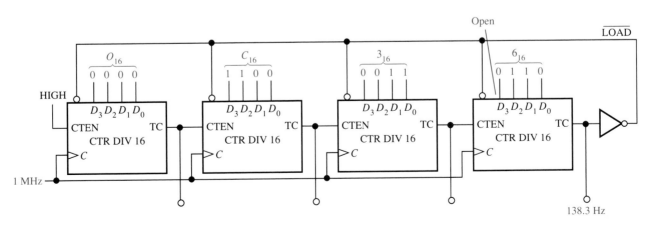

FIGURE 8–59
Cascaded counter with a truncated sequence.

For example, suppose the D_3 input of the most significant counter in Figure 8–59 is open and acts as a HIGH. Instead of 6_{16} being preset into the counter, E_{16} is preset in. So, instead of beginning with $63C0_{16}$ ($25,536_{10}$) each time the counter recycles, the sequence will begin with $E3C0_{16}$ ($58,304_{10}$). This changes the modulus of the counter from 40,000 to $65,536 - 58,304 = 7232$.

To check this counter, apply a known clock frequency, say 1 MHz, and measure the output frequency at the final terminal count output. If the counter is operating properly, the output frequency is

$$f = \frac{1 \text{ MHz}}{40,000} = 25 \text{ Hz}$$

In this case the specific failure described in the preceding paragraph will cause the output frequency to be

$$f = \frac{1 \text{ MHz}}{7232} = 138.3 \text{ Hz}$$

EXAMPLE 8–10

Frequency measurements are made on the truncated counter in Figure 8–60 as indicated. Determine if the counter is working properly, and if not, isolate the fault.

Solution First, we check to see if the frequency measured at TC4 is correct. If it is, the counter is working properly.

$$\text{Truncated modulus} = \text{full modulus} - \text{preset count}$$
$$= 16^4 - 82C0_{16}$$
$$= 65,536 - 33,472 = 32,064$$

The correct frequency at TC4 is

$$f_4 = \frac{10 \text{ MHz}}{32,064} = 311.88 \text{ Hz}$$

Uh oh! We have a problem. The measured frequency of 333.2 Hz does not agree with the correct frequency that we calculated. To find the problem, we will begin at the clock input and trace the frequency through the counter, one stage at a time.

Step 1 Make sure the clock frequency is 10 MHz.
Step 2 Check the frequency at the TC1 output of the first counter, and compare it with the correct calculated value. The truncated modulus of counter 1 is

$$16^1 - 0 = 16$$

The frequency at TC1 is

$$f_1 = \frac{10 \text{ MHz}}{16} = 625 \text{ kHz}$$

This agrees with the measured value.

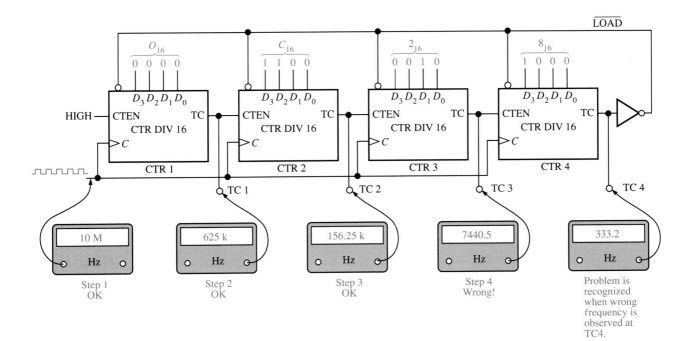

FIGURE 8–60

Step 3 Check the frequency at TC2, and compare it with the calculated value. The truncated modulus of counter 1 and counter 2 combined is

$$16^2 - C0_{16} = 256 - 192 = 64$$

The frequency at TC2 is

$$f_2 = \frac{10 \text{ MHz}}{64} = 156.25 \text{ kHz}$$

This agrees with the measured value.

Step 4 Check the frequency at TC3, and compare it with the calculated value. The truncated modulus of counter 1, counter 2, and counter 3 combined is

$$16^3 - 2C0_{16} = 4096 - 704 = 3392$$

The frequency at TC3 is

$$f_3 = \frac{10 \text{ MHz}}{3392} = 2948.1 \text{ Hz}$$

This does *not* agree with the measured frequency. So the measured frequency is wrong, and there is a problem with counter 3 (f_1 and f_2 are correct).

Step 5 Since the TC3 frequency is wrong, let us check to see what value of the overall modulus for the first three counters would produce the incorrect frequency of 7440.5 Hz:

$$\text{Modulus} = \frac{10 \text{ MHz}}{f_3} = \frac{10 \text{ MHz}}{7440.5 \text{ Hz}} = 1344$$

The number on the parallel data inputs to the counter must be wrong, because if $2C0_{16}$ is being loaded into the counter, the modulus has to be 3392, not 1344.

Step 6 Check the states of all the parallel inputs to counter 3. Assume that we find them to be correct (0010). This tells us that there must be an internal open in one of the parallel inputs, causing an incorrect count to be loaded. To find what is being loaded into the parallel inputs of counter 3, we can subtract the erroneous modulus (1344) from the full modulus of the first three counters.

$$16^3 - 1344 = 4096 - 1344 = 2752 = AC0_{16}$$

Ah ha! The number $AC0_{16}$, instead of $2C0_{16}$, is being loaded into the first three counters. So, A_{16} instead of 2_{16} is getting into counter 3. Counter 3 is actually seeing $A_{16} = 1010$ on its inputs. It should see $2_{16} = 0010$. Therefore, the MSB appears to be a 1 instead of a 0. This error strongly indicates that the D_3 input of counter 3 is open.

First check for an external open (bad connection, etc); if none can be found, replace the IC and the counter should work properly.

Discrete Flip-Flop Counters

Counters implemented with discrete flip-flops and gates are sometimes more difficult to troubleshoot because there are many more inputs and outputs with external connections than there are in an IC counter. The sequence of a counter can be altered by a single open or short on an input or output, as the next example illustrates.

EXAMPLE 8–11

Suppose that you observe the output waveforms that are indicated for the counter in Figure 8–61. Determine if there is a problem with the counter.

Solution The Q_2 waveform is incorrect. The correct waveform is shown in color. You can see that the Q_2 waveform looks exactly like the Q_1 waveform. So whatever is causing FF1 to toggle appears to also be controlling FF2.

Checking the *J-K* input to FF2, we find a waveform that looks like Q_0. This result indicates that Q_0 is somehow getting through the AND gate. The only way this can happen is if the Q_1 input to the AND gate is always HIGH. But we have seen that Q_1 has a correct waveform. This observation leads to the conclusion that the lower input to the AND gate must be internally open and acting as a HIGH. Replace the AND gate and retest the circuit.

FIGURE 8–61

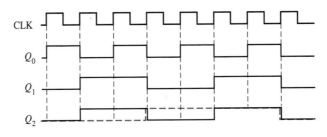

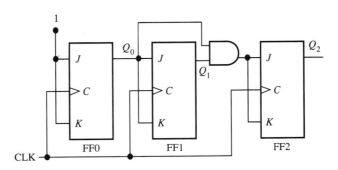

SECTION 8–8
REVIEW

Answers are found at the end of the chapter.

1. What failures can cause the counter in Figure 8–58 to have no pulse activity on any of the TC outputs?
2. What happens if the inverter in Figure 8–60 develops an open output?

LOGIC SYMBOLS WITH DEPENDENCY NOTATION

8–9

Up to this point the logic symbols with dependency notation specified in ANSI/IEEE Standard 91–1984 have been introduced on a limited basis. In many cases the new symbols do not deviate greatly from the traditional symbols. A significant departure from what we are accustomed to does occur, however, for some devices, including counters and other more complex devices.

Dependency notation is fundamental to the new ANSI/IEEE standard. Dependency notation is used in conjunction with the logic symbols to specify the relationships of inputs and outputs so that the logical operation of a given device can be determined entirely from its logic symbol without a prior knowledge of the details of its internal structure and without a detailed logic diagram for reference. Dependency notation was introduced in relation to specific flip-flops in Chapter 7.

Although we will continue to use primarily the more traditional and familiar symbols throughout this book, a brief coverage of logic symbols with dependency notation is provided in this section to prepare you for a projected increase in usage of these symbols in the future. This section can be treated as optional, and its omission will not affect the remainder of the book.

The 74LS163A four-bit binary counter is used for illustration. For comparison, Figure 8–62 shows a traditional block symbol and the ANSI/IEEE symbol with dependency notation. Basic descriptions of the symbol and the dependency notation follow.

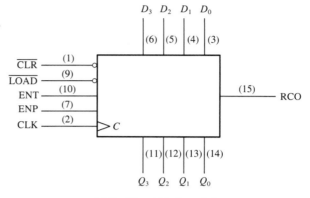

(a) Traditional block symbol

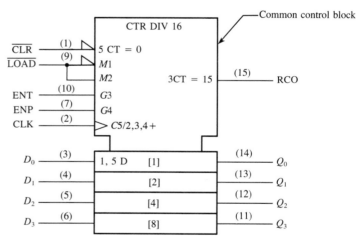

(b) ANSI/IEEE Std. 91–1984 logic symbol

Common Control Block

The upper block with notched corners in Figure 8–62(b) has inputs and an output that are considered common to all elements in the device and not unique to any one of the elements.

Individual Elements

The lower block, which is partitioned into four abutted sections, represents the four storage elements (D flip-flops) in the counter, with inputs D_0, D_1, D_2, and D_3 and outputs Q_0, Q_1, Q_2, and Q_3.

Qualifying Symbol

The label "CTR DIV 16" identifies the device as a counter (CTR) with sixteen states (DIV 16).

Control Dependency (*C*)

The letter *C* denotes control dependency. Control inputs usually enable or disable the data inputs (*D, J, K, S,* and *R*) of a storage element. The *C* input is usually the clock input. In this case the digit 5 following *C* (*C*5/2,3,4+) indicates that the inputs labeled with a 5 prefix are dependent on the clock (synchronous with the clock). For example, 5 CT = 0 on the $\overline{\text{CLR}}$ input indicates that the clear function is dependent on the clock; that is, it is a synchronous clear. When the $\overline{\text{CLR}}$ input is LOW (0), the counter is reset to zero (CT = 0) on the triggering edge of the clock pulse. Also, the 5 D label at the input of storage element [1] indicates that the data storage is dependent on (synchronous with) the clock. All labels in the [1] storage element apply to the [2], [4], and [8] elements below it, since they are not labeled differently.

Mode Dependency (*M*)

The letter *M* denotes mode dependency. This is used to indicate how the functions of various inputs or outputs depend on the mode in which the device is operating. In this case the device has two modes of operation. When the $\overline{\text{LOAD}}$ input is LOW (0), as indicated by the triangle input, the counter is in a preset mode (*M*1) in which the input data (*D*$_0$, *D*$_1$, *D*$_2$, and *D*$_3$) are synchronously loaded into the four flip-flops. The digit 1 following *M* in *M*1 and the 1 in 1, 5 D show a dependency relationship and indicate that input data are stored only when the device is in the preset mode (*M*1), in which $\overline{\text{LOAD}}$ = 0. When the $\overline{\text{LOAD}}$ input is HIGH (1), the counter advances through its normal binary sequence, as indicated by *M*2 and the 2 in *C*5/2,3,4+.

AND Dependency (*G*)

The letter *G* denotes AND dependency, indicating that an input designated with *G* followed by a digit is ANDed with any other input or output having the same digit as a prefix in its label. In this particular example, the *G*3 at the ENT input and the 3CT = 15 at the RCO output are related, as indicated by the 3, and that relationship is an AND dependency, indicated by the *G*. This tells us that ENT must be HIGH (no triangle on the input) AND the count must be fifteen (CT = 15) for the RCO output to be HIGH.

Also, the digits 2, 3, and 4 in the label *C*5/2,3,4+ indicate that the counter advances through its states when $\overline{\text{LOAD}}$ = 1, as indicated by the mode dependency label *M*2, and when ENT = 1 AND ENP = 1, as indicated by the AND dependency labels *G*3 and *G*4. The + indicates that the counter advances by one count when these conditions exist.

This coverage of a specific logic symbol with dependency notation is intended tc aid in the interpretation of other such symbols that you may encounter in the future.

SECTION 8–9 REVIEW

Answers are found at the end of the chapter.
1. In dependency notation, what do the letters *C, M,* and *G* stand for?
2. By what letter is data storage denoted?

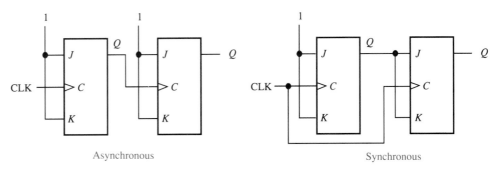

Asynchronous   Synchronous

FIGURE 8–63

Comparison of asynchronous and synchronous counters.

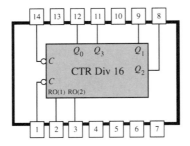

7493A asynchronous 4-bit binary counter

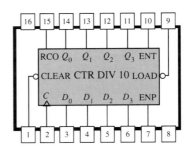

74LS160A synchronous decade
counter with asynchronous clear

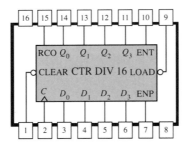

74LS161A synchronous 4-bit binary
counter with asynchronous clear

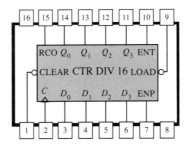

74LS163 A synchronous 4-bit binary
counter with synchronous clear

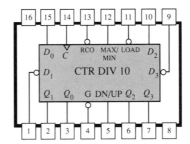

74190 synchronous up/down
decade counter

FIGURE 8–64

*Note that the labels (names of inputs and outputs) are consistent with text and may not agree
with the particular manufacturer's data book you are using.*

428

SUMMARY

☐ Asynchronous and synchronous counters differ only in the way in which they are clocked, as shown in Figure 8–63. Synchronous counters can run at faster clock rates than asynchronous counters.

☐ The maximum modulus of a counter is the maximum number of possible states and is a function of the number of stages (flip-flops). Thus,

$$\text{Maximum modulus} = 2^n$$

where n is the number of stages in the counter. The modulus of a counter is the *actual* number of states in its sequence and can be equal to or less than the maximum modulus.

☐ Counter design procedure:

1. Specify the required sequence of states with a *state diagram*.
2. Derive a *next-state table* from the state diagram.
3. Develop a *transition table*.
4. Transfer the *JK* states from the transition table to present-state Karnaugh maps. There must be a map for each input of each flip-flop.
5. Group the 1s and the ''don't cares'' on the maps and derive a Boolean expression in terms of the flip-flop outputs for each flip-flop input.
6. Implement the expressions with combinational logic, and connect the combinational logic circuits to the flip-flops.

☐ Connection diagrams for the IC counters used in this chapter are shown in Figure 8–64.

 Integrated Circuit Technologies (Chapter 13) may be covered after this chapter, covered after a later chapter, partially omitted, or completely omitted, depending on the needs and goals of your program. Coverage is not prerequisite or corequisite to any other material in this book. Inclusion is purely optional.

GLOSSARY

Asynchronous counter   A type of counter in which each stage is clocked from the output of the preceding stage.

Decade counter   A digital counter having ten states.

Decrement   To decrease the binary state of a counter by one.

Increment   To increase the binary state of a counter by one.

Modulus   The number of states in a counter sequence.

Recycle   To undergo transition (as a counter) from the final, or terminal, state back to the initial state.

Ripple counter   An asynchronous counter.

Sequential circuit   A digital circuit whose logic states depend on a specified time sequence.

State machine   Any sequential circuit exhibiting a specified sequence of states.

Synchronous counter   A type of counter in which each stage is clocked by the same pulse.

Terminal count   The final state in a counter's sequence.

Truncated sequence   A sequence that does not include all of the possible states of a counter.

Up/down counter   A counter that can progress in either direction through a certain sequence.

SELF-TEST

Answers and solutions are found at the end of the book.

1. What is the reason for the difference between asynchronous and synchronous counter performance?

2. What is the primary disadvantage of an asynchronous counter?

3. What is the maximum modulus for a counter with each of the following numbers of flip-flops?
   (a) 2   (b) 4   (c) 5   (d) 6   (e) 7   (f) 8

4. Determine the number of flip-flops in each of the following counters:
   (a) modulus-3   (b) modulus-8   (c) modulus-12
   (d) modulus-16   (e) modulus-18   (f) modulus-36
   (g) modulus-64   (h) modulus-144

5. A four-bit asynchronous (ripple) counter consists of flip-flops each of which has a clock-to-Q propagation delay of 12 ns. How long does it take the counter to recycle from 1111 to 0000 after the triggering edge of the clock pulse?

6. Answer question 5 for a synchronous counter.

7. Explain how a BCD decade counter can be used as a divide-by-10 device, and illustrate it with a timing diagram.

8. A four-bit decade counter does not use its maximum possible modulus, so there are several invalid states. List these states.

9. A 74LS163A four-bit synchronous counter (see Figure 8–19) is in the 1010 state, and the following input conditions exist: $\overline{CLR} = 1$, $\overline{LOAD} = 0$, $D_0 = 1$, $D_1 = 0$, $D_2 = 1$, $D_3 = 0$, ENT = 1, and ENP = 1. To what state does the counter go on the next clock pulse?

10. A four-bit binary up/down counter (modulus-16) is in the binary state of zero. What is its next state in the up mode? In the down mode?

11. A flip-flop is in the RESET state and must stay RESET at the next clock pulse. What must J and K be?

12. With general block diagrams, show how to obtain the following frequencies from a 10-MHz clock by using single flip-flops, modulus-5 counters, and decade counters:
    (a) 5 MHz   (b) 2.5 MHz   (c) 2 MHz   (d) 1 MHz   (e) 500 kHz
    (f) 250 kHz   (g) 62.5 kHz   (h) 40 kHz   (i) 10 kHz   (j) 1 kHz

13. Show how to decode the following states of a four-bit binary counter by using active-LOW decoding:
    (a) 1   (b) 5   (c) 10   (d) 14

14. In the circuit of Figure 8–52, explain how the recycling to count 1 is accomplished.

15. If the clock frequency is 10 kHz in Figure 8–55, how long does it take for two bytes to be transmitted in serial form?

PROBLEMS

Answers to odd-numbered problems are at the end of the book.

Section 8–1 Asynchronous Counters

8–1 For the ripple counter shown in Figure 8–65, draw the complete timing diagram for eight clock pulses, showing the clock, Q_0 and Q_1 waveforms.

8–2 For the ripple counter in Figure 8–66, draw the complete timing diagram for sixteen clock pulses. Show the clock, Q_0, Q_1, and Q_2 waveforms.

8–3 In the counter of Problem 8–2, assume that each flip-flop has a propagation delay from the triggering edge of the clock to a change in the Q output of 8 ns. Determine the worst-case (longest) delay time from a clock pulse to the arrival of the counter in a given state. Specify the state or states for which this worst-case delay occurs.

FIGURE 8–65

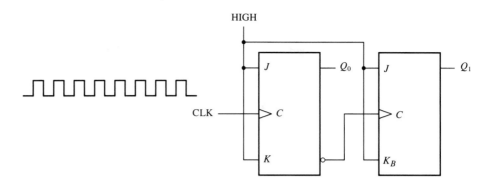

FIGURE 8–66

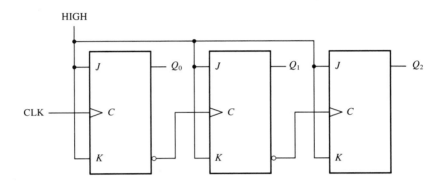

8–4 Show how to connect a 7493A four-bit asynchronous counter for each of the following moduli:

(a) 9 **(b)** 11 **(c)** 13 **(d)** 14 **(e)** 15

Section 8–2 Synchronous Counters

8–5 If the counter of Problem 8–3 were synchronous rather than asynchronous, what would be the longest delay time?

8–6 Draw the complete timing diagram for the five-stage synchronous binary counter in Figure 8–67. Verify that the waveforms of the Q outputs represent the proper binary number after each clock pulse.

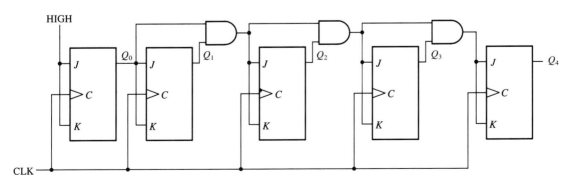

FIGURE 8–67

8–7 By analyzing the J and K inputs to each flip-flop prior to each clock pulse, prove that the decade counter in Figure 8–68 progresses through a BCD sequence. Explain how these conditions in each case cause the counter to go to the next proper state.

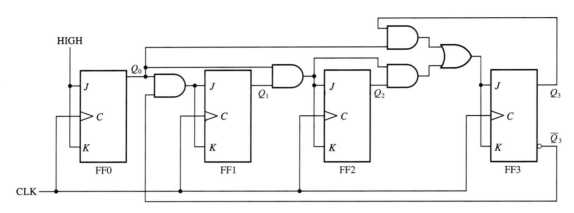

FIGURE 8–68

8–8 The waveforms in Figure 8–69 are applied to the count Enable, clear, and clock inputs as indicated. Sketch the counter output waveforms in proper relation to these inputs. The clear input is asynchronous.

8–9 A BCD decade counter is shown in Figure 8–70. The waveforms are applied to the clock and clear inputs as indicated. Determine the waveforms for each of the counter outputs (Q_0, Q_1, Q_2, and Q_3). The clear is synchronous, and the counter is initially in the binary 1000 state.

8–10 The waveforms in Figure 8–71 are applied to a 74LS163A counter. Determine the Q outputs and the RCO. The inputs are $D_0 = 1$, $D_1 = 1$, $D_2 = 0$, and $D_3 = 1$.

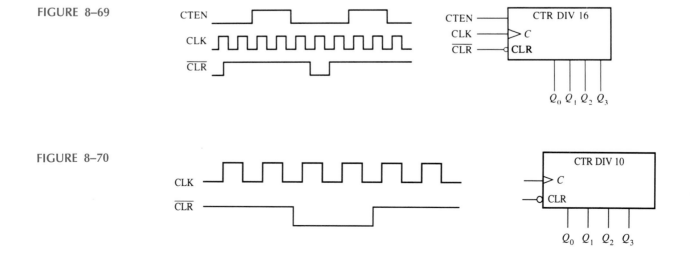

FIGURE 8–69

FIGURE 8–70

FIGURE 8–71

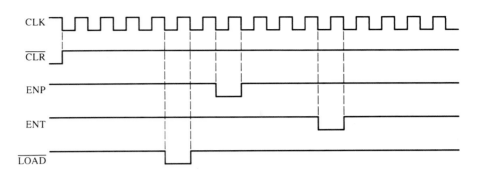

8–11 The waveforms in Figure 8–71 are applied to a 74LS160A counter. Determine the Q outputs and the RCO. The inputs are $D_0 = 1$, $D_1 = 0$, $D_2 = 0$, and $D_3 = 1$.

Section 8–3 Up/Down Synchronous Counters

8–12 Draw a complete timing diagram for a three-bit up/down counter that goes through the following sequence. Indicate when the counter is in the UP mode and when it is in the DOWN mode. Assume positive edge-triggering.

$$0, 1, 2, 3, 2, 1, 2, 3, 4, 5, 6, 5, 4, 3, 2, 1, 0$$

8–13 Sketch the Q output waveforms for a 74190 up/down counter with the input waveforms shown in Figure 8–72. A binary 0 is on the data inputs. Start with a count of 0000.

Section 8–4 Design of Sequential Circuits

8–14 Determine the sequence of the counter in Figure 8–73.

8–15 Determine the sequence of the counter in Figure 8–74. Begin with the counter cleared.

8–16 Design a counter to produce the following sequence. Use J-K flip-flops.

$$00, 10, 01, 11, 00, \ldots$$

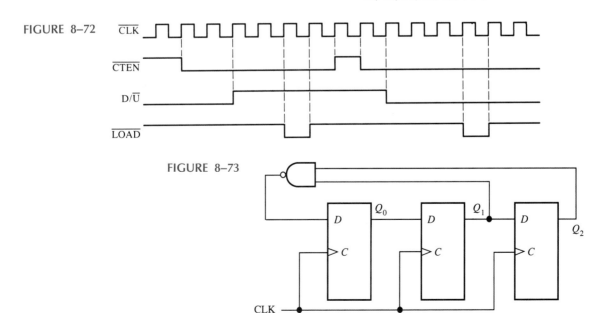

FIGURE 8–72

FIGURE 8–73

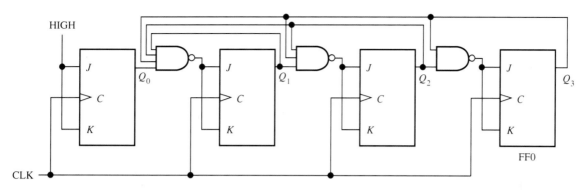

FIGURE 8–74

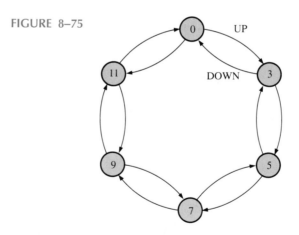

8–17 Design a counter to produce the following binary sequence. Use J-K flip-flops.

1, 4, 3, 5, 7, 6, 2, 1, . . .

8–18 Design a counter to produce the following binary sequence. Use J-K flip-flops.

0, 9, 1, 8, 2, 7, 3, 6, 4, 5, 0, . . .

8–19 Design a counter with the sequence shown in the state diagram of Figure 8–75.

FIGURE 8–75

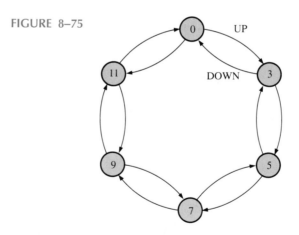

Section 8–5 Cascaded Counters

8–20 For each of the cascaded counter configurations in Figure 8–76, determine the frequency of the waveform at each point indicated by a circled number, and determine the overall modulus.

8–21 Expand the counter in Figure 8–40 to create a divide-by-10,000 counter and a divide-by-100,000 counter.

8–22 Design a modulus-1000 counter by using 74LS160A decade counters.

8–23 Modify the design of the counter in Figure 8–43 to achieve a modulus of 30,000.

8–24 Solve Problem 8–23 for a modulus of 50,000.

FIGURE 8–76

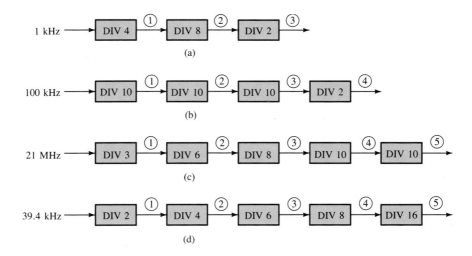

(a)

(b)

(c)

(d)

Section 8–6 Counter Decoding

8–25 Given a BCD decade counter with only the Q outputs available, show what decoding logic is required to decode each of the following states and how it should be connected to the counter. A HIGH output indication is required for each decoded state. The MSB is to the left.

(a) 0001 **(b)** 0011 **(c)** 0101 **(d)** 0111 **(e)** 1000

8–26 For the four-bit binary counter connected to the decoder in Figure 8–77, determine each of the decoder output waveforms in relation to the clock pulses.

8–27 If the counter in Figure 8–77 is asynchronous, determine where the decoding glitches occur on the decoder output waveforms.

8–28 Modify the circuit in Figure 8–77 to eliminate decoding glitches.

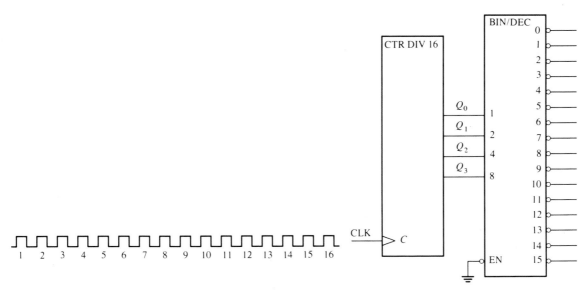

FIGURE 8–77

8–29 Analyze the counter in Figure 8–44 for the occurrence of glitches on the decode gate output. If glitches occur, suggest a way to eliminate them.

8–30 Analyze the counter in Figure 8–45 for the occurrence of glitches on the outputs of the decoding gates. If glitches occur, make a design change that will eliminate them.

Section 8–7 Counter Applications

8–31 Modify the digital clock in Figures 8–50, 8–51, and 8–52 so that it can be preset to any desired time.

8–32 Design an alarm circuit for the digital clock that can detect a predetermined time (hours and minutes only) and produce a signal to activate an audio alarm.

8–33 Modify the circuit in Figure 8–54 for a 1000-space parking garage and a 3000-space parking garage.

8–34 Implement the parallel-to-serial data conversion logic in Figure 8–55 with specific devices.

Section 8–8 Troubleshooting

8–35 For the counter in Figure 8–1 draw the timing diagram for the Q_0 and Q_1 waveforms for each of the following faults (assume Q_0 and Q_1 are initially LOW):
  (a) clock input to FF0 shorted to ground
  (b) Q_0 output open
  (c) clock input to FF1 open
  (d) J input to FF0 open
  (e) K input to FF1 shorted to ground

8–36 Solve Problem 8–35 for the counter in Figure 8–11.

8–37 Isolate the fault in the counter in Figure 8–3 by analyzing the waveforms in Figure 8–78.

FIGURE 8–78

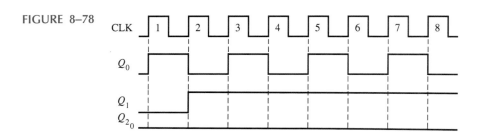

8–38 From the waveform diagram in Figure 8–79, determine the most likely fault in the counter of Figure 8–14.

8–39 Solve Problem 8–38 if the Q_2 output has the waveform observed in Figure 8–80. Outputs Q_0 and Q_1 are the same as in Figure 8–79.

8–40 In Problem 8–15 you found that the counter locks up and alternates between two states. It turns out that this operation is the result of a design flaw. Redesign the counter so that when it goes into the second of the lock-up states, it will recycle to the all-0s state on the next clock pulse.

FIGURE 8–79

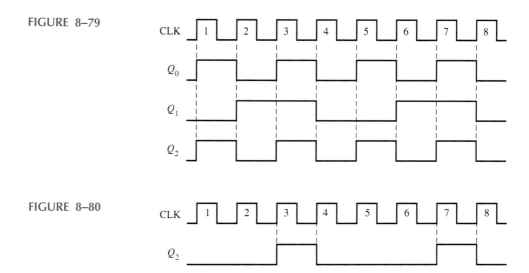

FIGURE 8–80

8–41 You apply a 5-MHz clock to the cascaded counter in Figure 8–43 and measure a frequency of 76.2939 Hz at the last RCO output. Is this correct, and if not, what is the most likely problem?

8–42 Develop a table for use in testing the counter in Figure 8–43 that will show the frequency at the final RCO output for all possible open failures of the parallel data inputs (D_0, D_1, D_2, and D_3) taken one at a time. Use 10 MHz as the test frequency for the clock.

8–43 The tens-of-hours seven-segment display in the digital clock system of Figure 8–50 continuously displays a 1. All the other digits work properly. What could be the problem?

8–44 What would be the visual indication of an open Q_1 output in the decade portion of the minutes counter in Figure 8–50? Also see Figure 8–51.

8–45 One day (perhaps a Monday) complaints begin flooding in from patrons of a parking garage that has the control system depicted in Figures 8–53 and 8–54. The patrons say that they enter the garage because the gate is up and the FULL sign is off but that, once in, they can find no empty space. As the technician in charge of this facility, what do you think the problem is, and how will you troubleshoot and repair the system as quickly as possible?

ANSWERS TO SECTION REVIEWS

Section 8–1   Review
1. Each flip-flop after the first one is clocked by the output of the preceding flip-flop.
2. fourteen

Section 8–2   Review
1. All flip-flops in a synchronous counter are clocked simultaneously.
2. The counter can be preset (initialized) to any given state.
3. Counter is enabled when ENP and ENT are both HIGH; RCO goes HIGH when final state in sequence is reached.

Section 8–3 Review

1. 1001 **2.** 1111, 0000, 1111

Section 8–4 Review

1. $J = 1$, $K = X$ (don't care)   **2.** $J = X$ (don't care), $K = 0$
3. **(a)** 1011
   (b) Q_3 (MSB): no-change or SET; Q_2: no-change or RESET; Q_1: no-change or SET; Q_0 (LSB): SET or toggle

Section 8–5 Review

1. $n = 10$; $n = 14$
2. **(a)** flip-flop and DIV 10   **(b)** flip-flop and DIV 16
   (c) DIV 16 and DIV 10   **(d)** DIV 16 and DIV 10 and flip-flop

Section 8–6 Review

1. **(a)** none, because there is a single bit change
   (b) 0010, 0000
   (c) none, because there is a single bit change
   (d) 1110, 1100, 1000

Section 8–7 Review

1. Gate G_1 resets flip-flop on first clock pulse after count 12. Gate G_2 decodes count 12 to preset counter to 0001.
2. The decade counter advances through each state from zero to nine, and as it recycles from nine back to zero, the flip-flop is toggled to the SET state. This produces a ten (10) on the display. In state 12 the decode NAND gate causes the decade counter to preset to one. The flip-flop RESETS. This results in a one (01) on the display.

Section 8–8 Review

1. CTEN of first counter shorted to ground or to a LOW; clock input of first counter open; clock line shorted to ground or to a LOW; TC output of first counter shorted to ground or to a LOW
2. The counter does not recycle at the preset count but acts as a full-modulus counter.

Section 8–9 Review

1. C: control, usually clock; M: mode; G: AND
2. D

Shift registers are a type of sequential logic circuit closely related to counters. They are used basically for the storage of digital data and typically do not possess a characteristic internal sequence as do counters.

In this chapter the basic types of shift registers are studied and several applications are presented. Also, a new troubleshooting method is introduced.

Specific devices introduced in this chapter are as follows:

1. 7491A eight-bit serial shift register
2. 74164 eight-bit parallel out serial shift register
3. 74165 eight-bit parallel load shift register
4. 74195 four-bit parallel access shift register
5. 74194 four-bit bidirectional universal shift register
6. 74199 eight-bit bidirectional universal shift register

After completing this chapter, you will be able to

☐ identify the basic forms of data movement in shift registers.
☐ control data movement in a shift register.
☐ explain how serial in–serial out, serial in–parallel out, parallel in–serial out, and parallel in–parallel out shift registers operate.
☐ describe how a bidirectional shift register operates.
☐ determine the sequence of a Johnson counter.
☐ set up a ring counter to produce a specified sequence.
☐ construct a ring counter from a shift register.
☐ use a shift register as a time-delay device.
☐ use a shift register to implement a serial-to-parallel data converter.
☐ implement a basic shift-register-controlled keyboard encoder.
☐ troubleshoot digital systems by ''exercising'' the system with a known test pattern.
☐ apply signature analysis to troubleshooting.
☐ interpret ANSI/IEEE Standard 91-1984 shift register symbols with dependency notation.

9

Shift Registers

SHIFT REGISTER FUNCTIONS

9–1

Shift registers are very important in applications involving the storage and transfer of data in a digital system. The basic difference between a register and a counter is that a register has no specified sequence of states, except in certain very specialized applications. A register, in general, is used solely for storing and shifting data (1s and 0s) entered into it from an external source and possesses no characteristic internal sequence of states.

The storage capability of a register is one of its two basic functional characteristics and makes it an important type of memory device. Figure 9–1 illustrates the concept of storing a 1 or a 0 in a flip-flop. A 1 is applied to the input as shown, and a clock pulse is applied that stores the 1 by setting the flip-flop. When the 1 on the input is removed, the flip-flop remains in the SET state, thereby storing the 1. The same procedure applies to the storage of a 0, as also illustrated in Figure 9–1.

FIGURE 9–1
The flip-flop as a storage element.

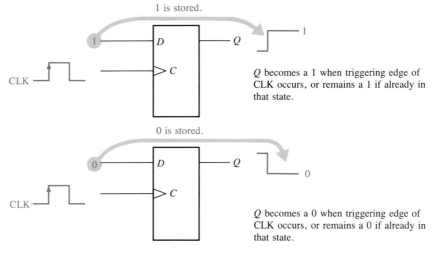

The *storage capacity* of a register is the number of bits (1s and 0s) of digital data it can retain. Each stage of a shift register represents one bit of storage capacity; therefore, the number of stages in a register determines its total storage capacity.

Registers are commonly used for the *temporary* storage of data within a digital system. Registers are implemented with flip-flops or other storage devices to be introduced later. The *shifting* capability of a register permits the movement of data from stage to stage within the register or into or out of the register upon application of clock pulses. Figure 9–2 shows symbolically the types of data movement in shift register operations. The block represents any arbitrary four-bit register, and the arrows indicate the direction and type of data movement.

SECTION 9–1
REVIEW

Answers are found at the end of the chapter.
1. Generally, what is the difference between a counter and a shift register?
2. What two principal functions are performed by a shift register?

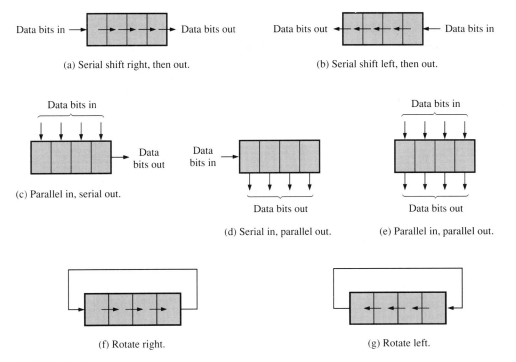

(a) Serial shift right, then out.

(b) Serial shift left, then out.

(c) Parallel in, serial out.

(d) Serial in, parallel out.

(e) Parallel in, parallel out.

(f) Rotate right.

(g) Rotate left.

FIGURE 9–2

Basic data movement in registers (four bits shown for illustration).

SERIAL IN–SERIAL OUT SHIFT REGISTERS

9–2

This type of shift register accepts data serially—that is, one bit at a time on a single line. It produces the stored information on its output also in serial form.

Let us first look at the serial entry of data into a typical shift register with the aid of Figure 9–3, which shows a four-bit device implemented with D flip-flops.

With four stages, this register can store up to four bits of data; its storage capacity is four bits. We will illustrate entry of the four-bit binary number 1010 into the register, beginning with the right-most bit. The 0 is put onto the data input line, making $D = 0$ for FF0. When the first clock pulse is applied, FF0 is RESET, thus storing the 0. Next the 1 is applied to the data input, making $D = 1$ for FF0 and $D = 0$ for FF1 because the D input of FF1 is connected to the Q_0 output.

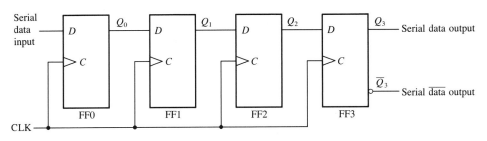

FIGURE 9–3

Serial in–serial out shift register.

When the second clock pulse occurs, the 1 on the data input is "shifted" into FF0 because FF0 SETS, and the 0 that was in FF0 is shifted into FF1. The next 0 in the binary number is now put onto the data-input line, and a clock pulse is applied. The 0 is entered into FF0, the 1 stored in FF0 is shifted into FF1, and the 0 stored in FF1 is shifted into FF2. The last bit in the binary number, a 1, is now applied to the data input, and a clock pulse is applied. This time the 1 is entered into FF0, the 0 stored in FF0 is shifted into FF1, the 1 stored in FF1 is shifted into FF2, and the 0 stored in FF2 is shifted into FF3. This completes the serial entry of the four-bit number into the shift

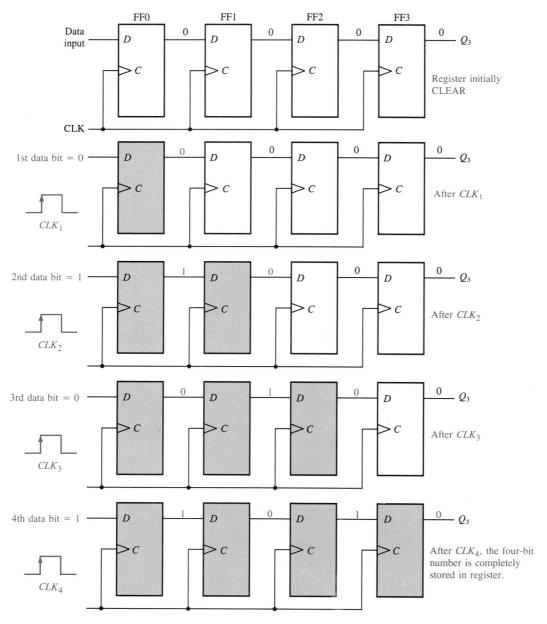

FIGURE 9–4
Four bits (1010) being entered serially into the register.

register, where it can be stored for any length of time. Figure 9–4 illustrates each step in the shifting of the four bits into the register.

If we want to get the data out of the register, they must be shifted out serially and taken off the Q_3 output. After CLK_4 in the data-entry operation just described, the right-most 0 in the number appears on the Q_3 output. When clock pulse CLK_5 is applied, the second bit appears on the Q_3 output. Clock pulse CLK_6 shifts the third bit to the output, and CLK_7 shifts the fourth bit to the output, as illustrated in Figure 9–5. Notice that while the original four bits are being shifted out, a new four-bit number can be shifted in. An all-0 number is shown being shifted in.

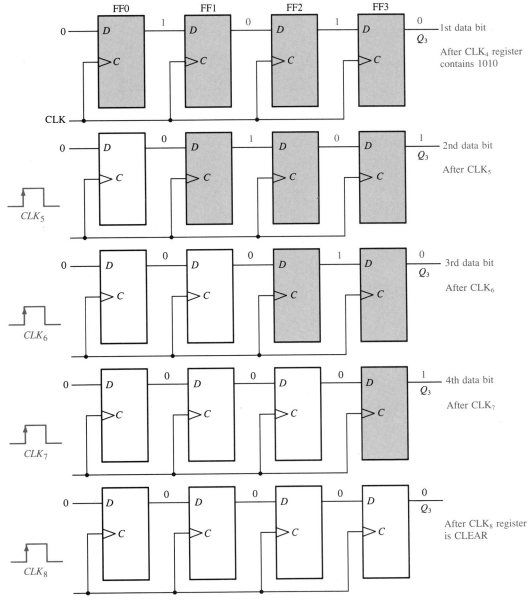

FIGURE 9–5
Four bits (1010) being serially shifted out of the register.

EXAMPLE 9–1   Show the states of the five-bit register in Figure 9–6(a) for the specified data input and clock waveforms. Assume that the register is initially cleared (all 0s).

FIGURE 9–6

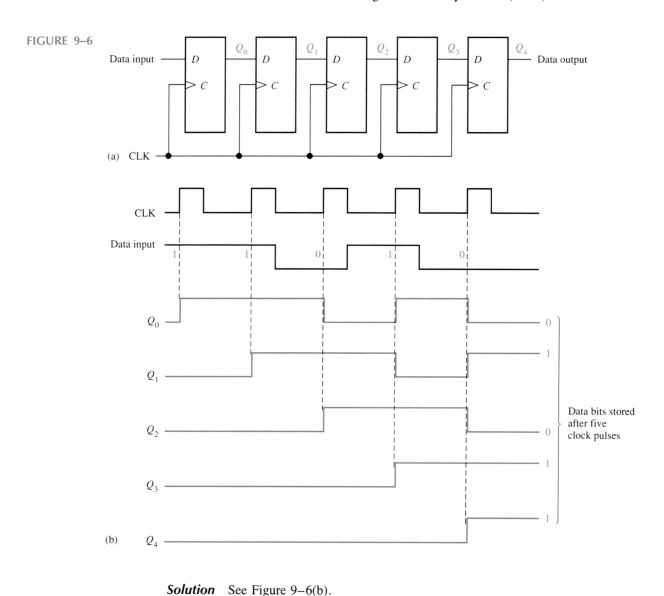

Solution   See Figure 9–6(b).

The 7491A Eight-Bit Shift Register

The 7491A is an example of an IC serial in–serial out shift register. The logic diagram is shown in Figure 9–7(a). As you can see, S-R flip-flops are used to implement this device. There are two gated data-input lines, *A* and *B*, for serial data entry. When data are entered on A, the B input must be HIGH, and vice versa. The serial data output is Q_7, and its complement is $\overline{Q_7}$.

A traditional logic block symbol is shown in Figure 9–7(b). The "SRG 8" designation means a shift register (SRG) with an eight-bit capacity.

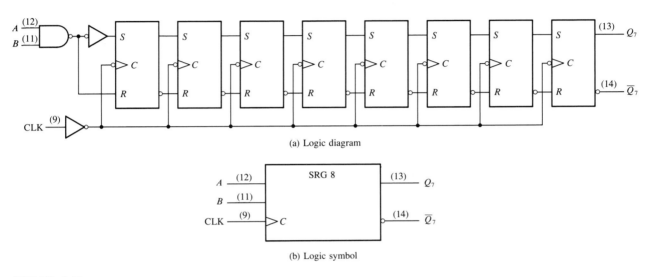

(a) Logic diagram

(b) Logic symbol

FIGURE 9–7
The 7491A eight-bit serial in–serial out shift register.

SECTION 9–2
REVIEW

Answers are found at the end of the chapter.
1. Draw the logic diagram for the register in Figure 9–3, using J-K flip-flops to replace the D flip-flops.
2. How many clock pulses are required to enter a byte of data serially into an eight-bit shift register?

SERIAL IN–PARALLEL OUT SHIFT REGISTERS

9–3

Data bits are entered serially into this type of register in the same manner as discussed in the last section. The difference is the way in which the data bits are taken out of the register; in the parallel output register, the output of each stage is available. Once the data are stored, each bit appears on its respective output line, and all bits are available simultaneously, rather than on a bit-by-bit basis as with the serial output.

Figure 9–8 shows a four-bit serial in–parallel out register and its logic block symbol.

FIGURE 9–8

A serial in–parallel out shift register.

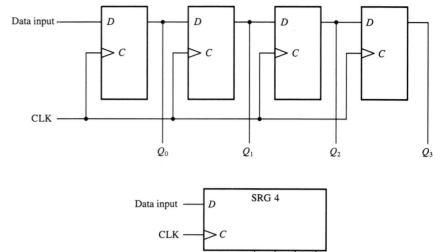

EXAMPLE 9–2

Show the states of the four-bit register for the data input and clock waveforms in Figure 9–9(a). The register initially contains all 1s.

FIGURE 9–9

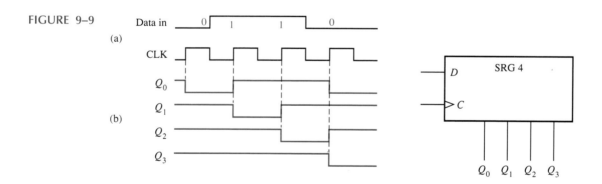

(a)

(b)

Solution The register contains 0110 after four clock pulses. See Figure 9–9(b).

The 74164 Eight-Bit Serial In–Parallel Out Shift Register

The 74164 is an example of an IC shift register having serial in–parallel out operation. The logic diagram is shown in Figure 9–10(a), and a typical logic block symbol is shown in part (b). Notice that this device has two gated serial inputs, A and B, and a clear ($\overline{CLR}$) input that is active-LOW. The parallel outputs are Q_0 through Q_7.

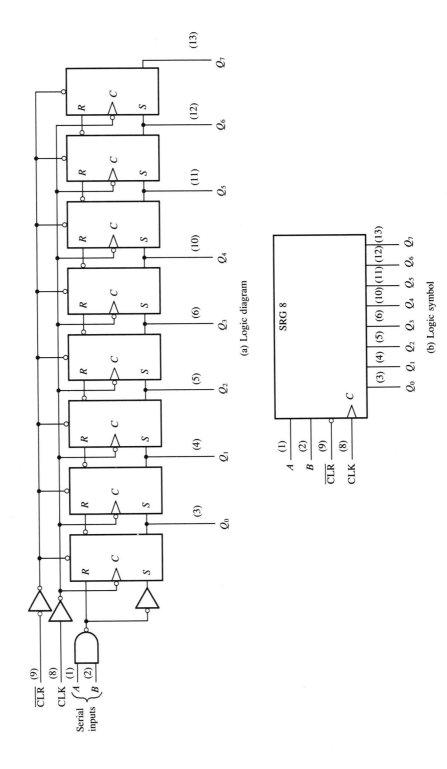

FIGURE 9–10
The 74164 eight-bit serial in–parallel out shift register.

449

A sample timing diagram for the 74164 is shown in Figure 9–11. Notice that the serial input data on input A are shifted into and through the register after input B goes HIGH.

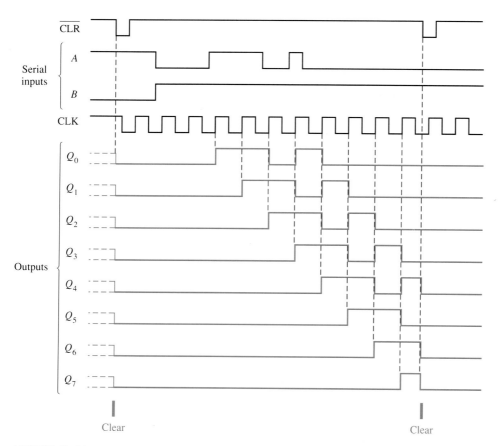

FIGURE 9–11
Sample timing diagram for a 74164 shift register.

SECTION 9–3
REVIEW

Answers are found at the end of the chapter.

1. The binary number 1101 is serially entered (right-most bit first) into a four-bit parallel out shift register that is initially clear. What are the Q outputs after two clock pulses?
2. How can a serial in–parallel out register be used as a serial in–serial out register?

PARALLEL IN–SERIAL OUT SHIFT REGISTERS

9–4

For a register with parallel data inputs, the bits are entered simultaneously into their respective stages on parallel lines rather than on a bit-by-bit basis on one line as with serial data inputs. The serial output is the same as described in Section 9–2, once the data are completely stored in the register.

Figure 9–12 illustrates a four-bit parallel in–serial out register. Notice that there are four data-input lines, D_0, D_1, D_2, and D_3, and a SHIFT/$\overline{\text{LOAD}}$ input, which allows four bits of data to be entered in parallel into the register. When SHIFT/$\overline{\text{LOAD}}$ is LOW, gates G_1 through G_3 are enabled, allowing each data bit to be applied to the D input of its respective flip-flop. When a clock pulse is applied, the flip-flops with $D = 1$ will SET and those with $D = 0$ will RESET, thereby storing all four bits simultaneously.

When SHIFT/$\overline{\text{LOAD}}$ is HIGH, gates G_1 through G_3 are disabled and gates G_4 through G_6 are enabled, allowing the data bits to shift right from one stage to the next. The OR gates allow either the normal shifting operation or the parallel data-entry operation, depending on which AND gates are enabled by the level on the SHIFT/$\overline{\text{LOAD}}$ input.

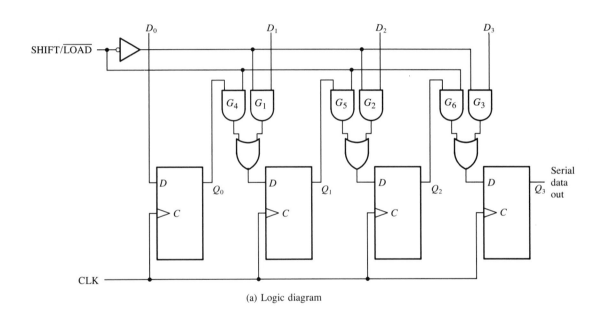

(a) Logic diagram

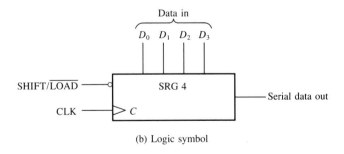

(b) Logic symbol

FIGURE 9–12
A four-bit parallel in–serial out shift register.

EXAMPLE 9–3

Show the data-output waveform for a four-bit register with the parallel input data and clock waveform given in Figure 9–13(a). Refer to Figure 9–12 for the logic diagram.

FIGURE 9–13

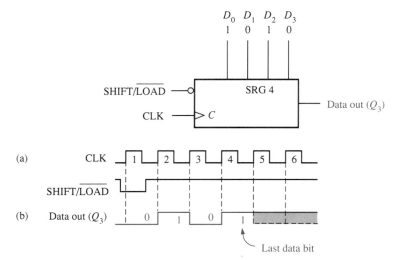

Solution On clock pulse 1, the parallel data ($D_0 D_1 D_2 D_3 = 1010$) are loaded into the register, making Q_3 a 0. On clock pulse 2 the 1 from Q_2 is shifted onto Q_3; on clock pulse 3 the 0 is shifted onto Q_3; on clock pulse 4 the last data bit (1) is shifted onto Q_3; and on clock pulse 5, all data bits have been shifted out, and only 1s remain in the register (assuming the D_0 input remains a 1). See Figure 9–13(b).

The 74165 Eight-Bit Parallel Load Shift Register

The 74165 is an example of an IC shift register that has a parallel in–serial out operation (it can also be operated as serial in–serial out). Figure 9–14 shows the internal logic diagram for this device and a typical logic block symbol.

A LOW on the SHIFT/LOAD (SH/$\overline{\text{LD}}$) input enables all the NAND gates for parallel loading. When an input data bit is a 1, the flip-flop is asynchronously SET by a LOW out of the upper gate. When an input data bit is a 0, the flip-flop is asynchronously RESET by a LOW out of the lower gate.

Additionally, data can be entered serially on the SER input. Also, the clock can be inhibited anytime with a HIGH on the CLK INH input. The serial data outputs of the register are Q_7 and its complement, $\overline{Q}_7$.

This implementation is different from the synchronous method of parallel loading previously discussed, demonstrating that there are usually several ways to accomplish the same function.

Figure 9–15 is a timing diagram showing an example of the operation of a 74165 shift register.

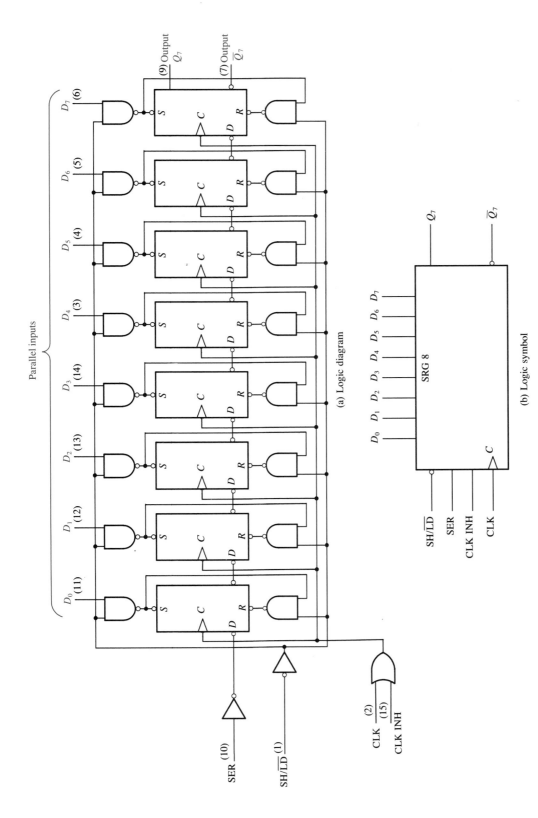

FIGURE 9–14

The 74165 eight-bit parallel load shift register.

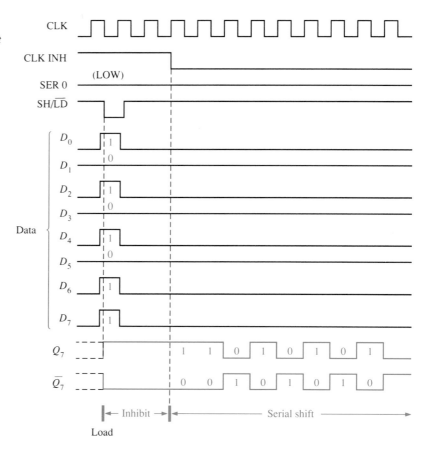

FIGURE 9–15
Sample timing diagram for a 74165 shift register.

SECTION 9–4
REVIEW

Answers are found at the end of the chapter.
1. Explain the function of the SHIFT/$\overline{\text{LOAD}}$ input.
2. Is the parallel load operation in a 74165 shift register synchronous or asynchronous? What does this mean?

PARALLEL IN–PARALLEL OUT REGISTERS

9–5

Parallel entry of data was described in Section 9–4, and parallel output of data has also been discussed previously. The parallel in–parallel out register employs both methods: Immediately following the simultaneous entry of all data bits, the bits appear on the parallel outputs. This type of register is shown in Figure 9–16.

The 74195 Four-Bit Parallel Access Shift Register

This device can be used for parallel in–parallel out operation. Since it also has a serial input, it can be used for serial in–serial out and serial in–parallel out operation. It can

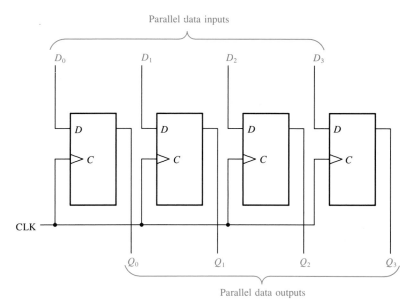

Parallel data inputs

FIGURE 9–16
A parallel in–parallel out register.

be used for parallel in–serial out operation by using Q_3 as the output. A typical logic block symbol is shown in Figure 9–17.

When the SHIFT/$\overline{\text{LOAD}}$ (SH/$\overline{\text{LD}}$) input is LOW, the data on the parallel inputs are entered synchronously on the positive transition of the clock. When the SH/$\overline{\text{LD}}$ input is HIGH, stored data will shift right (Q_0 to Q_3) synchronously with the clock. Inputs J and $\overline{K}$ are the serial data inputs to the first stage of the register (Q_0); Q_3 can be used for serial output data. The active-LOW clear is asynchronous.

The timing diagram in Figure 9–18 illustrates the operation of this register.

FIGURE 9–17
The 74195 four-bit parallel access shift register.

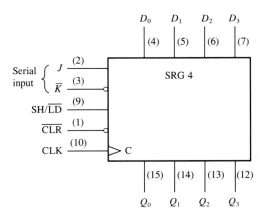

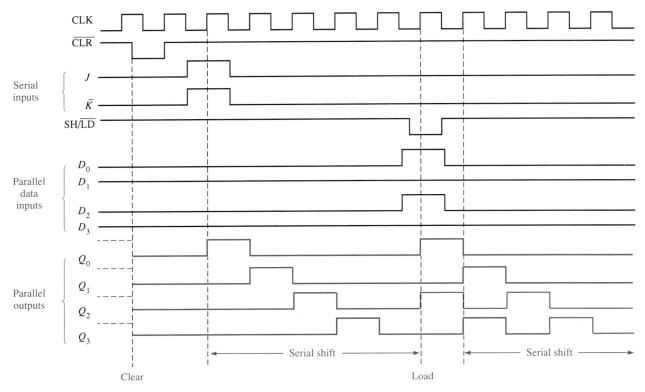

FIGURE 9–18
Example timing diagram for a 74195 shift register.

SECTION 9–5 REVIEW

Answers are found at the end of the chapter.

1. In Figure 9–16, $D_0 = 1$, $D_1 = 0$, $D_2 = 0$, and $D_3 = 1$. After three clock pulses, what are the data outputs?
2. For a 74195, $SH/\overline{LD} = 1$, $J = 1$, and $\overline{K} = 1$. What is Q_0 after one clock pulse?

BIDIRECTIONAL SHIFT REGISTERS

9–6

A bidirectional shift register is one in which the data can be shifted either left or right. It can be implemented by using gating logic that enables the transfer of a data bit from one stage to the next stage to the right or to the left, depending on the level of a control line.

A four-bit implementation is shown in Figure 9–19 for illustration. A HIGH on the RIGHT/$\overline{LEFT}$ control input allows data bits inside the register to be shifted to the right, and a LOW enables data bits inside the register to be shifted to the left. An examination of the gating logic should make the operation apparent. When the RIGHT/$\overline{LEFT}$ control is HIGH, gates G_1 through G_4 are enabled, and the state of the Q output of each flip-flop is passed through to the D input of the *following* flip-flop. When a clock pulse occurs, the data bits are then effectively shifted one place to the *right*. When the

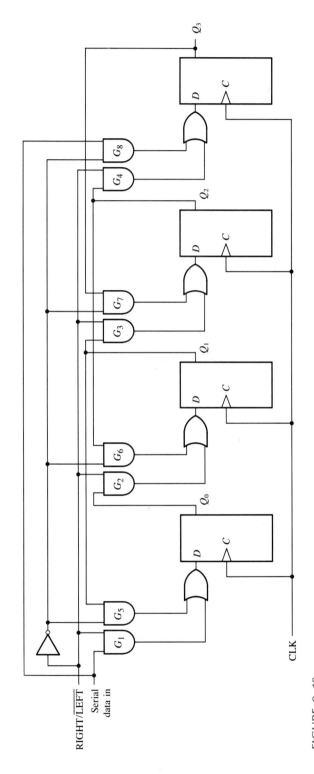

FIGURE 9–19
Four-bit bidirectional shift register.

RIGHT/$\overline{\text{LEFT}}$ control is LOW, gates G_5 through G_8 are enabled, and the Q output of each flip-flop is passed through to the D input of the *preceding* flip-flop. When a clock pulse occurs, the data bits are then effectively shifted one place to the *left*.

EXAMPLE 9–4

Determine the state of the shift register of Figure 9–19 after each clock pulse for the given RIGHT/$\overline{\text{LEFT}}$ control input waveform in Figure 9–20(a). Assume that $Q_0 = 1$, $Q_1 = 1$, $Q_2 = 0$, and $Q_3 = 1$ and that the serial data-input line is LOW.

FIGURE 9–20

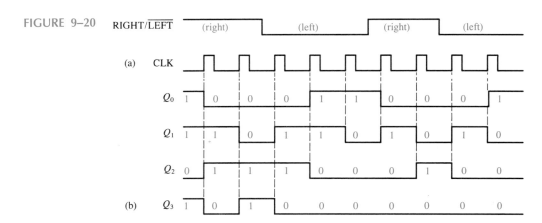

Solution See Figure 9–20(b).

The 74194 Four-Bit Bidirectional Universal Shift Register

The 74194 is an example of a bidirectional shift register in integrated circuit form. A logic block symbol is shown in Figure 9–21.

Parallel loading, which is synchronous with a positive transition of the clock, is accomplished by applying the four bits of data to the parallel inputs and a HIGH to the S_0 and S_1 inputs.

Shift right is accomplished synchronously with the positive edge of the clock when S_0 is HIGH and S_1 is LOW. Serial data in this mode are entered at the *shift-right*

FIGURE 9–21

The 74194 four-bit bidirectional universal shift register.

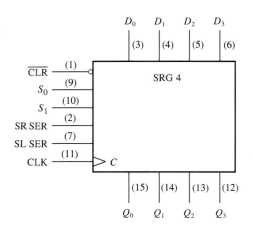

serial input (SR SER). When S_0 is LOW and S_1 is HIGH, data bits shift left synchronously with the clock, and new data are entered at the *shift-left serial input* (SL SER). A sample timing diagram is shown in Figure 9–22. Input SR SER goes into the Q_0 stage, and SL SER goes into the Q_3 stage.

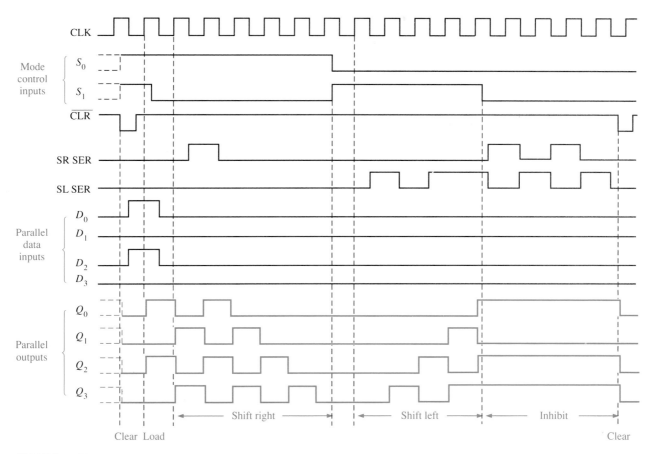

FIGURE 9–22
Sample timing diagram for a 74194 shift register.

SECTION 9–6
REVIEW

Answers are found at the end of the chapter.

1. Assume that the bidirectional shift register in Figure 9–19 has the following contents: $Q_0 = 1$, $Q_1 = 1$, $Q_2 = 0$, and $Q_3 = 0$. There is a 1 on the serial data-input line. If RIGHT/$\overline{\text{LEFT}}$ is HIGH for three clock pulses and LOW for two more clock pulses, what are the contents after the fifth clock pulse?

SHIFT REGISTER COUNTERS

9–7

A shift register counter is basically a shift register with the serial output connected back to the serial input to produce special sequences. These devices are often classified as counters because they exhibit a specified sequence of states. Two of the most common types of shift register counters are introduced in this section: The **Johnson counter** *and the* **ring counter.**

The Johnson Counter

In a Johnson counter the complement of the output of the last flip-flop is connected back to the D input of the first flip-flop (it can be implemented with other types of flip-flops as well). This *feedback* arrangement produces a characteristic sequence of states, as shown in Table 9–1 for a four-bit device and in Table 9–2 for a five-bit device. Notice that the four-bit sequence has a total of *eight* states, or bit patterns, and that the five-bit sequence has a total of *ten* states. In general, a Johnson counter will produce a modulus of $2n$, where n is the number of stages in the counter.

TABLE 9–1

Four-bit Johnson sequence.

Clock Pulse	Q_0	Q_1	Q_2	Q_3
0	0	0	0	0
1	1	0	0	0
2	1	1	0	0
3	1	1	1	0
4	1	1	1	1
5	0	1	1	1
6	0	0	1	1
7	0	0	0	1

TABLE 9–2

Five-bit Johnson sequence.

Clock Pulse	Q_0	Q_1	Q_2	Q_3	Q_4
0	0	0	0	0	0
1	1	0	0	0	0
2	1	1	0	0	0
3	1	1	1	0	0
4	1	1	1	1	0
5	1	1	1	1	1
6	0	1	1	1	1
7	0	0	1	1	1
8	0	0	0	1	1
9	0	0	0	0	1

The implementations of the four- and five-stage Johnson counters are shown in Figure 9–23. The implementation of a Johnson counter is very straightforward and is the same regardless of the number of stages. The Q output of each stage is connected to the D input of the next stage (assuming that D flip-flops are used). The single exception is that the $\overline{Q}$ output of the last stage is connected back to the D input of the first stage. As the sequences in Tables 9–1 and 9–2 show, the counter will "fill up" with 1s from left to right, and then it will "fill up" with 0s again.

Diagrams of the timing operations of the four- and five-bit counters are shown in Figures 9–24 and 9–25, respectively.

FIGURE 9–23
Johnson counters.

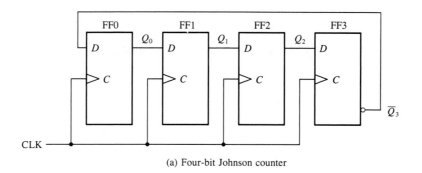

(a) Four-bit Johnson counter

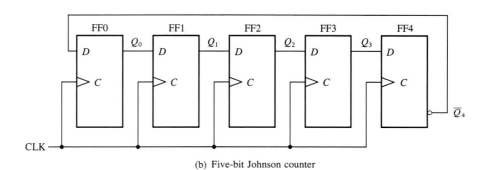

(b) Five-bit Johnson counter

FIGURE 9–24
Timing sequence for a four-bit Johnson counter.

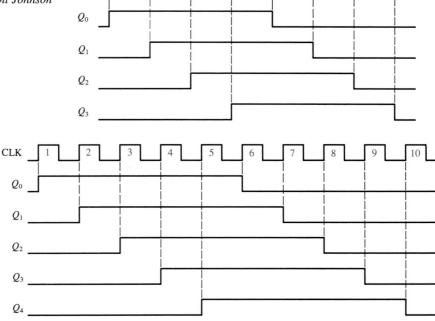

FIGURE 9–25
Timing sequence for a five-bit Johnson counter.

The Ring Counter

The ring counter utilizes one flip-flop for each state in its sequence. It has the advantage that decoding gates are not required for decimal conversion, because there is an output for each decimal number.

A logic diagram for a ten-bit ring counter is shown in Figure 9–26. The sequence for this ring counter is given in Table 9–3. Initially, a 1 is preset into the first flip-flop, and the rest of the flip-flops are cleared. Notice that the interstage connections are the same as those for a Johnson counter, except that Q rather than $\overline{Q}$ is fed back from the last stage. The ten outputs of the counter indicate directly the decimal count of the clock pulse. For instance, a 1 on Q_0 is a zero, a 1 on Q_1 is a one, a 1 on Q_2 is a two, a 1 on Q_3 is a three, and so on. You should verify for yourself that the 1 is always retained in the counter and simply shifted "around the ring," advancing one stage for each clock pulse.

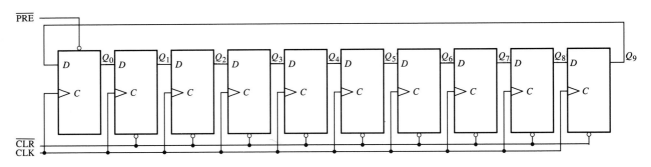

FIGURE 9–26
A ten-bit ring counter.

TABLE 9–3
Ring counter sequence (ten bits).

Clock Pulse	Q_0	Q_1	Q_2	Q_3	Q_4	Q_5	Q_6	Q_7	Q_8	Q_9
0	1	0	0	0	0	0	0	0	0	0
1	0	1	0	0	0	0	0	0	0	0
2	0	0	1	0	0	0	0	0	0	0
3	0	0	0	1	0	0	0	0	0	0
4	0	0	0	0	1	0	0	0	0	0
5	0	0	0	0	0	1	0	0	0	0
6	0	0	0	0	0	0	1	0	0	0
7	0	0	0	0	0	0	0	1	0	0
8	0	0	0	0	0	0	0	0	1	0
9	0	0	0	0	0	0	0	0	0	1

Modified sequences can be achieved by having more than a single 1 in the counter, as illustrated in the following example.

EXAMPLE 9–5   If a ten-bit ring counter similar to Figure 9–26 has the initial state 1010000000, determine the waveform for each of the Q outputs.

FIGURE 9–27

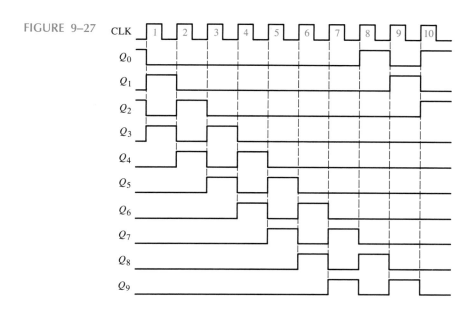

Solution   See Figure 9–27.

SECTION 9–7
REVIEW

Answers are found at the end of the chapter.
1. How many states are there in an eight-bit Johnson counter sequence?
2. Write the sequence of states for a three-bit Johnson counter starting with 000.

SHIFT REGISTER APPLICATIONS

9–8

Shift registers are found in an almost endless array of applications, a few of which are presented in this section.

Time Delay

The serial in–serial out shift register can be used to provide a time delay from input to output that is a function of both the number of stages (n) in the register and the clock frequency.

When a data pulse is applied to the serial input in Figure 9–28 (*A* and *B* connected together), it enters the first stage on the triggering edge of the clock pulse. It is then shifted from stage to stage on each successive clock pulse until it appears on the serial output n clock periods later. This time-delay operation is illustrated in Figure 9–28, in which a 7491A eight-bit shift register is used with a clock frequency of 1 MHz to achieve a time delay of 8 μs (8 × 1 μs). This time can be adjusted up or down by changing the clock frequency. The time delay can also be increased by cascading shift registers and decreased by taking the outputs from successively lower stages in the register if the outputs are available, as illustrated in Example 9–6.

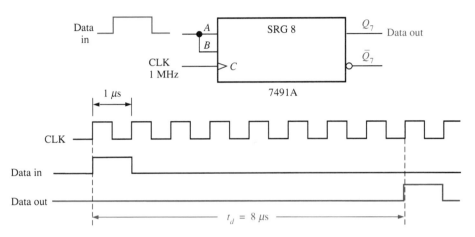

FIGURE 9–28
The shift register as a time-delay device.

EXAMPLE 9–6

Determine the amount of time delay between the serial input and each output in Figure 9–29. Show a timing diagram to illustrate.

FIGURE 9–29

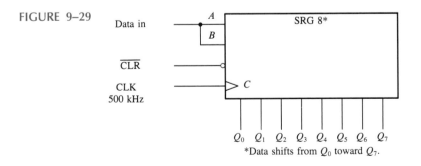

Solution The clock period is 2 μs. Thus, the time delay can be increased or decreased in 2-μs increments from a minimum of 2 μs to a maximum of 16 μs, as illustrated in Figure 9–30.

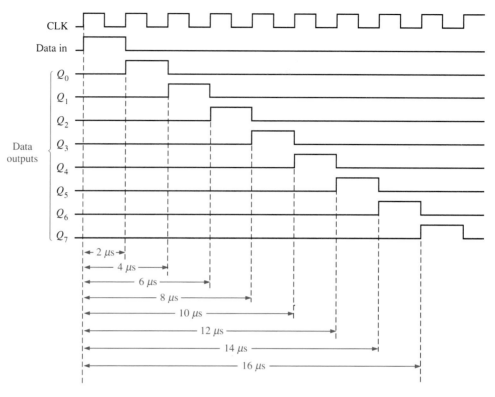

FIGURE 9–30
Timing diagram showing time delays for the register in Figure 9–29.

A Ring Counter Using a 74195 Shift Register

If the output is connected back to the serial input, a shift register can be used as a ring counter. This application is illustrated in Figure 9–31 with a 74195 four-bit shift register.

FIGURE 9–31
74195 connected as a ring counter.

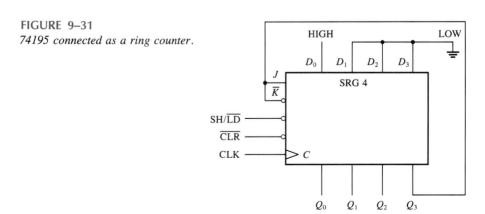

Initially, a 1000 bit pattern (or any other pattern) can be synchronously preset into the counter by applying the bit pattern to the parallel data inputs, taking the SH/$\overline{\text{LD}}$ input LOW, and applying a clock pulse. After this initialization, the 1 continues to circulate through the ring counter, as the timing diagram in Figure 9–32 shows.

FIGURE 9–32

Timing diagram showing two complete cycles of the ring counter in Figure 9–31 when it is initially preset to 1000.

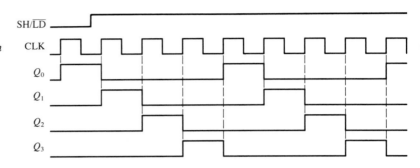

Serial-to-Parallel Data Converter

Serial data transmission from one digital system to another is commonly used to reduce the number of wires in the transmission line. For example, eight bits can be sent serially over one wire, but it takes eight wires to send the same data in parallel.

A computer or microprocessor-based system commonly requires incoming data to be in parallel format, thus the requirement for serial-to-parallel conversion.

A simplified serial-to-parallel data converter, in which two types of shift registers are used, is shown in Figure 9–33. In this case, use of specific devices is reserved as an end-of-chapter problem.

To illustrate the operation of this system, the serial data format shown in Figure 9–34 is used. It consists of eleven bits. The first bit (start bit) is always 0 and always begins with a HIGH-to-LOW transition. The next eight bits (D_0 through D_7) are the data bits (one of the bits can be parity), and the last two bits (stop bits) are always 1s. When no data are being sent, there is a continuous 1 on the serial data line.

The HIGH-to-LOW transition of the start bit SETS the control flip-flop, which enables the clock generator. After a fixed delay time, the clock generator begins producing a pulse waveform, which is applied to the data-input register and to the divide-by-8 counter. The clock has a frequency precisely equal to that of the incoming serial data, and the first clock pulse after the start bit is coincident with the first data bit.

The timing diagram in Figure 9–35 illustrates the following basic operation: The eight data bits (D_0 through D_7) are serially shifted into the data-input register. After the eighth clock pulse, a HIGH-to-LOW transition of the *terminal count* (TC) output of the counter ANDed with the clock loads the eight bits that are in the data-input register into the data-output register. This same transition also triggers the one-shot, which produces a short-duration pulse to clear the counter and RESET the control flip-flop and thus disable the clock generator. The system is now ready for the next group of eleven bits, and it waits for the next HIGH-to-LOW transition at the beginning of the start bit.

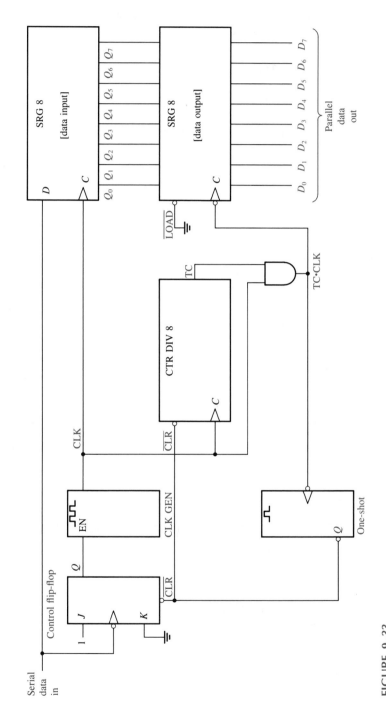

FIGURE 9–33
Simplified logic diagram of a serial-to-parallel converter.

467

FIGURE 9–34
Serial data format.

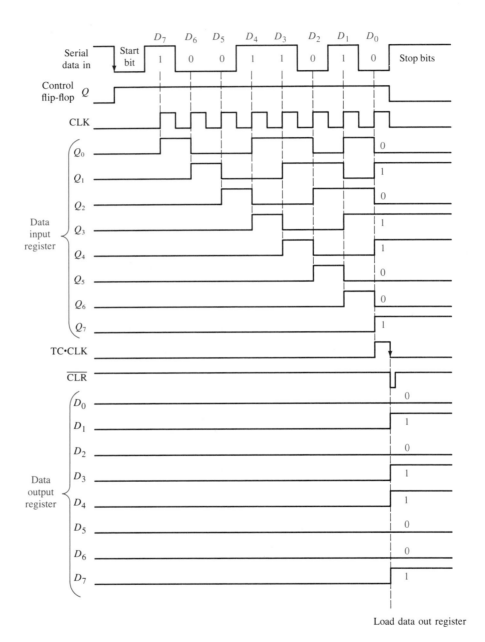

FIGURE 9–35
Timing diagram illustrating the operation of the serial-to-parallel data converter in Figure 9–33.

By reversing the process just covered, parallel-to-serial data conversion can be accomplished. However, since the serial data format must be produced, additional requirements must be taken into consideration. This is left as a problem at the end of the chapter.

Universal Asynchronous Receiver Transmitter (UART)

As mentioned, computers and microprocessor-based systems often send and receive data in a parallel format. Frequently, these systems must communicate with external devices that send and/or receive *serial* data. An interfacing device used to accomplish these conversions is the UART. Figure 9–36 illustrates the UART in a general microprocessor-based system application.

A UART includes a serial-to-parallel data converter such as we have discussed and a parallel-to-serial converter, as shown in Figure 9–37. The *data bus* is basically a set of parallel conductors along which data move between the UART and the microprocessor system. Buffers interface the data registers with the data bus.

The UART receives data in serial format, converts the data to parallel format, and places them on the data bus. The UART also accepts parallel data from the data bus, converts the data to serial format, and transmits them to an external device.

FIGURE 9–36
UART interface.

FIGURE 9–37
Basic UART block diagram.

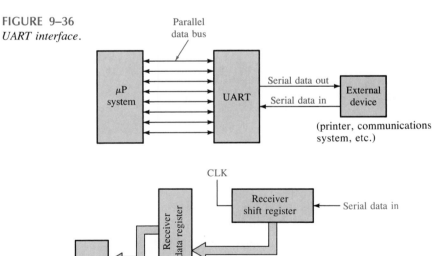

Keyboard Encoder

The keyboard encoder is a good example of the application of a shift register used as a ring counter in conjunction with other devices. Recall that a simplified calculator keyboard encoder without data storage was presented in Chapter 6.

Figure 9–38 shows a simplified keyboard encoder for encoding a key closure in 64-key matrix organized in eight rows and eight columns. A 74199 eight-bit universal shift register is connected as a ring counter (Q_0 to J, $\overline{K}$) with a fixed bit pattern of seven 1s and one 0 preset into it when the power is turned on. Two 74147 priority encoders

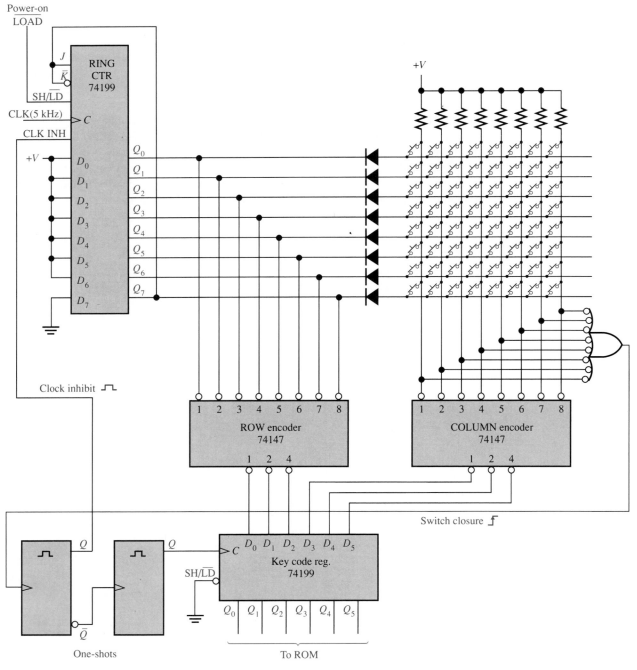

FIGURE 9–38
Simplified keyboard encoding circuit.

(introduced in Chapter 6) are used as eight-line-to-three-line encoders (9 input HIGH, 8 output unused) to encode the ROW and COLUMN lines of the keyboard matrix. Another 74199 shift register is used as a parallel in–parallel out register in which the ROW/COLUMN code from the priority encoders is stored.

The basic operation of the keyboard encoder in Figure 9–38 is as follows: The *ring counter* "scans" the rows for a key closure as the clock signal shifts the 0 around the counter at a 5-kHz rate. The 0 (LOW) is sequentially applied to each ROW line, while all other ROW lines are HIGH. All the ROW lines are connected to the *ROW encoder* inputs, so the three-bit output of the ROW encoder at any time is the binary representation of the ROW line that is LOW.

When there is a key closure, one COLUMN line is connected to one ROW line. When the ROW line is taken LOW by the ring counter, that particular COLUMN line is also pulled LOW. The *COLUMN encoder* produces a binary output corresponding to the COLUMN in which the key is closed. The three-bit ROW code plus the three-bit COLUMN code uniquely identifies the key that is closed. This six-bit code is applied to the inputs of the *key code register*. When a key is closed, the two one-shots produce a delayed clock pulse to parallel-load the six-bit code into the key code register. This delay allows the contact bounce to die out. Also, the first one-shot output inhibits the ring counter to prevent it from scanning while the data are being loaded into the key code register.

The six-bit code in the key code register is now applied to a ROM (read only memory) to be converted to ASCII or some other appropriate alphanumeric code that identifies the keyboard character. ROMs are studied in the next chapter.

SECTION 9–8
REVIEW

Answers are found at the end of the chapter.
1. In the keyboard encoder, how many times per second does the ring counter scan the keyboard?
2. What is the six-bit ROW/COLUMN code (key code) for the top row and the leftmost column in the keyboard encoder?
3. What is the purpose of the diodes in the keyboard encoder? What is the purpose of the resistors?

TROUBLESHOOTING

9–9

One basic method of troubleshooting sequential logic and other more complex digital systems uses a procedure of "exercising" the circuit under test with a known input waveform (stimulus) and then observing the output for the proper bit pattern.

The serial-to-parallel data converter in Figure 9–33 is used to illustrate this procedure. The main objective in "exercising" the circuit is to force all elements (flip-flops and gates) into all of their states to be certain that nothing is "stuck" in a given state as a result of a fault. The *input test pattern,* in this case, must be designed to force each flip-flop in the registers into both states, to clock the counter through all of its eight states, and to take the control flip-flop, the clock generator, the one-shot, and the AND gate through their paces.

The input test pattern that accomplishes this objective for the serial-to-parallel data converter is based on the serial data format in Figure 9–34. It consists of the pattern

10101010 in one serial group of data bits followed by 01010101 in the next group, as shown in Figure 9–39. These patterns are generated on a repetitive basis by a special test-pattern generator. The basic test setup is shown in Figure 9–40.

FIGURE 9–39
Sample test pattern.

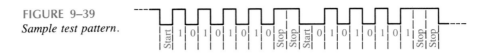

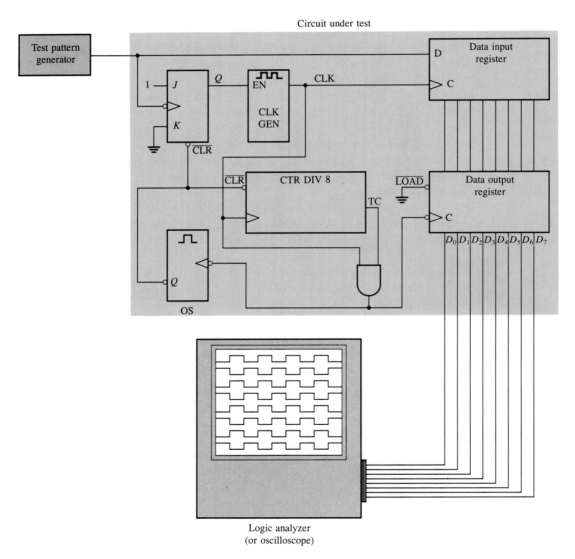

FIGURE 9–40
Basic test setup for the serial-to-parallel data converter of Figure 9–33.

After both patterns have been run through the circuit under test, all the flip-flops in the data-input register and in the data-output register have resided in both SET and RESET states, the counter has gone through its sequence (once for each bit pattern), and all the other devices have been exercised.

To check for proper operation, each of the parallel data outputs is observed for an alternating pattern of 1s and 0s as the input test patterns are repetitively shifted into the data-input register and then loaded into the data-output register. The proper timing diagram is shown in Figure 9–41. Each output can be observed individually, or the outputs can be observed in pairs with a dual-trace oscilloscope, or all eight outputs can be observed simultaneously with a logic analyzer configured for timing analysis.

If one or more outputs of the data-output register are incorrect, then we must back up to the outputs of the data-input register. If these outputs are correct, then probably the data-output register is defective. If the data-input register outputs are also incorrect, the fault could be with the register itself or with any of the other logic, and additional investigation is necessary to isolate the problem.

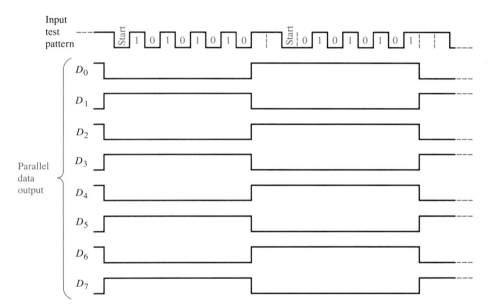

FIGURE 9–41
Proper outputs for the circuit under test in Figure 9–40. Two full cycles of the input test pattern are shown.

Signature Analysis

Signature analysis is another troubleshooting method. It is based on the comparison of measured bit patterns (called *signatures*) with documented signatures at various test points *(nodes)* in a system that is being tested.

During troubleshooting, measured signatures are compared against documented ones that appear on the schematic diagram or on a test point list for the system under test. There is no diagnostic information in the signature itself; thus, a signature display is useless without a known correct signature with which to compare it.

A typical signature analyzer is shown in Figure 9–42. The *measurement cycle* of the analyzer is controlled by an internal state called the *gate*. When the gate is open, the analyzer is taking a new signature measurement through the *data-probe input*. When

FIGURE 9–42

A typical signature analyzer.

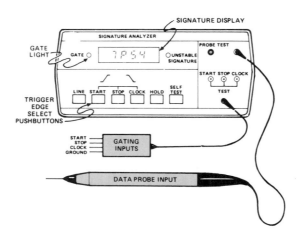

the gate closes, the analyzer stops taking the new signature measurement and displays it, replacing the previous measurement. The gate lamp on the front panel is on during a measurement cycle when the gate is open and is off when the gate is closed.

The analyzer gate is controlled by three inputs from the system or device under test: *start, stop,* and *clock*. Basic gate control for a signature measurement cycle is shown in Figure 9–43. The start and stop inputs are used to open and close the gate at

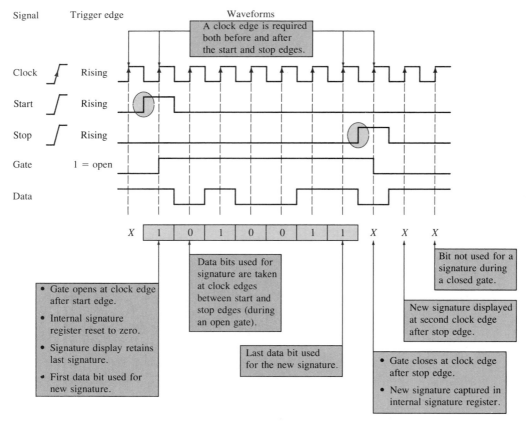

FIGURE 9–43

Example of the basic gate operation of a signature analyzer.

predetermined triggering edges, which are selected by front panel controls. The clock synchronizes the analyzer to the system or device being tested. The logic level of the waveform (bit pattern) measured by the data-probe input is sampled only at the clock pulse edges. These logic levels are used as the data bits that are compressed into the signature. The gate opens at the clock edge following the selected start edge and closes at the clock edge following the selected stop edge, as illustrated in Figure 9–43.

**SECTION 9–9
REVIEW**

Answers are found at the end of the chapter.
1. What is the purpose of providing a test input to a sequential logic circuit?
2. Generally, when an output waveform is found to be incorrect, what is the next step to be taken?
3. Define *signature analysis*.

LOGIC SYMBOLS WITH DEPENDENCY NOTATION

9–10

This section can be omitted without affecting other coverage. Two examples of ANSI/IEEE Standard 91-1984 symbols with dependency notation for shift registers are presented.

The logic symbol for a 74164 eight-bit parallel out serial shift register is shown in Figure 9–44. The common control inputs are shown on the notched block. The clear ($\overline{CLR}$) input is indicated by an R (for RESET) inside the block. Since there is no dependency prefix to link R with the clock ($C1$), the clear function is asynchronous. The right arrow symbol after $C1$ indicates data flow from Q_0 to Q_7. The A and B inputs are ANDed, as indicated by the imbedded AND symbol, to provide the synchronous data input, $1D$, to the first stage (Q_0). Note the dependency of D on C, as indicated by the 1 suffix on C and the 1 prefix on D.

FIGURE 9–44
Logic symbol for the 74164.

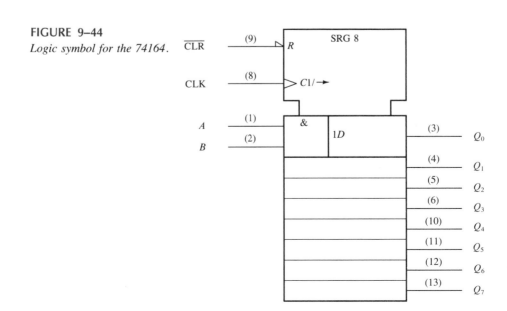

Figure 9–45 is the logic symbol for the 74194 four-bit bidirectional universal shift register. Starting at the top left side of the control block, note that the $\overline{\text{CLR}}$ input is active-LOW and is asynchronous (no prefix link with C). Inputs S_0 and S_1 are *mode* inputs that determine the *shift-right, shift-left,* and *parallel load* modes of operation, as indicated by the $\frac{0}{3}$ dependency designation following the M. The $\frac{0}{3}$ represents the binary states of 0, 1, 2, and 3 on the S_0 and S_1 inputs. When one of these digits is used as a prefix for another input, a dependency is established. The $1\rightarrow/2\leftarrow$ symbol on the clock input indicates the following: $1\rightarrow$ indicates that a right shift (Q_0 toward Q_3) occurs when the mode inputs (S_0, S_1) are in the binary 1 state ($S_0 = 1$, $S_1 = 0$), $2\leftarrow$ indicates that a left shift (Q_3 toward Q_0) occurs when the mode inputs are in the binary 2 state ($S_0 = 0$, $S_1 = 1$). The shift-right serial input (SR SER) is both mode- and clock-dependent, as indicated by 1, $4D$. The parallel inputs (D_0, D_1, D_2, and D_3) are all mode-dependent (prefix 3 indicates parallel load mode) and clock-dependent, as indicated by 3, $4D$. The shift-left serial input (SL SER) is both mode- and clock-dependent, as indicated by 2, $4D$.

FIGURE 9–45

Logic symbol for the 74194.

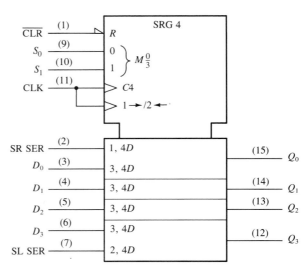

The four modes for this device are summarized below:

Do nothing: $S_0 = 0, S_1 = 0$ (mode 0)
Shift right: $S_0 = 1, S_1 = 0$ (mode 1, as in 1, $4D$)
Shift left: $S_0 = 0, S_1 = 1$ (mode 2, as in 2, $4D$)
Parallel load: $S_0 = 1, S_1 = 1$ (mode 3, as in 3, $4D$)

SECTION 9–10 REVIEW

Answers are found at the end of the chapter.
1. In Figure 9–45, are there any inputs that are dependent on the mode inputs being in the 0 state?
2. Is the parallel load synchronous with the clock?

SUMMARY

☐ The basic types of registers, classified by input and output, are

1. serial in–serial out
2. serial in–parallel out
3. parallel in–serial out
4. parallel in–parallel out

☐ The basic types of data movement in a register are

1. shift right
2. shift left
3. parallel shift in (load)
4. parallel shift out
5. recirculate (rotate) right
6. recirculate (rotate) left

☐ Shift register counters are shift registers with feedback that exhibit special sequences. Examples are the Johnson counter and the ring counter.
☐ The Johnson counter has $2n$ states in its sequence, where n is the number of stages.
☐ The ring counter has n states in its sequence.

 Integrated Circuit Technologies (Chapter 13) may be covered after this chapter, covered after a later chapter, partially omitted, or completely omitted, depending on the needs and goals of your program. Coverage is not prerequisite or corequisite to any other material in this book. Inclusion is purely optional.

GLOSSARY

Bidirectional   Having two directions. In a bidirectional shift register, the stored data can be shifted right or left.

Clear   To place a register in the state in which it contains all 0s.

Johnson counter   A type of register in which a specific pattern of 1s and 0s is shifted through the stages, creating a unique sequence of bit patterns.

Load   To enter data into a shift register.

Recirculate   To retain data in a register as it is shifted out.

Register   A digital circuit capable of storing binary data. Typically used as a temporary storage device.

Ring counter   A register in which a certain pattern of 1s and 0s is continuously recirculated.

Shift   To move binary data from stage to stage within a shift register or other storage device or to move binary data into or out of the device.

Stage   One storage element (flip-flop) in a register.

Universal shift register   A register that has both serial and parallel input and output capability.

SELF-TEST

Answers and solutions are at the end of the book.
1. Explain how a flip-flop can store a data bit.
2. How many clock pulses are required to shift a byte of data into and out of an eight-bit serial in–serial out shift register?

3. Answer question 2 for an eight-bit serial in–parallel out shift register.

4. Draw a logic diagram for the register in Figure 9–3, using J-K flip-flops to replace the D flip-flops.

5. Show how the shift register in Figure 9–8 can be used as a three-bit serial in–serial out shift register.

6. The binary number 10110101 is serially shifted, LSB first, into an eight-bit parallel out shift register that has an initial content of 11100100. What are the Q outputs after two clock pulses? After four clock pulses? After eight clock pulses?

7. An eight-bit bidirectional shift register contains 10111011. Zeros are applied to the shift-right and shift-left serial data inputs. The shift register is in the shift-right mode for three clock pulses, then in the shift-left mode for six clock pulses, then in the shift-right mode for two clock pulses. What is the final state of the register?

8. How many flip-flops are required to implement a divide-by-10 Johnson counter?

9. What is the basic difference between a Johnson counter and a ring counter?

10. Write the sequence of states for a four-bit Johnson counter.

11. Write the sequence of states for an eight-bit ring counter with 1s in the first and fourth stages and 0s in the rest.

12. In Figure 9–28, what clock frequency is required for a 24-μs delay?

13. Determine the sequence of states for the ring counter in Figure 9–31 if 1s are preset in D_3 and D_1 and 0s in D_2 and D_0.

14. What is the purpose of a UART?

15. Specify, by device number, a type of one-shot that can be used in the keyboard encoder of Figure 9–38.

16. Explain the purpose of the ring counter in Figure 9–38.

PROBLEMS

Answers to odd-numbered problems are at the end of the book.

Section 9–1 Shift Register Functions

9–1 Why are shift registers considered basic memory devices?

9–2 What is the storage capacity of a register than can retain two bytes of data?

Section 9–2 Serial In–Serial Out Shift Registers

9–3 For the data input and clock in Figure 9–46, determine the states of each flip-flop in the shift register of Figure 9–3, and draw the Q waveforms. Assume that the register contains all 1s initially.

9–4 Solve Problem 9–3 for the waveforms in Figure 9–47.

9–5 What is the state of the register in Figure 9–48 after each clock pulse if it starts in the 101001111000 state?

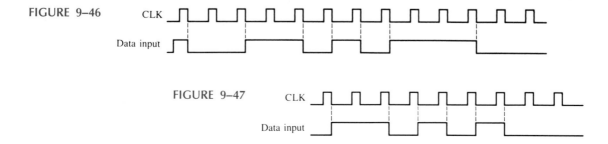

FIGURE 9–46

FIGURE 9–47

9–6 For the serial in–serial out shift register, determine the data-output waveform for the data-input and clock waveforms in Figure 9–49. Assume that the register is initially cleared.

9–7 Solve Problem 9–6 for the waveforms in Figure 9–50.

9–8 The data-output waveform in Figure 9–51 is related to the clock as indicated. What binary number is stored in an eight-bit serial in–serial out register if the first data bit out (left most) is the LSB?

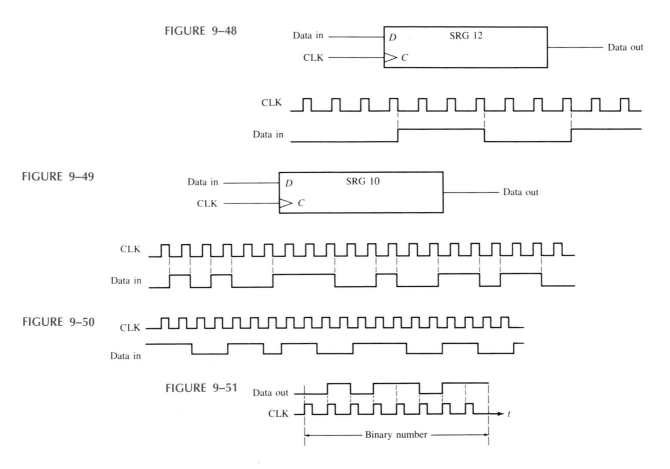

FIGURE 9–48

FIGURE 9–49

FIGURE 9–50

FIGURE 9–51

Section 9–3   Serial In–Parallel Out Shift Registers

9–9 Draw a complete timing diagram showing the parallel outputs for the shift register in Figure 9–8. Use the inputs in Figure 9–49 with the register initially clear.

9–10 Solve Problem 9–9 for the input waveforms in Figure 9–50.

9–11 Sketch the Q_0 through Q_7 outputs for a 74164 shift register with the input waveforms shown in Figure 9–52.

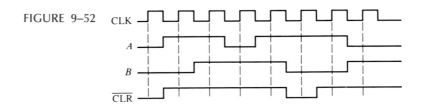

FIGURE 9–52

Section 9–4 Parallel In–Serial Out Shift Registers

9–12 The shift register in Figure 9–53(a) has SHIFT/$\overline{\text{LOAD}}$ and CLK inputs as shown in part (b). The serial data input (SER) is a 0. The parallel data inputs are $D_0 = 1$, $D_1 = 0$, $D_2 = 1$, and $D_3 = 0$. Sketch the data-output waveform in relation to the inputs.

9–13 The waveforms in Figure 9–54 are applied to a 74165 shift register. The parallel inputs are all 0. Determine the Q_7 waveform.

9–14 Solve Problem 9–13 if the parallel inputs are all 1.

9–15 Solve Problem 9–13 if the SER input is inverted.

FIGURE 9–53

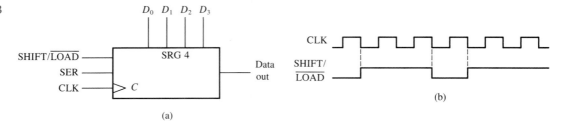

(a)

FIGURE 9–54

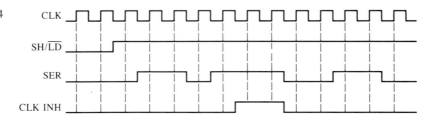

Section 9–5 Parallel In–Parallel Out Shift Registers

9–16 Determine all the Q output waveforms for a 74195 four-bit shift register when the inputs are as shown in Figure 9–55.

9–17 Solve Problem 9–16 if the SH/$\overline{\text{LD}}$ input is inverted and the register is initially clear.

9–18 Use two 74195 shift registers to form an eight-bit shift register. Show the required connections.

FIGURE 9–55

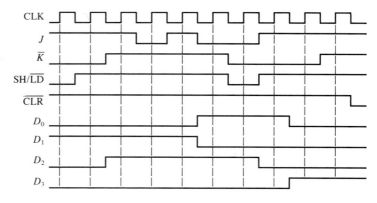

Section 9–6   Bidirectional Shift Registers

9–19   For the eight-bit bidirectional register in Figure 9–56, determine the state of the register after each clock pulse for the RIGHT/$\overline{\text{LEFT}}$ control waveform given. A HIGH on this input enables a shift to the right, and a LOW enables a shift to the left. Assume that the register is initially storing the decimal number seventy-six in binary, with the right-most position being the LSB. There is a LOW on the data-input line.

9–20   Solve Problem 9–19 for the waveforms in Figure 9–57.

9–21   Use two 74194 four-bit bidirectional shift registers to create an eight-bit bidirectional shift register. Show the connections.

9–22   Determine the Q outputs of a 74194 with the inputs shown in Figure 9–58. Inputs D_0, D_1, D_2, and D_3 are all HIGH.

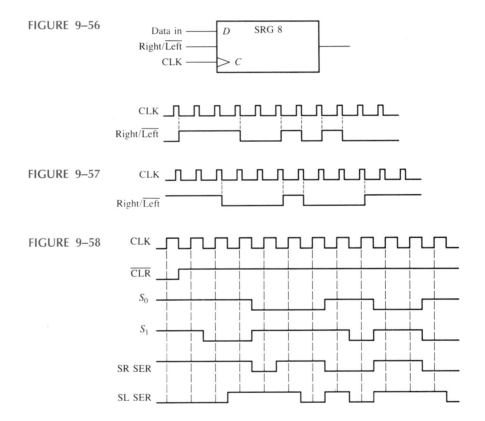

FIGURE 9–56

FIGURE 9–57

FIGURE 9–58

Section 9–7   Shift Register Counters

9–23   How many flip-flops are required to implement each of the following in a Johnson counter configuration:

    (a) divide-by-6   **(b)** divide-by-10   **(c)** divide-by-14

    (d) divide-by-16   **(e)** divide-by-20   **(f)** divide-by-24

    (g) divide-by-36

9–24   Draw the logic diagram for a divide-by-18 Johnson counter. Sketch the timing diagram and write the sequence in tabular form.

9–25 For the ring counter in Figure 9–59, draw the waveforms for each flip-flop output with respect to the clock. Assume that FF0 is initially SET and that the rest are RESET. Show at least ten clock pulses.

9–26 The waveform pattern in Figure 9–60 is required. Devise a ring counter, and indicate how it can be preset to produce this waveform on its final output. At CLK_{16} the pattern begins to repeat.

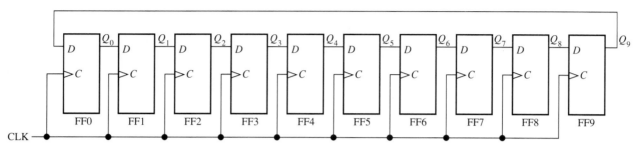

FIGURE 9–59

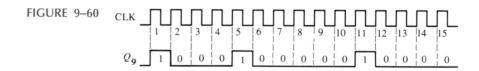

FIGURE 9–60

Section 9–8   Shift Register Applications

9–27 Use 74195 four-bit shift registers to implement a sixteen-bit ring counter. Show the connections.

9–28 Specify the devices that can be used to implement the serial-to-parallel data converter in Figure 9–33. Draw the complete logic diagram, showing any modifications necessary to accommodate the specific devices used.

9–29 Modify the serial-to-parallel converter in Figure 9–33 to provide sixteen-bit conversion.

9–30 Design an eight-bit parallel-to-serial data converter that produces the data format in Figure 9–34. Show a logic diagram and specify the devices.

9–31 What is the purpose of the power-on $\overline{\text{LOAD}}$ input in Figure 9–38?

9–32 Develop a power-on $\overline{\text{LOAD}}$ circuit for the keyboard encoder in Figure 9–38. This circuit must generate a short-duration LOW pulse when the power switch is turned on.

9–33 What happens when two keys are pressed simultaneously in Figure 9–38?

Section 9–9   Troubleshooting

9–34 Based on the waveforms in Figure 9–61(a), determine the most likely problem with the register in part (b) of the figure.

9–35 Refer to the parallel in–serial out shift register in Figure 9–12. The register is in the state where $Q_0 Q_1 Q_2 Q_3 = 1001$, and $D_0 D_1 D_2 D_3 = 1010$ is loaded in. When the SHIFT/LOAD input is taken HIGH, the data shown in Figure 9–62 are shifted out. Is this operation correct? If not, what is the most likely problem?

9–36 You have found that the bidirectional register in Figure 9–19 will shift data right but not left. What is the most likely fault?

FIGURE 9–61

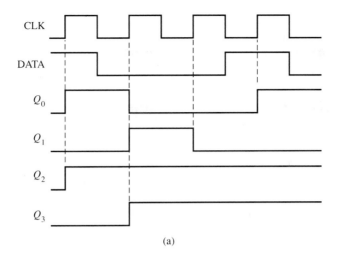

(a)

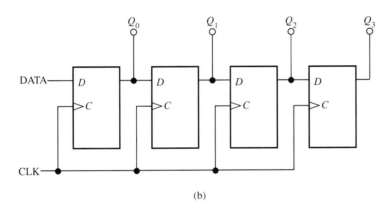

(b)

FIGURE 9–62

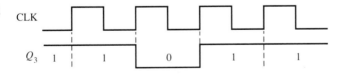

9–37 For the keyboard encoder in Figure 9–38, list the possible faults for each of the following symptoms:

(a) The state of the key code register does not change for any key closure.

(b) The state of the key code register does not change when any key in the third row is closed. A proper code occurs for all other key closures.

(c) The state of the key code register does not change when any key in the first column is closed. A proper code occurs for all other key closures.

(d) When any key in the second column is closed, the left three bits of the key code ($Q_0Q_1Q_2$) are correct, but the right three bits are all 1s.

9–38 Develop a test procedure for exercising the keyboard encoder in Figure 9–38. Specify the procedure on a step-by-step basis, indicating the output code from the key code register that should be observed at each step in the test.

9–39 Implement the test pattern generator used in Figure 9–40 to troubleshoot the serial-to-parallel converter.

9–40 What symptoms are observed for the following failures in the serial-to-parallel converter in Figure 9–33:
(a) AND gate output stuck in HIGH state
(b) clock generator output stuck in LOW state
(c) third stage of data-input register stuck in SET state
(d) terminal count output of counter stuck in HIGH state

ANSWERS TO SECTION REVIEWS

Section 9–1 Review

1. A counter has a specified sequence of states, but a register does not.
2. storage, data movement (shifting)

Section 9–2 Review

1. FF0: data input to J, $\overline{\text{data input}}$ to K; FF1: Q_0 to J, $\overline{Q_0}$ to K; FF2: Q_1 to J, $\overline{Q_1}$ to K; FF3: Q_2 to J, $\overline{Q_2}$ to K
2. Eight

Section 9–3 Review

1. 0100 **2.** Take the serial output from the right-most flip-flop.

Section 9–4 Review

1. When $\overline{\text{SHIFT/LOAD}}$ is HIGH, the data are shifted right one bit per clock pulse. When $\overline{\text{SHIFT/LOAD}}$ is LOW, the data on the parallel inputs are loaded into the register.
2. Asynchronous. The parallel load operation is not dependent on the clock.

Section 9–5 Review

1. 1001 **2.** $Q_0 = 1$

Section 9–6 Review

1. 1111

Section 9–7 Review

1. sixteen **2.** 000, 100, 110, 111, 011, 001, 000

Section 9–8 Review

1. 625 scans/second **2.** $Q_5Q_4Q_3Q_2Q_1Q_0 = 110110$
3. The diodes provide unidirectional paths for pulling the ROWs LOW and preventing HIGHs on the ROW lines from being connected to the switch matrix. The resistors pull the COLUMN lines HIGH.

Section 9–9 Review

1. to sequence the circuit through all of its states
2. Check the input to that portion of the circuit. If the signal on that input is correct, the fault is isolated to the circuitry between the good input and the bad output.
3. Signature analysis is a troubleshooting technique whereby measured bit patterns are compared with documented bit patterns (signatures).

Section 9–10 Review

1. no **2.** yes, as indicated by the $4D$ label

The last chapter covered shift registers, which are a type of storage device; in fact, a shift register is basically a memory. The memory devices covered in this chapter are generally used for longer-term storage of larger amounts of data than registers can provide.

Modern data-processing systems require the permanent or semipermanent storage of large amounts of data. Microprocessor-based systems rely on memories for their operation because memories are used to store programs and data during processing.

In this chapter both semiconductor and magnetic memories and their applications are covered. Programmable logic devices are also thoroughly covered. Specific memory devices introduced are the following:

1. 74184 ROM BCD-to-binary converter
2. 74185 ROM binary-to-BCD converter
3. 7488 ROM
4. TMS47256 ROM
5. TMS2516 16K-bit EPROM
6. TMS4016 16K-bit byte-oriented static RAM
7. 74189 64-bit static RAM
8. TMS4164 64K-bit dynamic RAM
9. 82S100 programmable logic array
10. PAL16R4 programmable logic sequencer.

After completing this chapter, you will be able to

☐ see the need for memories in digital systems.
☐ understand what a ROM is and how it works.
☐ explain how a ROM is programmed.
☐ make the distinction between ROMs and the various types of PROMs.
☐ understand what a RAM is and how it works.
☐ explain how ROMs and RAMs are used in microprocessor-based systems.
☐ explain the difference between static RAMs (SRAMs) and dynamic RAMs (DRAMs).
☐ expand ROMs and RAMs to increase word length and word capacity.
☐ describe the basic organization of a floppy disk.
☐ describe the components of a magnetic bubble memory.
☐ program FPLAs, FPALs, and FPLSs.
☐ describe basic methods for memory testing.
☐ develop flowcharts for memory testing.

10

Memories and Programmable Logic Devices

MEMORY CONCEPTS

10–1

Data are stored in a memory by a process called **write** *and are retrieved from the memory by a process called* **read**. *Memories are made up of storage locations in which data can be stored. Each location is identified by an* **address**. *The number of storage locations can vary from a few in some memories to hundreds of thousands in others.*

Each storage location can accommodate one or more bits. Generally, the total number of bits that a memory can store is its **capacity**. *Often the capacity is specified in bytes.*

Memories are made up of storage elements (flip-flops or capacitors in semiconductor memories and magnetic domains in magnetic storage), each of which stores one bit of data. A storage element is called a **cell**.

Figure 10–1 illustrates in a very simplified way the concepts of *write, read, address,* and *storage capacity* for a generalized memory.

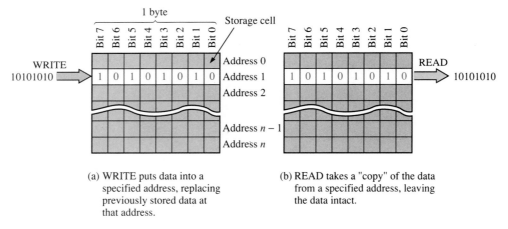

(a) WRITE puts data into a specified address, replacing previously stored data at that address.

(b) READ takes a "copy" of the data from a specified address, leaving the data intact.

FIGURE 10–1
A memory with n *addresses and a capacity of* n *bytes or 8* × n *bits.*

Types of Semiconductor Memories

The two basic types of semiconductor memories are the *ROM* and the *RAM*. The ROM is a *read only memory*. Data are permanently or semipermanently stored and can be read from the memory at any time. In a ROM in which the data are permanently stored, there is no write operation because specified data are either manufactured into the device or programmed into the device by the user and cannot be altered. In a ROM in which the data are semipermanently stored, the data can be altered by special methods, but there is no write operation. A ROM is a *nonvolatile* memory; that is, the stored data are retained even when power is removed. Examples of ROM applications are look-up tables, conversions, and programmed instructions.

The RAM is a *random access memory* that has both read and write capability. More precisely, the RAM is a read/write random access memory, and the ROM is a read only random access memory. The term *random access* means that all storage locations in the memory are equally accessible and they do not require sequential access. Actually, ROMs are also random access devices, so the terminology is somewhat misleading. A RAM is a *volatile* memory, so data are lost if the power is removed.

The ROM family Semiconductor ROMs are manufactured with bipolar technology (such as TTL) or with MOS (metal oxide semiconductor) technology. Figure 10–2 shows how ROMs are categorized. The *mask* ROM is the type in which data are permanently stored in the memory during the manufacturing process. The PROM, or *programmable* ROM, is the type in which the data are electrically stored by the user with the aid of specialized equipment. Notice that both the mask ROM and the PROM can be of either technology. The EPROM, or *erasable* PROM, is strictly an MOS device.

The EPROM is electrically programmable by the user, but the stored data must be erased by exposure to ultraviolet light over a period of several minutes. The *electrically erasable* PROM (EEPROM) or *electrically alterable* PROM (EAPROM) can be erased in a few milliseconds.

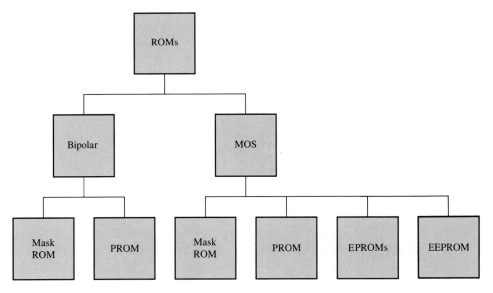

FIGURE 10–2
The semiconductor ROM family.

The RAM family Semiconductor RAMs are also manufactured with either bipolar or MOS technologies. Bipolar RAMs are all *static* RAMs (SRAM); that is, the storage elements used in the memory are latches, so data can be stored for an indefinite period of time as long as the power is on. Some MOS RAMs are of the static type and some are *dynamic*. A dynamic RAM (DRAM) memory is one in which data are stored on capacitors, which require periodic recharging *(refreshing)* to retain the data. Figure 10–3 shows the categories of RAMs.

FIGURE 10–3
The semiconductor RAM family.

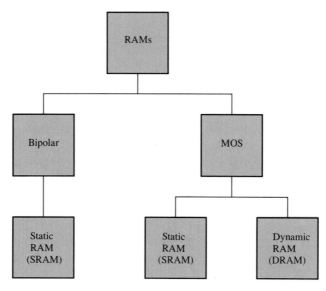

Types of Magnetic Memories

Magnetic storage devices are used in computer systems primarily for *mass data storage*. These devices are sometimes referred to as *external* or *secondary* memories. Magnetic memories have a much slower access time (time to read data) than semiconductor memories; thus their applications are limited to permanent or semipermanent storage of large amounts of data. Magnetic memories fall into three basic categories: *disk, tape,* and *bubble*.

Two basic types of disks are used in computer systems: *floppy,* or *flexible,* disks and *hard,* or *fixed,* disks. Basically, the floppy disk provides less storage capacity than the hard disk, but the floppy is transportable, and the 3.5-in. variety can be carried in a shirt pocket. The hard disk provides a very large storage capacity that can be accessed relatively fast.

Two basic types of tape storage that are sometimes used in computer systems are the *audiocassette tape* and the *streaming tape*. Although they are not used as widely as they once were, audiocassettes provide an inexpensive method for mass data storage. However, access time is extremely long. Streaming tape systems are more commonly used for mass storage and for backup of disk-based data because data can be accessed much faster than with the audiocassette.

The magnetic bubble memory is a more recent development than either disk or tape and is much faster. However, because of its expense, its use is limited at this time. Figure 10–4 shows the categories of magnetic memory.

How Memories Are Used in Microprocessor-Based Systems

Figure 10–5 shows a generalized basic block diagram of a typical microcomputer, which forms the heart of microprocessor-based systems. As you can see, the diagram consists of four functional blocks: *microprocessor, memory, input/output (I/O) inter-face,* and *I/O device*. The microprocessor unit within the microcomputer is interconnected to the memory and the I/O interface with an *address bus,* a *data bus,* and a *control bus*.

FIGURE 10–4
The magnetic memory family.

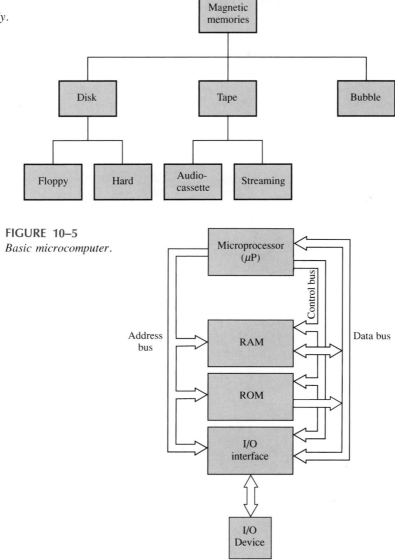

FIGURE 10–5
Basic microcomputer.

The function of the internal semiconductor memory is to store binary data that are to be used or processed by the microprocessor. The function of the I/O interface is to get information into and out of the memory or the microprocessor from the I/O device. The I/O device can be a keyboard, a video monitor, a printer, a magnetic disk or tape unit, a bubble memory, or another type of equipment. A given computer system typically has several I/O devices.

The address bus provides a path from microprocessor to memory and I/O interface. It allows the microprocessor to select the memory address from which to acquire data or in which data are to be stored. It also provides for communication with an I/O device for inputting or outputting data.

The data bus provides a path over which data are transferred between the microprocessor, the memory, and the I/O interface.

As mentioned, the microprocessor is connected to memory (RAM and ROM) with the address bus and the data bus. In addition, there are certain control signals that must be sent from the microprocessor to the memory, such as the read and write controls.

As illustrated in Figure 10–6, the address bus is unidirectional; that is, the address bits go only one way—from the microprocessor to the memory. The data bus is bidirectional. This provides for data bits to be transferred from the RAM and ROM to the microprocessor or from the microprocessor to the RAM.

Words and bytes   A *complete* unit of binary information or data is called a *word*. Most microprocessors handle bits in eight-bit groups called *bytes*. A byte may be a word or only part of a word. For instance, an *eight-bit* word consists of *one byte*. However, a *sixteen-bit word* consists of *two bytes*.

In the following example of the read and write operations, a memory capacity of 65,536 bits is used because it represents a common memory size used in many microprocessor-based systems. This memory capacity is traditionally designated as a 64K.

Read Operation

To transfer a byte of data from the memory to the microprocessor, a read operation must be performed, as illustrated in Figure 10–7. To begin, a *program counter* contains the 16-bit address of the byte to be read from the memory. This address is loaded into the *address register* and put onto the address bus.

Once the address code is on the bus, the microprocessor control unit sends a read signal to the memory. At the memory the address bits are decoded, and the desired memory location is selected. The read signal causes the contents of the selected address to be placed on the data bus. The data byte is then loaded into the data register to be used by the microprocessor. This loading completes the read operation.

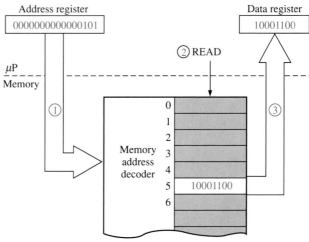

(1) Address 5_{10} placed on address bus.

(2) READ signal applied.

(3) Contents of address 5_{10} in memory placed on data bus and stored by data register.

FIGURE 10–7

Illustration of the read operation in a typical microcomputer.

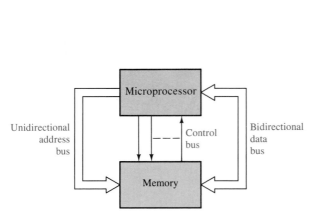

FIGURE 10–6

A microprocessor with memory.

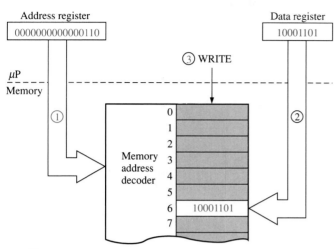

FIGURE 10–8
Illustration of the write operation.

Address (hexadecimal)	Contents
0000	
0001	
0002	
0003	
0004	
0005	
0006	
0007	
FFFB	
FFFC	
FFFD	
FFFE	
FFFF	

FIGURE 10–9
Representation of a 64K memory.

Note that each memory location contains one byte of data. When a byte is read from the memory, it is not destroyed but remains in the memory. This process of "copying" the contents of a memory location without destroying those contents is called *nondestructive readout*.

Write Operation

To transfer a byte of data from the microprocessor to the memory, a write operation is required, as illustrated in Figure 10–8.

The memory is addressed in the same way as during a read operation. A data byte being held in the data register is put onto the data bus, and the microprocessor sends the memory a write signal. This causes the byte on the data bus to be stored at the selected location in the memory, as specified by the sixteen-bit address code. The existing contents of that particular memory location are *replaced* by the new data. This replacement completes the write operation.

Hexadecimal Representation of Address and Data

The only things a microprocessor recognizes are combinations of 1s and 0s. However, most literature on microprocessors uses the hexadecimal number system to simplify the representation of binary quantities.

For instance, the binary address 0000000000001111 can be written as 000F in hexadecimal. A sixteen-bit address can have a *minimum* hexadecimal value of 0000_{16} and a *maximum* value of $FFFF_{16}$. With this notation, a 64K memory can be shown in block form as in Figure 10–9. The lowest memory address is 0000_{16} and the highest address is $FFFF_{16}$.

A data byte can also be represented in hexadecimal. A data byte is eight bits and can represent decimal numbers from 0_{10} to 255_{10}, or it can represent up to 256_{10} instructions. For example, a microprocessor code that is 10001100 in binary is written as 8C in hexadecimal.

SECTION 10–1
REVIEW

Answers are found at the end of the chapter.
1. Explain the basic difference between ROM and RAM.
2. What does the term *nonvolatile* mean?
3. Explain the difference between static and dynamic memories.

READ ONLY MEMORIES (ROMs)

10–2

As mentioned before, a ROM contains permanently or semipermanently stored data, which can be read from the memory but either cannot be changed at all or cannot be changed without specialized equipment. A ROM is used to store data that are used repeatedly in system applications, such as tables, conversions, or programmed instructions for system initialization and operation.

The Mask ROM

The mask ROM is referred to simply as a ROM. It is permanently programmed during the manufacturing process to provide widely used standard functions, such as popular conversions, or to provide user-specified functions. Once the memory is programmed, it cannot be changed. Most IC ROMs utilize the presence or absence of a transistor connection at a ROW/COLUMN junction to represent a 1 or a 0. A ROM can be either bipolar or MOS.

Figure 10–10(a) shows bipolar ROM cells. The presence of a connection from a ROW line to the *base* of a transistor represents a 1 at that location because when the ROW line is taken HIGH, all transistors with a base connection to that ROW line turn on and connect the HIGH (1) to the associated COLUMN lines. At ROW/COLUMN junctions where there are no base connections, the COLUMN lines remain LOW (0) when the ROW is addressed.

Figure 10–10(b) illustrates MOS ROM cells. They are basically the same as the bipolar cells, except they are made with MOSFETs (metal oxide semiconductor field-effect transistors). The presence or absence of a *gate* connection at a junction permanently stores a 1 or a 0, as shown.

FIGURE 10–10
ROM cells.

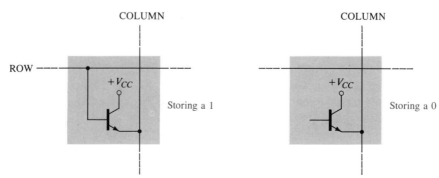

(a) Bipolar cells

FIGURE 10–10
(Continued)

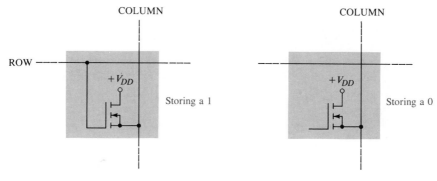

(b) MOS cells

A Simple ROM

To illustrate the concept, Figure 10–11 shows a small, simplified ROM array. The colored squares represent stored 1s, and the gray squares represent stored 0s. The basic read operation is as follows: When a binary address code is applied to the *address input,* the corresponding ROW line goes HIGH. This HIGH is connected to the COLUMN lines through the transistors at each junction (cell) where a 1 is stored. At each cell

FIGURE 10–11
A 16 × 8-bit ROM array.

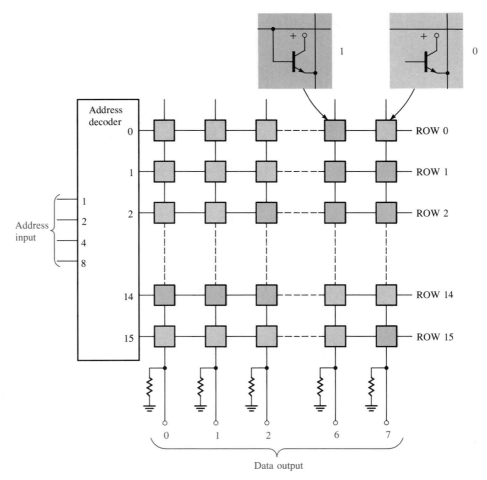

where a 0 is stored, the COLUMN line stays LOW because of the terminating resistor. The COLUMN lines form the *data output*. The eight data bits stored in the selected ROW appear on the output lines.

As you can see, this example ROM is organized into 16 addresses, each of which stores 8 data bits. Thus, it is a 16×8 (16-by-8) ROM, and its total capacity is 128 bits.

EXAMPLE 10–1

Show a basic ROM, similar to that shown in Figure 10–11, programmed for four-bit binary-to-Gray conversion.

Solution After reference to Chapter 2 for the Gray code, Table 10–1 is developed for use in programming the ROM.

TABLE 10–1

Binary				Gray			
B_3	B_2	B_1	B_0	G_3	G_2	G_1	G_0
0	0	0	0	0	0	0	0
0	0	0	1	0	0	0	1
0	0	1	0	0	0	1	1
0	0	1	1	0	0	1	0
0	1	0	0	0	1	1	0
0	1	0	1	0	1	1	1
0	1	1	0	0	1	0	1
0	1	1	1	0	1	0	0
1	0	0	0	1	1	0	0
1	0	0	1	1	1	0	1
1	0	1	0	1	1	1	1
1	0	1	1	1	1	1	0
1	1	0	0	1	0	1	0
1	1	0	1	1	0	1	1
1	1	1	0	1	0	0	1
1	1	1	1	1	0	0	0

The resulting ROM array is shown in Figure 10–12. You can see that a binary code on the address inputs produces the corresponding Gray code on the output lines (COLUMNs). For example, a condition in which there are 0s on all the address inputs represents the binary number 0000. This selects the zero address, in which are stored four 0s, representing the corresponding Gray code, 0000.

Internal ROM Structure

Most IC ROMs have a somewhat more complex internal structure than that in the basic simplified example just covered. To illustrate how an IC ROM is structured, a 1024-bit device with a 256×4 organization is used. The logic symbol is shown in Figure 10–13. When any one of 256 binary codes (eight bits) is applied to the address inputs, four data bits appear on the outputs if the chip-select inputs are LOW.

FIGURE 10–12
*A ROM programmed as a binary-to-Gray
code converter.*

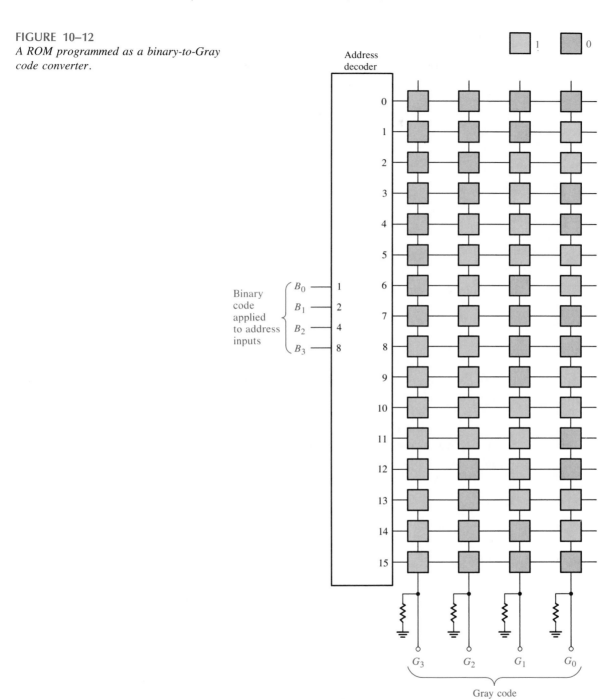

FIGURE 10–13
A 256×4 ROM logic symbol.

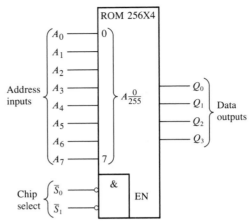

Although the 256×4 organization of this device implies that there are 256 rows and 4 columns in the memory array, this is not the case. The memory cell array is actually a 32×32 matrix (32 rows and 32 columns), as shown in the block diagram in Figure 10–14.

The ROM works as follows: Five of the eight address lines (A_0 through A_4) are decoded by the ROW decoder (often called the Y decoder) to select one of the 32 rows. Three of the eight address lines (A_5 through A_7) are decoded by the column decoder (often called the X decoder) to select four of the 32 columns. Actually, the X decoder consists of four 1-of-8 decoders (data selectors), as shown.

The result of this structure is that when an eight-bit address code (A_0 through A_7) is applied, a four-bit data word appears on the data outputs when the chip-select lines (S_0 and S_1) are LOW to enable the output buffers.

This type of internal structure (architecture) is typical of IC ROMs of various capacities. In fact, the 74187 ROM has this exact configuration.

A Small-Capacity ROM

Figure 10–15 shows the logic symbol for a 7488 as an example of a bipolar ROM. It is organized as a 32×8 memory; that is, it has 32 address locations, each of which has eight bits of storage. A five-bit address on inputs A_0 through A_4 selects one of the 32 locations (0 through 31). A LOW on the $\overline{S}$ input enables (EN) the device and places the selected data byte on the outputs (Q_0 through Q_7). This Enable input is called the *chip select*.

This ROM can be programmed to user specifications by the manufacturer.

A Large-Capacity ROM

As a comparison with the small bipolar ROM just discussed, the TMS47256 (TI) is an example of a large-capacity MOS ROM. It can store 262,144 bits with a $32,768 \times 8$ ($32K \times 8$) organization. That is, it has 32,768 address locations with eight bits at each location.

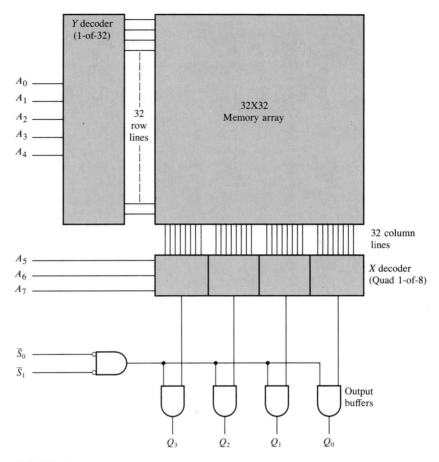

FIGURE 10–14
Internal structure of a typical ROM. This particular example is a 1024-bit ROM with a 256 × 4 organization.

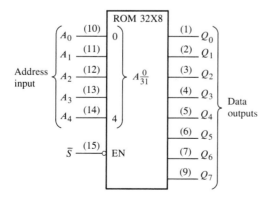

FIGURE 10–15
Logic symbol for the 7488 256-bit bipolar ROM.

FIGURE 10–16

Logic symbol for the TMS47256 MOS static ROM.

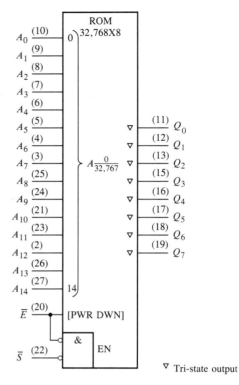

▽ Tri-state output

Figure 10–16 is a logic symbol for this device. Notice that it has 15 address lines. These are required to address the 32,768 locations (2^{15} = 32,768). There is an active-LOW chip select, $\overline{S}$, and, on this particular device, a *chip enable/power down* input, $\overline{E}$. These inputs work as follows: The $\overline{S}$ and $\overline{E}$ inputs must both be LOW to enable the memory output. When $\overline{E}$ is HIGH, the device is put into a low-power standby mode that reduces the current drain on the dc power supply.

Tristate Outputs and Buses

At this point, it is convenient to introduce the concept of the tristate output, which is commonly found on memory devices such as the ROM just presented. The tristate outputs are indicated on logic symbols by a small inverted triangle (▽), as shown in Figure 10–16, and are used for compatibility with bus structures such as those found in microprocessor-based systems.

Physically, a bus is a set of conductive paths that serves to interconnect two or more functional components of a system or several diverse systems. Electrically, a bus is a collection of specified voltage levels and/or current levels and signals that allow the various devices connected to the bus to work properly together.

For example, a microprocessor is connected to memories and input/output devices by certain bus structures. This was illustrated by the block diagram in Figure 10–5. An address bus allows the microprocessor to address the memories, and the data bus provides for transfer of data between the microprocessor, the memories, and the input/output devices. The control bus allows the microprocessor to control data flow and timing for the various components.

The physical bus is symbolically represented by wide lines with arrowheads indicating direction of data movement. Buses can also be used to interconnect various test instruments or other types of electronic systems.

Tristate interface to the bus In a typical application, several devices are connected to one bus, for example, a microprocessor, a RAM, and a ROM. For this reason, *tristate logic* circuits are used to interface digital devices such as memories to a bus. Figure 10–17(a) shows the logic symbol for a noninverting tristate buffer with an active-HIGH *Enable*. Part (b) of the figure shows one with an active-LOW *Enable*.

The basic operation of a tristate buffer can be understood in terms of switching action as illustrated in Figure 10–18. When the Enable input is active, the gate operates as a normal noninverting circuit. That is, the output is HIGH when the input is HIGH and LOW when the input is LOW, as shown in parts (a) and (b). The HIGH and LOW levels represent two of the states. The buffer operates in its third state when the Enable input is not active. In this state the circuit acts as an open switch, and the output is completely disconnected from the input, as shown in part (c). This is sometimes called the *high-impedance* or *high-Z* state. A more detailed coverage of tristate circuitry is found in Chapter 13.

Many microprocessors, memories, and other integrated circuit functions have tristate buffers that serve to interface with the buses. Such buffers are necessary when two or more devices are connected to a common bus. To prevent the devices from interfering with each other, the tristate buffers are used to disconnect all devices except the ones that are communicating at any given time.

FIGURE 10–17
Tristate buffer symbols.

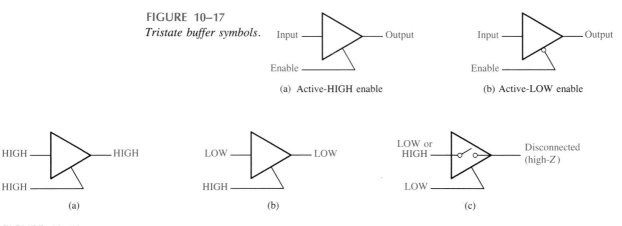

FIGURE 10–18
Tristate buffer operation.

ROM Access Time

A typical timing diagram that illustrates ROM access time is shown in Figure 10–19. The access time, t_a, of a ROM is the time from the application of a valid address code on the inputs until the appearance of valid output data. Access time can also be measured from the activation of the chip-select ($\overline{S}$) input to the occurrence of valid output data when a valid address is already on the inputs.

FIGURE 10–19
ROM access time (t_a) from address change to data output with chip Enable already active.

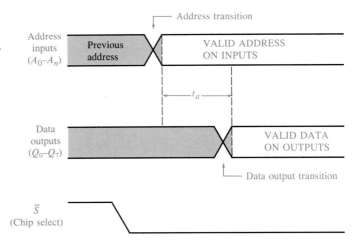

ROM as a Code Converter

As mentioned before, some ROMs are programmed to perform widely used functions and are available ''off the shelf.'' Examples are the 74184, a ROM device programmed as a BCD-to-binary converter, and the 74185, a ROM device programmed as a binary-to-BCD converter. Logic symbols for these devices used as six-bit converters are shown in Figure 10–20.

ROM in Computer Applications

ROM is used in the IBM personal computer, for example, to store what is referred to as the BIOS (Basic Input/Output Services). These are programs that are used to perform

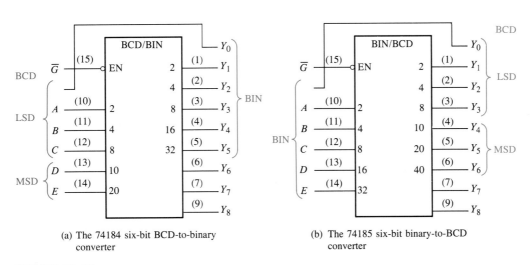

(a) The 74184 six-bit BCD-to-binary converter

(b) The 74185 six-bit binary-to-BCD converter

FIGURE 10–20
Two ROM devices programmed for specific conversions.

fundamental supervisory and support functions for the computer. For example, BIOS programs stored in the ROM control certain video monitor functions, provide for disk formatting, scan the keyboard for inputs, and control certain printer functions.

Another use of ROMs in computer systems is to store the language *interpreter* programs, for BASIC or PASCAL, for example.

SECTION 10–2
REVIEW

Answers are found at the end of the chapter.
1. What is the bit storage capacity of a ROM with a 512×4 organization?
2. Explain the purpose of tristate outputs.
3. How many address bits are required for a 2048-bit memory?

PROGRAMMABLE ROMs (PROMs AND EPROMs)

10–3

PROMs are basically the same as mask ROMs, once they have been programmed. The difference is that PROMs come from the manufacturer unprogrammed and are custom programmed in the field to meet the user's needs. There are two general categories: PROMs and EPROMs.

PROMs

PROMs are available in both bipolar and MOS technologies and generally have four-bit or eight-bit output word formats and bit capacities ranging in excess of 250,000. PROMs employ some type of fusing process to store bits, whereby a memory link is fused open or left intact to represent a 0 or a 1. The fusing process is irreversible. Once a PROM is programmed, it cannot be changed.

Figure 10–21 illustrates a bipolar PROM array with *fusible links*. The fusible links are manufactured into the PROM between the emitter of each cell's transistor and its column line. In the programming process a sufficient current is injected through the fusible link to burn it open to create a stored 0. The link is left intact for a stored 1.

Three basic fuse technologies are used in PROMs: metal links, silicon links, and PN junctions. A brief description of each of these follows:

1. *Metal links* are made of a material such as nichrome. Each bit in the memory array is represented by a separate link. During programming, the link is either "blown" open or left intact. This is done basically by first addressing a given cell and then forcing a sufficient amount of current through the link to cause it to open.

2. *Silicon links* are formed by narrow, notched strips of polycrystalline silicon. Programming of these fuses requires melting of the links by passing a sufficient amount of current through them. This amount of current causes a high temperature at the fuse location that oxidizes the silicon and forms an insulation around the now-open link.

3. *Shorted junction,* or *avalanche-induced migration,* technology consists basically of two pn junctions arranged back-to-back. During programming, one

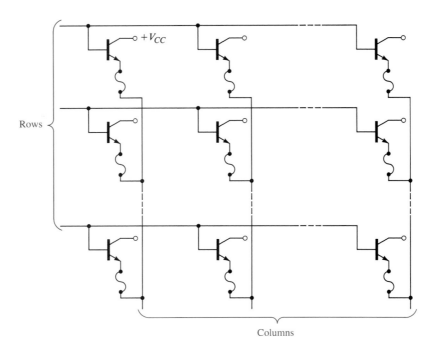

FIGURE 10–21
Bipolar PROM array with fusible links. (All collectors are commonly connected to V_{CC}.)

of the diode junctions is avalanched, and the resulting voltage and heat cause aluminum ions to migrate and short the junction. The remaining junction is then used as a forward-biased diode to represent a data bit.

PROM Programming

A PROM is normally programmed by plugging it into a special instrument called a *PROM programmer*. Basically, the programming is accomplished as shown by the simplified setup in Figure 10–22. An address is selected by the switch settings on the address lines, and then a pulse is applied to those output lines corresponding to bit locations where 0s are to be stored (the PROM starts out with all 1s). These pulses *blow* the fusible links, thus creating the desired bit pattern. The next address is then selected and the process is repeated. This sequence is done automatically by the PROM programmer.

EPROMS

An EPROM is an *erasable* PROM. Unlike an ordinary PROM, an EPROM can be reprogrammed if an existing program in the memory array is erased first.

An EPROM uses an NMOSFET array with an isolated-gate structure. The isolated transistor gate has no electrical connections and can store an electrical charge for indef-

FIGURE 10-22
Simplified PROM programming setup.

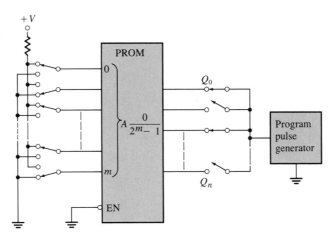

inite periods of time. The data bits in this type of array are represented by the presence or absence of a stored gate charge. Erasure of a data bit is a process that removes the gate charge.

There are two basic types of erasable PROMs: the *ultraviolet light erasable PROM (UV EPROM)* and the *electrically erasable PROM (EEPROM)*.

UV EPROMs   You can recognize the UV EPROM device by the transparent quartz lid on the package, as shown in Figure 10-23. The isolated gate in the FET of an ultraviolet EPROM is "floating" within an oxide insulating material. The programming process causes electrons to be removed from the floating gate. Erasure is done by exposure of the memory array chip to high-intensity ultraviolet radiation through the quartz window on top of the package. The positive charge stored on the gate is neutralized after several minutes to an hour of exposure time.

EEPROMs   Electrically erasable PROMs can be both erased and programmed with electrical pulses. This type of device is also known as an *electrically alterable ROM (EAROM)*. Since it can be both electrically written into and electrically erased, the EEPROM can be rapidly programmed and erased in circuit for reprogramming.

There are two types of EEPROMs: floating-gate MOS and metal-nitride-oxide-silicon (MNOS). The application of a voltage on the control gate in the floating-gate structure permits the storage and removal of charge from the floating gate.

FIGURE 10-23

Ultraviolet erasable PROM package. (Courtesy of Motorola Semiconductor Products)

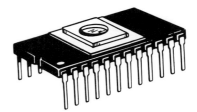

A Specific PROM

The TMS2516 is an example of a MOS EPROM device. Its operation is representative of that of other typical EPROMs. As the logic symbol in Figure 10–24 shows, this device has 2048 addresses (2^{11} = 2048), each with eight bits. Notice that the eight outputs are tristate ($\triangledown$).

To *read* from the memory, the select input ($\overline{S}$) must be LOW and the power-down/program (PD/PGM) input LOW. (The power-down mode was mentioned earlier in relation to the TMS47256 ROM.) To *erase* the stored data, the device is exposed to high-intensity ultraviolet light through the transparent lid. A typical 12 mW/cm² filterless UV lamp will erase the data in about 20 to 25 minutes. As in most EPROMs after erasure, all bits are 1s. Normal ambient light contains the correct wavelength of UV light for erasure. Therefore, the transparent lid must be kept covered.

To *program* the device, +25 V dc is applied to V_{PP} (which is normally +5 V), and $\overline{S}$ is HIGH. The eight data bits to be programmed into a given address are applied to the outputs (Q_0 through Q_7), and the address is selected on inputs A_0 through A_{10}. Next, a 10-ms to 55-ms HIGH level pulse is applied to the PD/PGM input. The addresses can be programmed in any order.

A timing diagram for the programming mode is shown in Figure 10–25. These signals are normally produced by an EPROM programmer.

FIGURE 10–24
The logic symbol for a TMS2516 PROM.

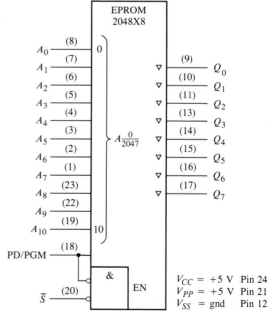

Answers are found at the end of the chapter.
1. How do PROMs differ from ROMs?
2. After erasure, all bits are (1s, 0s) in a typical EPROM.
3. What is the *normal* mode of operation for a PROM?

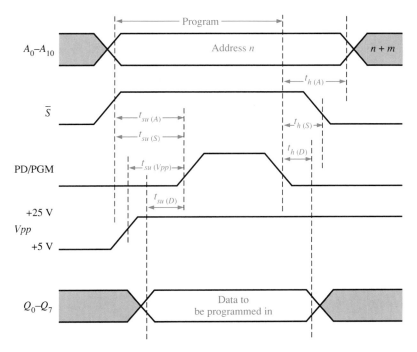

FIGURE 10–25

Timing diagram for TMS2516 PROM programming cycle, with critical setup times (t_s) and hold times (t_h) indicated.

READ/WRITE RANDOM ACCESS MEMORIES (RAMs)

10–4
Data can be readily written into and read from a RAM at any selected address in any sequence. When data are written into a given address in the RAM, the data previously stored at that address are destroyed and are replaced by the new data. When data are read from a given address in the RAM, the data at that address are not destroyed. This nondestructive read operation can be thought of as copying the contents of an address while leaving the contents intact. A RAM is used for short-term data storage.

The Static RAM Cell

As mentioned before, the storage cells in a *static RAM (SRAM)* are either bipolar or MOS latches. Once a data bit is stored in a cell, it remains indefinitely (unless power is lost or a new data bit is written in). Because data are lost if the power to the memory is lost, a static RAM is a type of *volatile* memory.

Figure 10–26 shows a functional logic diagram of a static RAM cell. The basic operation is as follows: The cell (or a group of cells) is selected by HIGHs on the ROW and COLUMN lines. When the READ/$\overline{\text{WRITE}}$ line is LOW (write), the input data bit is written into the cell by setting the latch for a 1 or resetting the latch for a 0. When the READ/$\overline{\text{WRITE}}$ line is HIGH (read), the latch is unaffected, but the stored data bit (Q) is gated to the data-output line.

FIGURE 10–26

A generalized logic diagram for a static RAM cell.

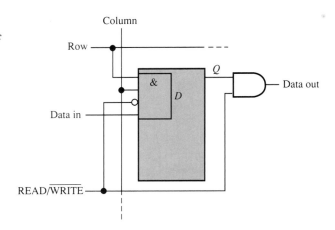

Basic Structure of a Static RAM

A RAM is addressed in basically the same way as a ROM. The main difference between the structures of RAMs and ROMs is that RAMs have data inputs and a read/write control.

To illustrate the generalized structure of a RAM, a 1024-bit device with a 256×4 organization is used. The logic symbol is shown in Figure 10–27.

In the read mode (R/$\overline{W}$ HIGH), four data bits from the selected address appear on the data outputs when the chip select ($\overline{CS}$) is LOW. In the write mode (R/$\overline{W}$ LOW), the four data bits that are applied to the data inputs are stored at the selected address.

Although the 256×4 organization of this device implies that there are 256 rows and 4 columns, the memory cell array actually is a 32×32 matrix (32 rows and 32 columns), as shown in Figure 10–28.

The static RAM works as follows: Five of the eight address lines (A_0 through A_4) are decoded by the ROW decoder to select one of the 32 rows. Three of the eight address lines (A_5 through A_7) are decoded by the output column decoder. In the read

FIGURE 10–27

Logic diagram for a 256×4 static RAM.

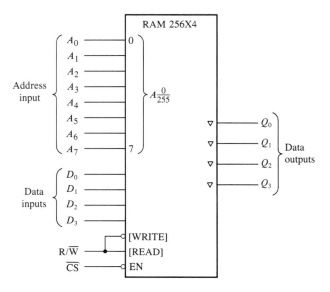

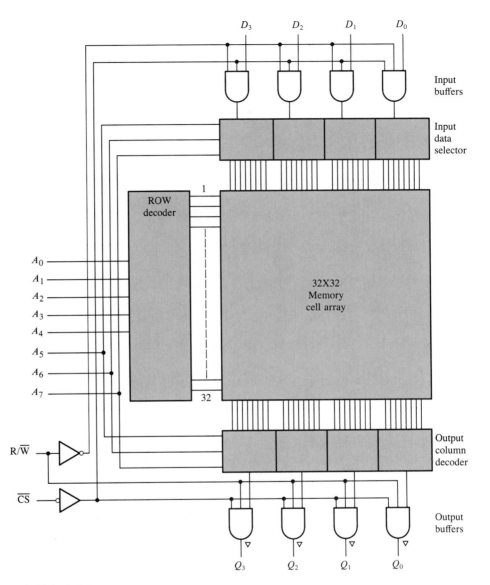

FIGURE 10–28
Basic structure of a 256×4 static RAM.

mode, the output buffers are enabled (the input buffers are disabled), and the four data bits from the selected address appear on the outputs. In the write mode, the input data buffers are enabled (the output buffers are disabled), and the four input data bits are routed through the input data selector by the address bits, A_5 through A_7, to the selected address for storage. The chip select must be LOW during read and write.

Variations of this basic structure and other RAM structures are possible, but this illustrates the basic static RAM operation and provides a comparison with the basic ROM structure in Figure 10–14.

A typical timing diagram for a read/write cycle is shown in Figure 10–29.

FIGURE 10–29

Basic read/write cycle timing for the static RAM in Figure 10–28.

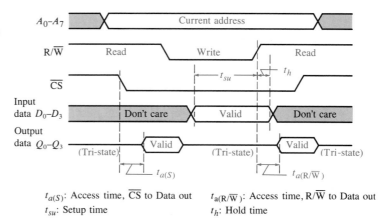

$t_{a(S)}$: Access time, $\overline{CS}$ to Data out $t_{a(R/\overline{W})}$: Access time, $R/\overline{W}$ to Data out
t_{su}: Setup time t_h: Hold time

Specific Static RAMs

As an example of a small-capacity static RAM, the 74189 TTL RAM is introduced. A logic symbol is shown in Figure 10–30(a).

This device has a 16×4 organization. The $R/\overline{W}$ input is HIGH for read and LOW for write. The four address inputs select one of the 16 four-bit locations. The chip select, $\overline{S}$, must be LOW for memory operation.

Because its capacity is small, the internal structure of this particular device is simpler than that discussed for the generalized RAM. For example, it does not have both ROW and COLUMN decoders. It has a ROW decoder (16 rows) only, because the array is actually a 16×4 matrix.

Another example of a static RAM is the TMS4016, which is an MOS device. As shown by the logic symbol in Figure 10–30(b), this memory has a 2048×8 organization; that is, there is storage capacity for 2048 eight-bit words. Since each memory location consists of eight bits, this is an example of a byte-organized static RAM (sometimes called *byte-wide*).

There are eleven address inputs to select the 2048 locations. The data-in/data-out (DQ_0 through DQ_7) are bidirectional terminals that provide for both data input during write and data output during read. The operation of this device is specified in the table in Figure 10–30(c).

The Dynamic RAM (DRAM) Cell

Dynamic memory cells store a data bit in a small capacitor rather than in a latch. The advantage of this type of cell is that it is very simple, thus allowing very large memory arrays to be constructed on a chip at a lower cost per bit than in static memories. The disadvantage is that the storage capacitor cannot hold its charge over an extended period of time and will lose the stored data bit unless its charge is refreshed periodically. This process of refreshing requires additional memory circuitry and complicates the operation of the dynamic RAM. Figure 10–31 shows a typical dynamic cell consisting of a single MOS transistor and a capacitor.

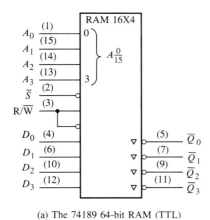

(a) The 74189 64-bit RAM (TTL)

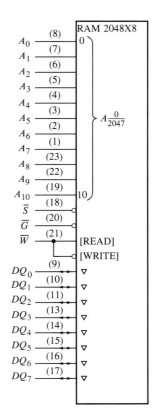

$\overline{W}$	$\overline{S}$	$\overline{G}$	$DQ_0 - DQ_7$	Mode
0	0	X	Valid data	WRITE
1	0	0	Data output	READ
X	1	X	High-impedance*	Device disabled
1	0	1	High-impedance*	Output disabled

*Tri-state output

(c) Truth table for the TMS4016

(b) The TMS4016 16,384-bit RAM (MOS)

FIGURE 10–30
Specific static RAMS.

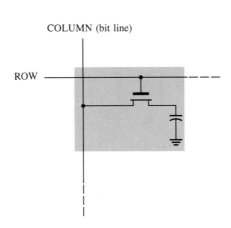

FIGURE 10–31
A dynamic MOS RAM cell.

In this type of cell, the transistor acts as a switch. The basic simplified operation is illustrated in Figure 10–32 and is as follows: A LOW on the R/$\overline{\text{W}}$ line (write mode) enables the tristate input buffer and disables the output buffer. For a 1 to be written into the cell, the D_{IN} line must be HIGH, and the transistor must be turned on by a HIGH on the ROW line. The transistor acts as a closed switch connecting the capacitor to the BIT line. This connection allows the capacitor to charge to a positive voltage, as shown in Figure 10–32(a). When a 0 is to be stored, a LOW is applied to the D_{IN} line. If the capacitor is storing a 0, it remains uncharged, or if it is storing a 1, it discharges as indicated in Figure 10–32(b). When the ROW line is taken back LOW, the transistor turns off and disconnects the capacitor from the BIT line, thus "trapping" the charge (1 or 0) on the capacitor.

To read from the cell, the R/$\overline{\text{W}}$ line is HIGH, enabling the output buffer and disabling the input buffer. When the ROW line is taken HIGH, the transistor turns on and connects the capacitor to the BIT line and thus to the output buffer (sense amplifier), so the data bit appears on the data-output line (D_{OUT}). This process is illustrated in Figure 10–32(c).

For refreshing the memory cell, the R/$\overline{\text{W}}$ line is HIGH, the ROW line is HIGH, and the REFRESH line is HIGH. The transistor turns on, connecting the capacitor to the BIT line. The output buffer is enabled, and the stored data bit is applied to the input of the *refresh buffer,* which is enabled by the HIGH on the REFRESH input. This produces a voltage on the BIT line corresponding to the stored bit, thus replenishing the capacitor as illustrated in Figure 10–32(d).

Basic Structure of a Dynamic RAM

As you have seen, the main difference between dynamic RAMs and static RAMs is the type of memory cell. Because the dynamic memory cell requires refreshing to retain data, additional circuitry is necessary. Several specific features common to most DRAMs are now discussed.

Address multiplexing   Most DRAMs employ a technique called *address multiplexing* to reduce the number of address lines and thus the number of input/output pins on the package. Figure 10–33 shows a block diagram of a 16K (16,384) DRAM. The diagram has been simplified to illustrate address multiplexing. This example memory has a 16K × 1 organization (it stores 16,384 one-bit words).

The fourteen-bit address (2^{14} = 16,384) is applied seven bits at a time to the address inputs. First, the seven-bit ROW address is applied, and the RAS (row address strobe) latches the seven bits into the *row address latch*. Next, the seven-bit COLUMN address is applied to the address inputs, and the CAS (column address strobe) latches these seven bits into the *column address latch*. The seven-bit ROW address and the seven-bit COLUMN address are then decoded to select the appropriate memory cell for a read or write operation. The basic timing for this address multiplexing operation is shown in Figure 10–34.

Memory refresh   All DRAMs require a refresh operation. The refresh logic can be either internal or external to the memory chip. Figure 10–35 shows a block diagram for the same basic generalized 16K memory just discussed, with the addition of refresh logic and somewhat greater detail.

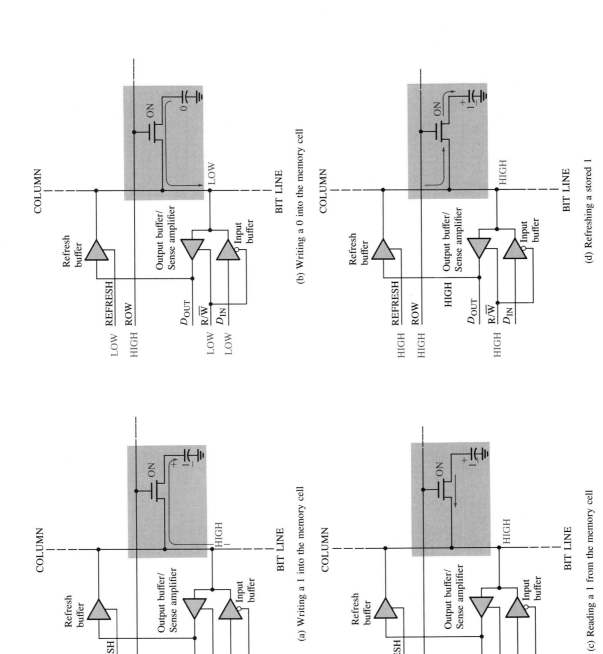

(a) Writing a 1 into the memory cell

(b) Writing a 0 into the memory cell

(c) Reading a 1 from the memory cell

(d) Refreshing a stored 1

FIGURE 10–32

Basic operation of a dynamic memory cell.

513

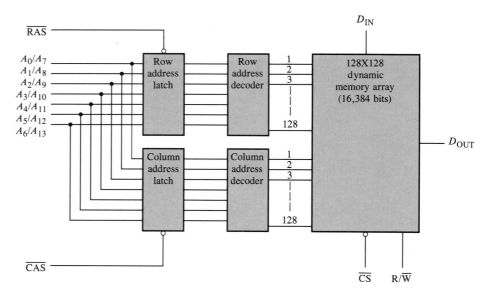

FIGURE 10–33

Simplified block diagram of a 16K dynamic RAM. The refresh circuitry is omitted for simplicity.

FIGURE 10–34

Timing for address multiplexing in Figure 10–33.

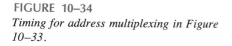

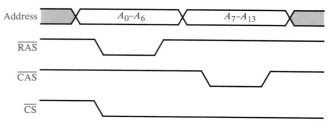

In its simplest form the memory refresh operation sequentially refreshes each row of cells, one row after the other, until all rows are refreshed. This process is called *burst refresh* and must be repeated every 2 ms to 4 ms in a typical DRAM. During refresh, data cannot be written into or read from the memory.

Another basic refresh method distributes the row refresh operations between the read and write operations, rather than doing them all at once in a burst. Still, each row must be refreshed within the specified refresh period.

Now let us see how refresh works in the memory of Figure 10–35. When a refresh operation is initiated, the seven output lines from the *refresh counter* go into the data selector (MUX) and are selected by the *refresh control* and applied to the row decoder in place of the row address lines.

The refresh counter is started and sequences through all its states (0 through 127). Each row is sequentially selected, and the $\overline{RAS}$ signal causes each cell in each row to be refreshed, as previously discussed. Meanwhile, the $\overline{CAS}$ signal is HIGH, disabling the output buffer so that the changing data do not appear on the outputs as the rows are refreshed. When the refresh counter reaches its terminal count, the refresh operation terminates, and the memory is available for normal read/write cycling. A basic timing diagram illustrating the refresh operation is shown in Figure 10–36.

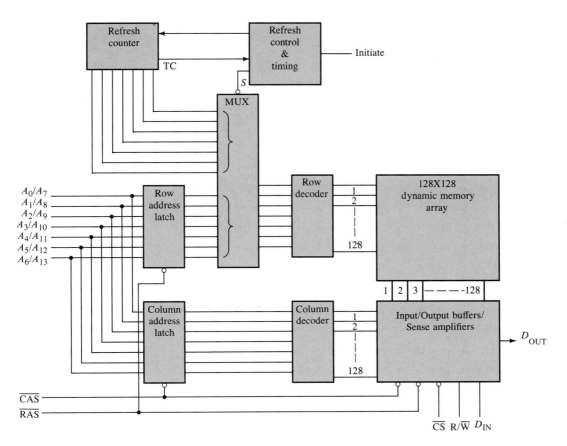

FIGURE 10–35
A typical dynamic RAM with basic refresh circuitry.

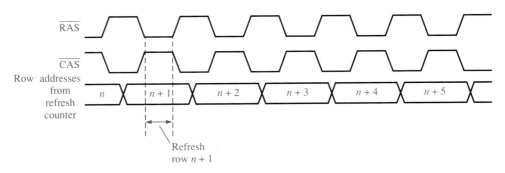

FIGURE 10–36
Basic refresh timing diagram for a portion of the complete memory refresh operation.

A Specific Dynamic RAM

The TMS4164 is a 65,536-bit (64K) DRAM. It is organized as a 64K × 1 memory. The logic symbol is shown in Figure 10–37. There are eight address lines (A_0 through A_7), a row address strobe ($\overline{RAS}$), a column address strobe ($\overline{CAS}$), and the read/write input ($\overline{W}$). The data input is D, and the data output is Q. This device uses a single $+5$-V supply.

FIGURE 10–37
Logic symbol for a TMS4164 64K × 1 dynamic RAM.

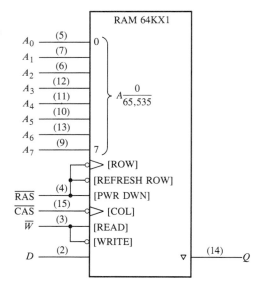

When $\overline{RAS}$ goes LOW, the eight row address bits are latched into the RAM. Next the column bits are latched in when $\overline{CAS}$ goes LOW. A read cycle is initiated when $\overline{W}$ is HIGH, and a write cycle is initiated when $\overline{W}$ is LOW.

Larger-capacity dynamic RAMs with 256K bits, such as the TMS41256, are also commonly used. Also, 1M-bit dynamic RAMs are available.

**SECTION 10–4
REVIEW**

Answers are found at the end of the chapter.
1. Explain the basic difference between static RAMs and dynamic RAMs.
2. What is the reason for the refresh operation in dynamic RAMs?

MEMORY EXPANSION

10–5

*Available memory can be expanded to increase the **word length** (number of bits in each address) or the **word capacity** (number of addresses) or both. Memory expansion is accomplished by adding an appropriate number of memory chips to the address, data, and control buses as explained in this section.*

Word-Length Expansion

To increase the word length of a memory, the number of bits in the data bus must be increased. For example, an eight-bit word length can be achieved by using two memories, each with four-bit words as illustrated in Figure 10–38(a). As you can see in part (b), the eight-bit address bus is commonly connected to both memories so that the combination memory still has the same number of addresses ($2^8 = 256$) as each individual memory. The four-bit data buses from the two memories are combined to form an eight-bit data bus. Now when an address is selected, eight bits are produced on the data bus—four from each memory.

The following example shows the details of 256×4-to-256×8 expansion.

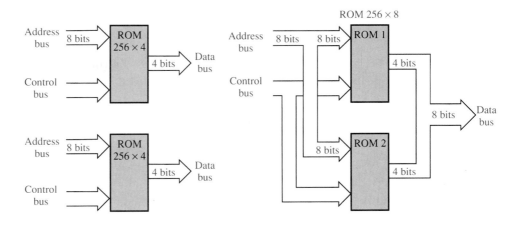

(a) Two separate 256 × 4 ROMs   (b) A 256 × 8 ROM from two 256 × 4 ROMs

FIGURE 10–38
Expansion of two 256 × 4 ROMs into a 256 × 8 ROM to illustrate word-length expansion.

EXAMPLE 10–2   Expand the 256 × 4 ROM in Figure 10–39 to form a 256 × 8 ROM.

FIGURE 10–39
A 256 × 4 ROM.

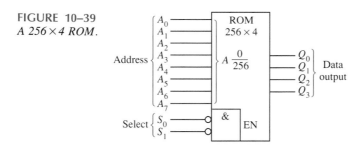

FIGURE 10–40

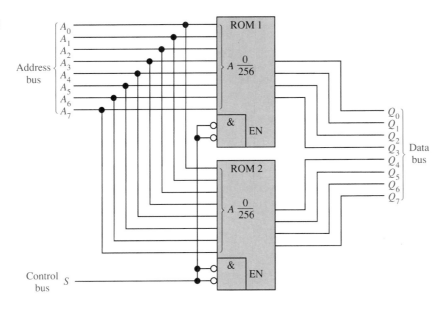

Solution Two 256×4 ROMs are connected as shown in Figure 10–40. Notice that a specific address is accessed in ROM 1 and ROM 2 at the same time. The four bits from a selected address in ROM 1 and the four bits from the corresponding address in ROM 2 go out in parallel to form an eight-bit word on the data bus. Also notice that a LOW on the select line, S_0, which forms a simple control bus, enables *both* memories.

EXAMPLE 10–3   Use the memories in Example 10–2 to form a 256×16 ROM.

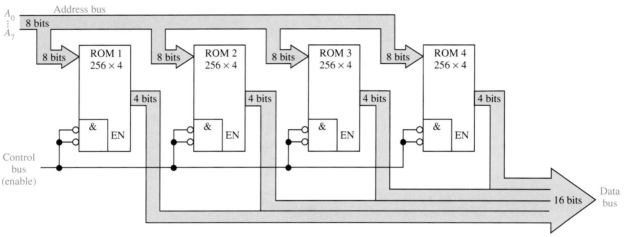

FIGURE 10–41

Solution In this case we need a memory that stores 256 sixteen-bit words. Four 256×4 ROMs are required to do the job, as shown in Figure 10–41.

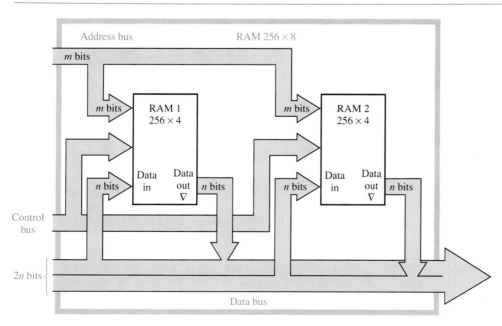

FIGURE 10–42
Illustration of word-length expansion with RAMs.

A ROM has only data outputs, but a RAM has both data inputs and data outputs. For word-length expansion in a RAM, the data inputs *and* data outputs form the data bus. Because data input lines and the corresponding data output lines must be connected together, tristate buffers are required. Most available IC RAMs provide internal tristate circuitry. Figure 10–42 illustrates RAM expansion to increase word length.

EXAMPLE 10–4

Use 74189 16×4 SRAMs to create a 16×8 SRAM. See Figure 10–30(a).

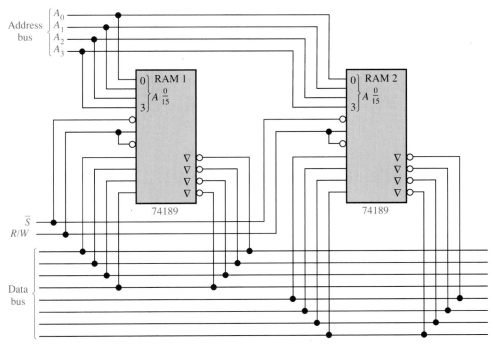

FIGURE 10–43

Solution Two 74189s are connected as shown in Figure 10–43. When the $R/\overline{W}$ input is LOW, the tristate data outputs are disabled for the write operation.

Word-Capacity Expansion

When memories are expanded to increase the word capacity, the number of addresses is increased. To achieve this increase, the number of address bits must be increased, as illustrated in Figure 10–44, where two 256×4 ROMs are expanded to form a 512×4 memory.

Each individual memory has eight address bits to select its 256 addresses, as shown in part (a). The expanded memory has 512 addresses and therefore requires *nine* address bits, as shown in part (b). The ninth address bit is used to enable the appropriate memory chip. The data bus for the expanded memory remains four bits wide. Details of this expansion are illustrated in Example 10–5.

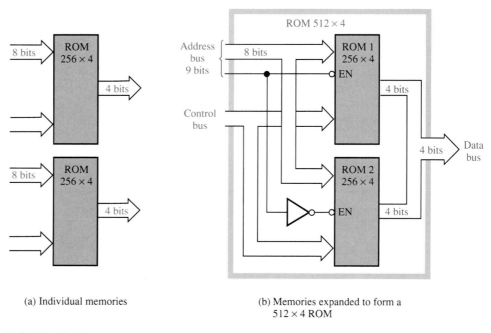

(a) Individual memories

(b) Memories expanded to form a
512 × 4 ROM

FIGURE 10–44
Illustration of word-capacity expansion.

EXAMPLE 10–5

Use 256 × 4 ROMs like the one in Figure 10–39 to implement a 512 × 4 memory.

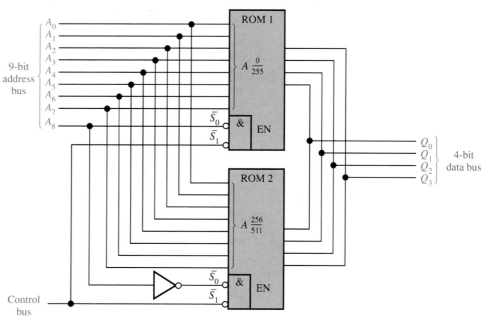

FIGURE 10–45

Solution The expanded addressing is achieved by connecting the chip-select ($\overline{S}_0$) input to the ninth address bit (A_8), as shown in Figure 10–45. Input $\overline{S}_1$ is used as an Enable input common to both memories. When the ninth address bit (A_8) is LOW, ROM 1 is selected (ROM 2 is disabled), and the eight lower-order address bits (A_0–A_7) access each of the addresses in ROM 1 (0 through 255). When the ninth address bit (A_8) is HIGH, ROM 2 is enabled by a LOW on the inverter output (ROM 1 is disabled), and the eight lower-order address bits (A_0–A_7) access each of the ROM 2 addresses (256 through 511).

SECTION 10–5 REVIEW

Answers are found at the end of the chapter.

1. How many $16K \times 1$ RAMs are required to achieve a memory with a word capacity of 16K and a word length of eight bits?
2. To expand the $16K \times 8$ memory in question 1 to a $32K \times 8$ organization, how many more $16K \times 1$ RAMs are required?

MAGNETIC MEMORIES

10–6

In this section, we will cover the fundamentals of disks, tapes, and bubble memories. These storage media are very important, particularly in computer applications, where they are used for mass storage and backup.

The Floppy Disk

The floppy disk (diskette) is a small, flexible Mylar® disk with a magnetic surface. It is permanently housed in a flexible square jacket for protective purposes, as shown in Figure 10–46(a). The 5.25-in. version is shown, but there is also a 3.5-in. disk that is widely used. The 3.5-in. disk differs from the 5.25-in. in that it is housed in a *rigid* plastic case.

The surface of the disk is coated with a thin magnetic film in which binary data are stored in the form of minute magnetized regions. There are cutout areas in the jacket for the drive spindle, the read/write head, and the index position sensor. The index hole establishes a reference point for all the tracks on the disk. As the disk rotates within the stationary jacket, the read/write head makes contact through the access slot.

Floppy disks are organized into *tracks* and *sectors,* which determine how much data can be stored. Figure 10–46(b) illustrates a typical track and sector arrangement. Standard 5.25-in. disks have 40 tracks (IBM) on each side. Double-sided 3.5-in. disks have 77 tracks per side. Some variations exist, such as the standard Apple disk with 35 tracks on one side and the AT 5.25-in. quad-density disk with 80 tracks per side.

The number of sectors per track also varies. The IBM standard is nine sectors per track. Apple disks have sixteen sectors per track.

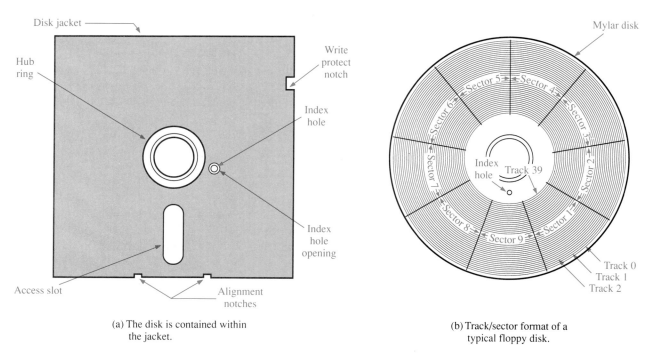

(a) The disk is contained within the jacket.

(b) Track/sector format of a typical floppy disk.

FIGURE 10–46

Floppy disk construction and format.

Each sector in a standard IBM disk can store 512 bytes of data. (Typically, 1 byte represents one character or numeral.) The total storage capacity of a double-sided disk is therefore

$$(512 \text{ bytes/sector})(9 \text{ sectors/track})(80 \text{ tracks}) = 368,640 \text{ bytes}$$

A typical sector format is shown in Figure 10–47, where each sector is divided into fields. The address mark passes the read/write head first and identifies the upcoming areas of the sector as the ID field. The ID field identifies the data field by sector and track number. The data mark indicates whether the upcoming data field contains a good record or a deleted record. The data field is the portion of the sector that contains the 512 data bytes. The disk is a random access device, and the average access time to a given sector is about 20 to 40 ms. This is much faster than for magnetic tape but much slower than the semiconductor memories.

Both the ID field and the data field typically contain two bytes for *cyclic redundancy check (CRC)*. These bytes provide for error detection. The CRC is computed by the disk controller with the recorded data, using a special algorithm. Then it is compared to the CRC recorded with the data. If they are not the same, there is an error.

FIGURE 10–47

A typical sector format for one track of a floppy disk.

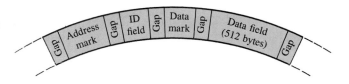

The Hard Disk

Hard disks are contained in a sealed drive package such as the one shown in Figure 10–48. The read/write head, unlike that of a floppy, floats above the disk surface on a layer of air. Because of the extremely small spacing between the disk and the head (about 10^{-6} in.), a very clean environment is essential to reliable operation. This requirement is the reason for using a sealed enclosure.

The storage capacity of a typical hard disk is about 20 megabytes, although 30 and 40 megabyte disks are also used, compared with a little over 360 kilobytes for a floppy disk. Also, in part because of a higher rotational speed, the data transfer rates are much faster than for a typical floppy disk drive.

Because of the type of arm mechanism that carries the read/write head, hard disk drives are often referred to as *Winchester drives*. A newer mechanism, called the Whitney arm, promises improved performance and may compete strongly with the Winchester drive.

The standard size for a hard disk is 5.25 inches. A 3.5-inch version, however, is becoming widely used. These disks are formatted into tracks and sections just as floppy disks are. The typical hard disk has over 300 tracks, and the number of sectors varies depending on the system.

FIGURE 10–48
A typical hard disk drive unit.

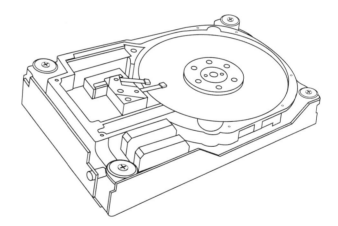

Magnetic Tape

Compared with disks, magnetic tape is a very slow storage medium in terms of access time because data are accessed *serially* rather than by random selection. That is, to get to a given block of data, all data preceding it must be read through.

At one time, before the advent of floppy disks, the audiocassette tape was widely used for mass storage in microcomputer systems. Today audiocassettes have largely been discarded as a means for mass data storage.

The *streaming tape* cartridge is the most widely used form of magnetic tape in computer applications and is used basically for backing up important data stored on disk. The standard streaming tape is $\frac{1}{4}$ inch wide and comes in lengths of 300 to 600 feet. Storage capacities range from about 10 Mbytes to 40 Mbytes and higher.

A typical tape format is shown in Figure 10–49. The standard nine-track cartridge stores backup data by recording front-to-back and then beginning at the back of the next track and recording to the front, as indicated in part (a). Typically, a tape is organized into records or blocks as indicated in part (b) of the figure. The marker indicates the beginning of a record, the ID specifically identifies the record, and the stored data then appear, followed by a checksum.

The *checksum* is the arithmetic sum of all the bytes in a record and is used by the computer to check for errors in the data as they are being taken from the tape. This check is made basically by adding the data bytes as they are read from the tape and comparing this figure to the checksum.

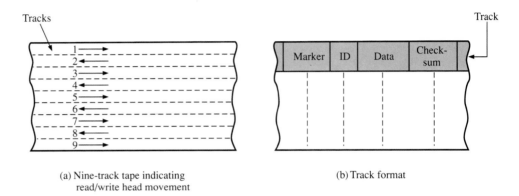

(a) Nine-track tape indicating read/write head movement

(b) Track format

FIGURE 10–49
Standard nine-track streaming tape.

Read/Write Head Principles

A simplified diagram of the magnetic surface read/write operation is shown in Figure 10–50. A data bit (1 or 0) is written on the magnetic surface by the magnetization of a small segment of the surface as it moves by the *write head*. The direction of the magnetic flux lines is controlled by the direction of the current pulse in the winding, as shown in Figure 10–50(a). At the air gap in the write head, the magnetic flux takes a path through the surface of the storage device. This magnetizes a small spot on the surface in the direction of the field. A magnetized spot of one polarity represents a binary 1, and one of the opposite polarity represents a binary 0. Once a spot on the surface is magnetized, it remains until written over with an opposite magnetic field.

When the magnetic surface passes a *read head,* the magnetized spots produce magnetic fields in the read head, which induce voltage pulses in the winding. The polarity of these pulses depends on the direction of the magnetized spot and indicates whether the stored bit is a 1 or a 0. This process is illustrated in Figure 10–50(b). Very often the read and write heads are combined into a single unit as shown in part (c).

Magnetic Bubble Memories (MBMs)

In an MBM, data bits are stored in the form of magnetic ''bubbles'' moving in thin films of magnetic material. The bubbles are actually cylindrical magnetic domains whose polarization is opposite that of the thin magnetic film in which they are embedded.

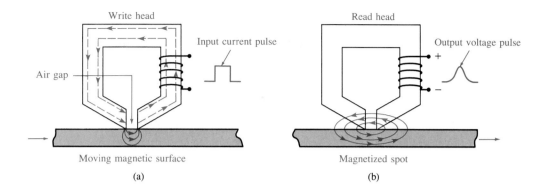

(a)

(b)

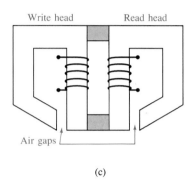

(c)

FIGURE 10–50

Read/write function on a magnetic surface.

The magnetic film   When a thin film of magnetic garnet is viewed by polarized light through a microscope, a pattern of wavy strips of magnetic domains can be seen. In one set of strips, the tiny internal magnets point up, and in the other areas, they point down. As a result, one set of strips appear bright and the other dark when exposed to the polarized light. This is illustrated graphically in Figure 10–51(a).

Now, if an *external magnetic field* is applied *perpendicular* to the film and slowly increased in strength, the wavy domain strips whose magnetization is opposite that of the external field begin to narrow. This effect is illustrated in Figure 10–51(b). At a certain magnitude of external field strength, all these domains suddenly contract into small circular areas called ''bubbles,'' as shown in part (c). These bubbles typically are only a few micrometers in diameter and act as tiny magnets floating in the external field. The bubbles can be easily moved and controlled within the film by rotating magnetic fields in the plane of the film or by current-carrying conductive elements.

Moving the bubbles   The bubbles can be made to move laterally in the film by the application of a magnetic field *parallel* to the film. To control the direction of movement, magnetic ''paths'' are created by deposits of magnetically conductive material in a specific pattern on the surface of the thin film.

For example, Figure 10–52 shows an asymmetrical chevron pattern. Various other patterns are also feasible. Imagine a magnetic field rotating counterclockwise, parallel with the thin garnet film. At successive points in its rotation, the magnetic field is

FIGURE 10–51

FIGURE 10–51
Creation of magnetic bubbles in a thin magnetic film by application of an external magnetic field.

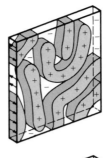

(a) No external magnetic field—wide magnetic domains.

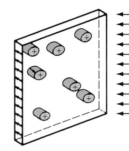

(b) Small magnetic field—positive magnetic domains shrink.

(c) Large magnetic field—bubbles form.

FIGURE 10–52
Chevron pattern in a rotating magnetic field.

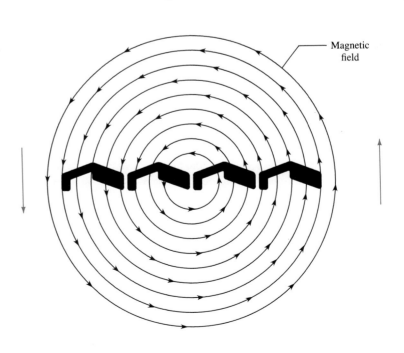

Magnetic field

pointing right, down, left, and up, as illustrated. As this happens, the chevrons are polarized in the direction of rotation of the magnetic field.

When a bubble is introduced at the right end of the chevron pattern, it moves step by step to the left as the field rotates. In one complete revolution of the field, a bubble moves from a position between two chevrons to a corresponding position between the next two chevrons, as illustrated in Figure 10–53. So, as the magnetic field continues to rotate, the bubbles continue to move along the chevron paths.

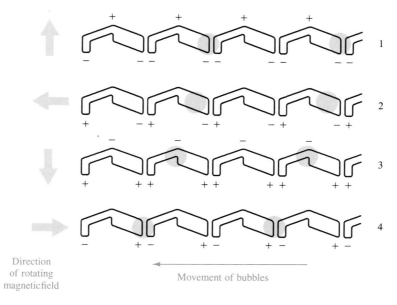

Direction
of rotating
magnetic field

Movement of bubbles

FIGURE 10–53
Propagation of magnetic bubbles along a chevron pattern.

Physical structure of an MBM   A typical MBM assembly is shown in Figure 10–54. The main components are the *thin film memory chip,* the *drive coils (orthogonal coils),* the *permanent magnets,* and the *control electronics.* These components are assembled into a case, which serves as a shield to protect the device from disruptive external magnetic fields.

Major/minor loop architecture   Loops in an MBM are in the form of continuous, elongated paths of chevron patterns. The major/minor loop arrangement consists basically of one *major loop* and many *minor loops,* as shown in Figure 10–55.

In an MBM, 1s and 0s are represented by the presence or absence of bubbles. The minor loops are essentially the memory cell arrays that store the data bits. The major loop is primarily a path to get data from the minor loops where they are stored to the output during read, and to get data from the input to the minor loops during write.

Five control functions are used in the read/write cycles of an MBM. These are *generation, transfer, replication, annihilation,* and *detection.*

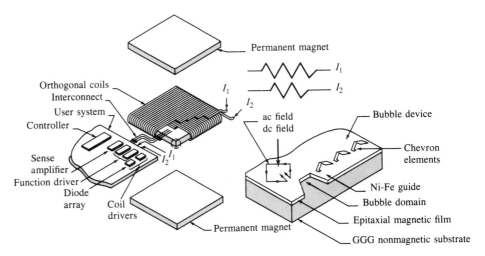

FIGURE 10–54
Construction of magnetic bubble memory. (Courtesy of Texas Instruments, Inc.)

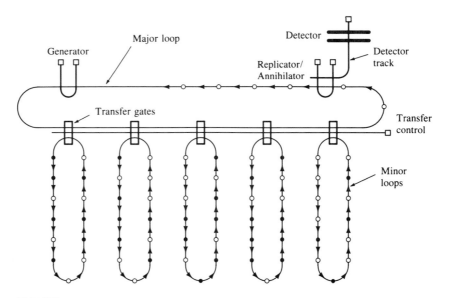

FIGURE 10–55
Major/minor loop architecture (greatly simplified).

The read cycle Data are read from the MBM in blocks called *pages*. Basically, a page consists of a number of bits equal to the number of minor loops. For example, a typical MBM may have 150 minor loops, each of which is capable of storing 600 bits. In this case, the page size is 150 bits.

Each bit in a given page occupies the same relative location in each minor loop. During read, all of the bits in a page of data are shifted to the transfer gates and onto the major loop at the same time. Then they are serially shifted around the major loop to the *replicator/annihilator* where each bubble is "stretched" by the replicator until it "splits" into two bubbles. One of the replicated bubbles (or no bubble, as the case may be) is transferred to the *detector,* where the presence or absence of the bubble is sensed and translated to the appropriate logic level to represent a 1 or a 0. The other replicated bubble continues along the major loop and is transferred back onto the appropriate minor loop for storage. This process continues until each bit in the page has been read. The replication/detection process results in a nondestructive readout.

The write cycle   Before a new page of data can be written into an address, the data currently stored at that address must be annihilated. Annihilation is done by a *destructive* readout, in which the bubbles are not replicated. Now the new page of data is produced by the *generator* one bit at a time. Each bit is injected onto the major loop until the entire page of data has been entered. The page is then moved serially into position and transferred onto the minor loops for storage.

Applications   An MBM is capable of storing large amounts of data. An MBM is also nonvolatile, so data are not lost if the power goes off. Because of these features, MBMs are competitive with semiconductor memories in many applications. The main disadvantage is that it takes much longer to get data into and out of an MBM than a semiconductor memory because MBMs are serially accessed rather than randomly accessed.

SECTION 10–6 REVIEW

Answers are found at the end of the chapter.
1. How are data bits stored on a magnetic surface?
2. How many tracks are there on each side of a typical $5\frac{1}{4}$-inch floppy disk?
3. What does *CRC* mean?
4. Name the two types of loops in an MBM.
5. How are 1s and 0s represented in an MBM?

MISCELLANEOUS MEMORIES

10–7

First In–First Out (FIFO) Memories

This type of memory is formed by an arrangement of shift registers. The term *FIFO* refers to the basic operation of this type of memory, in which the first data bit written into the memory is the first to be read out.

There is one important difference between a conventional shift register and a FIFO register: in a conventional register a data bit moves through the register only as new data bits are entered; in a FIFO register a data bit immediately goes through the register to the right-most bit location that is empty. This difference is illustrated in Figure 10–56.

Figure 10–57 is a block diagram of a FIFO serial memory. This particular memory has four serial 64-bit data registers and a 64-bit control register (marker register). When data are entered by a shift-in pulse, they move automatically under control of the marker register to the empty location closest to the output. Data cannot advance into occupied positions. However, when a data bit is shifted out by a shift-out pulse, the data bits remaining in the registers automatically move to the next position toward the output. In an *asynchronous* FIFO, data are shifted out independent of data entry, with the use of two separate clocks.

FIGURE 10–56
Comparison of conventional and FIFO register operation.

Conventional shift register

Input	X	X	X	X	Output
0	0	X	X	X	⟶
1	1	0	X	X	⟶
1	1	1	0	X	⟶
0	0	1	1	0	⟶

X = unknown data bits.

In a conventional shift register, data stay to the left until "forced" through by additional data.

FIFO shift register

Input	—	—	—	—	Output
0	—	—	—	0	⟶
1	—	—	1	0	⟶
1	—	1	1	0	⟶
0	0	1	1	0	⟶

— = empty positions.

In a FIFO shift register, data "fall" through (go right).

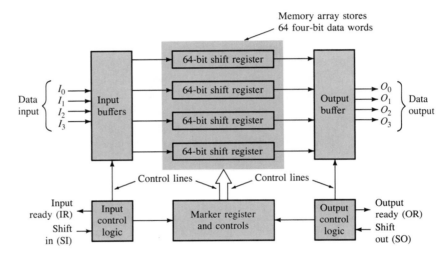

FIGURE 10–57
Block diagram of a typical FIFO serial memory.

FIFO Applications

One important application area for the FIFO register is the case in which two systems of differing data rates must communicate. Data can be entered into a FIFO register at one rate and taken out at another rate. Figure 10–58 illustrates how a FIFO register might be used in these situations.

Last In–First Out (LIFO) Memories

The last in–first out memory is found in applications involving microprocessors and other computing systems. It allows data to be stored and then recalled in reverse order; that is, the last data byte to be stored is the first data byte to be retrieved.

Stacks   A LIFO memory is commonly referred to as a *push-down stack*. In some systems, it is implemented with a group of registers as shown in Figure 10–59. A stack can consist of any number of registers, but the register at the top is called the *top of the stack*.

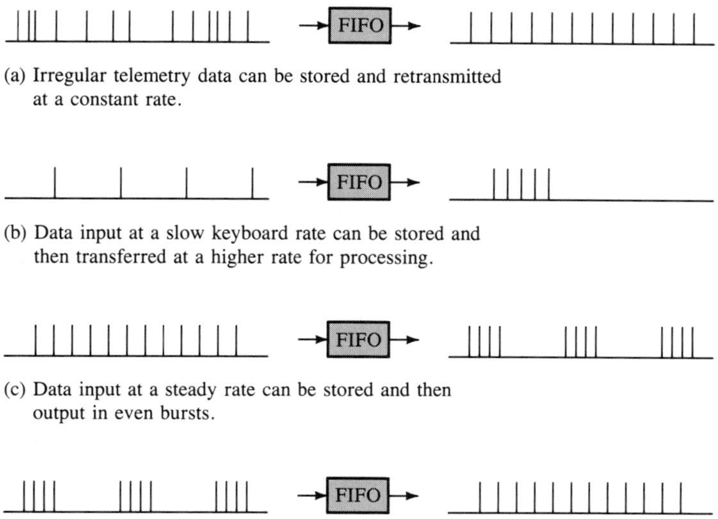

(a) Irregular telemetry data can be stored and retransmitted
    at a constant rate.

(b) Data input at a slow keyboard rate can be stored and
    then transferred at a higher rate for processing.

(c) Data input at a steady rate can be stored and then
    output in even bursts.

(d) Data in bursts can be stored and reformatted into a
    steady-rate output.

FIGURE 10–58
The FIFO register in data-rate buffering applications.

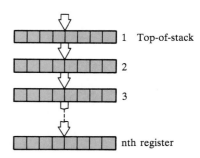

FIGURE 10–59
Register stack.

To illustrate the principle, a byte of data is loaded in parallel onto the top of the stack. Each successive byte "pushes" the previous one down into the next register. This process is illustrated in Figure 10–60. Notice that the new data byte is always loaded into the top register and the previously stored bytes are pushed deeper into the stack. The name *push-down* stack comes from this characteristic.

Data bytes are retrieved in the reverse order. The last byte entered is always at the top of the stack, so when it is "pulled" from the stack, the other bytes "pop" up into the next higher locations. This process is illustrated in Figure 10–61.

RAM stack   Another approach to LIFO used in some microprocessor-based systems is the allocation of a section of RAM as the stack rather than the use of a dedicated set of registers.

FIGURE 10–60
Pushing data onto the stack.

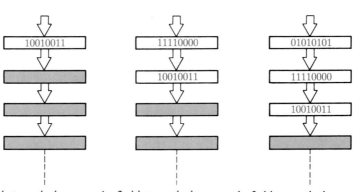

1st byte pushed onto stack.   2nd byte pushed onto stack.   3rd byte pushed onto stack.

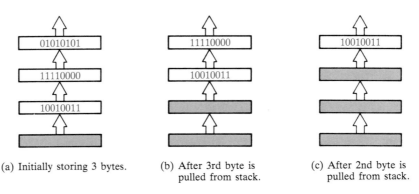

(a) Initially storing 3 bytes.

(b) After 3rd byte is pulled from stack.

(c) After 2nd byte is pulled from stack.

FIGURE 10–61
Pulling data from the stack.

Consider a random access memory that is *byte* oriented—that is, one in which each address contains eight bits—as illustrated in Figure 10–62.

Next, consider a section of RAM set aside for use as a stack. A special separate register called a *stack pointer* contains the address of the top of the stack, as illustrated in Figure 10–63. A four-digit hexadecimal representation is used for the binary addresses (we are assuming a sixteen-bit address code). In the figure the addresses are chosen arbitrarily for purposes of illustration.

Now let us see how data are pushed onto the stack. A data byte is stored by a normal memory write operation at address 00FF, which is the top of the stack. The stack pointer is then decremented (decreased by 1) to 00FE. This moves the top of the stack to the next lower memory address, as shown in Figure 10–63(b).

Notice that the top of the stack is not stationary as in the fixed register stack but moves *downward* (to lower addresses) in the RAM as data bytes are stored. Figure 10–63(c) and (d) show two more bytes being pushed onto the stack. After the third byte is stored, the top of the stack is at 00FC.

To retrieve data from the stack, the stack pointer is first incremented (increased by 1) and then a RAM read operation is performed. The first data byte to be pulled from the stack is at address 00FD. When this byte is read, the stack pointer is again

FIGURE 10–62
Representation of a byte-organized memory.

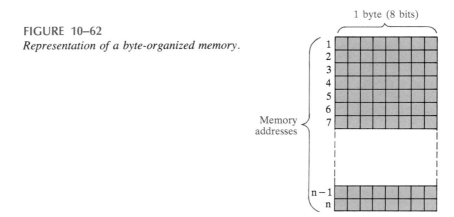

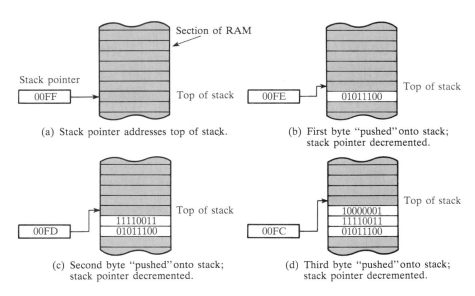

FIGURE 10–63
Illustration of the pushing of data onto a RAM stack.

incremented to 00FE. Another read operation pulls the next byte, and so on, as shown in Figure 10–64. Keep in mind that RAMs are nondestructive when read, so the data byte still remains in the memory after a read operation. A data byte is destroyed only when a new byte is written over it.

A RAM stack can be of any depth, depending on the number of continuous memory addresses not needed for any other use.

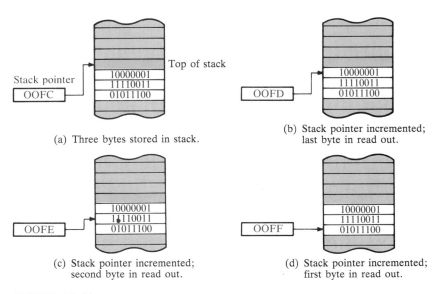

FIGURE 10–64
Illustration of the pulling of data out of the stack.

CCD Memories

The *CCD* (charge-coupled device) *memory* stores data as charges on capacitors. Unlike the dynamic RAM, however, the storage cell does not include a transistor. High density is the main advantage of CCDs.

The CCD memory consists of long rows of semiconductor capacitors, called *channels*. Data are entered into a channel serially by depositing a small charge for a 0 and a large charge for a 1 on the capacitors. These charge *packets* are then shifted along the channel by clock signals as more data are entered.

As with the dynamic RAM, the charges must be refreshed periodically. This process is done by shifting the charge packets serially through a refresh circuit. Figure 10–65 shows the basic concept of a CCD channel. Because data are shifted serially through the channels, the CCD memory has a relatively long access time.

FIGURE 10–65
A CCD (charge-coupled device) channel.

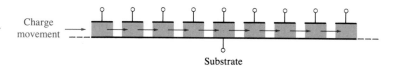

Charge
movement

Substrate

SECTION 10–7
REVIEW

Answers are found at the end of the chapter.
1. What is a FIFO memory?
2. What is a LIFO memory?
3. What does the term *CCD* stand for?

PROGRAMMABLE LOGIC DEVICES (PLDs)

10–8

Programmable logic devices (PLDs) *are similar to PROMs because they are fuse-programmable. However, they are quite different from PROMs in their applications. A PLD is used to implement Boolean logic functions and can replace several individual gate or flip-flop ICs in many situations. In practical terms, PLDs can save space and assembly costs on printed circuit boards, and they can also cut inventory costs because fewer parts are required to implement a given function.*

There are three basic categories of PLDs: the **PLA** *(programmable logic array), the* **PAL** *(programmable array logic), and the* **PLS** *(programmable logic sequencer). Normally these designations are preceded by an* **F,** *to indicate that they are fuse-programmable.*

The FPLA

The basic FPLA consists of a programmable *AND plane* and a programmable *OR plane*. The AND plane can be programmed to produce product terms of the input variables. The OR plane can be programmed to combine the product terms to form sum-of-products functions. A basic FPLA block diagram is shown in Figure 10–66.

FIGURE 10–66

A basic FPLA block diagram.

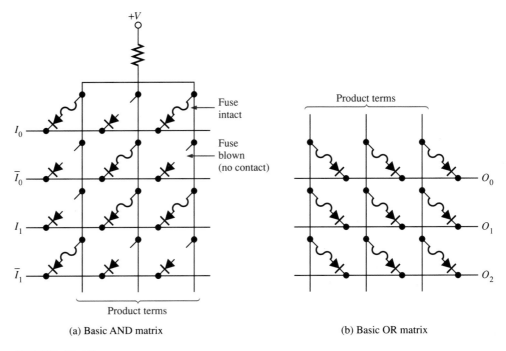

The AND plane and the OR plane consist of a matrix of conductors that are connected at each cross point by a fusible contact, as illustrated in Figure 10–67. Programming is accomplished by blowing the fuses to eliminate the connections at specified cross points and leaving the connections at other cross points. It is done with a PROM programmer.

A small FPLA is shown in Figure 10–68 to illustrate the principle. Actual FPLAs have much larger matrices. This particular device has three inputs, eight product terms, and four outputs. Notice that there are actually six rows in the AND plane, for the three input variables and their complements.

Starting out, all cross points are connections. To program the device, all cross point fuses are blown except those that are necessary to implement the required logic functions, and these are left intact.

(a) Basic AND matrix

(b) Basic OR matrix

FIGURE 10–67

Basic concept of an FPLA matrix.

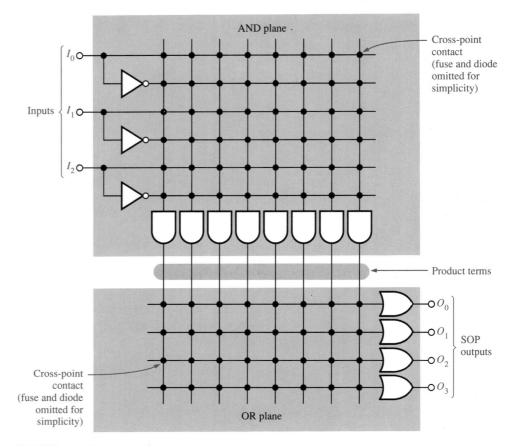

FIGURE 10–68
Simplified example of an FPLA.

Suppose we want to program this device to produce two output logic functions as follows:

$$O_0 = I_2 I_1 I_0 + \bar{I}_2 I_1 \bar{I}_0 + I_2 \bar{I}_1 I_0$$
$$O_1 = \bar{I}_2 \bar{I}_1 \bar{I}_0 + \bar{I}_2 I_1 \bar{I}_0$$

In Figure 10–69 the colored dots represent the cross point contacts that remain intact; the rest are blown.

The AND gates are such that *all* the input rows connected to the column or product line for a given gate must be HIGH for the AND gate output to be HIGH. The OR gates are such that when one or more columns that are connected to the row line for a given gate are HIGH, the OR gate output is HIGH.

Additional features   Most IC FPLAs provide XOR gates at each output so that the output functions can be inverted if desired. As Figure 10–70 shows, the *A* input of an

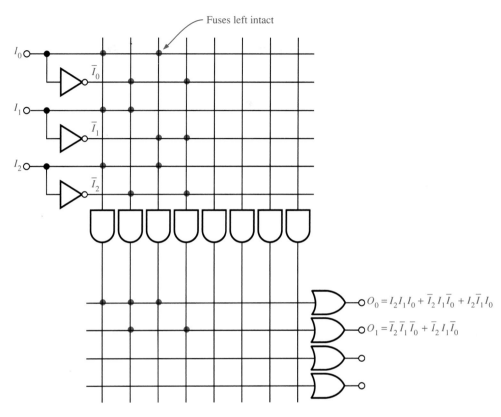

FIGURE 10–69

The FPLA programmed for the specified functions.

FIGURE 10–70

The XOR as a programmable inverter.

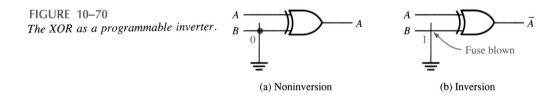

(a) Noninversion   (b) Inversion

XOR is uncomplemented (noninverted) when the *B* input is a 0, and it is complemented (inverted) when the *B* input is a 1. The FPLA outputs are programmed for noninversion or inversion by retaining or blowing the fusible contact at the cross point as indicated.

Many FPLAs also provide tristate outputs for connection to a bus. Example 10–6 further illustrates FPLA programming.

EXAMPLE 10–6

Show the cross point connections that are to be left intact to make the FPLA in Figure 10–71 a BCD-to-seven-segment decoder. The segment outputs are to be active-LOW.

FIGURE 10–71

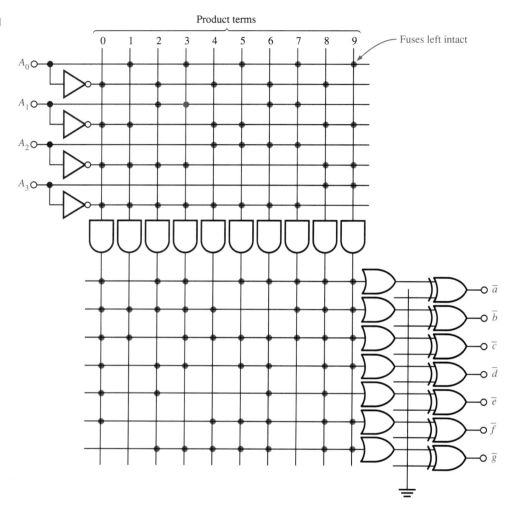

Solution   Refer back to Chapter 6 for the logic equation for each of the seven segments. These functions are summarized in the following equations by expressing each segment output in terms of the digits in which it is used in a seven segment display.

$$a = 0 + 2 + 3 + 5 + 6 + 7 + 8 + 9$$
$$b = 0 + 1 + 2 + 3 + 4 + 7 + 8 + 9$$
$$c = 0 + 1 + 3 + 4 + 5 + 6 + 7 + 8 + 9$$
$$d = 0 + 2 + 3 + 5 + 6 + 8 + 9$$
$$e = 0 + 2 + 6 + 8$$
$$f = 0 + 4 + 5 + 6 + 8 + 9$$
$$g = 2 + 3 + 4 + 5 + 6 + 8 + 9$$

Each column in the AND plane is a product term representing a decimal digit. For example, the 0 column is programmed for $\overline{A_3}\overline{A_2}\overline{A_1}\overline{A_0}$, which is the BCD code for decimal zero. The 0-column AND gate output is then connected to the rows in the OR plane for segments a through f, since a through f are all used in a 0 display. Next, the

1 column is programmed in the AND plane for the product term $\overline{A}_3\overline{A}_2\overline{A}_1A_0$ and connected in the OR plane to the row for segments b and c. This procedure is repeated for each product term and segment. A colored dot indicates that there is a contact at the cross point. The cross points for the XOR output gates are blown, so the functions are inverted and are thus active-LOW.

A specific FPLA The 82S100, shown in Figure 10–72, is a good example of a commercially available FPLA. It has 16 input variables, 48 product terms, and 8 outputs, and it contains approximately 2000 cross-point fuses.

FIGURE 10–72
A typical FPLA.

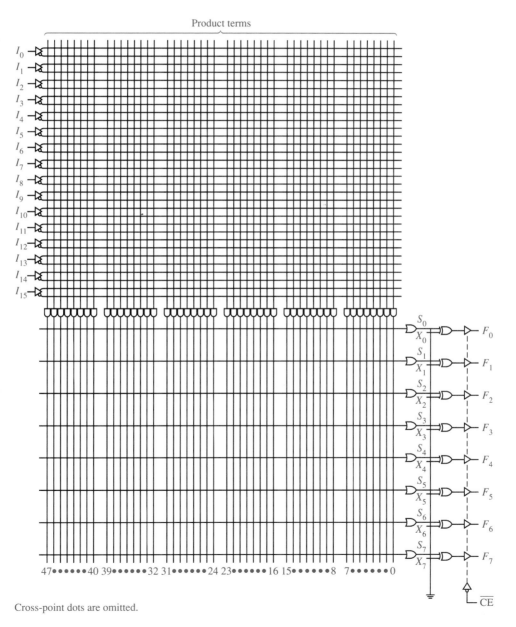

Cross-point dots are omitted.

The FPAL

The FPAL (fuse-programmable array logic) is another PLD. It differs from the FPLA in that it has a programmable AND plane with a fixed OR array. It does not have a programmable OR plane.

A simplified FPAL is shown in Figure 10–73 as an example. This device is programmed to implement the logic function

$$O_0 = \bar{I}_2\bar{I}_1\bar{I}_0 + \bar{I}_2I_1 + I_1\bar{I}_0$$

Notice that in this sample FPAL, each AND-OR output is limited to four product terms, since there are four AND gates for each OR gate. If we need a function that has more than four product terms, the output from one AND-OR array can be connected to the AND input of another AND-OR array, as Example 10–7 illustrates.

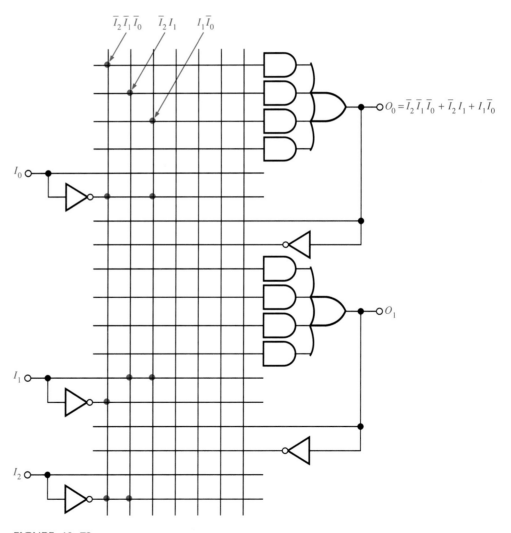

FIGURE 10–73
Simplified example of an FPAL.

EXAMPLE 10–7   Use an FPAL like the one in Figure 10–73 to implement the following function:

$$O_1 = I_2I_1I_0 + I_2I_1\bar{I}_0 + I_2\bar{I}_1I_0 + \bar{I}_2I_1I_0 + \bar{I}_2\bar{I}_1\bar{I}_0 + \bar{I}_2\bar{I}_1I_0$$

FIGURE 10–74

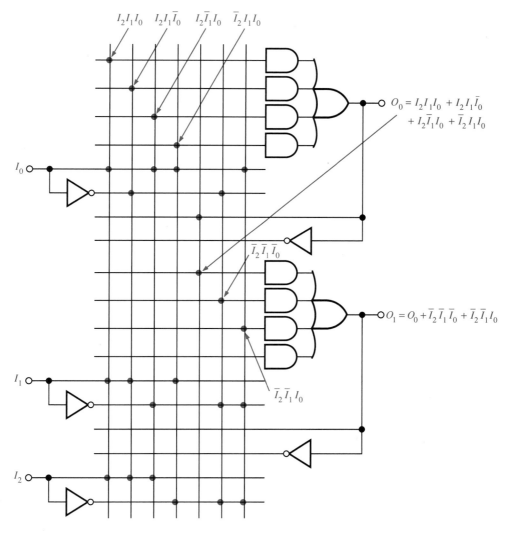

Solution   The function is programmed into the FPAL in Figure 10–74.

The FPLS

The FPLS (fuse-programmable logic sequencer) is essentially the same as the FPAL except that flip-flops are incorporated in the device. An example is the PAL16R4 shown in Figure 10–75.

This device is actually a sixteen-input (variables plus complements) FPAL with four D flip-flops and with tristate outputs. With this device, sequential logic such as counters and registers as well as combinational logic can be implemented.

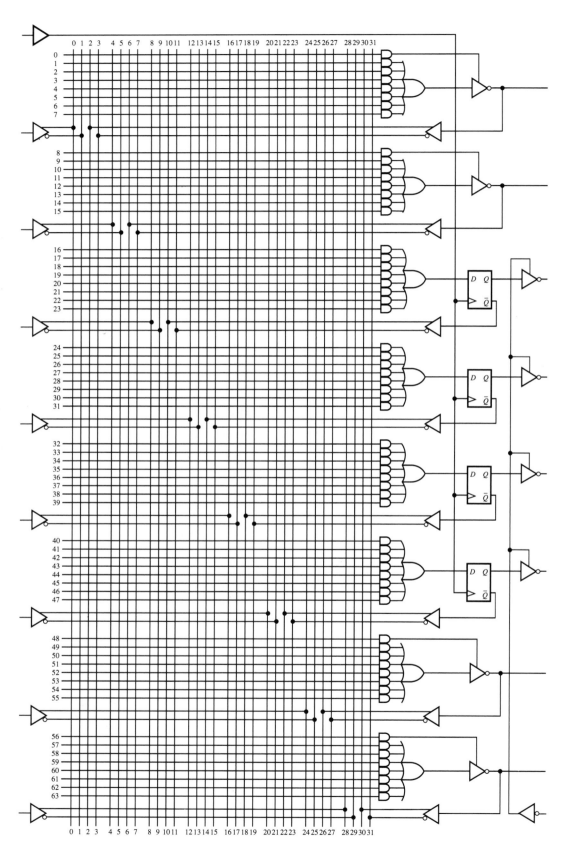

FIGURE 10–75
The PAL16R4 programmable logic sequencer.

SECTION 10–8
REVIEW

Answers are found at the end of the chapter.
1. List three types of PLDs, and explain the differences.
2. What can you do with a PLD?

TESTING AND TROUBLESHOOTING

10–9

Because memories can contain large numbers of storage cells, testing each cell can be a very lengthy and frustrating process. Fortunately, memory testing is usually an automated process performed with a programmable test instrument or with the aid of software for in-system testing. Most microprocessor-based systems provide automatic memory testing as part of their system software.

ROM Testing

Since ROMs contain known data, they can be checked for the correctness of the stored data by reading each data word from the memory and comparing it with a data word that is known to be correct. One way of doing this is illustrated in Figure 10–76. This process requires a *reference* ROM that contains the same data as the ROM to be tested. A special test instrument is programmed to read each address in both ROMs simultaneously and to compare the contents. A flowchart in Figure 10–77 illustrates the basic sequence.

FIGURE 10–76
Block diagram for a complete-contents check of a ROM.

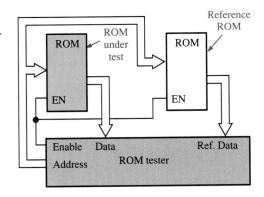

Checksum method Although the previous method checks each ROM address for correct data, it has the disadvantage of requiring a reference ROM for each different ROM to be tested. Also, a failure in the reference ROM can produce a fault indication.

In the *checksum* method a number, the sum of the contents of all the ROM addresses, is stored in a designated ROM address when the ROM is programmed. To test the ROM, the *contents* of all the addresses except the checksum are added, and the result is compared with the checksum stored in the ROM. If there is a difference, there is definitely a fault. If the checksums compare, the ROM is most likely good. However, there is a remote possibility that a combination of bad memory cells could cause the checksums to compare.

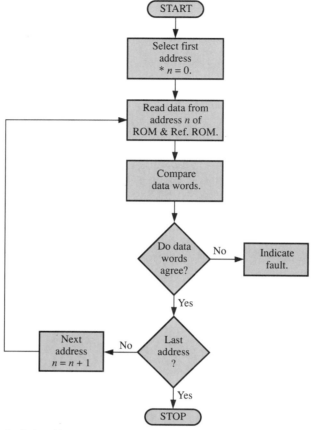

* n is the address number.

FIGURE 10–77
Flowchart for a complete-contents check of a ROM.

This process is illustrated in Figure 10–78 with a simple example. The checksum in this case is produced by taking the sum of each column of data bits and discarding the carries. This is actually an XOR sum of each column. The flowchart in Figure 10–79 illustrates the basic checksum test.

The checksum test can be implemented with a special test instrument, or it can be incorporated as a test routine in the built-in (system) software of microprocessor-based systems. In that case the ROM test routine is automatically run on system start-up.

FIGURE 10–78
Illustration of a programmed ROM with the checksum stored at a designated address.

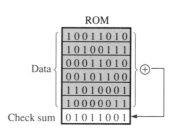

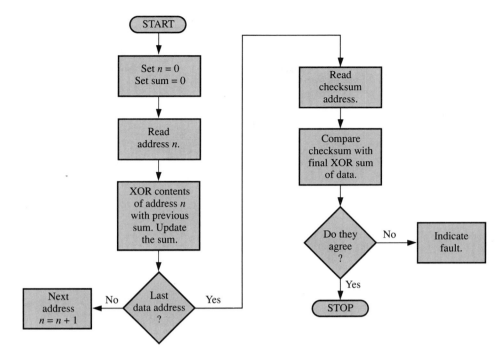

FIGURE 10–79
Flowchart for a basic checksum test.

RAM Testing

To test a RAM for its ability to store both 0s and 1s in each cell, first 0s are written into all the cells in each address and then read out and checked. Next, 1s are written into all the cells in each address and then read out and checked. This basic test will detect a cell that is stuck in either a 1 state or a 0 state.

Some memory faults cannot be detected with the all-0s–all-1s test. For example, if two adjacent memory cells are shorted, they will always be in the same state, both 0s or both 1s. Also, the all-0s–all-1s test is ineffective if there are internal noise problems such that the contents of one or more addresses are altered by a change in the contents of another address.

The checkerboard pattern test One way to more fully test a RAM is by using a checkerboard pattern of 1s and 0s, as illustrated in Figure 10–80. Notice that all adjacent cells have opposite bits. This pattern checks for a short between two adjacent cells, because if there is a short, both cells will be in the same state.

After the RAM is checked with the pattern in Figure 10–80(a), the pattern is reversed, as shown in part (b). This reversal checks the ability of all cells to store both 1s and 0s.

A further test is to alternate the pattern one address at a time and check all the other addresses for the proper pattern. This test will catch a problem in which the contents of an address are dynamically altered when the contents of another address change.

RAM

```
1 0 1 0 1 0 1 0
0 1 0 1 0 1 0 1
1 0 1 0 1 0 1 0
0 1 0 1 0 1 0 1
1 0 1 0 1 0 1 0
0 1 0 1 0 1 0 1

1 0 1 0 1 0 1 0
0 1 0 1 0 1 0 1
1 0 1 0 1 0 1 0
```

(a)

RAM

```
0 1 0 1 0 1 0 1
1 0 1 0 1 0 1 0
0 1 0 1 0 1 0 1
1 0 1 0 1 0 1 0
0 1 0 1 0 1 0 1
1 0 1 0 1 0 1 0

0 1 0 1 0 1 0 1
1 0 1 0 1 0 1 0
0 1 0 1 0 1 0 1
```

(b)

FIGURE 10–80
The RAM checkerboard test pattern.

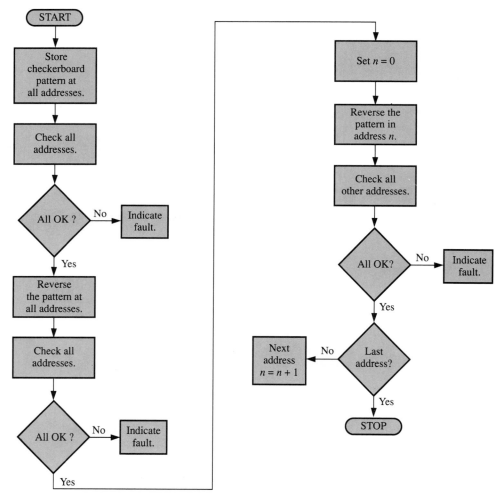

FIGURE 10–81
Flowchart for basic RAM checkerboard test.

A basic procedure for the checkerboard test is illustrated by the flowchart in Figure 10–81. The procedure can be implemented with the system software in microprocessor-based systems so that either the tests are automatic when the system is powered up or they can be initiated from the keyboard.

SECTION 10–9
REVIEW

Answers are found at the end of the chapter.
1. Describe the checksum method of ROM testing.
2. Why can the checksum method not be applied to RAM testing?
3. List the three basic faults that the checkerboard pattern test can detect in a RAM.

SUMMARY

☐ Semiconductor memories:

ROM
RAM
CCD

☐ Magnetic memories:

streaming tape
floppy disk
hard disk
bubble

☐ Random access memories:

ROM
RAM
disk

☐ Serial access memories:

CCD
tape
bubble

☐ Read/write memories:

RAM
CCD
tape
disk
bubble

☐ Special function memories:

FIFO
LIFO
PLD

Integrated Circuit Technologies (Chapter 13) may be covered after this chapter, covered after a later chapter, partially omitted, or completely omitted, depending on the needs and goals of your program. Coverage is not prerequisite or corequisite to any other material in this book. Inclusion is purely optional.

GLOSSARY

Address   The location of a given storage cell or group of cells in a memory.

Bubble memory   A type of memory that uses tiny magnetic bubbles to store 1s and 0s.

Capacity   The total number of bits that a memory can store.

CCD   Charge-coupled device. A type of semiconductor memory that stores data in the form of charge packets.

Cell   A single storage element in a memory.

DRAM   Dynamic random access memory.

Dynamic memory   A memory having cells that tend to lose stored data over a period of time and, therefore, must be refreshed. Typically the storage elements are capacitors.

EAPROM   Electrically alterable programmable read only memory.

EEPROM   Electrically erasable programmable read only memory

EPROM   Erasable programmable read only memory.

FIFO   First in–first out memory.

Floppy disk   A magnetic storage device. Typically a 5.25-in. flexible Mylar® disk enclosed in a flexible jacket. More recently, a 3.5-in. flexible disk enclosed in a rigid case.

FPAL   Fuse-programmable array logic.

FPLA   Fuse-programmable logic array.

FPLS   Fuse-programmable logic sequencer.

LIFO   Last in–first out memory. A memory stack.

Memory array   An array of memory cells.

Nondestructive readout   The process of ''copying'' the contents of a memory location without destroying them.

PLD   Programmable logic device.

RAM   Random access memory. A read/write type of memory.

Read   The process of retrieving data from a memory.

Refresh   To renew the contents of a dynamic memory.

ROM   Read only memory, accessed randomly.

SRAM   Static random access memory.

Static memory   A memory composed of storage elements such as latches or magnetic elements and capable of retaining data indefinitely without refreshing.

Tristate logic   A type of logic circuit having the normal two-state output and, in addition, an open or high-impedance state in which the circuit is disconnected from its load.

UV EPROM   Ultraviolet erasable programmable ROM.

Volatility   The characteristic of a memory whereby it loses stored data if power is removed.

Winchester drive   A type of hard or fixed disk system which is totally sealed to maintain a clean environment. The term *Winchester* applies to the design of the arm mechanism.

Word capacity   The number of words in a memory.

Word length   The number of bits in a word.

Write   The process of storing data in a memory.

SELF-TEST

Answers and solutions are found at the end of the book.

1. What is the bit capacity of a memory that has 512 addresses and can store 8 bits at each address?

2. To retrieve data from a memory, a _____ operation must be performed.

3. (a) List three types of memories in the MOS ROM category.
   (b) List two types of memories in the bipolar ROM category.

4. List two types of memories in **(a)** the MOS RAM category and **(b)** the bipolar RAM category.

5. How many bytes does a 32-bit data word contain?

6. What is the purpose of the program counter in a microprocessor system?

7. What is the purpose of the address buffer in a microprocessor system?

8. How many bits can actually be stored in a 64K memory?

9. What is the total capacity of a ROM with 1024 rows and 4 columns?

10. How many address bits are required for a 512×4 memory?

11. What does the chip-select input of a memory do?

12. Explain the power-down mode that is available on some memory devices.

13. For a certain ROM, the access time from address to data output is 450 ns, and the access time from chip select is 200 ns. The chip select is activated 100 ns before the valid address is applied. How long after chip select do valid output data appear?

14. A certain 256×4 RAM has an access time of 300 ns from address to data output. How long does it take to read data from all memory addresses if the address transition times are neglected?

15. Explain the purpose of the address latches in a dynamic RAM such as in Figure 10–33.

16. What is the function of the refresh counter in Figure 10–35?

17. How many $64K \times 4$ RAMs are needed to form a $64K \times 8$ memory? Is this an example of word-capacity expansion or word-length expansion?

18. What are the storage sections in an MBM called?

19. What is the purpose of checksum in a ROM or magnetic tape?

20. How does a PLD differ from a ROM?

PROBLEMS

Answers to odd-numbered problems are found at the end of the book.

Section 10–1 Memory Concepts

10–1 Identify the ROM and the RAM in Figure 10–82.

10–2 Explain why RAMs and ROMs are both random access memories.

10–3 Explain the purposes of the address bus, the data bus, and the control bus in a typical microprocessor-based system.

10–4 What memory address (0 through 256) is represented by each of the following hexadecimal numbers:
   (a) $0A_{16}$ **(b)** $3F_{16}$ **(c)** CD_{16}

FIGURE 10–82

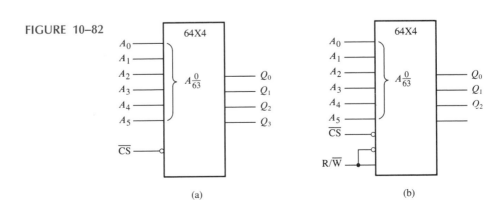

(a)   (b)

Section 10–2 Read Only Memories (ROMs)

10–5 For the ROM array in Figure 10–83, determine the outputs for all possible input combinations, and summarize them in tabular form (light cell is a 1, dark cell is a 0).

10–6 Determine the truth table for the ROM in Figure 10–84.

10–7 Using a procedure similar to that in Example 10–1, design a ROM for conversion of single-digit BCD to Excess-3 code.

10–8 What is the total bit capacity of a ROM that has 14 address lines, 8 data inputs, and 8 data outputs?

FIGURE 10–83

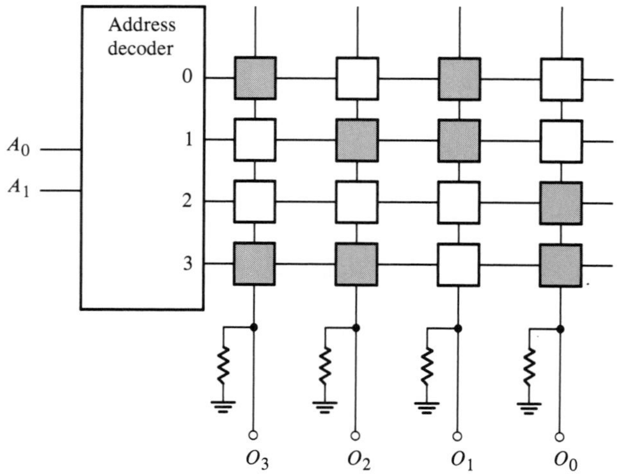

FIGURE 10–84

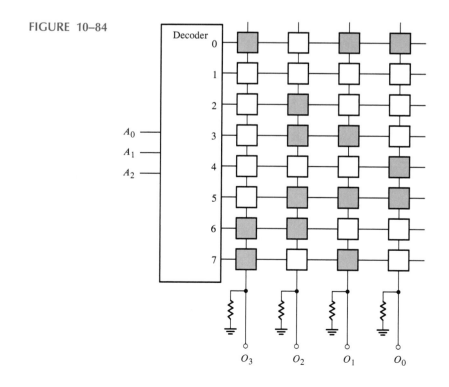

10–9 Determine the output of the 74185 binary-to-BCD converter in Figure 10–85 for each of the following binary inputs:

(a) 001101 **(b)** 101110 **(c)** 110111

10–10 Expand the binary-to-BCD converter in Figure 10–20 to handle a nine-bit binary input.

Section 10–3 Programmable ROMs (PROMs and EPROMs)

10–11 Assuming that the PROM matrix in Figure 10–86 is programmed by blowing a fuse link to create a 0, indicate the links to be blown to program an x^3 look-up table, where x is a number from 0 through 7.

FIGURE 10–85

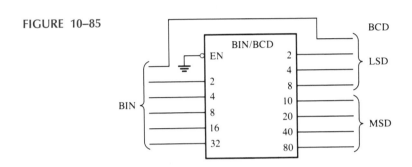

FIGURE 10–86

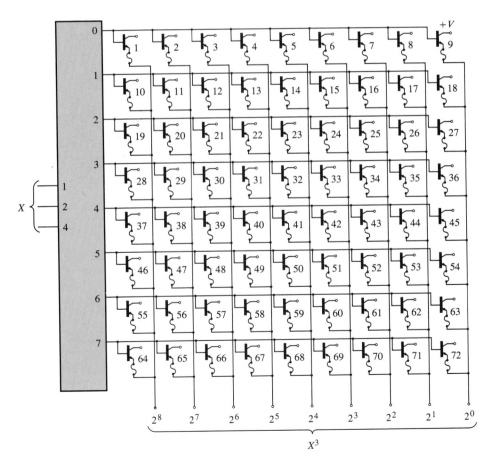

10–12 Determine the addresses that are programmed and the contents of each address after the programming sequence in Figure 10–87 has been applied to a TMS2516 PROM.

Section 10–4 Read/Write Random Access Memories (RAMs)

10–13 A static memory cell such as the one in Figure 10–26 is storing a 0 (RESET). What is its state after each of the following conditions?
   (a) ROW = 1, COLUMN = 1, DATA IN = 1, READ/$\overline{\text{WRITE}}$ = 1
   (b) ROW = 0, COLUMN = 1, DATA IN = 1, READ/$\overline{\text{WRITE}}$ = 1
   (c) ROW = 1, COLUMN = 1, DATA IN = 1, READ/$\overline{\text{WRITE}}$ = 0

10–14 Draw a basic logic diagram for a 512×8-bit static RAM, showing all the inputs and outputs.

10–15 Assuming that a 512×8 static RAM has a structure similar to that of the 256×4 RAM in Figure 10–28, determine the size of its memory cell array.

10–16 Redraw the block diagram in Figure 10–35 for a 32K memory.

FIGURE 10–87

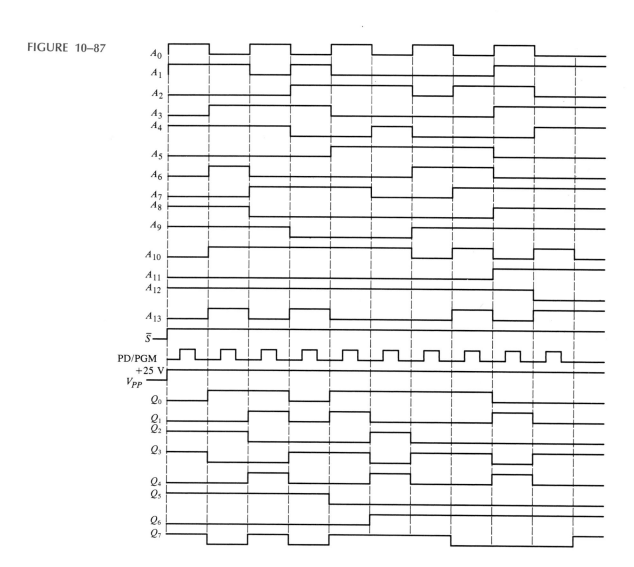

Section 10–5 Memory Expansion

10–17 Use 74189 16×4 RAMs to build a 64×8 RAM. Show the logic diagram.

10–18 Use TMS4164 $64K \times 1$ dynamic RAMs to build a $256K \times 4$ RAM.

10–19 What is the word length and the word capacity of the memory of Problem 10–17? Problem 10–18?

Section 10–6 Magnetic Memories

10–20 How do floppy disks compare with magnetic tape in access time?

10–21 How many bytes of data can be stored on a double-sided 3.5-in. disk with sixteen sectors per track if each sector stores 128 bytes?

10–22 A certain MBM has 250 minor loops, each of which stores 1000 bits. What is the total bit capacity? What is the page size?

10–23 What mechanism in an MBM permits nondestructive readout?

Section 10–7 Miscellaneous Memories

10–24 Complete the timing diagram in Figure 10–88 by showing the output waveforms for a FIFO serial memory like that shown in Figure 10–57.

10–25 Consider a 4096×8 RAM in which the last 64 addresses are used as a LIFO stack. If the first address in the RAM is 000_{16}, designate the 64 addresses used for the stack.

10–26 In the memory of Problem 10–25, sixteen bytes are pushed into the stack. At what address is the first byte in located? At what address is the last byte in located?

FIGURE 10–88

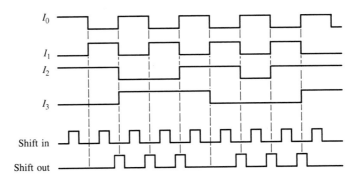

Section 10–8 Programmable Logic Devices (PLDs)

10–27 Program the FPLA in Figure 10–68 for the following output functions by indicating the cross-points that are to be left intact. All others are to be blown.

$$O_0 = \bar{I}_2 I_1 \bar{I}_0 + I_2 \bar{I}_1 I_0 + I_2 \bar{I}_1 \bar{I}_0 + \bar{I}_2 \bar{I}_1 I_0$$
$$O_1 = I_2 I_1 I_0 + I_2 \bar{I}_1 + \bar{I}_1 \bar{I}_0$$
$$O_2 = \bar{I}_2 \bar{I}_1 \bar{I}_0 + I_2 I_1 I_0 + I_2 \bar{I}_1 \bar{I}_0$$

10–28 Program the FPLA in Figure 10–72 as a BCD-to-binary converter. Indicate the cross points that are to be left intact. Limit the binary output to four bits.

10–29 Solve Problem 10–28 for a five-bit binary output.

10–30 Program the FPAL in Figure 10–73 to produce the following output functions:

$$O_0 = \bar{I}_2 \bar{I}_1 I_0 + I_2 \bar{I}_1 \bar{I}_0 + \bar{I}_2 I_1 I_0 + \bar{I}_2 I_1 I_0 + I_2 I_1 \bar{I}_0$$
$$O_1 = I_2 I_1 I_0 + \bar{I}_2 \bar{I}_1 \bar{I}_0 + I_2 \bar{I}_1 I_0$$

10–31 Program the FPLS in Figure 10–75 to operate as a four-bit binary counter.

10–32 Program the FPLS in Figure 10–75 to operate as a four-bit serial input–parallel output shift register.

10–33 Implement the three-bit Gray code counter in Chapter 8 (Figure 8–30) with the FPLS in Figure 10–75.

Section 10–9   Testing and Troubleshooting

10–34 Develop a basic block diagram for the ROM tester for the setup in Figure 10–76 to test 7488 ROMs (see Figure 10–15).

10–35 Repeat Problem 10–34 for a TMS47256 ROM.

10–36 Determine if the contents of the ROM in Figure 10–89 are correct.

10–37 A 128×8 ROM is implemented with four 7488 ROMs as shown in Figure 10–90. The 74139 decodes the two most significant address bits to enable the ROMs one at a time, depending on the address selected.

   (a) Express the lowest address and the highest address of each ROM as hexadecimal numbers.

   (b) Assume that a *single* checksum is used for the *entire* memory and it is stored at the highest address. Develop a flowchart for testing the complete memory system.

FIGURE 10–89

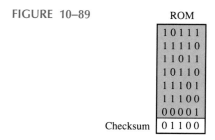

FIGURE 10–90

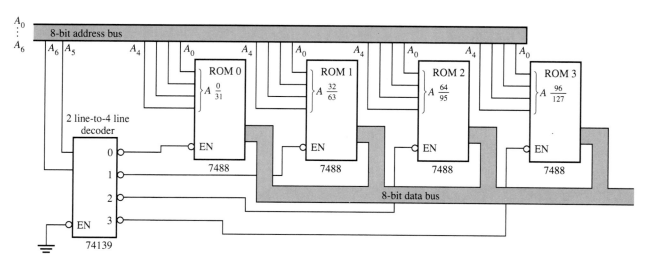

(c) Assume that *each* 7488 ROM has a checksum stored at its highest address. Modify the flowchart developed in part (b) to accommodate this change.

(d) What is the disadvantage of using a single checksum for the entire memory rather than a checksum for each individual ROM?

10–38 Suppose that a checksum test is run on the memory in Figure 10–90 and each individual ROM has a checksum at its highest address. What IC or ICs will you replace for each of the following error messages that appear on the system's video monitor?

(a) ADDRESSES 40 − 5F FAULTY

(b) ADDRESSES 20 − 3F FAULTY

(c) ADDRESSES 00 − 7F FAULTY

10–39 An FPLA, a portion of which is shown in Figure 10–91(a), is programmed as indicated (dotted cross points represent intact fuses). The test waveforms shown in part (b) are applied to the inputs, and the given output waveforms are observed. Is the device functioning properly, or is there a fault? If there is a fault, what is the correct output?

FIGURE 10–91

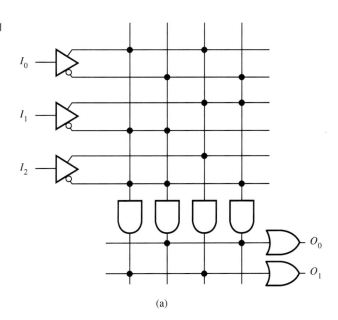

(a)

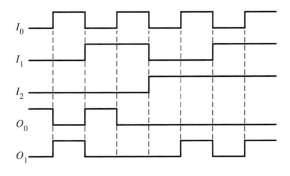

(b)

ANSWERS TO SECTION REVIEWS

Section 10–1   Review

1. A ROM has only a read operation; A RAM has both read and write operations.
2. Data are not lost when power is removed.
3. In a static memory the storage cells are latches and can retain data indefinitely. In a dynamic memory the storage cells are capacitors and must be refreshed periodically.

Section 10–2   Review

1. 2048 bits   **2.**   Tristate logic allows outputs to be disconnected from bus lines.
3. eleven bits

Section 10–3   Review

1. PROMs are field-programmable; ROMs are not.   **2.** 1s   **3.** read

Section 10–4   Review

1. Static RAMs have latch storage cells that can retain data indefinitely. Dynamic RAMs have capacitive storage cells that must be periodically refreshed.
2. The refresh operation prevents data from being lost because of capacitive discharge. A stored bit is restored periodically by recharging the capacitor.

Section 10–5   Review

1. Eight   **2.**   Eight

Section 10–6   Review

1. as magnetized spots with specified polarities
2. 40   **3.**   cyclic redundancy check
3. major loop, minor loop
4. 1 = presence of magnetic bubble; 0 = absence of magnetic bubble

Section 10–7   Review

1. In a FIFO the *first* bit (or word) *in* is the *first* bit (or word) *out.*
2. In a LIFO the *last* bit (or word) *in* is the *first* bit (or word) *out.* A stack is a LIFO.
3. charge-coupled device

Section 10–8   Review

1. An *FPLA* (a fuse-programmable logic array) has an AND plane and an OR plane that are programmable. An *FPAL* (a fuse-programmable array logic) has a programmable AND plane and a fixed OR array. An *FPLS* (a fuse-programmable logic sequence) is an FPAL with flip-flops.
2. Combinational and sequential logic functions can be implemented by programming PLDS.

Section 10–9   Review

1. The contents of the ROM are added and compared with a prestored checksum.
2. The contents of a RAM are not fixed.
3. (1) a short between adjacent cells; (2) an inability of some cells to store both 1s and 0s; (3) dynamic altering of the contents of one address when the contents of another address change.

Interfacing is the process of making two or more electronic devices or systems operationally compatible with each other so that they function together as required.

After completing this chapter, you will be able to

☐ describe applications of digital-to-analog and analog-to-digital conversion.

☐ explain D/A conversion by the binary-weighted input method and by the *R*/2*R* ladder method.

☐ explain A/D conversion by any of the following processes: flash (simultaneous), stairstep-ramp, tracking, single-slope, dual-slope, and successive approximation.

☐ recognize in D/A and A/D converters such conversion errors as nonmonotonicity, differential nonlinearity, offset, improper gain, and missing or incorrect codes.

☐ troubleshoot D/A and A/D converter systems.

☐ explain how digital devices are interfaced by the use of bus systems.

☐ define the basic characteristics and applications of the Multibus bus standard.

☐ define the basic characteristics and applications of the GPIB and EIA-232-D/RS-232-C bus standards.

☐ explain how to expand GPIB systems.

11

System Interfacing

INTERFACING THE ANALOG AND DIGITAL WORLDS

11–1

*In Chapter 1 we briefly described analog and digital quantities and pointed out the difference between them. Analog quantities are sometimes called **real-world** quantities because most physical quantities are analog in nature. Many applications of computers and other digital systems require the input of real-world quantities, such as temperature, speed, position, pressure, and force. Real-world quantities can even include graphic images. Also, digital systems often must produce outputs to control real-world quantities.*

Digital and Analog Signals

You already know how a quantity can be expressed in digital form, but what about an *analog* quantity? As you know, an analog quantity is one that has a *continuous* set of values over a given range, as contrasted with *discrete* values for the digital case.

Almost any measurable quantity is analog in nature, such as temperature, pressure, speed, and time. To further illustrate the difference between an analog and a digital representation of a quantity, let us take the case of a voltage that varies over a range from 0 V to +15 V. The analog representation of this quantity takes in *all* values between 0 and +15 V of which there are an infinite number.

In the case of a digital representation using a four-bit binary code, only sixteen values can be defined. More values between 0 and +15 can be represented by using more bits in the digital code. So an analog quantity can be represented to some degree of accuracy with a digital code that specifies discrete values within the range. This concept is illustrated in Figure 11–1, where the analog function shown is a smoothly changing curve that takes on values between 0 V and +15 V. If a four-bit code is used to represent this curve, each binary number represents a discrete point on the curve.

In Figure 11–1 the voltage on the analog curve is measured, or *sampled,* at each of thirty-five equal intervals. The voltage at each of these intervals is represented by a four-bit code as indicated. At this point, we have a series of binary numbers representing various voltage values along the analog curve. This is the basic idea of *analog-to-digital (A/D) conversion.*

An approximation of the analog function in Figure 11–1 can be reconstructed from the sequence of digital numbers that has been generated. Obviously, there will be some error in the reconstruction because only certain values are represented (thirty-six in this example) and not the continuous set of values. If the digital values at all of the thirty-six points are graphed as shown in Figure 11–2, we have a reconstructed function. As you can see, the graph only approximates the original curve, because values between the points are not known.

Real-World Interfacing

To interface between the digital and analog worlds, two basic processes are required. These are *analog-to-digital (A/D) conversion* and *digital-to-analog (D/A) conversion.* The following two system examples illustrate the application of these conversion processes.

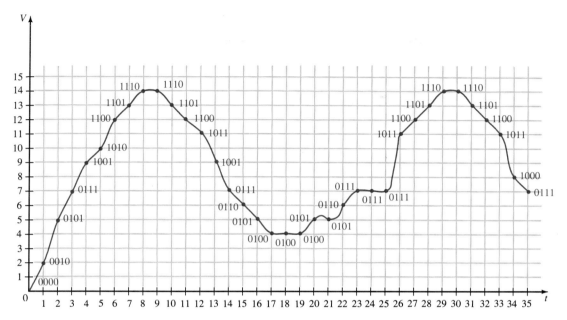

FIGURE 11–1
Discrete (digital) points on an analog curve.

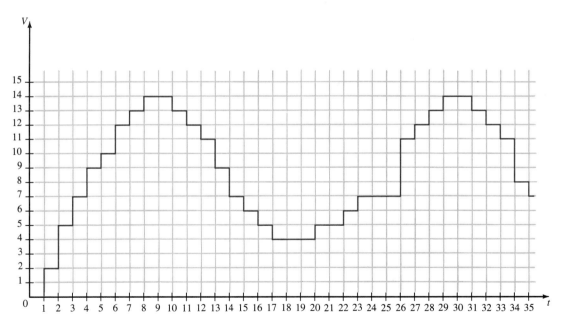

FIGURE 11–2
A rough digital reproduction of an analog curve.

An electronic thermostat A simplified block diagram of a microprocessor-based electronic thermostat is shown in Figure 11–3. The room temperature sensor produces an analog voltage that is proportional to the temperature. This voltage is increased by the linear amplifier and applied to the A/D converter, where it is converted to a digital code and periodically sampled by the microprocessor. For example, suppose the room temperature is 67° F. A specific voltage value corresponding to this temperature appears on the A/D converter input and is converted to an eight-bit binary number, 01000011.

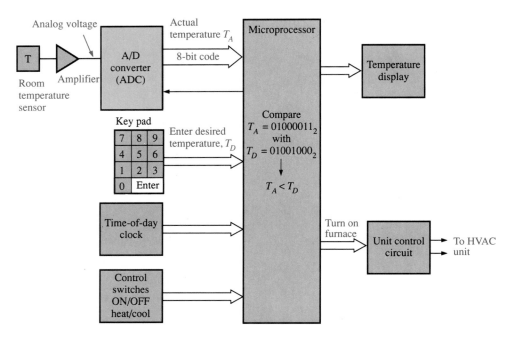

FIGURE 11–3

An electronic thermostat that uses an A/D converter.

Internally, the microprocessor compares this binary number with a binary number representing the *desired* temperature (say 01001000 for 72° F). This desired value has been previously entered from the keypad and stored in a register. The comparison shows that the actual room temperature is less than the desired temperature. As a result, the microprocessor instructs the unit control circuit to turn the furnace on. As the furnace runs, the microprocessor continues to monitor the actual temperature via the A/D converter (ADC). When the actual temperature equals or exceeds the desired temperature, the microprocessor turns the furnace off.

A digital audiotape (DAT) player/recorder Another system example that includes both A/D and D/A conversion is the *DAT* player/recorder. A basic block diagram is shown in Figure 11–4.

An audio signal, of course, is an analog quantity. In the *record mode,* sound is picked up, amplified, and converted to digital form by the A/D converter. The digital codes representing the audio signal are processed and recorded on the tape.

In the *play mode,* the digitized audio signal is read from the tape, processed, and converted back to analog form by the D/A converter (DAC). It is then amplified and sent to the speaker system.

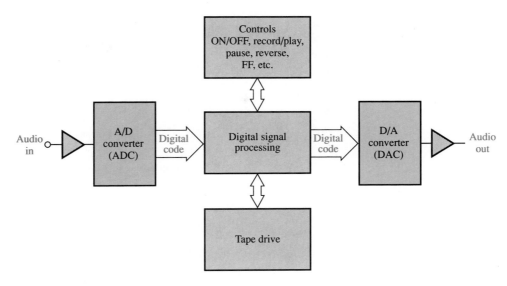

FIGURE 11–4
Basic block diagram of a DAT system.

Answers are found at the end of the chapter.
1. In the real world, quantities normally appear in what form?
2. Explain the basic purpose of A/D conversion.
3. Explain the basic purpose of D/A conversion.

DIGITAL-TO-ANALOG (D/A) CONVERSION

11–2

As you have seen, D/A conversion is an important part of many systems. In this section we will examine two basic types of D/A converters and learn about their performance characteristics.

The Operational Amplifier

Before getting into D/A converters, let us look briefly at an element that is common to most types of D/A and A/D converters. This element is the *operational amplifier,* or *op-amp* for short.

An op-amp is a linear amplifier that has two inputs (inverting and noninverting) and one output. It has a very high voltage gain and a very high input impedance, as well as a very low output impedance. The op-amp symbol is shown in Figure 11–5(a). When used as an *inverting amplifier,* the op-amp is configured as shown in part (b). The feedback resistor, R_F, and the input resistor, R_{IN}, control the voltage gain according to the formula in Equation (11–1), where V_{OUT}/V_{IN} is the closed-loop voltage gain (closed loop refers to the feedback from output to input provided by R_F). The negative sign indicates inversion.

$$\frac{V_{OUT}}{V_{IN}} = -\frac{R_F}{R_{IN}} \tag{11–1}$$

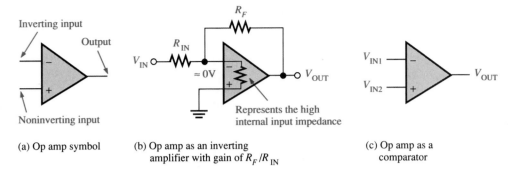

(a) Op amp symbol

(b) Op amp as an inverting amplifier with gain of R_F/R_{IN}

(c) Op amp as a comparator

FIGURE 11–5
The operational amplifier (op amp).

In the inverting amplifier configuration, the inverting input of the op-amp is approximately at ground potential (0 V) because feedback and the extremely high open-loop gain make the differential voltage between the two inputs extremely small. Since the noninverting input is grounded, the inverting input is at approximately 0 V, which is called *virtual ground*.

When the op-amp is used as a *comparator,* as shown in part (c) of the figure, two voltages are applied to the inputs. When these input voltages differ by a very small amount, the op-amp is driven into one of its saturated output states, depending on which input voltage is greater.

Binary-Weighted-Input D/A Converter

One method of D/A conversion uses a resistor network with resistance values that represent the binary weights of the input bits of the digital code. Figure 11–6 shows a four-bit D/A converter of this type. Each of the input resistors will either have current or have no current, depending on the input voltage level. If the input voltage is zero (binary 0), the current is also zero. If the input voltage is HIGH (binary 1), the amount of current depends on the input resistor value and is different for each input resistor, as indicated in the figure.

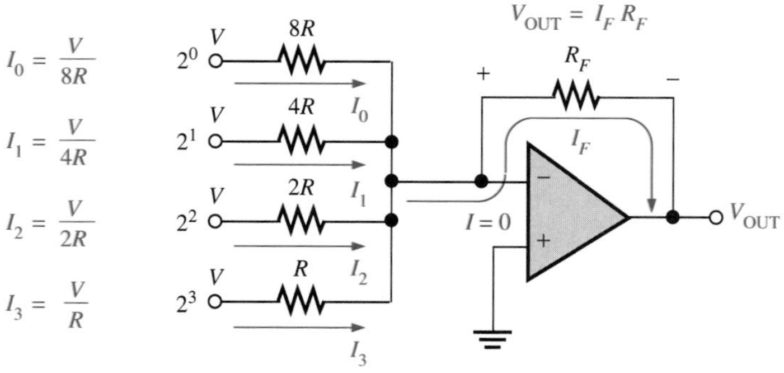

$$I_0 = \frac{V}{8R}$$

$$I_1 = \frac{V}{4R}$$

$$I_2 = \frac{V}{2R}$$

$$I_3 = \frac{V}{R}$$

$$V_{OUT} = I_F R_F$$

FIGURE 11–6
A four-bit D/A converter with binary-weighted inputs.

Since there is practically no current into the op-amp inverting input, all of the input currents sum together and flow through R_F. Since the inverting input is at 0 V (virtual ground), the drop across R_F is equal to the output voltage, so $V_{OUT} = I_F R_F$.

The values of the input resistors are chosen to be inversely proportional to the binary weights of the corresponding input bits. The lowest-value resistor (R) corresponds to the highest binary-weighted input (2^3). The other resistors are multiples of R: $2R$, $4R$, and $8R$, corresponding to the binary weights 2^2, 2^1, and 2^0, respectively. The input currents are also proportional to the binary weights. Thus, the output voltage is proportional to the sum of the binary weights, since the sum of the currents flows through R_F.

One of the disadvantages of this type of D/A converter is the number of different resistor values. For example, an eight-bit converter requires eight resistors, ranging from some value of R to $128R$ in binary-weighted steps. This range of resistors requires tolerances of one part in 255 (less than 0.5%) to accurately convert the input, making this type of D/A converter very difficult to mass-produce.

EXAMPLE 11–1

Determine the output of the D/A converter in Figure 11–7(a) if the waveforms representing a sequence of four-bit binary numbers in Figure 11–7(b) are applied to the inputs. Input D_0 is the LSB.

FIGURE 11–7

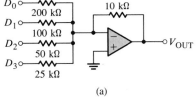
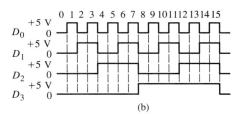

(a)

(b)

Solution First, we determine the current for each of the weighted inputs. Since the inverting input $(-)$ of the op-amp is at 0 V (virtual ground) and a binary 1 corresponds to $+5$ V, the current through any of the input resistors is 5 V divided by the resistance value:

$$I_0 = 5 \text{ V}/200 \text{ k}\Omega = 0.025 \text{ mA}$$

$$I_1 = 5 \text{ V}/100 \text{ k}\Omega = 0.05 \text{ mA}$$

$$I_2 = 5 \text{ V}/50 \text{ k}\Omega = 0.1 \text{ mA}$$

$$I_3 = 5 \text{ V}/25 \text{ k}\Omega = 0.2 \text{ mA}$$

Almost no current goes into the inverting op-amp input, because of its extremely high impedance. Therefore, we assume that all of the current goes through the feedback resistor R_F. Since one end of R_F is at 0 V (virtual ground), the drop across R_F equals the output voltage, which is negative with respect to virtual ground.

$$V_{OUT(D0)} = (10 \text{ k}\Omega)(-0.025 \text{ mA}) = -0.25 \text{ V}$$

$$V_{OUT(D1)} = (10 \text{ k}\Omega)(-0.05 \text{ mA}) = -0.5 \text{ V}$$

$$V_{OUT(D2)} = (10 \text{ k}\Omega)(-0.1 \text{ mA}) = -1 \text{ V}$$

$$V_{OUT(D3)} = (10 \text{ k}\Omega)(-0.2 \text{ mA}) = -2 \text{ V}$$

From Figure 11–7(b), the first binary input code is 0000, which produces an output voltage of 0 V. The next input code is 0001, which produces an output voltage of −0.25 V. For this, the output voltage is −0.25 V. The next code is 0010 which produces an output voltage of −0.5 V. The next code is 0011 which produces an output voltage of −0.25 V + −0.5 V = −0.75 V. Each successive binary code increases the output voltage by −0.25 V, so for this particular straight binary sequence on the inputs, the output is a *stairstep* waveform going from 0 V to −3.75 V in −0.25-V steps. This is shown in Figure 11–8.

FIGURE 11–8

Output of D/A converter in Figure 11–7.

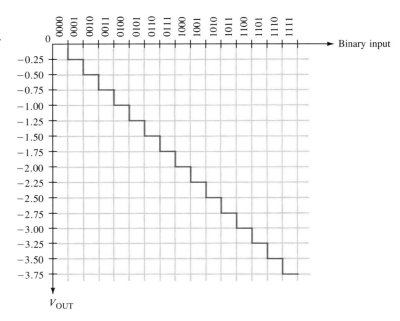

The *R/2R* Ladder D/A Converter

Another method of D/A conversion is the *R/2R ladder,* as shown in Figure 11–9 for four bits. It overcomes one of the problems in the previous type in that it requires only two resistor values.

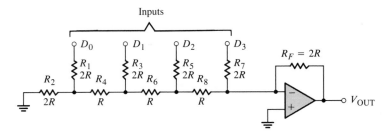

FIGURE 11–9

An R/2R ladder D/A converter.

Start by assuming that the D_3 input is HIGH ($+5$ V) and the others are LOW (ground, 0 V). This condition represents the binary number 1000. A circuit analysis will show that this reduces to the equivalent form shown in Figure 11–10(a) (p. 568). There is essentially no current through the $2R$ *equivalent* resistance because the inverting input is at virtual ground. Thus, all of the current ($I = 5$ V/2R) through R_7 also goes through R_F, and the output voltage is -5 V.

Part (b) of the figure shows the equivalent circuit when the D_2 input is at $+5$ V and the others are at ground. This condition represents 0100. If we thevenize looking from R_8, we get 2.5 V in series with R, as shown in part (b). This results in a current through R_F of $I = 2.5$ V/2R, which gives an output voltage of -2.5 V. Keep in mind that there is no current into the op-amp inverting input and that there is no current through the equivalent resistance to ground because it has 0 V across it, due to the virtual ground.

Part (c) of the figure shows the equivalent circuit when the D_1 input is at $+5$ V and the others are at ground. Again thevenizing looking from R_8, we get 1.25 V in series with R as shown. This results in a current through R_F of $I = 1.25$ V/2R, which gives an output voltage of -1.25 V.

In part (d) of the figure, the equivalent circuit representing the case where D_0 is at $+5$ V and the other inputs are at ground is shown. Thevenizing from R_8 gives an equivalent of 0.625 V in series with R as shown. The resulting current through R_F is $I = 0.625$ V/2R, which gives an output voltage of -0.625 V.

Notice that each successively lower weighted input produces an output voltage that is halved, so that the output voltage is proportional to the binary weight of the input bits.

Performance Characteristics of D/A Converters

The performance characteristics of a D/A converter include resolution, accuracy, linearity, monotonicity, and settling time, each of which is discussed in the following list:

1. *Resolution.* The resolution of a D/A converter is the reciprocal of the number of discrete steps in the D/A output. This, of course, is dependent on the number of input bits. For example, a four-bit D/A converter has a resolution of one part in $2^4 - 1$ (one part in fifteen). Expressed as a percentage, this is $(1/15)100 = 6.67\%$. The total number of discrete steps equals $2^n - 1$, where n is the number of bits. Resolution can also be expressed as the number of bits that are converted.

2. *Accuracy.* Accuracy is a comparison of the *actual* output of a D/A converter with the *expected* output. It is expressed as a percentage of a full-scale, or maximum, output voltage. For example, if a converter has a full-scale output of 10 V and the accuracy is $\pm 0.1\%$, then the maximum error for any output voltage is $(10$ V$)(0.001) = 10$ mV. Ideally, the accuracy should be, at most, $\pm \frac{1}{2}$ of an LSB. For an eight-bit converter, 1 LSB is $1/256 = 0.0039$ (0.39% of full scale). The accuracy should be approximately $\pm 0.2\%$.

3. *Linearity.* A linear error is a deviation from the ideal straight-line output of a D/A converter. A special case is an *offset error,* which is the amount of output voltage when the input bits are all zeros.

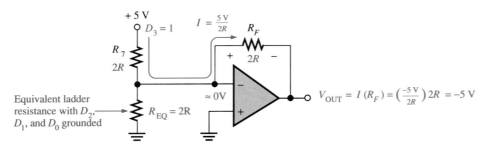

(a) Equivalent circuit for $D_3 = 1, D_2 = 0, D_1 = 0, D_0 = 0$

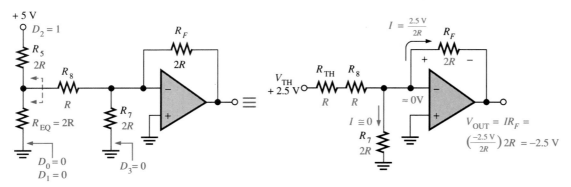

(b) Equivalent circuit for $D_3 = 0, D_2 = 1, D_1 = 0, D_0 = 0$

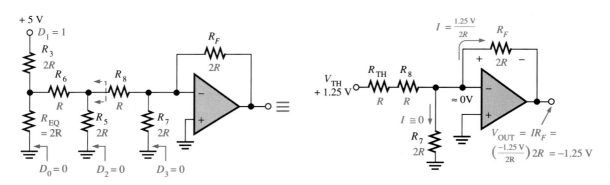

(c) Equivalent circuit for $D_3 = 0, D_2 = 0, D_1 = 1, D_0 = 0$

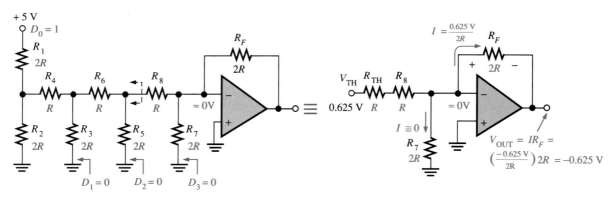

(d) Equivalent circuit for $D_3 = 0, D_2 = 0, D_1 = 0, D_0 = 1$

FIGURE 11–10 (Opposite)
Analysis of the R/2R ladder D/A converter. (The operational amplifier keeps the inverting (−)
input near zero volts (≈0 V) because of negative feedback. Therefore, all current goes through
R_F rather than into the inverting input.)

    4. *Monotonicity.* A D/A converter is *monotonic* if it does not take any reverse
        steps when it is sequenced over its entire range of input bits.
    5. *Settling time.* This is normally defined as the time it takes a D/A converter to
        settle within ± ½ LSB of its final value when a change occurs in the input
        code.

EXAMPLE 11–2

Determine the resolution, expressed as a percentage, of **(a)** an eight-bit and **(b)** a twelve-bit D/A converter.

Solutions
(a) For the eight-bit converter,

$$\frac{1}{2^8 - 1} \times 100 = \frac{1}{255} \times 100 = 0.392\%$$

(b) For the 12-bit converter,

$$\frac{1}{2^{12} - 1} \times 100 = \frac{1}{4095} \times 100 = 0.0244\%$$

SECTION 11–2
REVIEW

Answers are found at the end of the chapter.
1. What is the disadvantage of the D/A converter with binary weighted inputs?
2. What is the resolution of a four-bit D/A converter?

ANALOG-TO-DIGITAL (A/D) CONVERSION

11–3

As you have seen, analog-to-digital conversion is the process by which an analog
quantity is converted to digital form. It is necessary when measured quantities must
be in digital form for processing in a computer or for display or storage. Several
A/D conversion methods are now examined.

Flash (Simultaneous) A/D Converter

This method utilizes *comparators* that compare reference voltages with the analog input voltage. When the analog voltage exceeds the reference voltage for a given comparator, a HIGH is generated. Figure 11–11 shows a three-bit converter that uses seven comparator circuits; a comparator is not needed for the all-0s condition. A four-bit converter of this type requires fifteen comparators. In general, $2^n - 1$ comparators are required for conversion to an *n*-bit binary code. The large number of comparators necessary for a reasonable-sized binary number is one of the disadvantages of the flash A/D converter. Its chief advantage is that it provides a *short conversion time*.

FIGURE 11–11
A three-bit flash A/D converter.

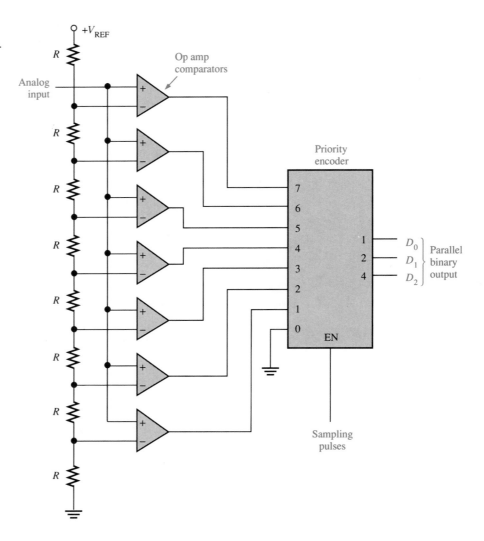

The reference voltage for each comparator is set by the resistive voltage-divider network. The output of each comparator is connected to an input of the priority encoder. The encoder is sampled by a pulse on the Enable input, and a three-bit binary code representing the value of the analog input appears on the encoder's outputs. The binary code is determined by the highest-order input having a HIGH level.

The *sampling rate* determines the accuracy with which the sequence of digital codes represents the analog input of the A/D converter. The more samples taken in a given unit of time, the more accurately the analog signal is represented in digital form.

The following example illustrates the basic operation of the flash A/D converter in Figure 11–11.

EXAMPLE 11–3

Determine the binary code output of the three-bit flash A/D converter for the analog input signal in Figure 11–12 and the sampling pulses (encoder Enable) shown. For this example, $V_{REF} = + 8$ V.

FIGURE 11–12

Sampling of values on an analog waveform for conversion to digital form.

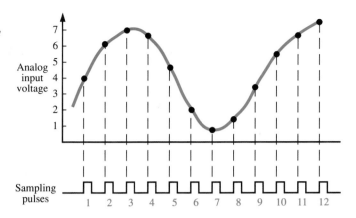

Solution   The resulting A/D output sequence is listed as follows and shown in the waveform diagram of Figure 11–13 in relation to the sampling pulses:

$$100, \ 101, \ 110, \ 110, \ 100, \ 010, \ 000, \ 001, \ 011, \ 101, \ 110, \ 111$$

FIGURE 11–13

Resulting digital outputs for sampled values. Output D_0 is the LSB.

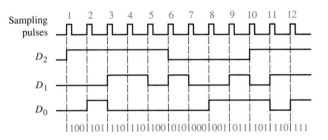

Stairstep-Ramp A/D Converter

This method of A/D conversion is also known as the *digital-ramp* or the *counter* method. It employs a D/A converter and a binary counter to generate the digital value of an analog input. Figure 11–14 shows a diagram of this type of converter.

Assume that the counter begins RESET and the output of the D/A converter is zero. Now assume that an analog voltage is applied to the input. When it exceeds the reference voltage (output of D/A), the comparator switches to a HIGH output state and enables the AND gate. The clock pulses begin advancing the counter through its binary states, producing a stairstep reference voltage from the D/A converter. The counter continues to advance from one binary state to the next, producing successively higher steps in the reference voltage. When the stairstep reference voltage reaches the analog input voltage, the comparator output will go LOW and disable the AND gate, thus cutting off the clock pulses to stop the counter. The binary state of the counter at this point equals the number of steps in the reference voltage required to make the reference equal to or greater than the analog input. This binary number, of course, represents the value of the analog input. The control logic loads the binary count into the latches and resets the counter, thus beginning another count sequence to sample the input value.

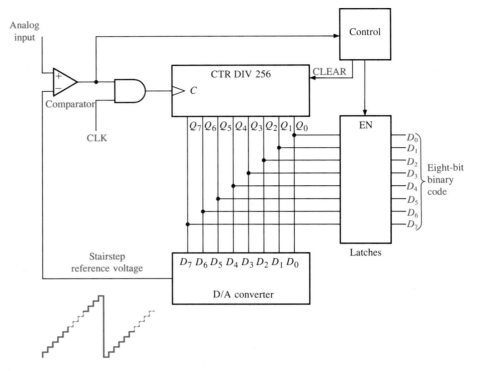

FIGURE 11–14
Stairstep-ramp A/D converter (eight bits).

This method is slower than the flash method because, in the worst case of maximum input, the counter must sequence through its maximum number of states before a conversion occurs. For an eight-bit conversion, this means a maximum of 256 counter states. Figure 11–15 illustrates a conversion sequence for a four-bit conversion. Notice that for each sample, the counter must count from *zero* up to the point at which the stairstep reference voltage reaches the analog input voltage. The conversion time varies, depending on the analog voltage.

Tracking A/D Converter

This method uses an *up/down counter* and is faster than the stairstep-ramp method because the counter is not reset after each sample, but rather tends to *track* the analog input. Figure 11–16 shows a typical eight-bit tracking A/D converter.

As long as the D/A output reference voltage is less than the analog input, the comparator output is HIGH, putting the counter in the UP mode, which causes it to produce an up sequence of binary counts. This causes an increasing stairstep reference voltage out of the D/A converter, which continues until the stairstep reaches the value of the input voltage.

When the reference voltage equals the analog input, the comparator's output switches LOW and puts the counter in the DOWN mode, causing it to back up one count. If the analog input is decreasing, the counter will continue to back down in its sequence and effectively track the input. If the input is increasing, the counter will back down one count after the comparison occurs and then will begin counting up again.

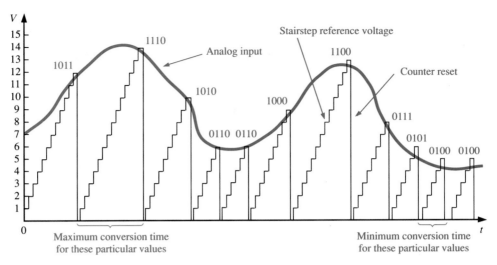

FIGURE 11–15
Example of a four-bit conversion, showing an analog input and the stairstep reference voltage.

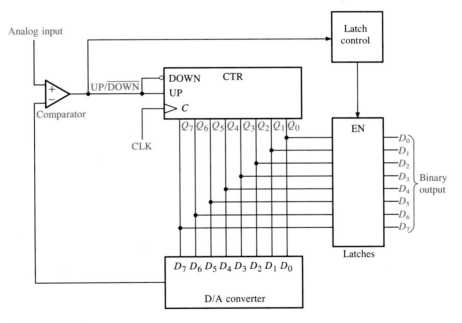

FIGURE 11–16
An eight-bit tracking A/D converter.

When the input is constant, the counter backs down one count when a comparison occurs. The reference output is now less than the analog input, and the comparator output goes HIGH, causing the counter to count up. As soon as the counter increases one state, the reference voltage becomes greater than the input, switching the comparator to its LOW state. This causes the counter to back down one count. This back-and-forth action continues as long as the analog input is a constant value, thus causing an oscillation between two binary states in the A/D output. This is a disadvantage of this type of converter.

Figure 11–17 illustrates the tracking action of this type of A/D converter for a four-bit conversion.

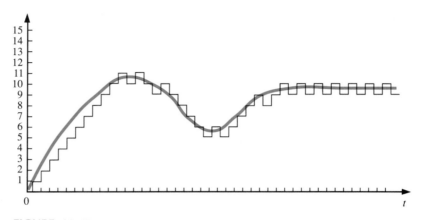

FIGURE 11–17
Tracking action of an A/D converter.

Single-Slope A/D Converter

Unlike the previous two methods, this type of converter does not require a D/A converter. It uses a linear ramp generator to produce a constant-slope reference voltage. A diagram is shown in Figure 11–18.

At the beginning of a conversion cycle, the counter is RESET and the ramp generator output is 0 V. The analog input is greater than the reference voltage at this point and therefore produces a HIGH output from the comparator. This HIGH enables the clock to the counter and starts the ramp generator.

Assume that the slope of the ramp is 1 V/ms. The ramp will increase until it equals the analog input; at this point the ramp is RESET, and the binary or BCD count is stored in the latches by the control logic. Let us assume that the analog input is 2 V at the point of comparison. This means that the ramp is also 2 V and has been running for 2 ms. Since the comparator output has been HIGH for 2 ms, 200 clock pulses have been allowed to pass through the gate to the counter (assuming a clock frequency of 100 kHz). At the point of comparison, the counter is in the binary state that represents decimal 200. With proper scaling and decoding, this binary number can be displayed as 2.00 V. This basic concept is used in some digital voltmeters.

Dual-Slope A/D Converter

The operation of this type of A/D converter is similar to that of the single-slope type except that a variable-slope ramp and a fixed-slope ramp are both used. This type of converter is common in digital voltmeters and other types of measurement instruments.

A ramp generator (integrator), A_1, is used to produce the dual-slope characteristic. A block diagram of a dual-slope A/D converter is shown in Figure 11–19 for reference.

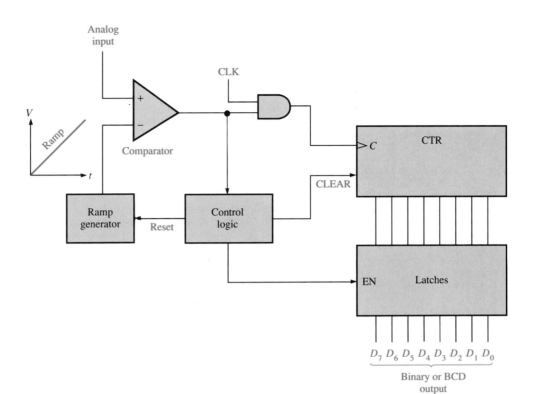

FIGURE 11–18
Single-slope A/D converter.

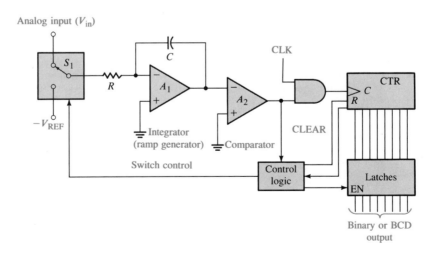

FIGURE 11–19
Dual-slope A/D converter.

We will start by assuming that the counter is RESET and the output of the integrator is zero. Now assume that a positive input voltage is applied to the input through the switch (S_1) as selected by the control logic. Since the inverting input of A_1 is at virtual ground, and assuming that V_{IN} is constant for a period of time, there will be constant current through the input resistor R and therefore through the capacitor C. Capacitor C will charge linearly because the current is constant, and as a result, there will be a negative-going linear voltage ramp on the output of A_1, as illustrated in Figure 11–20(a).

When the counter reaches a specified count, it will be reset, and the control logic will switch the negative reference voltage ($-V_{REF}$) to input A_1 as shown in Figure 11–20(b). At this point the capacitor is charged to a negative voltage ($-V$) proportional to the input analog voltage.

Now the capacitor discharges linearly because of the constant current from the $-V_{REF}$ as shown in Figure 11–20(c). This linear discharge produces a positive-going ramp on the A_1 output, starting at $-V$ and having a constant slope that is independent of the charge voltage.

As the capacitor discharges, the counter advances from its reset state. The time it takes the capacitor to discharge to zero depends on the initial voltage $-V$ (proportional to V_{IN}) because the discharge rate (slope) is constant. When the integrator (A_1) output voltage reaches zero, the comparator (A_2) switches to the LOW state and disables the clock to the counter. The binary count is latched, thus completing one conversion cycle. The binary count is proportional to V_{IN} because the time it takes the capacitor to discharge depends only on $-V$, and the counter records this interval of time.

Successive-Approximation A/D Converter

This is perhaps the most widely used method of A/D conversion. It has a much shorter conversion time than the other methods with the exception of the flash method. It also has a fixed conversion time that is the same for any value of the analog input.

Figure 11–21 shows a basic block diagram of a four-bit successive-approximation A/D converter. It consists of a D/A converter, a successive-approximation register (SAR), and a comparator. The basic operation is as follows: The bits of the D/A converter are enabled one at a time, starting with the MSB. As each bit is enabled, the comparator produces an output that indicates whether the analog input voltage is greater or less than the output of the D/A. If the D/A output is greater than the analog input, the comparator's output is LOW, causing the bit in the register to RESET. If the D/A output is less than the analog input, the bit is retained in the register.

The system does this with the MSB first, then the next most significant bit, then the next, and so on. After all the bits of the D/A have been tried, the conversion cycle is complete.

In order to better understand the operation of this type of A/D converter, we will take a specific example of a four-bit conversion. Figure 11–22 illustrates the step-by-step conversion of a given analog input voltage (5 V in this case). We will assume that the D/A converter has the following output characteristic: $V_{OUT} = 8$ V for the 2^3 bit (MSB), $V_{OUT} = 4$ V for the 2^2 bit, $V_{OUT} = 2$ V for the 2^1 bit, and $V_{OUT} = 1$ V for the 2^0 bit (LSB).

Figure 11–22(a) shows the first step in the conversion cycle with the MSB = 1. The output of the D/A is 8 V. Since this is *greater* than the analog input of 5 V, the output of the comparator is LOW, causing the MSB in the SAR to be RESET to a 0.

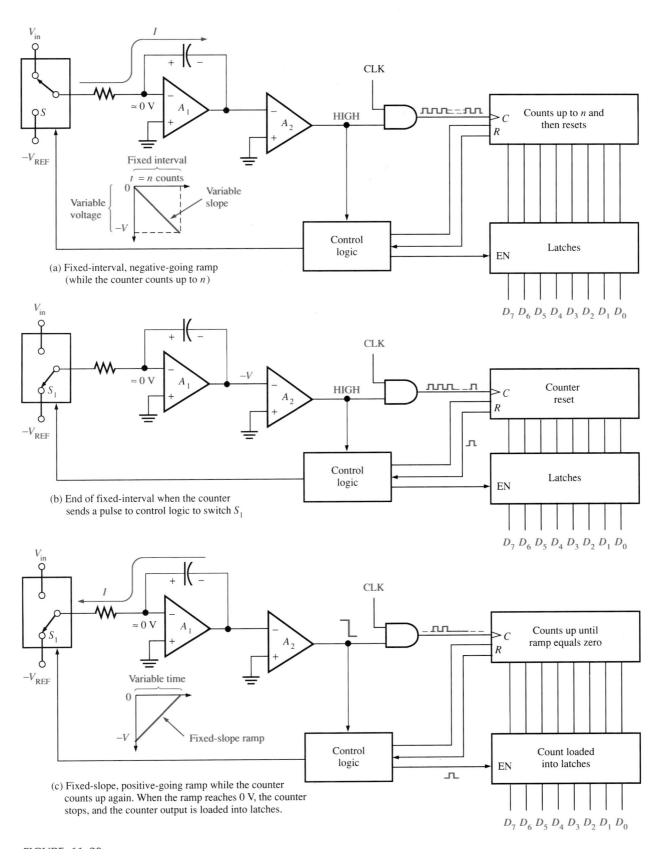

(a) Fixed-interval, negative-going ramp
(while the counter counts up to n)

(b) End of fixed-interval when the counter
sends a pulse to control logic to switch S_1

(c) Fixed-slope, positive-going ramp while the counter
counts up again. When the ramp reaches 0 V, the counter
stops, and the counter output is loaded into latches.

FIGURE 11–20
Dual-slope conversion.

FIGURE 11–21
Successive-approximation A/D converter.

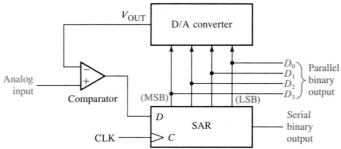

FIGURE 11–22
Successive-approximation conversion process.

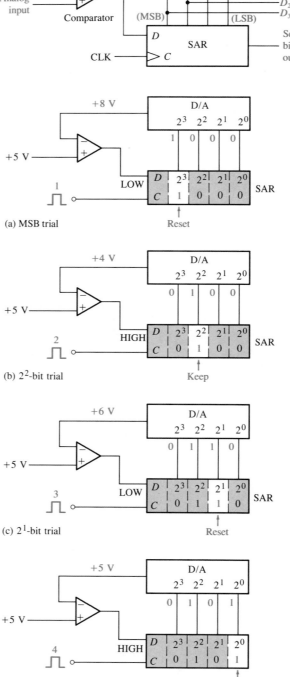

(a) MSB trial

(b) 2^2-bit trial

(c) 2^1-bit trial

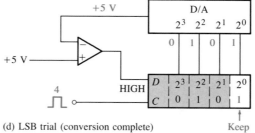

(d) LSB trial (conversion complete)

Figure 11–22(b) shows the second step in the conversion cycle with the 2^2 bit equal to a 1. The output of the D/A is 4 V. Since this is *less* than the analog input of 5 V, the output of the comparator switches to a HIGH, causing this bit to be retained in the SAR.

Figure 11–22(c) shows the third step in the conversion cycle with the 2^1 bit equal to a 1. The output of the D/A is 6 V because there is a 1 on the 2^2 bit input and on the 2^1 bit input and 4 V + 2 V = 6 V. Since this is *greater* than the analog input of 5 V, the output of the comparator switches to a LOW, causing this bit to be RESET to a 0.

Figure 11–22(d) shows the fourth and final step in the conversion cycle with the 2^0 bit equal to a 1. The output of the D/A is 5 V because there is a 1 on the 2^2 bit input and on the 2^0 bit input and 4 V + 1 V = 5 V.

The four bits have all been tried, thus completing the conversion cycle. At this point the binary code in the register is 0101, which is the binary value of the analog input of 5 V. Another conversion cycle now begins, and the basic process is repeated. The SAR is cleared at the beginning of each cycle.

A Specific A/D Converter

The ADC0801 is an example of a successive-approximation analog-to-digital converter. A block diagram is shown in Figure 11–23. This device operates from a +5 V supply and has a resolution of eight bits with a conversion time of 100 μs. Also, it has guaranteed monotonicity and an on-chip clock generator. The data outputs are tristate so that it can be interfaced with a microprocessor bus system.

FIGURE 11–23
The ADC0801 analog-to-digital converter.

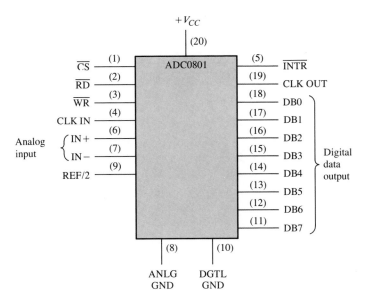

A detailed logic diagram of the ADC0801 is shown in Figure 11–24, and the basic operation of the device is as follows: The ADC0801 contains the equivalent of a 256-resistor D/A converter network. The successive-approximation logic sequences the network to match the analog differential input voltage ($V_{IN+} - V_{IN-}$) with an output from the resistive network. The MSB is tested first. After eight comparisons (sixty-four clock periods), an eight-bit binary code is transferred to an output latch, and the interrupt ($\overline{\text{INTR}}$) output goes LOW. The device can be operated in a free-running mode by connecting the $\overline{\text{INTR}}$ output to the write ($\overline{\text{WR}}$) input and holding the conversion start ($\overline{\text{CS}}$) LOW. To ensure start-up under all conditions, a LOW $\overline{\text{WR}}$ input is required during the power-up cycle. Taking $\overline{\text{CS}}$ low anytime after that will interrupt a conversion in process.

When the $\overline{\text{WR}}$ input goes LOW, the internal successive-approximation register (SAR) and the eight-bit shift register are RESET. As long as both $\overline{\text{CS}}$ and $\overline{\text{WR}}$ remain LOW, the analog-to-digital converter remains in a RESET state. One to eight clock periods after $\overline{\text{CS}}$ or $\overline{\text{WR}}$ makes a LOW-to-HIGH transition, conversion starts.

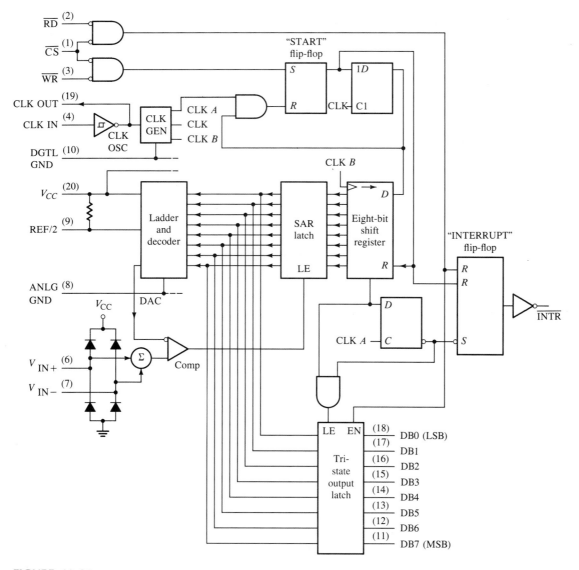

FIGURE 11–24
Logic diagram of the ADC0801 A/D converter.

When the $\overline{CS}$ and $\overline{WR}$ inputs are LOW, the start flip-flop is SET, and the interrupt flip-flop and eight-bit register are RESET. The HIGH is ANDed with the next clock pulse, which puts a HIGH on the RESET input of the start flip-flop. If either $\overline{CS}$ or $\overline{WR}$ has gone HIGH, the SET signal to the start flip-flop is removed, causing it to be RESET. A HIGH is placed on the D input of the eight-bit shift register, and the conversion process is started. If the $\overline{CS}$ and $\overline{WR}$ inputs are still LOW, the start flip-flop, the eight-bit shift register, and the SAR remain RESET. This action allows for wide $\overline{CS}$ and $\overline{WR}$ inputs, with conversion starting from one to eight clock periods after one of the inputs has gone HIGH.

When the HIGH input has been clocked through the eight-bit shift register, completing the SAR search, it is applied to an AND gate controlling the output latches and to the D input of a flip-flop. On the next clock pulse, the digital word is transferred to the tristate output latches, and the interrupt flip-flop is SET. The output of the interrupt flip-flop is inverted to provide an $\overline{INTR}$ output that is HIGH during conversion and LOW when conversion is complete.

When a LOW is at both the $\overline{CS}$ and $\overline{RD}$ inputs, the tristate output latch is enabled, the output code is applied to the DB0 through DB7 lines, and the interrupt flip-flop is RESET. When either the $\overline{CS}$ or the $\overline{RD}$ input returns to a HIGH, the DB0 through DB7 outputs are disabled. The interrupt flip-flop remains RESET.

SECTION 11–3
REVIEW

Answers are found at the end of the chapter.
1. What is the fastest method of analog-to-digital conversion?
2. Which A/D conversion method uses an up/down counter?
3. True or false: The successive-approximation converter has a fixed conversion time.

TROUBLESHOOTING

11–4

Basic testing of D/A and A/D converters includes checking their performance characteristics, such as monotonicity, offset, linearity, and gain, and checking for missing or incorrect codes. In this section the fundamentals of testing these analog interfaces are introduced.

Testing D/A Converters

The concept of D/A converter testing is illustrated in Figure 11–25. In this basic method a sequence of binary codes is applied to the inputs, and the resulting output is observed. The binary code sequence extends over the full range of values from 0 to $2^n - 1$ in ascending order, where n is the number of bits.

FIGURE 11–25
Basic test setup for a D/A converter.

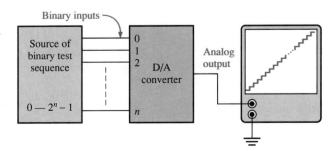

The ideal output is a straight-line stairstep as indicated. As the number of bits in the binary code is increased, the resolution is improved. That is, the number of discrete steps increases, and the output approaches a straight-line linear ramp.

D/A Conversion Errors

Several D/A conversion errors to be checked for are shown in Figure 11–26, which uses a four-bit conversion for illustration purposes. A four-bit conversion produces fifteen discrete steps. Each graph in the figure includes an ideal stairstep ramp for comparison with the faulty outputs.

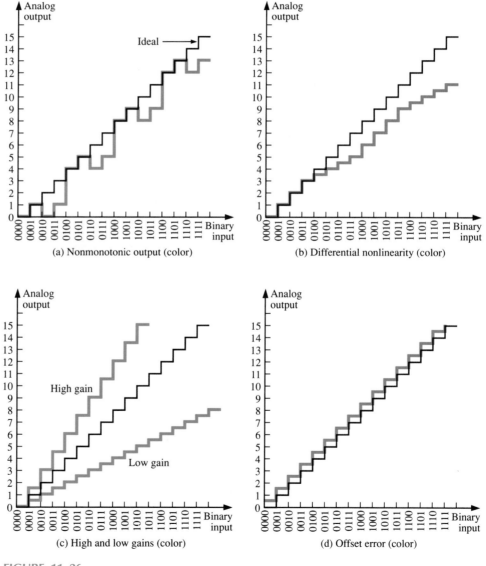

FIGURE 11–26

Illustrations of several D/A conversion errors.

Nonmonotonicity   The step reversals in Figure 11–26(a) indicate nonmonotonic performance, which is a form of nonlinearity. In this particular case the error occurs because the 2^1 bit in the binary code is interpreted as a constant 0. That is, a short is causing the bit input line to be stuck LOW.

Differential nonlinearity   Figure 11–26(b) illustrates differential nonlinearity in which the step amplitude is less than it should be for certain input codes. This particular output could be caused by the 2^2 bit's having an insufficient weight, perhaps because of a faulty input resistor. We could also see steps with amplitudes greater than normal, if a particular binary weight were greater than it should be.

Low or high gain   Output errors caused by low or high gain are illustrated in Figure 11–26(c). In the case of low gain, all of the step amplitudes are less than ideal. In the case of high gain, all of the step amplitudes are greater than ideal. This situation may be caused by a faulty feedback resistor in the op amp circuit.

Offset error   An offset error is illustrated in Figure 11–26(d). Notice that when the binary input is 0000, the output voltage is nonzero and that this amount of offset is the same for all steps in the conversion. A faulty op amp may be the culprit in this situation.

EXAMPLE 11–4

The D/A converter output in Figure 11–27 is observed when a straight four-bit binary sequence is applied to the inputs. Identify the type of error, and suggest an approach to isolate the fault.

FIGURE 11–27

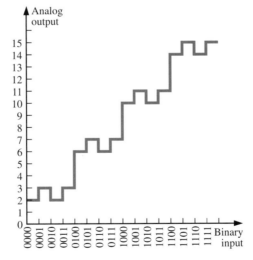

Solution   The D/A converter in this case is nonmonotonic. Analysis of the output reveals that the device is converting the following sequence, rather than the actual binary sequence applied to the inputs.

0010, 0011, 0010, 0011, 0110, 0111, 0110, 0111, 1010, 1011, 1010, 1011, 1110, 1111, 1110, 1111

Apparently, the 2^1 bit is stuck in the HIGH (1) state. To find the problem, first monitor the bit input pin to the device. If it is changing states, the fault is internal, most likely an open. If the external pin is not changing states and is always HIGH, check for an external short to $+V$ that may be caused by a solder bridge somewhere on the circuit board. If no problem is found here, disconnect the source output from the D/A input pin, and see if the output signal is correct. If these checks produce no results, the fault is most likely internal to the D/A converter, perhaps a short to the supply voltage.

Testing A/D Converters

One method for testing A/D converters is shown in Figure 11–28. A D/A converter is used as part of the test setup to convert the A/D output back to analog form for comparison with the test input.

A test input in the form of a linear ramp is applied to the input of the A/D converter. The resulting binary output sequence is then applied to the D/A test unit and converted to a stairstep ramp. The input and output ramps are compared for any deviation.

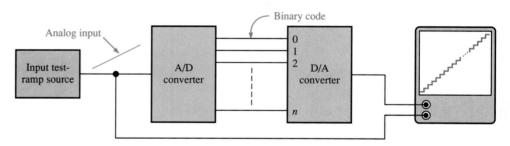

FIGURE 11–28
A method for testing A/D converters.

A/D Conversion Errors

Again, a four-bit conversion is used to illustrate the principles. We assume that the test input is an ideal linear ramp.

Missing code   The stairstep output in Figure 11–29(a) indicates that the binary code 1001 does not appear on the output of the A/D converter. Notice that the 1000 value stays for two intervals and then the output jumps to the 1010 value.

In a flash A/D converter, for example, a failure of one of the comparators can cause a missing-code error.

Incorrect codes   The stairstep output in Figure 11–29(b) indicates that several of the binary code words coming out of the A/D converter are incorrect. Analysis indicates that the 2^1-bit line is stuck in the LOW (0) state in this particular case.

Offset   Offset conditions are shown in 11–29(c). In this situation the A/D converter interprets the analog input voltage as greater than its actual value. This error is probably due to a faulty comparator circuit.

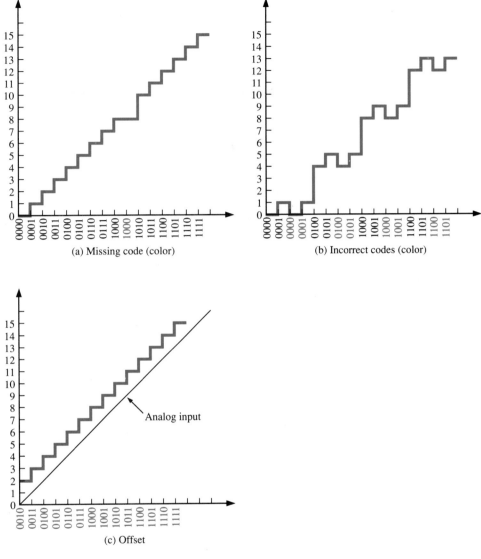

(a) Missing code (color)

(b) Incorrect codes (color)

(c) Offset

FIGURE 11–29
Illustrations of A/D conversion errors.

EXAMPLE 11–5

A four-bit flash A/D converter is shown in Figure 11–30(a). It is tested with a setup like the one in Figure 11–28. The resulting reconstructed analog output is shown in Figure 11–30(b). Identify the problem and the most probable fault.

Solution The binary code 0011 is missing from the A/D output, as indicated by the missing step.

Most likely, the output of comparator 3 is stuck in its inactive state (LOW).

FIGURE 11–30

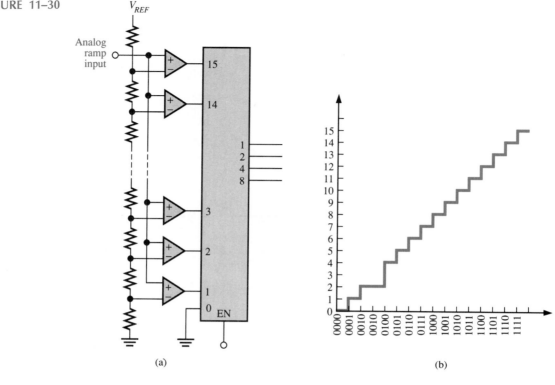

(a)

(b)

SECTION 11–4
REVIEW

Answers are found at the end of the chapter.
1. How do you detect nonmonotonic behavior in a D/A converter?
2. What effect does low gain have on a D/A output?
3. Name two types of output errors in an A/D converter.

INTERNAL DEVICE INTERFACING

11–5

The concept of the bus was introduced in Chapter 10 in relation to memories. All the devices within a given digital system are interconnected by buses, which serve as communication paths.

Multiplexed Buses

Most digital systems these days are microprocessor-based. In such systems the microprocessor controls and communicates with other devices, such as memories and input/output (I/O) devices, via the internal bus structure, as indicated in Figure 11–31.

A bus is *multiplexed* so that any of the devices connected to it can either send or receive data to or from one of the other devices. A sending device is called a *source*, and a receiving device is called an *acceptor*. At any given time, there is only one source active. For example, the RAM may be sending data to the I/O under control of the microprocessor.

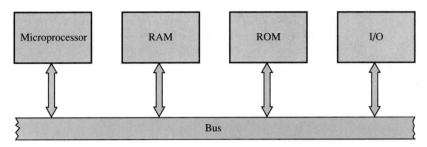

FIGURE 11–31
The interconnection of digital system components by a bidirectional, multiplexed bus.

Bus Signals

With *synchronous* bus control, the microprocessor usually originates all control and timing signals. The other devices then synchronize their operations to those control and timing signals. With *asynchronous* bus control, the control and timing signals are generated jointly by a source and an acceptor. The process of jointly establishing communication is called *handshaking*. A simple example of a handshaking sequence is given in Figure 11–32.

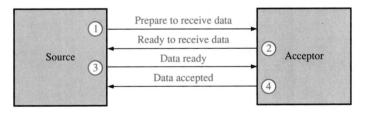

FIGURE 11–32
An example of a handshaking sequence.

An important control function is called *bus arbitration*. Arbitration prevents two sources from trying to use the bus at the same time.

Connecting Devices to the Bus

Tristate or open-collector drivers (see Chapter 13) are normally used to interface the outputs of a source device to the bus. Usually more than one source is connected to a bus, but only one can have access at any given time. All the other sources must be disconnected from the bus to prevent interference.

Typically, tristate circuits are used to connect a source to the bus or disconnect it from the bus, as illustrated in Figure 11–33(a) for the case of two sources. The select input is used to connect either source A or source B but not both at the same time to the bus. When $S = 0$, source A is connected and source B is disconnected. When $S = 1$, source B is connected and source A is disconnected. A switch equivalent of this action is shown in part (b) of the figure.

Recall from Chapter 10 that when the Enable input of a tristate circuit is not active, the device is in a high-impedance (high-Z) state and acts like an open switch. Many digital ICs provide internal tristate drivers for the output lines. A tristate output is indicated by a $\triangledown$ symbol as shown in Figure 11–34.

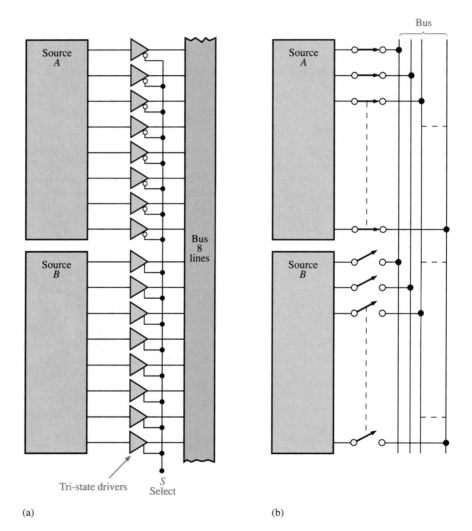

(a)   (b)

FIGURE 11–33
Tristate driver interface.

FIGURE 11–34

A device that has tristate outputs.

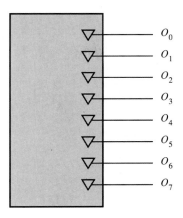

O_0

O_1

O_2

O_3

O_4

O_5

O_6

O_7

Multiplexed I/Os

Some devices that send and receive data have combined input and output lines, called I/O ports, that must be multiplexed onto the data bus. *Bidirectional* tristate drivers interface this type of device with the bus as illustrated in Figure 11–35(a).

Each I/O port has a pair of tristate drivers. When the $\overline{\text{SND}}/\text{RCV}$ (Send/Receive) line is LOW, the upper tristate driver in each pair is enabled and the lower one disabled. In this state the device is acting as a source and sending data to the bus. When the $\overline{\text{SND}}/\text{RCV}$ line is HIGH, the lower tristate driver in each pair is enabled so that the device is acting as an acceptor and receiving data from the bus. This operation is illustrated in Figure 11–35(b). Some devices provide for multiplexed I/O operation with internal circuitry.

Standard Buses

There are several so-called standard buses used to provide interfacing for the various internal components of digital systems. These standard buses assure compatibility of printed circuit board size, pin numbers, type of signal on each pin, and input and output characteristics. The standard bus allows for system expansion by specifying the conditions under which expansion or replacement units or modules must operate in order to interface properly with existing modules.

To introduce the concept of standard buses, two examples will be discussed briefly. A full, detailed coverage of these buses is beyond the scope of this book.

The S-100 bus   The S-100 bus was originally developed around the 8080A microprocessor and is one of the older bus standards. It is discussed here only to introduce the concept of internal bus standards. This bus is characterized by 100 lines, divided into 8 data output lines, 8 data-input lines, 16 address lines, 4 power-supply lines, 8 interrupt lines, 2 grounds, and 38 control lines. The remaining 16 lines are either unused or are spares. A given system may or may not use all of the functional lines.

FIGURE 11–35
Multiplexed I/O operation.

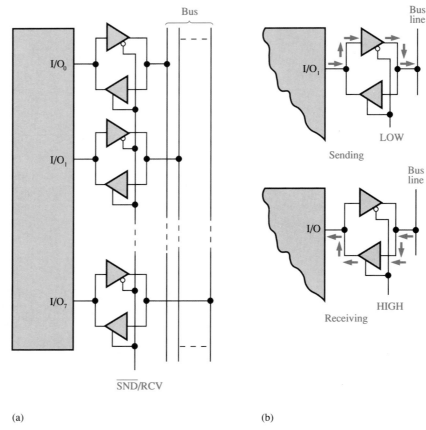

(a)

(b)

The S-100 is implemented physically with a motherboard arrangement consisting of 100 conductive strips that interconnect several 100-pin receptacles. Individual printed circuit boards plug into these receptacles with a card-edge connector, as illustrated in Figure 11–36.

FIGURE 11–36
Example of an S-100 physical configuration.

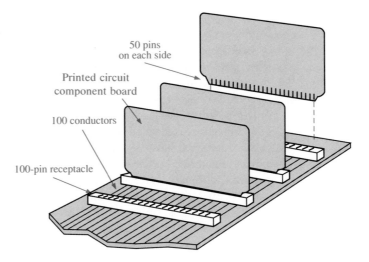

A simple example of a microcomputer that uses a portion of the S-100 bus is shown by the block diagram in Figure 11–37. The microprocessor has available 16 address lines (A0–A15), 8 data-output lines (DO0–DO7), 8 data-input lines (DI0–DI7), and a number of control and status lines, which are defined as follows:

MWRITE   Memory write. Data are to be placed in the RAM.
SMEMR   Memory read. A status signal indicating that a read operation will occur.
SOUT   Output status signal indicating that the output device address is on the address bus.
$\overline{\text{PWR}}$ Output signal for memory write or I/O control.
PDBIN   Output signal indicating that the bidirectional data bus is in the input mode.
SINP   Input status signal indicating that the input device address is on the address bus.

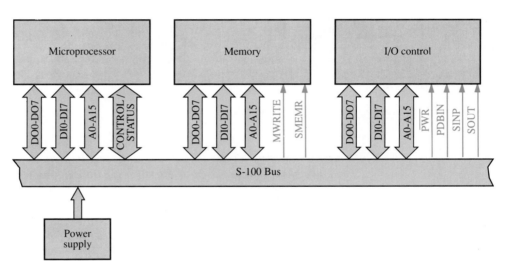

FIGURE 11–37
Example of a microcomputer using an S-100 bus.

The Multibus The Multibus is a general-purpose bus system developed by Intel but widely used in the industry. Many manufacturers offer products that are compatible with this particular bus system. The Multibus provides a flexible interface that can be used to interconnect a wide variety of microcomputer devices or modules. The modules in a Multibus system are designated as *masters* or *slaves*. Masters may obtain use of the bus and initiate data transfers. Slaves are devices that cannot transfer data themselves or control the bus. A major feature of the Multibus is that several processors (masters) can be connected to the bus at the same time to implement a *multiprocessing* operation, in which each processor performs a dedicated task.

The Multibus provides a total of 86 lines. There are 16 bidirectional data lines, 20 address lines, 8 interrupt lines, and various control lines, command lines, and power lines. The Multibus standard defines the physical and electrical parameters for devices using the bus. A generalized Multibus system configuration is shown in Figure 11–38.

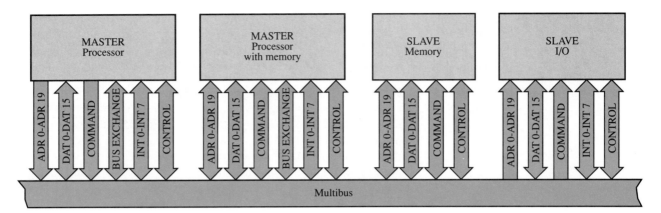

FIGURE 11–38
A Multibus system configuration.

SECTION 11–5
REVIEW
Answers are found at the end of the chapter.
1. Why are tristate drivers required to interface digital devices to a bus?
2. What is the purpose of a bus system?

DIGITAL EQUIPMENT INTERFACING

11–6

In the last section the idea of internal buses was introduced. These buses provide interconnections for the modules or units within a single digital system. In this section buses that are used to interface separate instruments are introduced. These buses can be considered external buses. External buses can be used to interface a microcomputer with test instruments or a computer terminal with a printer.

The General-Purpose Interface Bus (GPIB)

The GPIB is a bus system that provides an orderly and predictable way to transfer *parallel* data and maintain control between various digital instruments. Other designations, such as IEEE-488 or HPIB, also refer to the GPIB. This bus system is as close to a universal standard as you can get. A common application of the GPIB is the interfacing of test and measurement instruments with a computer to create an automated test system.

GPIB specifications One type of GPIB connector is a 24-pin miniature ribbon type, as shown in Figure 11–39. The signal name is indicated for each pin.

In a typical GPIB setup, there is a designated *controller* (usually a computer) and one or more *controlled devices* (test instruments, for example). When the controller issues a command for a controlled device to perform a specified operation, such as a frequency measurement, it is said that the controller "talks" and the controlled device "listens."

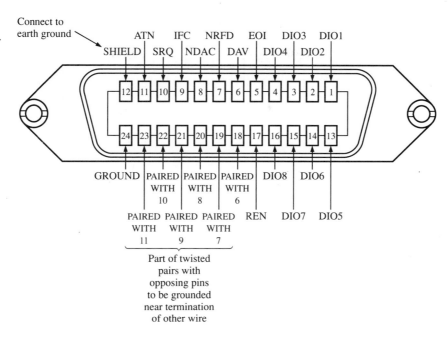

FIGURE 11–39
The GPIB connector.

A *listener* is an instrument capable of receiving data over the parallel interface bus when it is addressed by the controller. Examples of listeners are printers, monitors, programmable power supplies, and programmable signal generators. The GPIB allows up to fifteen active listeners on the bus at one time.

A *talker* is an instrument capable of transmitting data over the bus. Examples are DMMs that output data and frequency counters that output data, etc. The GPIB permits only one active talker on the bus at any given time.

Some instruments can transmit and receive and are called *talker/listeners*. An example is the *modem,* a device for interfacing digital systems to the telephone line. The term *modem* is a contraction of *modulator/demodulator*.

The *controller* is an instrument that can specify each of the other instruments on the bus as either a talker or a listener for the purpose of data transfer. A controller generally can be both a talker and a listener itself. A properly configured microcomputer is an example of a controller.

As mentioned, there are a total of twenty-four lines in the GPIB, including both signal lines and ground lines. The signal lines are divided into three functional groups: the *data bus,* the *data transfer control bus,* and the *interface management bus.* Figure 11–40 illustrates the three bus groupings that make up the GPIB in a typical arrangement.

The data bus and the data transfer control bus   Eight data bits are transferred in parallel on the bidirectional data bus (DIO1–DIO8). This type of transfer is a *serial-byte–parallel-bit* format. Every byte that is transferred undergoes a handshaking operation via the transfer bus. The three active-low handshaking lines of the transfer bus indicate if data are valid (DAV), if the addressed instrument is not ready for data

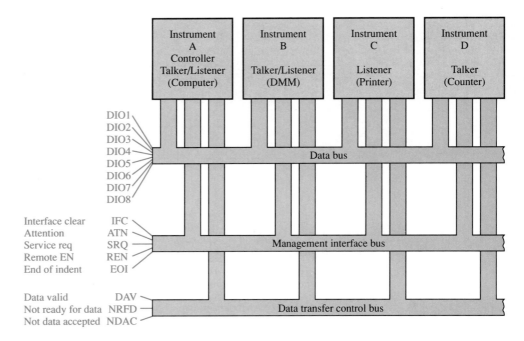

FIGURE 11–40

A typical GPIB interface arrangement showing the three bus groups.

(NRFD), or if the data are not accepted (NDAC). More than one instrument can accept data at the same time, and the slowest instrument sets the rate of transfer. Table 11–1 describes the handshaking signals, and Figure 11–41 shows the timing diagram for the GPIB handshaking sequence.

TABLE 11–1

The GPIB handshaking signals.

Name	Description
DAV	**Data Valid:** LOW is placed on this line by the talker when the data on its I/O are settled and valid, after the talker detects a HIGH on the NRFD line.
NRFD	**Not Ready for Data:** The listener places a LOW on this line to indicate that it is not ready for data. A HIGH indicates that it is ready. The NRFD line will not go HIGH until all addressed listeners are ready to accept data.
NDAC	**Not Data Accepted:** The listener places a LOW on this line to indicate that it has not accepted data. When it accepts data from its I/O, it releases its NDAC line. The NDAC line to the talker does not go HIGH until the last listener has accepted data.

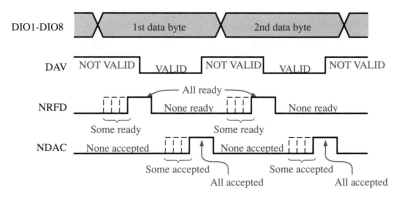

FIGURE 11–41
Timing diagram for the GPIB handshaking sequence.

Interface management bus These five lines control the orderly flow of data. The ATN (attention) line is monitored by all instruments connected to the bus. When the ATN line is active, the management bus is placed in the *command mode*. In this mode, the controller selects the specific interface operation, designates the talkers and the listeners, and provides specific addressing for the listeners. Each instrument designed to the GPIB standard has a specific identifying address that is used by the controller in the command mode. Table 11–2 describes the GPIB management lines and their functions.

TABLE 11–2
The GPIB management lines.

Name	Description
ATN	**Attention:** Causes all the devices on the bus to interpret data, as a controller command or address and activates the handshaking function.
IFC	**Interface Clear:** Initializes the bus.
SRQ	**Service Request:** Alerts the controller that a device needs to communicate.
REN	**Remote Enable:** Enables devices to respond to remote program control.
EOI	**End or Identify:** Indicates the last byte of data to be transferred.

Limitations of the GPIB There are three basic limitations on the use of the GPIB: (1) the cable length cannot exceed 15 meters, (2) there can be no more than one device per meter, and (3) the capacitive loading cannot exceed 50 pF per device.

The cable-length limitation can be overcome by the use of *bus extenders* and modems. A bus extender provides for cable-interfacing of instruments that are separated by a distance greater than that allowed by the GPIB cable limitation or for communicating over great distances via modem-interfaced telephone lines. The use of bus extenders is illustrated in Figure 11–42.

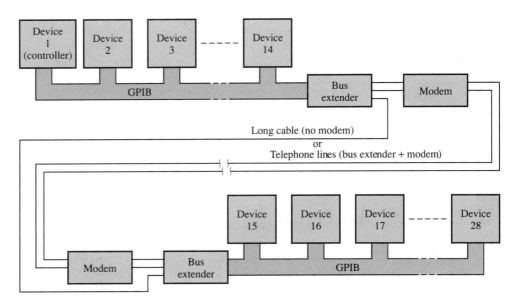

FIGURE 11–42
A bus extender and a modem can be used for interfacing remote GPIB systems.

The EIA-232-D (RS-232-C) Serial Interface

The EIA-232-D is an updated version of the RS-232-C interface standard for *serial* transfer of data bits.

This standard is commonly used for interfacing *data terminal equipment (DTE)* with *data communications equipment (DCE)*. An example is a serial computer terminal (DTE) interfaced with a modem (DCE).

This standard specifies twenty-five lines, but in many applications only a few of the lines are used. For example, at its simplest extreme, the EIA-232-D interface can be implemented with only two wires: a signal and a ground. Figure 11–43 illustrates a more common interface configuration, consisting of eight lines: transmitted data (TD), received data (RD), request to send (RTS), clear to send (CTS), data set ready (DSR), data terminal ready (DTR), protective ground, and signal ground.

FIGURE 11–43
A typical EIA-232-D/RS-232-C interface.

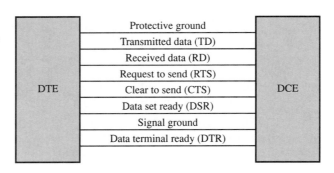

FIGURE 11–44

A full EIA-232-D/RS-232-C interface.

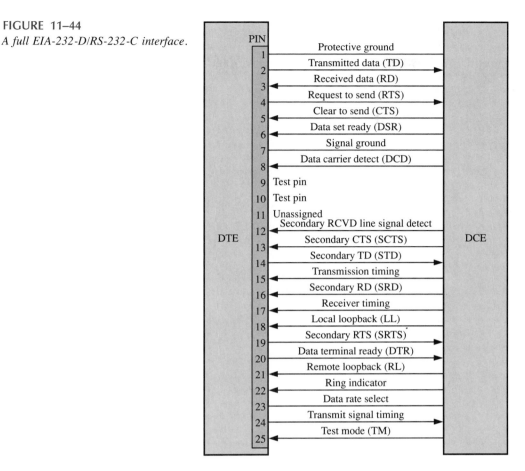

A full twenty-five-line interface showing all specified signals with connector pin numbers is illustrated in Figure 11–44.

There are three signals that make the EIA-232-D different from the RS-232-C standard. These signals are as follows: pin 18 is now local loopback (LL), pin 21 is now remote loopback (RL), and pin 25 is now test mode (TM). These new functions provide for loopback testing, in which a signal is sent out over part of the system, returned to the origin, and compared to check for system faults.

Electrical and mechanical specifications for the EIA-232-D interface standard are listed in Table 11–3.

Troubleshooting Buses

Troubleshooting bus systems is very difficult, if not impossible, without specialized equipment, such as a GPIB bus-system analyzer (the HP 59401A is an example).

A GPIB analyzer allows you to see the status of all bus lines as well as the actual codes by completely exercising the talkers, the listeners, or the controller. Normally, the system can be single-stepped for software debugging or run at the normal rate to test for system-related faults. The analyzer can operate as a talker or a listener, and data can be examined and compared at any preselected point in a transfer sequence to check for timing-related problems.

Also, RS-232 bus analyzers are available to ease the problem of troubleshooting these bus systems.

TABLE 11–3
Specifications for the EIA-232-D.

Parameter	Specification
DTE connector	DB-25 male
DCE connector	DB-25 female
Maximum cable length	capacitance limited (2500 pF)
Maximum data rate	20 kbits/s
Number of drivers on line	1 driver
Number of receivers on line	1 receiver
Driver output swing	± 5 V min, ± 15 V max
Driver load	3 kΩ to 7 kΩ
Driver slew rate	30 V/μs max
Receiver input resistance	3 kΩ to 7 kΩ
Receiver input threshold	± 3 V
Receiver input range	± 30 V max

SECTION 11–6 REVIEW

Answers are found at the end of the chapter.
1. True or false: The GPIB provides for parallel data transfer.
2. True or false: The EIA-232-C provides for serial data transfer.
3. Give an example of a GPIB application.
4. Give an example of a EIA-232-D or RS-232-C application.

SUMMARY

☐ D/A converters:

Binary-weighted input
$R/2R$ ladder (most common)

☐ A/D converters:

Flash (fastest)
Stairstep-ramp
Tracking
Single-slope
Dual-slope
Successive-approximation (most common)

☐ S-100 bus:

Data-input lines	8
Data-output lines	8
Address lines	16
Control lines	38
Interrupt lines	8
Power lines	4
Ground lines	2
Total lines	100

☐ Multibus

Bidirectional data lines	16
Address lines	20
Control/command lines	17
Interrupt lines	8
Power lines	14
Ground lines	8
Unused/reserved lines	3
Total lines	86

☐ General Purpose Interface Bus (GPIB)/IEEE-488/HPIB

Parallel data lines	8
Transfer lines	3
Management interface lines	5
Ground/shield lines	8
Total lines	24

☐ EIA-232-D/RS-232-C (serial interface)

Serial data lines	2
Timing and control lines	18
Ground lines	2
Unassigned lines	1
Reserved lines	2
Total lines	25

Integrated Circuit Technologies (Chapter 13) may be covered after this chapter, covered after a later chapter, partially omitted, or completely omitted, depending on the needs and goals of your program. Coverage is not prerequisite or corequisite to any other material in this book. Inclusion is purely optional.

GLOSSARY

Acceptor   A receiving device on a bus.

A/D conversion   The process of converting an analog signal to a sequence of digital codes.

Bus   A set of interconnections that interface components of a digital system or systems based on standardized specifications.

Bus arbitration   The process that prevents two sources from using a bus at the same time.

DAT   Digital audiotape.

D/A conversion   The process of converting a sequence of digital codes to an analog form.

DCE   Data communications equipment.

DTE   Data terminal equipment.

GPIB   General-purpose interface bus.

Handshaking   The method or process of signal interchange by which two digital devices or systems jointly establish communication.

Interfacing   The process of making two or more electronic devices or systems operationally compatible with each other so that they function properly together.

Listener   An instrument capable of receiving data on a GPIB.

Modem   A modulator/demodulator for interfacing digital devices to analog transmission systems such as telephone lines.

Monotonicity   The characteristic of a D/A converter defined by the absence of any incorrect step reversals. One type of D/A linearity.

Operational amplifier (op-amp)   A linear device with two differential inputs that has very high gain, very high input impedance, and very low output impedance.

Talker   An instrument capable of transmitting data on a GPIB.

SELF-TEST

Answers and solutions are found at the end of the book.

1. Define the term *interfacing*.
2. Distinguish between analog and digital signals.
3. What does it mean when we say a D/A converter is monotonic? Is this good or bad?
4. What is the resolution, expressed as a percentage, of a five-bit D/A converter?
5. In what other way can the resolution of the D/A converter in question 4 be expressed?
6. What is the advantage of the tracking A/D converter over the stairstep-ramp converter?
7. Why are tristate drivers required when a device is interfaced with a bus?
8. If one digital system is based on the S-100 bus and another on the Multibus, which has the greatest addressing capability?
9. Distinguish between internal and external bus systems. Give examples of each.
10. Name the three types of devices defined to operate on the GPIB.
11. What is a common application of the GPIB?
12. What is the main difference between the GPIB and the EIA-232-D standard interface?

PROBLEMS

Answers to odd-numbered problems are found at the end of the book.

Section 11–1   Interfacing the Analog and Digital Worlds

11–1   The analog curve in Figure 11–45 is sampled at 1-ms intervals. Represent the total curve by a series of four-bit binary numbers.

FIGURE 11–45

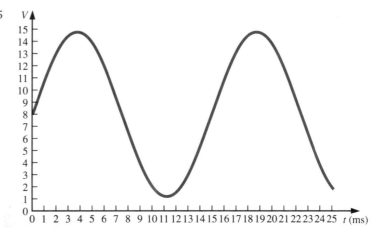

11–2 Sketch the digital reproduction of the curve in Problem 11–1.

11–3 Graph the analog function represented by the following sequence of binary numbers: 1111, 1110, 1101, 1100, 1010, 1001, 1000, 0111, 0110, 0101, 0100, 0101, 0110, 0111, 1000, 1001, 1010, 1011, 1100, 1100, 1100, 1011, 1010, 1010.

Section 11–2 Digital-to-Analog (D/A) Conversion

11–4 The input voltage to a certain op amp inverting amplifier is 10 mV, and the output is 2 V. What is the closed-loop voltage gain?

11–5 To achieve a closed-loop voltage gain of 330 with an inverting amplifier, what value of feedback resistor do you use if $R_{IN} = 1$ kΩ?

11–6 In the four-bit D/A converter in Figure 11–6, the lowest-weighted resistor has a value of 10 kΩ. What should the values of the other input resistors be?

11–7 Determine the output of the D/A converter in Figure 11–46(a) if the sequence of four-bit numbers in part (b) is applied to the inputs.

11–8 Repeat Problem 11–7 for the inputs in Figure 11–47.

11–9 Determine the resolution expressed as a percentage, for each of the following D/A converters:
   (a) three-bit   **(b)** ten-bit   **(c)** eighteen-bit

FIGURE 11–46

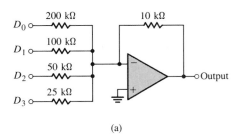
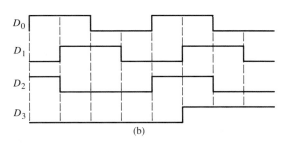

(a)

(b)

FIGURE 11–47

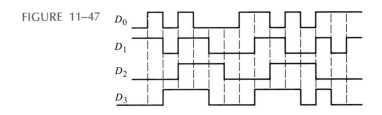

Section 11–3 Analog-to-Digital (A/D) Conversion

11–10 Determine the binary output code of a three-bit flash A/D converter for the analog input signal in Figure 11–48. The sampling rate is 100 kHz.

FIGURE 11–48

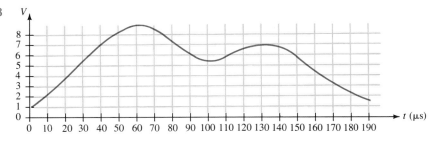

11–11 Repeat Problem 11–10 for the analog waveform in Figure 11–49.

11–12 For a three-bit stairstep-ramp A/D converter, the reference voltage advances one step every microsecond. Determine the encoded binary sequence for the analog signal in Figure 11–50.

11–13 For a four-bit stairstep-ramp A/D converter, assume that the clock period is 1 μs. Determine the binary sequence on the output for the input signal in Figure 11–51.

11–14 Repeat Problem 11–12 for a tracking A/D converter.

11–15 For a certain four-bit successive-approximation A/D converter, the maximum ladder output is +8 V. If a constant +6 V is applied to the analog input, determine the sequence of binary states for the SAR.

FIGURE 11–49

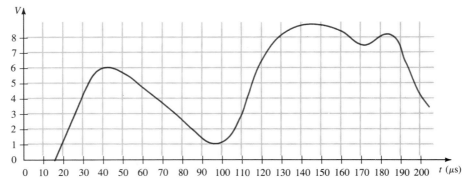

FIGURE 11–50

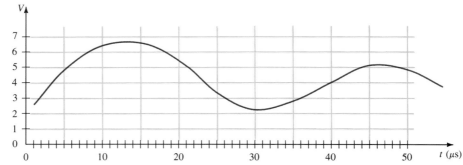

FIGURE 11–51

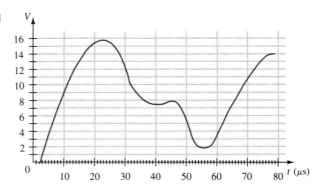

Section 11–4   Troubleshooting

11–16 Design a circuit for generating an 8-bit binary test sequence for the test setup in Figure 11–25.

11–17 A four-bit D/A converter has failed in such a way that the MSB is stuck in the 0 state. Draw the analog output when a straight binary sequence is applied to the inputs.

11–18 A straight binary sequence is applied to a four-bit D/A converter, and the output in Figure 11–52 is observed. What is the problem?

11–19 An A/D converter produces the following sequence of binary numbers when an analog signal is applied to its input: 0000, 0001, 0010, 0011, 0100, 0101, 0110, 0111, 0110, 0101, 0100, 0011, 0010, 0001, 0000.

   (a)   Reconstruct the input digitally.

   (b)   If the A/D failed so that the code 0111 were missing, what would the reconstructed output look like?

FIGURE 11–52

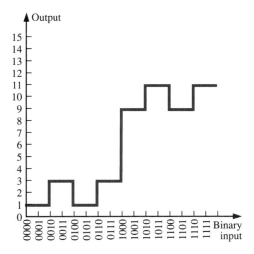

Section 11–5   Internal Device Interfacing

11–20 In a simple serial transfer of eight data bits from a source device to an acceptor device, the handshaking sequence in Figure 11–53 is observed on the four generic bus lines. By analyzing the time relationships, identify the function of each signal, and indicate if it originates at the source or at the acceptor.

11–21 Determine the signal on the bus line in Figure 11–54 for the data-input and Enable waveforms shown.

11–22 In Figure 11–55(a) data from the two sources are being placed on the data bus under control of the select line. The select waveform is shown in Figure 11–55(b). Determine the data-bus waveforms for the device output codes indicated.

FIGURE 11–53

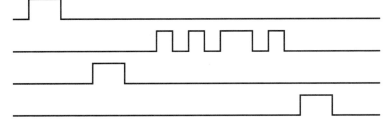

FIGURE 11–54

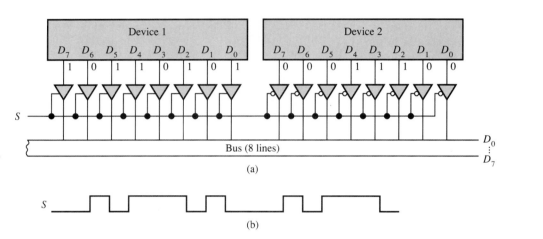

FIGURE 11–55

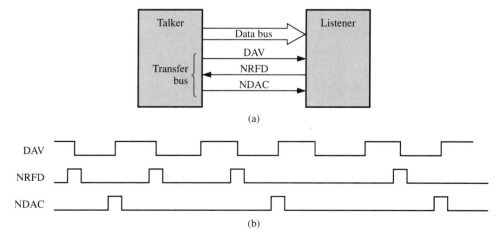

Section 11–6 Digital Equipment Interfacing

11–23 Eight GPIB-compatible instruments are connected to the bus. How many more can be added without exceeding the specifications?

11–24 Consider the GPIB interface between a talker and a listener as shown in Figure 11–56(a). From the handshaking timing diagram in part (b), determine how many data bytes are actually transferred to the listening device.

FIGURE 11–56

11–25 Describe the operations depicted in the GPIB timing diagram of Figure 11–57(a). Sketch a basic block diagram of the system involved in this operation.

11–26 A talker sends a data byte to a listener in a GPIB system. Simultaneously, a DTE sends a data byte to a DCE on an EIA-232-D interface. Which system will receive the complete data byte first? Why?

FIGURE 11–57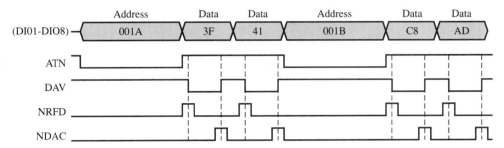

ANSWERS TO
SECTION REVIEWS

Section 11–1   Review

1. analog
2. to convert an analog quantity to digital form
3. to convert a digital quantity to analog form

Section 11–2   Review

1. Each resistor has a different value.
2. 6.67%

Section 11–3   Review

1. simultaneous (flash) method
2. tracking
3. true

Section 11–4   Review

1. A step reversal indicates nonmonotonic behavior.
2. Step amplitudes are less than ideal.
3. missing code; incorrect code

Section 11–5   Review

1. Tristate drivers allow devices to be completely disconnected from the bus when not in use, thus preventing interference with other devices.
2. A bus interconnects all the devices in a system and makes communication between devices possible.

Section 11–6   Review

1. true
2. true
3. automated test system
4. data communications

The purpose of this chapter is to provide a brief coverage of microprocessor-based systems. The Intel 8088-based microcomputer is used to illustrate the fundamental concepts and to provide some insight into the operation of systems that use microprocessors.

Naturally, a single chapter's coverage is very limited. One or more chapters could easily be devoted to *each* of the section topics. Keep in mind, however, that the purpose here is only to give you some basic familiarity with the subject before you proceed to full microprocessor and microcomputer courses. Of course, this chapter is an option, and some instructors may not wish to cover it at all; but for those who do, it will be helpful in preparing students for further study.

Although the 8088 is used for illustration, the concepts can easily be transferred to other types of microprocessors, such as the 68000 family. The 8086/8088 family is very widely used, and the newer versions—the 80286, the 80386, and the 80486—although more powerful, are closely related in architecture and other basic features.

After completing this chapter, you will be able to

☐ name the five basic units of a computer.

☐ explain the basic operation of an 8088 CPU.

☐ see how the CPU, memory, and I/O work together as a system.

☐ explain the multiplexed bus operation of the 8088 microprocessor.

☐ describe the function of the bus controller in an 8088 CPU.

☐ analyze a timing diagram for memory read and write cycles.

☐ distinguish between a dedicated I/O port and a memory-mapped I/O port.

☐ compare polled I/O to interrupt-driven I/O.

☐ describe the functions of PIC and PPI devices.

☐ define and explain the advantage of DMA.

☐ explain the basic architecture of the 8086/8088 microprocessors.

☐ distinguish between assembly language and machine language.

12

Introduction to Microprocessor-Based Systems

THE MICROCOMPUTER

12–1

In Chapter 10 memories were studied, and in Chapter 11 bus structures were introduced. In those chapters you saw generally how memories and buses are essential parts of microprocessor-based systems. Now, in this and several following sections, a particular microprocessor-based system, the **microcomputer,** *will be discussed.*

Basic Parts of a Microcomputer

The five basic parts that make up a microcomputer, or any computer for that matter, are

1. the arithmetic logic unit (ALU),
2. the control unit,
3. the memory unit,
4. the input unit,
5. the output unit.

These basic parts are shown in the block diagram of Figure 12–1. The arithmetic logic unit and the control unit are actually part of the *central processing unit (CPU)*, as indicated. The memory unit contains both RAM and ROM.

FIGURE 12–1

Block diagram of the basic parts of a microcomputer.

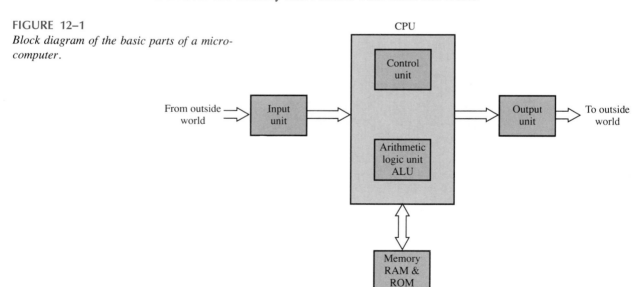

The Bus Structure of a Microcomputer

As you have learned in previous chapters, microcomputers contain three separate buses that interconnect all the basic parts. These buses are the *address bus,* the *data bus,* and the *control bus.* This structure is shown in Figure 12–2.

Addresses, data, and control signals are transferred between the CPU and the other units along these buses. In a typical microcomputer the address bus consists of twenty or more lines. Having twenty lines permits up to $2^{20} = 1,048,576$ memory

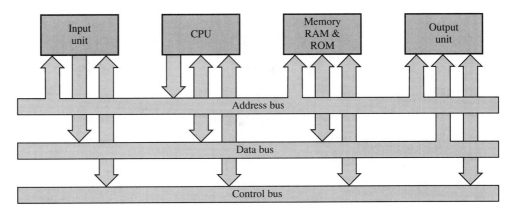

FIGURE 12–2
Basic bus structure of a microcomputer.

locations to be addressed. The data bus typically has either eight or sixteen lines, depending on the system. Some microprocessors have data buses with thirty-two lines. With an eight-line data bus, one data byte can be transferred at a time and with sixteen lines, two data bytes can be transferred. Thus, a sixteen-bit microcomputer operates faster than an eight-bit machine because more data can be transferred in a given time. The number of lines in the control bus varies depending on the requirements and the design of the specific microcomputer.

A Typical Microcomputer System

For a microcomputer to accomplish a given task, it must communicate with the "outside world." To do so, it must interface with human operators, sensing devices, and devices to be controlled. A typical microcomputer system configuration is shown in Figure 12–3. Usually there are one or more *disk drives* (hard disk and/or floppy disk) for mass storage, a *keyboard* for data entry by an operator, a *video monitor* for visual output, and a *printer* for hard-copy output. These external input and output devices are called *peripherals*.

FIGURE 12–3
A typical microcomputer system.

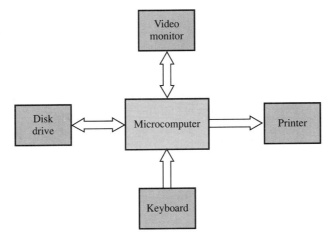

Other types of peripherals, such as a modem, a tape drive, a scanner, a graphics tablet, a mouse, a joystick, and a voice input, can also be connected to most microcomputers.

Now let us look at the basic purpose of each of the units within a microcomputer system.

Central processing unit (CPU)   In a microcomputer, a microprocessor and some support circuitry make up the CPU. For example, the IBM PC is based on the Intel 8088 microprocessor, which is part of the 8086 family. This microprocessor has an eight-bit data bus and a twenty-bit address bus.

Basically, a microprocessor-based CPU addresses a memory location, *fetches* a program instruction, and carries out *(executes)* the instruction. After completing one instruction, it moves on to the next one. This fetch-and-execute process is repeated until all the instructions in a specific program have been completed. A simple example of an application program is a set of instructions that directs the CPU to add a series of numbers and produces a sum.

Memory unit   As you know, the memory unit consists of RAM and ROM. The RAM is used to store data and programs *temporarily* during processing. Data are the numbers and other information that the computer works with, and *programs* are lists of instructions that tell the computer what to do. The RAM functions basically as a ''scratch pad'' for the computer. Everything stored in RAM is lost when the power is turned off, unless it has been transferred to disk storage.

The contents of ROM are *permanently* stored and are retained even when the power is turned off. Typically, the ROM contains the *system programs,* such as ROM BIOS (Basic Input/Output System). These are routines for handling video display graphics, printer communications, and the power-up self-test and other routines for servicing peripherals, as well as high-level language interpreter programs such as a BASIC interpreter. Since these types of programs (software) are stored in ROM (hardware), they are called *firmware*.

Input unit   We communicate with the computer via the input unit, which can generally handle several peripherals. This communication is often referred to as *man-machine interface*. For example, the keyboard is an input device that interfaces with the input unit and allows the operator to enter data and commands into the computer, and the mouse is an input device that allows the cursor to be easily moved about the video screen by the operator.

Output unit   The computer communicates with us via the output unit, which also can handle several peripherals. This communication is referred to as *machine-man interface*. For example, the output unit allows information from the computer to be displayed on a video monitor or to be printed by a printer.

Some peripherals function as both input and output devices. Examples are disk drives, tape drives, video monitors, and modems.

SECTION 12–1 REVIEW

Answers are found at the end of the chapter.
1. List the five basic parts of a microcomputer.
2. How are data moved from one unit to another in a microcomputer?
3. What does the term *peripheral* mean in relation to computer systems?

THE CENTRAL PROCESSING UNIT (CPU)

12–2

A typical CPU consists of a microprocessor and various associated ICs called support circuits. First we will look at the general operation of a CPU, and then we will examine a specific implementation using the 8088 microprocessor. Keep in mind that this coverage is intended to present a general look at the implementation and operation of a microcomputer. A thorough, detailed study must, by necessity, be left to later courses. The CPU is also known as the MPU (microprocessing unit).

The CPU is connected to the rest of the microcomputer units by three buses: the address bus, the data bus, and the control bus, as shown in Figure 12–4.

FIGURE 12–4

The CPU interfaces and communicates with the other parts of the system via its buses.

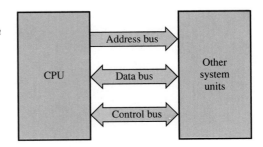

The Buses

Many microcomputers, including the IBM PC, have a twenty-bit address bus and an eight-bit data bus. Some systems have fewer than twenty address lines, and some have more. Also, some systems operate with a sixteen-bit data bus, and some even have thirty-two bits as previously mentioned. More address lines allow more memory locations to be addressed, and more data lines allow a faster transfer of data within the computer. The sizes of the address and data buses are determined by the type of microprocessor that is used.

Regardless of the bus sizes, the basic principles of operation are essentially the same for all microcomputers. The *address bus* is a "one-way street" on which the CPU sends an address code to select a specific memory location or a specific input/output (I/O) device. The *data bus* is a "two-way street" on which the CPU can receive an instruction or data for processing from the memory or from an input device and then return the result of a processing operation back to the memory for storage or to an output device. In general, some of the lines in the *control bus* are one-way and some are two-way. The CPU uses the control bus to communicate with a selected device to move data back and forth in an orderly manner.

An 8088-based CPU

As mentioned before, a typical CPU is implemented with a microprocessor and some additional support circuits. Figure 12–5 shows an 8088-based CPU similar to that used in the IBM PC.

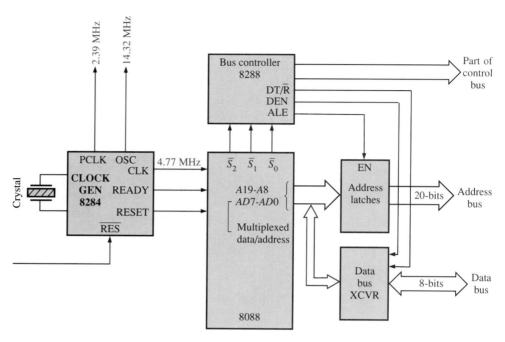

FIGURE 12–5
A simplified 8088-based CPU.

The 8088 microprocessor The 8088 is basically the same as the 8086 microprocessor except that the 8088 has an eight-bit external data bus rather than a sixteen-bit data bus. We will look at the internal structure (architecture) of the 8086/8088 microprocessors in a later section. For now, just a brief introduction to the device is adequate for the purpose of this coverage.

As mentioned, the 8088 has an eight-bit data bus, making it compatible with many eight-bit peripheral devices. Internally, however, the 8088 can handle data in sixteen-bit words, or it can handle data one byte at a time, depending on how it is programmed. Also, the 8088 has a twenty-bit address bus, which allows up to 1 megabyte (1 MB) of memory (actually 1,048,576 bytes) to be addressed.

Actually, in the 8088 the address bus and the data bus are combined. There are a total of twenty lines, AD0–AD7 and A8–A19. The eight lines designated AD0–AD7 serve as both data and address lines, using a multiplexed operation. The twelve lines designated A8–A19 are used strictly for addressing.

When a twenty-bit address is sent out by the microprocessor, all twenty lines are used as the address bus, as illustrated in Figure 12–6(a). When data are sent or received by the microprocessor, the lower eight lines, AD0–AD7, are used as a bidirectional data bus, as shown in Figure 12–6(b). The bus multiplexing is controlled by the bus controller, as you will see later. The bus controller is one of the support circuits. Other microprocessor input and output lines will be described in association with the appropriate support circuits.

The clock generator A clock generator circuit, such as the 8284, provides the basic timing signals to the 8088 and to other parts of the system. It also provides a RESET signal to the 8088 to initialize the internal circuitry and a READY signal to synchronize

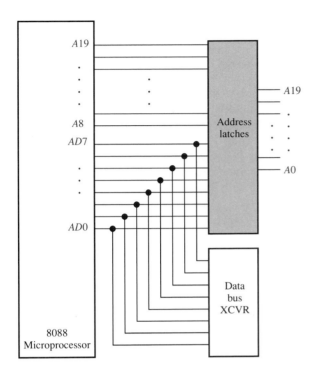

(a) Address latches enabled

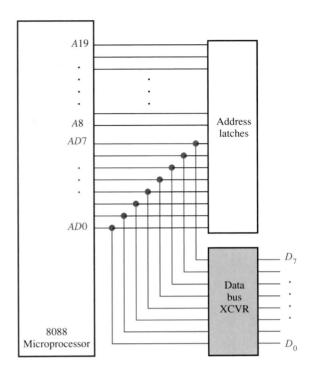

(b) Data bus transceiver enabled

FIGURE 12–6
Multiplexed bus operation of the 8088.

the microprocessor with the rest of the system. An external resonant crystal is used to establish the frequency of oscillation, as shown in Figure 12–7. In the PC a crystal frequency of 14.31818 MHz is used and is available on the OSC output. A divide-by-three counter inside the clock generator produces a 4.772727-MHz clock with a duty cycle of 33% to run the microprocessor. This signal is available on the CLK output.

FIGURE 12–7
The 8284 clock generator.

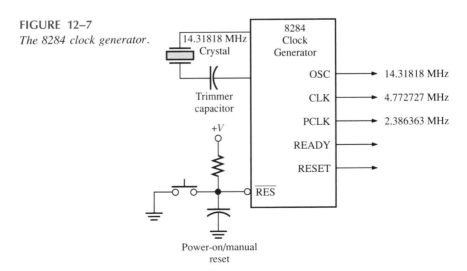

Also, a 2.386363-MHz signal is derived from the CLK frequency and is available on the PCLK output of the clock generator.

The bus controller   The 8288 bus controller is used to relieve the 8088 microprocessor of most bus control functions in order to maximize the processing efficiency. One function of the bus controller in the CPU is to provide the signals to multiplex the AD0–AD7 bus lines coming from the microprocessor. The bus controller generates these control signals based on the codes on the status lines, $\overline{S}_0$, $\overline{S}_1$, and $\overline{S}_2$, from the microprocessor.

When the microprocessor needs to communicate with memory or I/O, it sends a code on the status lines to the bus controller. The bus controller then generates an ALE (address latch Enable) signal, which latches the address placed on AD0–AD7 and A8–A19 by the microprocessor into the *address latches*. There are twenty latches (A_0–A_{19}), whose tristate outputs form the system address bus. During a memory read/write cycle, a valid address is available on this address bus, and, therefore, the lower eight bits (AD0–AD7) of the microprocessor's combined bus are freed up to send or receive data. The steps in this operation are illustrated in Figure 12–8.

After the address has been latched, the microprocessor signals the bus controller via the status lines ($\overline{S}_2$, $\overline{S}_1$, and $\overline{S}_0$) that it wants to read data or write data to the memory location or I/O that is specified by the address that is held on the address bus. The status line codes for the four possible read/write operations are given in Table 12–1.

When the bus controller receives one of the status codes, it sends a read command ($\overline{\text{IORC}}$ or $\overline{\text{MRDC}}$) or a write command ($\overline{\text{IOWC}}$ or $\overline{\text{MWTC}}$) to the appropriate device,

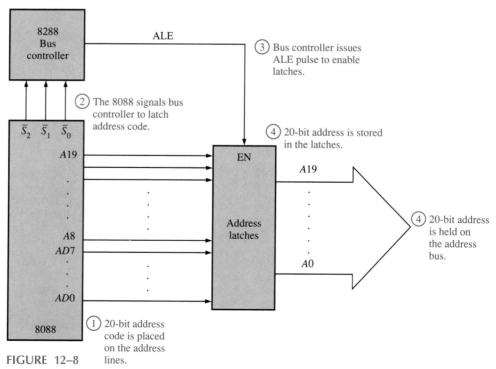

FIGURE 12–8

Steps in placing a valid address on the bus.

TABLE 12–1

The 8088 read/write status codes.

Status Code				
$\bar{S}_2$	$\bar{S}_1$	$\bar{S}_0$	Operation	Active Bus Controller Output Command
0	0	1	Read I/O port	$\overline{IORC}$
0	1	0	Write I/O port	$\overline{IOWC}$
1	0	1	Read memory	$\overline{MRDC}$
1	1	0	Write to memory	$\overline{MWTC}$

and it also sends two signals to the tristate *data bus transceiver*. These two signals are DT/$\overline{R}$ (data transmit/$\overline{receive}$) and DEN (data Enable). The DEN signal enables the data bus transceiver, and the DT/$\overline{R}$ signal selects the direction of data on the data bus (D_0–D_7). The basic operation is illustrated in Figure 12–9 for both read and write. Timing diagrams for the memory read and write cycles are given in the next section.

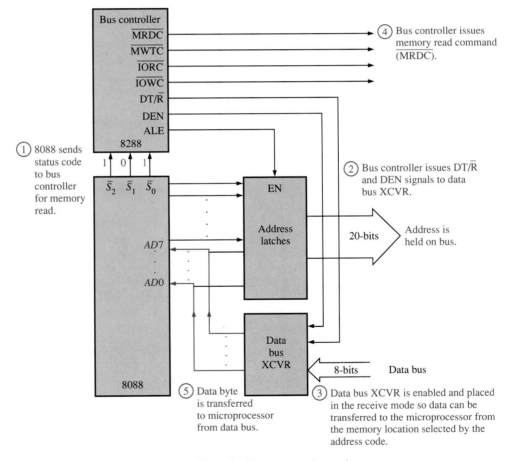

(a) Example of a memory read operation

FIGURE 12–9

Examples of CPU read and write operations.

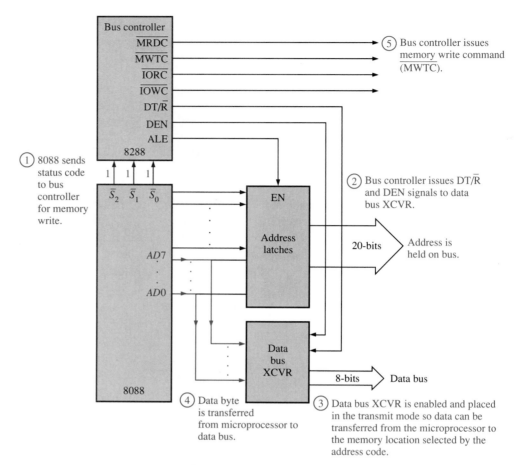

(b) Example of a memory write operation

FIGURE 12–9
(*Continued*)

In the IBM PC the address latches are implemented with three 74LS373 tristate octal-latch ICs. The data bus transceiver is implemented with three 74LS245 octal-bus-transceiver ICs.

SECTION 12–2 REVIEW

Answers are found at the end of the chapter.
1. What is the purpose of the clock generator in the CPU?
2. List two functions of the bus controller.
3. Explain why bus multiplexing is necessary in an 8088-based CPU.

THE MEMORY

12–3

Now that the basic operation of the CPU has been discussed, we will move on to the memory portion of a typical microcomputer to see how it is used by the CPU.

Address Allocation

As you have seen, the CPU can address up to 1,048,576 locations with its twenty-bit address bus (A_0–A_{19}). In a typical microcomputer system, a portion of the total address space is allocated to RAM and ROM and a portion to I/O. Input and output ports have unique addresses that are treated like memory locations, as you will see in the next section.

In the IBM PC, for example, the RAM is allocated addresses 00000_{16} through $BFFFF_{16}$ (a total of 786,431 locations), although not all of this RAM memory space is actually used. In the maximum configuration, 640 kilobytes (640 KB) of available memory are utilized. The ROM is allocated addresses $C0000_{16}$ through $FFFFF_{16}$ (a total of 262,143 locations). This particular memory allocation is shown in Figure 12–10. Within the RAM and the ROM, the addresses are segmented into areas for various purposes, such as system programs, user programs, data, and various types of system-related information. This segmenting is discussed in more detail later.

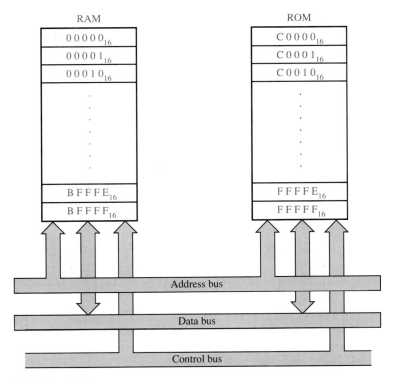

FIGURE 12–10
Example of memory address allocation in a microcomputer.

CPU-Memory Operation

The CPU handles two types of memory transfers, read and write. Basically, during the read operation, the CPU fetches either a program instruction, another address, or data. During the write operation, the CPU sends data resulting from a computation or some other operation back to RAM. A diagram of the CPU and memory portions of a microcomputer is shown in Figure 12–11.

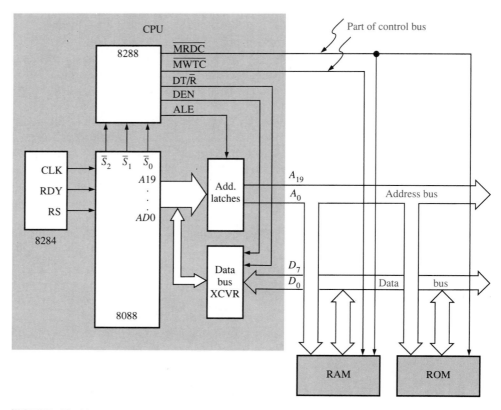

FIGURE 12–11
Basic CPU-memory organization.

The memory read cycle The microprocessor initiates a memory read cycle by sending a status code of $\overline{S}_2\overline{S}_1\overline{S}_0 = 101$ to the bus controller and placing a binary address on the AD0–AD7 and A8–A19 bus lines. The bus controller then issues an ALE signal, which latches the address onto the address bus (A_0–A_{19}), freeing up the AD0–AD7 lines for data transfer. Also, the bus controller issues a low DT/$\overline{R}$ signal to the bus transceiver to set it up for receiving data from the memory and passing it on to the microprocessor.

Next the bus controller issues a memory read command ($\overline{MRDC}$) to the memories. This instructs the memory to place the contents of the selected address on the data bus. The DEN signal from the bus controller then enables the bus transceiver, and a byte of data flows into the microprocessor. The basic timing diagram for a memory read cycle is shown in Figure 12–12.

Memory write cycle The microprocessor initiates a memory write cycle by sending status code $\overline{S}_2\overline{S}_1\overline{S}_0 = 110$ to the bus controller and placing a binary address on the bus lines. The bus controller then issues an ALE signal to latch the address onto the address bus (A_0–A_{19}) and a HIGH DT/$\overline{R}$ signal to set up the bus transceiver for sending data to the memory. The microprocessor then places the data byte on the AD0–AD7 lines. The DEN signal from the bus controller then enables the bus transceiver to place the data byte on the data bus (D_0–D_7).

FIGURE 12–12

Basic timing diagram for an 8088-based CPU memory read cycle.

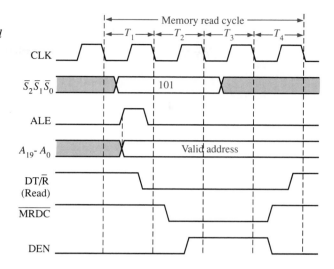

Next the bus controller issues a memory write command ($\overline{\text{MWTC}}$) to the memories. This command instructs the RAM to store the data byte in the selected address. A basic write-cycle timing diagram is shown in Figure 12–13.

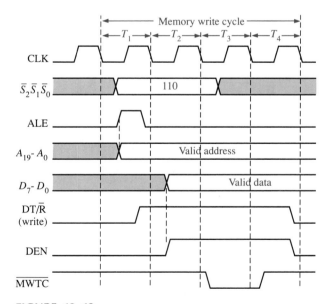

FIGURE 12–13

Basic timing diagram for an 8088-based CPU memory write cycle.

SECTION 12–3
REVIEW

Answers are found at the end of the chapter.

1. Name the two cycles used by the CPU to transfer data to and from the memory.
2. How many clock cycles does it take to read data from the memory in an 8088-based system?
3. When does a valid address appear on the address bus (A_0–A_{19}) during a write cycle?

THE INPUT/OUTPUT (I/O) PORT

12–4

In this section, you will see generally how the 8088-based CPU communicates with input and output devices via I/O ports. An I/O port can be thought of as a "window" between the computer and the outside world, through which information is passed.

A block diagram of a basic microcomputer system, including I/O ports and typical peripherals, is shown in Figure 12–14. Notice that a port can be strictly for input, strictly for output, or bidirectional for both input and output. Although only five I/O ports are shown in the figure, the CPU is capable of handling many more than that.

As mentioned, the purpose of an I/O port is to provide an interface between the computer and the "outside world" of peripheral equipment. Depending on the system

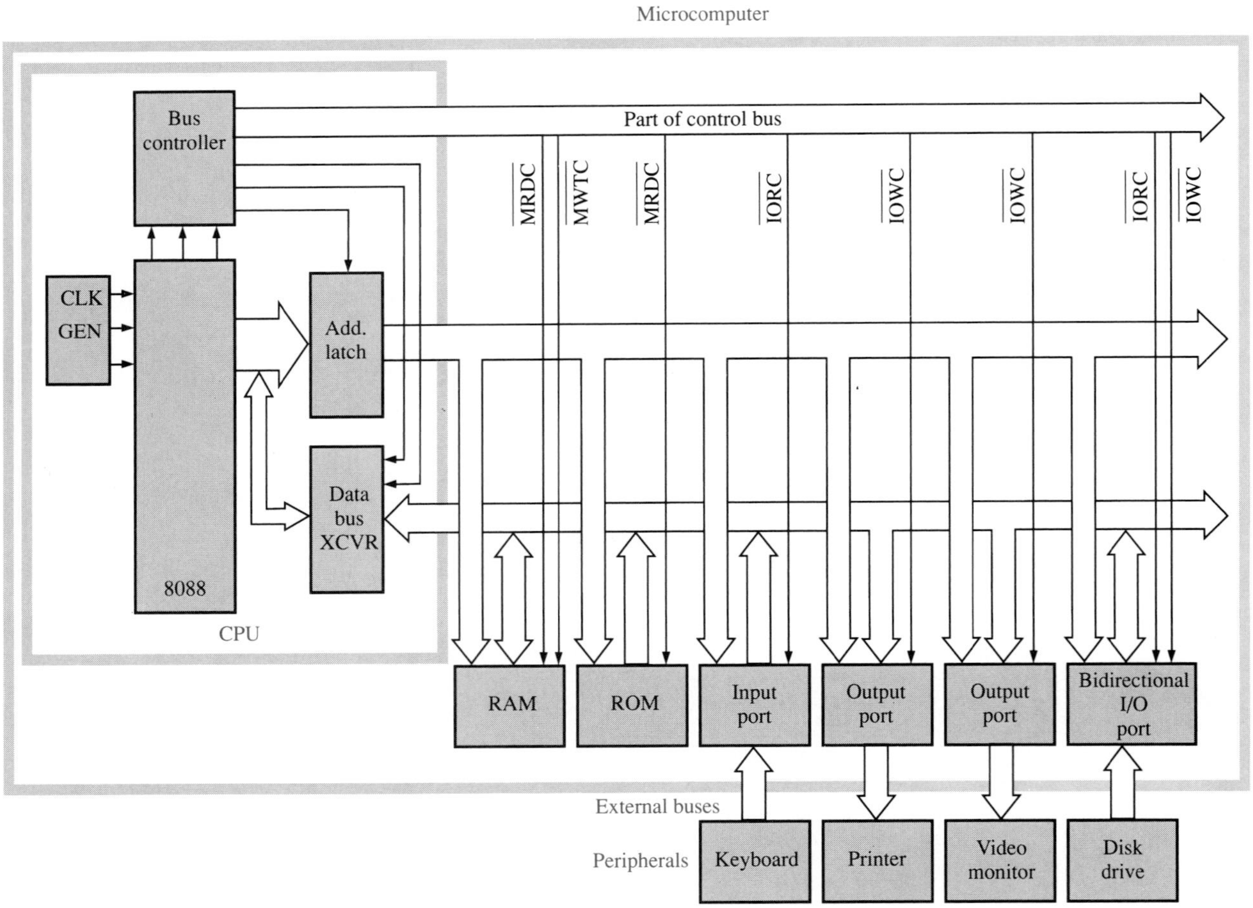

FIGURE 12–14
Basic microcomputer system with I/O ports and peripherals.

design, the I/O ports can be either *dedicated* or *memory-mapped*. These terms describe the ways in which the ports are accessed by the CPU.

Dedicated I/O Ports

A dedicated I/O port is assigned a unique address within the I/O address space of the computer, and dedicated I/O commands are used for communication. Dedicated I/O ports are accessed by the I/O read and write commands, $\overline{IORC}$ and $\overline{IOWC}$ in an 8088-based system. Figure 12–15 shows how a dedicated output port might be implemented.

FIGURE 12–15
A basic dedicated output port.

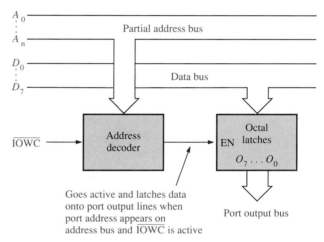

Output port   The basic operation of the output port is as follows: The port occupies one location within the I/O address space of the system; that is, it has a unique address different from the addresses of other ports. Typically, only a portion of the twenty-bit address bus is used for port addresses, because the actual number of I/O ports required is small compared with the number of memory locations.

During a CPU write cycle, the port address is placed on the address bus, and the CPU issues a low $\overline{IOWC}$ (I/O write command) signal to enable the *address decoder*. The decoder then produces a signal to latch the data byte onto the port output lines.

Input port   A dedicated input port is not quite as simple as an output port, because the port output lines are connected to the system data bus. This arrangement requires that the port output lines be disabled when the port is not in use, to prevent interference with other activity on the data bus. An example of a basic input port implementation is shown in Figure 12–16.

Bidirectional port   The third type of dedicated I/O port is the bidirectional port, which is basically a combination of the input and output ports with additional enabling controls. A bidirectional port allows data to flow to or from a peripheral device, such as a disk drive.

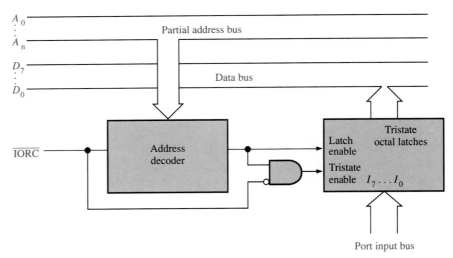

FIGURE 12–16
A basic dedicated input port.

Memory-Mapped I/O Ports

A second approach to accessing I/O ports is the memory-mapped port. The port implementation is essentially the same as for the dedicated ports in Figures 12–15 and 12–16 except for two things:

1. The memory-mapped ports are assigned addresses within the computer memory address space and are actually viewed as memory locations by the CPU. All twenty address lines are used for memory-mapped I/O.
2. The CPU accesses a memory-mapped I/O port with the $\overline{MRDC}$ and $\overline{MWTC}$ commands rather than the $\overline{IORC}$ and $\overline{IOWC}$ commands. In other words, the memory-mapped I/O ports are treated exactly as memory locations.

Programmable Peripheral Interface

In many systems, including the IBM PC, special ICs are used to implement the I/O ports. One such device is the 8255A programmable peripheral interface (*PPI*), shown in Figure 12–17. The address decoder is a separate device for enabling the PPI when a port address is decoded.

This particular PPI provides three I/O ports, labeled *A, B,* and *C*. Ports *A* and *B* have 8 lines each, and port *C* is split into two 4-line groups. The ports can be programmed for various combinations of input ports, output ports, and bidirectional ports and for control and handshaking lines. The PPI can be used for either dedicated or memory-mapped ports.

**SECTION 12–4
REVIEW**

Answers are found at the end of the chapter.
1. What is an I/O port?
2. Explain the difference between a *dedicated* I/O port and a *memory-mapped* I/O port.
3. What is the purpose of a PPI?

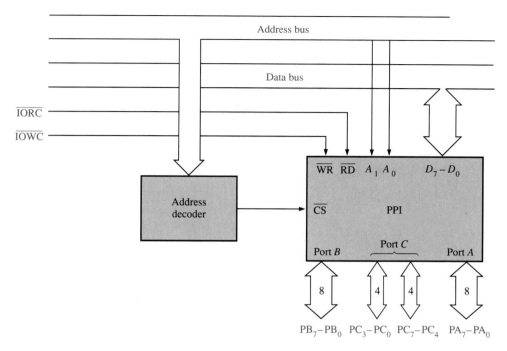

FIGURE 12–17
A basic programmable peripheral interface (PPI) configuration.

I/O INTERRUPTS

12–5

In the last section the concept of an I/O port was presented, and both dedicated and memory-mapped ports were introduced. In this section the coverage of I/O is continued with a basic introduction to the concepts of the establishment of communications between a peripheral and the CPU. Two methods are discussed: **polled I/O** *and* **interrupt-driven I/O.**

In microprocessor-based systems, such as the microcomputer, peripheral devices require periodic service from the CPU. The term *service* generally means sending data to or taking data from the device or performing some updating process.

In general, peripheral devices are very slow compared with the CPU. For example, the 8088 CPU can perform about 1,000,000 read or write operations in one second (based on a 4.77-MHz clock frequency). A printer may average only a few characters per second (one character is represented by eight bits), depending on the type of material being printed. A keyboard input may be about one or two characters per second, depending on the operator. So, in between the times that the CPU is required to service a peripheral, it can do a lot of processing, and in most systems this processing time must be maximized by using an efficient method for servicing the peripherals.

Polled I/O

One method of servicing the peripherals is called *polling*. In this method the CPU must test each peripheral device in sequence at certain intervals to see if it needs or is ready for servicing. The basic polled I/O method is illustrated in Figure 12–18.

The CPU sequentially selects each peripheral device via the multiplexer to see if it needs service by checking the state of its *ready* line. Certain peripherals may need service at irregular and unpredictable intervals, that is, more frequently on some occasions than on others. Nevertheless, the CPU must poll the device at the *highest* rate. For example, let us say that a certain peripheral *occasionally* needs service every 1000 μs but most of the time requires service only once every 100 ms. As you can see, precious processing time is wasted if the CPU polls the device, as it must, at its maximum rate (every 1000 μs), because most of the time the device will not need service when it is polled.

Each time the CPU polls a device, it must stop the program that it is currently processing, go through the polling sequence, provide service if needed, and then return to the point where it left off in its current program.

Another problem with the sequentially polled I/O approach is that if two or more devices need service at the same time, the first one polled will be serviced first, and the other devices will have to wait, although they may need servicing much more urgently than the first.

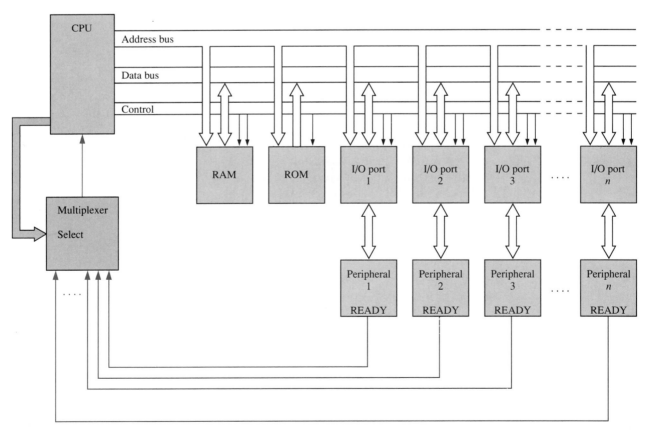

FIGURE 12–18
The basic polled I/O configuration.

As you can see, polling is suitable only for devices that can be serviced at regular and predictable intervals and only in situations in which there are no priority considerations.

Interrupt-driven I/O

This approach overcomes the disadvantages of the polling method. In the interrupt method the CPU responds to a need for service only when service is requested by a peripheral device. Thus, the CPU can concentrate on running the current program without having to break away unnecessarily to see if a device needs service.

A request for service from a peripheral device is called an *interrupt*. When the CPU receives an I/O interrupt, it temporarily stops its current program and fetches a special program (service routine) from memory for the particular device that has issued the interrupt. When the service routine is complete, the CPU returns where it left off.

A device called a *programmable interrupt controller* (*PIC*) handles the interrupts on a priority basis. The 8259A is an example of a PIC that can handle up to eight interrupts.

The PIC accepts service requests from the peripherals. If two or more devices request service at the same time, the one assigned the highest priority is serviced first, then one with the next highest priority, and so on. After issuing an interrupt (INTR) signal to the CPU, the PIC provides the CPU with information that ''points'' the CPU to the beginning memory address of the appropriate service routine. This process is called ''vectoring.'' A basic interrupt-driven I/O configuration is shown in Figure 12–19.

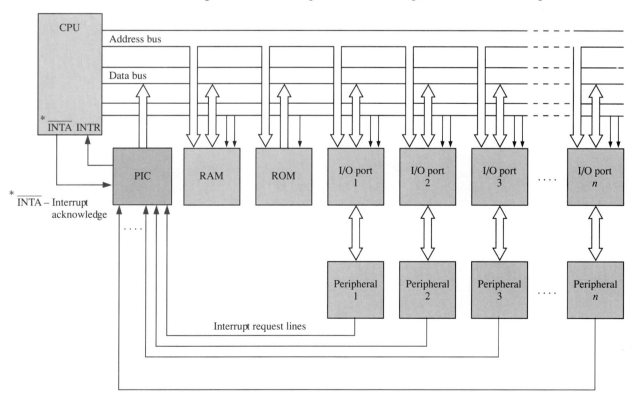

FIGURE 12–19
A basic interrupt-driven I/O configuration.

SECTION 12–5
REVIEW

Answers are found at the end of the chapter.
1. How does an interrupt-driven I/O differ from a polled I/O?
2. What is the main advantage of an interrupt-driven I/O?

DIRECT MEMORY ACCESS (DMA)

12–6

All I/O data transfers discussed so far have passed through the CPU. For example, when data are to be transferred from RAM to a peripheral device, the CPU reads the first data byte from the memory and loads it into an internal register within the microprocessor. Then the CPU writes the data byte to the appropriate I/O port. This read/write operation is repeated for each byte in the group of data to be transferred. This process is illustrated in Figure 12–20(a).

For large blocks of data, intermediate stops in the microprocessor consume a lot of time. For this reason, many systems use a technique called *direct memory access* (*DMA*) to speed up data transfers between RAM and certain peripheral devices. Basically, DMA bypasses the CPU for certain types of data transfers, thus eliminating the time consumed by the normal fetch and execute cycles required for each read or write operation.

For direct memory transfers, a device called the *DMA controller* takes control of the system buses and allows data to flow directly between RAM and the peripheral device, as indicated in Figure 12–20(b).

Transfers between the disk drive and RAM are particularly suited for DMA because of the large amounts of data involved and the serial nature of the transfers. The DMA controller can handle data transfers several times faster than the CPU.

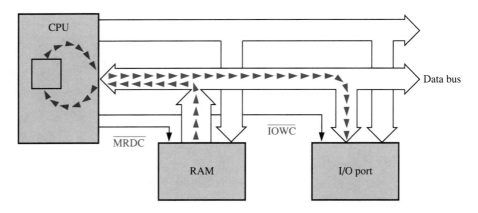

(a) A memory / I/O transfer handled by the CPU

FIGURE 12–20
A simplified comparison of a CPU-handled transfer and a DMA transfer.

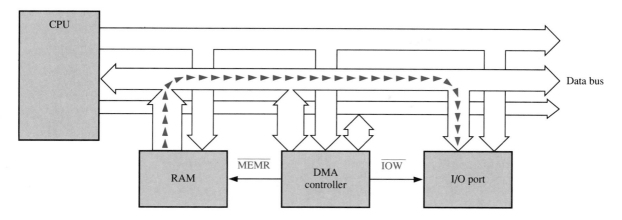

(b) A DMA transfer

FIGURE 12–20
(*Continued*)

SECTION 12–6
REVIEW

Answers are found at the end of the chapter.
1. What does DMA stand for?
2. Discuss the advantage of DMA, and give an example of a type of transfer for which it is often used.

THE MICROPROCESSOR

12–7

Now that you have been introduced to some basic microprocessor-based-system concepts and are familiar with some aspects of the 8088 microprocessor, we will look a bit more closely at the microprocessor itself. In this section we will discuss the arrangement and function of the internal parts (**architecture**) *of the 8086 and 8088 microprocessors. Again, this coverage is brief and is intended only to introduce you to basic concepts.*

The 8086 and 8088 microprocessors are third-generation devices; they were preceded by the 8080 and the 8085. More recent improvements in the performance and operational features of the 8086/8088 family have led to the 80286, the 80386, and the 80486 microprocessors.

The 8086 and 8088 are functionally identical except that the 8088 has an eight-bit external data bus and the 8086 has a sixteen-bit bus.

Basic Operation

A microprocessor is a VLSI device that executes a program (list of instructions) by repeatedly cycling through the following basic steps:

1. Fetch the next instruction from memory.

2. Read an *operand* if required by the instruction (an operand is a quantity to be operated on as directed by its associated instruction).

3. Execute the instruction (do what the instruction says).

4. Write the result back into memory (if required by the instruction).

These basic processing steps are performed in the 8086/8088 by two separate internal units, the *execution unit* (EU), which executes instructions, and the *bus interface unit* (BIU), which interfaces with the system buses and fetches instructions, reads operands, and writes results. These are shown in Figure 12–21.

8086/8088 Microprocessor

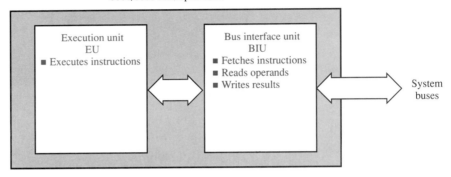

FIGURE 12–21
The 8086/8088 has two separate internal units, the EU and the BIU.

The BIU performs all the bus operations for the EU, such as data transfers from memory or I/O. While the EU is executing instructions, the BIU "looks ahead" and fetches more instructions from memory. This action is called *prefetching*. The prefetched instructions are stored in an internal memory called the instruction queue (pronounced "Q"). This queue allows the BIU to keep the EU supplied with instructions so that the EU does not have to wait for the next instruction to be fetched each time it completes the previous one. The queue thus speeds up the processing by effectively overlapping fetch and execute operations, which, in earlier microprocessors, were handled serially (fetch, then execute, fetch, then execute, etc.). Figure 12–22 compares the serial fetch/execute cycle with the overlapping fetch/execute operation performed by the 8086/8088.

Basic 8086/8088 Architecture

Figure 12–23 is a block diagram of the internal organization of an 8088 microprocessor. The 8086 is identical except that it has a sixteen-bit data bus and a six-byte instruction queue.

The Bus Interface Unit (BIU)

The major parts of the BIU are the *segment registers* (ES, CS, SS, and DS), the *instruction pointer* (IP), the *address summing block* (Σ), and the four-byte *instruction queue*. The internal data buses and the Q bus interconnect the BIU and the EU.

	Serial Fetch/Execute							
Serial Fetch/Execute	Fetch 1st inst	Execute 1st inst	Write result	Fetch 2nd inst	Execute 2nd inst	Fetch 3rd inst	Read operand	Execute 3rd inst

Elapsed time

Overlapped Fetch/Execute						
EU		Execute 1st inst	Execute 2nd inst			Execute 3rd inst
BIU	Fetch 1st inst	Fetch 2nd inst	Write result	Fetch 3rd inst	Read operand	Fetch 4th inst

FIGURE 12–22
A comparison of serial and overlapped fetch/execute cycles.

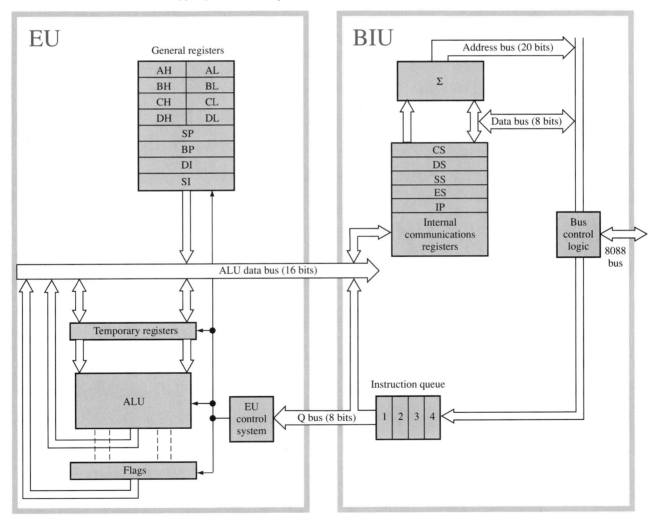

FIGURE 12–23
The internal organization of the 8088 microprocessor.

Instruction queue The instruction queue increases the average speed with which a program is executed (called the *throughput*) by storing up to four prefetched instructions (six in the 8086). As described earlier, this technique allows the 8088 essentially to do two things (fetch and execute) at one time.

Segment registers The segment registers are all sixteen-bit registers used for addressing the 1 MB (1,048,576 bytes) of memory space.

The memory space is divided into groups of 64-KB (65,536-byte) segments. Every segment is assigned, by program, a *base address,* which is its starting location in the memory space. The segment registers point to (contain the base addresses of) four currently addressable segments. These registers can be changed by the program to point to other desired segments.

Currently addressable memory segments are those defined by the base addresses contained in the CS (code segment) register, the DS (data segment) register, the SS (stack segment) register, and the ES (extra segment) register.

Instruction pointer (IP) and address summing block The sixteen-bit IP is analogous to the program counter in other microprocessors; it points to the next instruction in memory. The IP contains the *offset address* of the next instruction, which is the distance in bytes from the beginning, or base, address of the current code segment (in the CS register).

To achieve the twenty-bit physical memory address that goes out on the address bus, the sixteen-bit offset address in the IP is added to the segment base address, which has been shifted four bits to the left, as indicated in Figure 12–24. This addition is done by the address summing block. Figure 12–25 illustrates the addressing of a location in memory by the segmented method. In this example, $A000_{16}$ is in the segment register and $A0B0_{16}$ is in the IP. When the segment register is shifted and added to the IP, we get $A0000_{16} + A0B0_{16} = AA0B0_{16}$.

FIGURE 12–24
Formation of the twenty-bit address from the segment base address and the offset address.

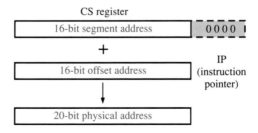

FIGURE 12–25
Illustration of the segmented addressing method.

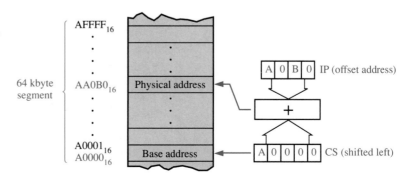

EXAMPLE 12–1   The hexadecimal contents of the CS register and the IP are shown in Figure 12–26. Determine the physical address in memory of the next instruction.

FIGURE 12–26

A034$_{16}$ CS

0FF2$_{16}$ IP

Solution   Shifting the CS base address left four bits (one hex digit), effectively places a 0_{16} in the LSD position, as shown in Figure 12–27. The shifted base address and the offset address are added to produce the twenty-bit physical address.

FIGURE 12–27

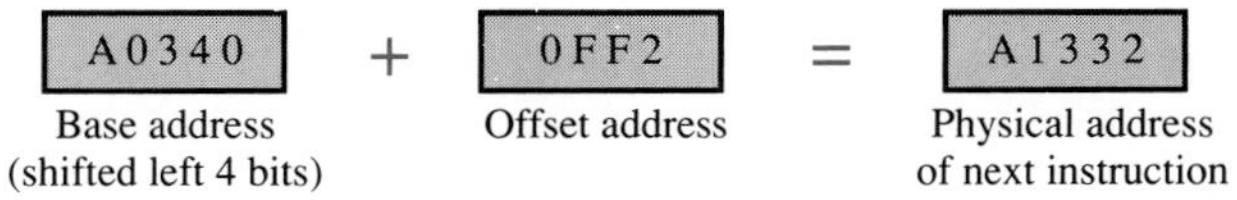

A0340 $+$ 0FF2 $=$ A1332

Base address (shifted left 4 bits)   Offset address   Physical address of next instruction

The Execution Unit (EU)

The EU decodes instructions fetched by the BIU, generates appropriate control signals, and executes the instructions. The main parts of the EU are the *arithmetic logic unit (ALU)*, the *general registers,* and the *flags*.

The ALU   This unit does all the computational and logic operations, working with either eight- or sixteen-bit operands.

The general registers   This set of sixteen-bit registers is divided into two sets of four registers each, as shown in Figure 12–28. One set consists of the *data registers,* and the other set consists of the *pointer* and *index registers*.

FIGURE 12–28
The 8088 general register set.

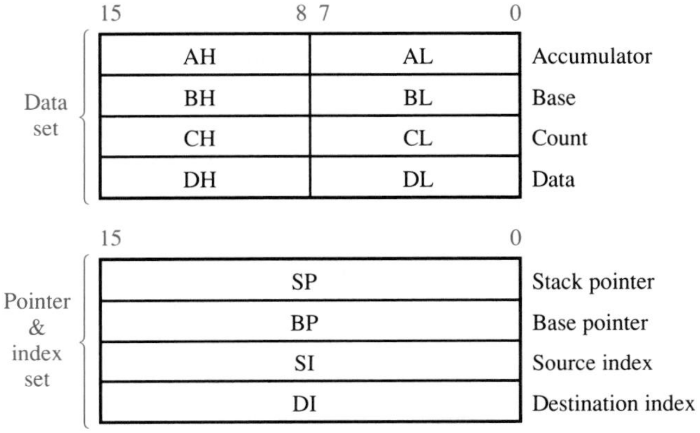

	15	8 7	0	
Data set	AH	AL		Accumulator
	BH	BL		Base
	CH	CL		Count
	DH	DL		Data

	15	0	
Pointer & index set	SP		Stack pointer
	BP		Base pointer
	SI		Source index
	DI		Destination index

Each of the sixteen-bit data registers has two separately accessible eight-bit sections and can be used either as a sixteen-bit register or as two eight-bit registers. The pointer and index registers are used only as sixteen-bit registers.

The low-order bytes of the data registers are designated as *AL, BL, CL,* and *DL.* The high-order bytes are designated as *AH, BH, CH,* and *DH.* These registers can be used in most arithmetic and logic operations in any manner specified by the programmer for storing data prior to and after processing. Also, some of these registers are used specifically by certain program instructions.

The pointer and index registers are the *stack pointer (SP)*, the *base pointer (BP)*, the *source index (SI)*, and the *destination index (DI)*. These registers are used in various forms of memory addressing under control of the EU.

The flags   The flag register contains nine independent status and control bits *(flags)*, as shown in Figure 12–29. A *status flag* is a one-bit indicator used to reflect a certain condition after an arithmetic or logic operation by the ALU, such as a carry (CF), a zero result (ZF), or the sign of a result (SF), among others. The control flags are used to alter processor operations in certain situations.

FIGURE 12–29
The 8088 status and control flags.

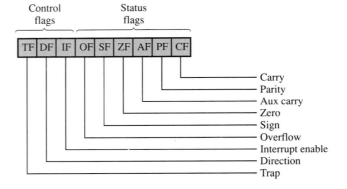

SECTION 12–7 REVIEW

Answers are found at the end of the chapter.
1. Name the two operational units within an 8086 or 8088 microprocessor.
2. What is the purpose of the BIU?
3. Does the EU interface with the system buses?
4. What is the function of the instruction queue?
5. What is the advantage of prefetched instructions?

MICROPROCESSOR PROGRAMMING

12–8

*The **hardware** (circuitry and physical components) of a microcomputer can do nothing without **software** (programs). All computers must be programmed to perform even the most elementary tasks. This section gives a brief introduction to the concept of programming in microprocessor-based systems. A detailed study of programming is left for later courses.*

A program is a list of computer instructions arranged to achieve a specific result. To give you a basic idea of how a computer carries out a task under program control, we will look at how a computer runs a very simple program. The steps in our example task are as follows:

1. Input a number from port 2.
2. Add 5 to this number.
3. Output the sum to port 3.

Each step in the task must be implemented with a *program statement* that instructs the microprocessor what to do. Every microprocessor has an *instruction set* from which a programmer can choose the most appropriate instructions to accomplish a given task.

For *step 1* in our example task, an input instruction, IN, is used to tell the microprocessor to transfer the data byte from the specified input port to the AL register. Let us assume that the IN instruction is represented by the eight-bit code 11100100, or $E4_{16}$. This is the *op code,* or operation code, for this particular instruction, and it is stored in RAM at an address determined by the programmer. The RAM address *following* the IN op code contains the input port number.

For *step 2* an addition instruction, ADD, is used to tell the microprocessor to add an operand (in this case 5) to the contents of the AL register (which now contains the number from the input port) and then place the sum back into the AL register, replacing the previous number. The op code for the ADD instruction is 00000100, or 04_{16}, and is stored at the next RAM address. The address following the ADD op code contains the operand 5.

For *step 3* an output instruction, OUT, is used to tell the microprocessor to transfer the contents of the AL register (which is now the sum) to the specified output port. The op code for OUT is 11100110, or $E6_{16}$. The RAM address following the op code address contains the number of the output port.

Figure 12–30 shows how this sample program "looks" in memory. A beginning address of 00000_{16} is assumed for illustration.

FIGURE 12–30
Example of a simple program stored in RAM.

Description of code	RAM	Physical address (hexadecimal)
Op code for IN	1 1 1 0 0 1 0 0	00000
Input port number	0 0 0 0 0 0 1 0	00001
Op code for ADD	0 0 0 0 0 1 0 0	00002
Operand "5"	0 0 0 0 0 1 0 1	00003
Op code for OUT	1 1 1 0 0 1 1 0	00004
Output port number	0 0 0 0 0 0 1 1	00005

When a program is written with the *mnemonic* designations for the instructions, such as IN, ADD, and OUT, it is an *assembly-language* program. When a program is written with the binary codes, it is a *machine-language* program. The sample program is written in both forms as follows:

Assembly-Language Program	*Comments*
IN 02_{16}	Input byte from port 2 and store in AL register.
ADD 05_{16}	Add 5 to AL register and store sum back in register.
OUT 03_{16}	Output sum to port 3.

Machine-Language Program	*Comments*
11100100 ($E4_{16}$)	IN op code
00000010 (02_{16})	Port number
00000100 (04_{16})	ADD op code
00000101 (05_{16})	Operand
11100110 ($E6_{16}$)	OUT op code
00000011 (03_{16})	Port number

Running the Program

When the microcomputer is commanded to *run* the program, it goes through the following sequence:

1. Fetch the op code for the IN instruction from memory.
2. Decode the IN op code. This tells the microprocessor to go to the next memory address and get the input port number.
3. Read the input port number from the memory.
4. Access the port with an I/O read operation, and transfer the number that is on the input port lines to the AL register. Let us say the number is binary seven (00000111).
5. Fetch the op code for ADD from the memory.
6. Decode the ADD op code. This tells the microprocessor to go to the next memory address and get the operand (5_{10}) and add it to the contents of the AL register (7_{10}).
7. Read the operand (5_{10}) from the memory, add it to the number 7_{10}, which is in the AL register, and store the sum (12_{10}) in the AL register (replacing the previous number, 7_{10}).
8. Fetch the op code for the OUT instruction from the memory.
9. Decode the OUT op code. This tells the microprocessor to go to the next memory address and get the output port number.
10. Read the output port number from the memory.
11. Access the port with an I/O write operation, and transfer the sum (12_{10}), which is in the AL register, to the output port.

This simple program should give you some feel for microcomputer software operation. Although only three instructions were used, most microprocessors have a large number of instructions in their instruction sets. For example, the 8086/8088 has about 100 instructions. This number allows enormous flexibility and programming power.

A coverage of the complete instruction set is obviously beyond the scope of a single section, or even a chapter. Only the *categories* of 8086/8088 instructions and a brief discussion of a few selected instructions are given here.

Data Transfer Instructions

Two data transfer instructions, IN and OUT, were used in the example program. Others include MOV (move), PUSH (push onto stack), POP (take off of stack), and XCHG (exchange). The MOV instruction, for example, can be used in several ways to move a byte or a word (sixteen bits) between various sources and destinations, such as registers, memory, and I/O ports.

Arithmetic Instructions

There are a number of instructions for addition, subtraction, multiplication, and division. The ADD instruction was used in the sample program. Others include INC (increment), DEC (decrement), CMP (compare), SUB (subtract), MUL (multiply), and DIV (divide). These instructions allow for specification of operands located in memory, registers, and I/O.

Bit Manipulation Instructions

This group of instructions includes those used for three classes of operations: *logical (Boolean) operations, shifts,* and *rotations.* The logical instructions are NOT, AND, OR, XOR, and TEST. An example of a shift instruction is SAL (shift arithmetic right). An example of a rotate instruction is ROL (rotate left). When bits are shifted out of an operand, they are lost, but when bits are rotated out of an operand, they are looped back into the other end. These logical, shift, and rotate instructions can operate on bytes or words in registers or memory.

Other categories of instructions are *string instructions, program transfer instructions,* and *processor control instructions.*

**SECTION 12–8
REVIEW**

Answers are found at the end of the chapter.
1. Define *program.*
2. What is an instruction set?
3. What is an op code?
4. What is an operand?

SUMMARY

☐ Basic units of a microcomputer are shown in Figure 12–31.

FIGURE 12–31

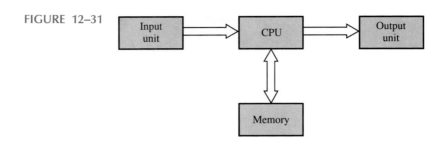

☐ Microcomputer buses:
Address bus
Data bus
Control bus
☐ Typical peripherals:
Keyboard
Video monitor
Disk drive
Printer
Modem
☐ 8086/8088 microprocessors:
8086: sixteen-bit data bus, twenty-bit address bus
8088: eight-bit data bus, twenty-bit address bus
Execution unit (EU): Executes instructions
Bus interface unit (BIU): Fetches instructions, reads operands, and writes results
General registers:

Accumulator (AH, AL)	Stack pointer (SP)
Base (BH, BL)	Base pointer (BP)
Count (CH, CL)	Source index (SI)
Data (DH, DL)	Destination index (DI)

Flags:

Trap (TF)	Zero (ZF)
Direction (DF)	Auxiliary carry (AF)
Interrupt enable (IF)	Parity (PF)
Overflow (OF)	Carry (CF)
Sign (SF)	

☐ Ports:
Physical types: Input, output, input/output (bidirectional)
Methods of access: Dedicated, memory-mapped
Methods of servicing: Polled, interrupt-driven

Integrated Circuit Technologies (Chapter 13) may be covered after this chapter, partially omitted, or completely omitted, depending on the needs and goals of your program. Coverage is not prerequisite or corequisite to any other material in this book. Inclusion is purely optional.

GLOSSARY

Address allocation   The arrangement of memory space (addresses) in a computer.

ALU   Arithmetic logic unit. The part of a computer that processes information by using arithmetic and logic operations.

Architecture   In computers, the internal organization and design.

Assembly language   A form of computer program language in which statements expressed in mnemonics represent the instructions.

Base address   The beginning address of a segment of memory.

CPU   Central processing unit. A major component of all computers that controls the internal operations and processes data.

DMA   Direct memory access. A technique in which the CPU is bypassed for certain types of data transfers.

Execute   A CPU process in which an instruction is carried out.

Fetch   A CPU process in which an instruction is obtained from the memory.

Firmware   Programs stored in ROM.

Flag   A bit used to control or indicate the status of an operation in the CPU.

Hardware   The circuitry and mechanical components of an electronic system.

Instruction   One step in a computer program. A unit of information that tells the CPU what to do.

Interrupt   A request for service from a peripheral device.

Machine language   The representation of computer program instructions by a binary code.

Microcomputer   A computer in which the CPU is built around a microprocessor.

Mnemonic   In a computer programming language, the ''shorthand'' representation of an instruction.

Offset address   The distance in bytes of a physical address from the base address.

Op code   Operation code. The code representing a particular microprocessor instruction.

Operand   A quantity or element operated on during the execution of an instruction.

Peripheral   A device or instrument that provides communication with a computer or provides auxiliary services or functions for the computer.

Physical address   The actual location of a byte in memory.

PIC   Programmable interrupt controller.

Polling   The process in which the CPU periodically checks peripheral devices to see if they need service.

Port   The physical interface between a computer and a peripheral.

PPI   Programmable peripheral interface.

Program   A list of computer instructions arranged to achieve a specific result. Software.

Queue   A file or lineup of elements, for example, prefetched instructions.

Software   Computer programs.

Throughput   The average speed with which a program is executed.

SELF-TEST

Answers and solutions are found at the end of the book.

1. Describe the functions of the CPU in a microcomputer.
2. What is an application program?
3. What unit acts as a ''scratch pad'' for the computer?

4. How many memory locations can be addressed with a twenty-four-bit address bus?
5. What is the advantage of a sixteen-bit data bus over an eight-bit data bus?
6. List the basic support circuits in the CPU of a PC.
7. What sets the frequency of the 8284 clock generator?
8. Define the function of each of the following signals, and name their source.
   (a) $\overline{\text{MRDC}}$ **(b)** $\overline{\text{MWTC}}$ **(c)** $\overline{\text{IORC}}$ **(d)** $\overline{\text{IOWC}}$
9. What is the purpose of the status code output of the 8088?
10. Explain the purpose of the ALE signal from the bus controller.
11. What is the capacity of a byte-oriented memory whose low address is 0000_{16} and whose high address is FFFF_{16}?
12. What is the elapsed time for an 8088 read or write cycle? The clock frequency is 4.77 MHz.
13. What does it take to enable the octal latch in the dedicated output port of Figure 12–15?
14. What is the purpose of interrupts in a computer?
15. Define "vectoring."
16. How does the 8088 instruction queue increase throughput?
17. What is the instruction pointer, and what does it do?
18. A segment address is $\text{C}400_{16}$, and the offset address is $35\text{A}1_{16}$. What is the physical address?

ANSWERS TO SECTION REVIEWS

Section 12–1   Review

1. ALU, control unit, input unit, output unit, memory
2. along the data bus
3. a device external to the computer used for inputting or outputting information

Section 12–2   Review

1. to provide basic timing signals for the microcomputer
2. It issues memory read and write commands, issues I/O read and write commands, and enables the address latches and the data bus transceiver.
3. The twenty bus lines available must be used for both the twenty-bit address bus and the eight-bit data bus.

Section 12–3   Review

1. write and read
2. four
3. after the leading edge of the ALE pulse

Section 12–4   Review

1. the interface between the internal data bus and a peripheral
2. A dedicated I/O port has an address within the I/O address space. A memory-mapped I/O port has an address within the memory address space.
3. A PIC provides I/O ports.

Section 12–5   Review

1. For an interrupt-driven I/O, the CPU provides service to a peripheral only when requested to do so by the peripheral. For a polled I/O, the CPU periodically checks a peripheral to see if it needs service.
2. Interrupt-driven I/Os save CPU time.

Section 12–6   Review

1. direct memory access
2. A DMA transfer of data from memory to I/O or vice versa saves CPU time. Direct memory access is often used in transferring data between RAM and a disk drive.

Section 12–7   Review

1. bus interface unit (BIU) and execution unit (EU)
2. It provides addressing and data interface.
3. no
4. to store prefetched instructions for the EU
5. They increase throughput.

Section 12–8   Review

1. a list of computer instructions arranged to achieve a specific result
2. all the instructions available to a particular microprocessor
3. the code for an instruction
4. the quantity or element operated on by an instruction

This chapter can be used as a "floating" chapter. It can be covered at several appropriate points throughout the book or completely omitted, depending on course objectives and personal preference. Suggested points that seem most appropriate are after Chapters 3, 6, 9, or 12, but it can be placed after other chapters if desired. You may wish to divide the coverage so that selected topics are introduced at various points.

Other approaches are to use this chapter as an optional reading assignment or to use it strictly for reference.

After completing this chapter, you will be able to

☐ determine the noise margin of a device from data sheet parameters.

☐ calculate the power dissipation of a device.

☐ explain how propagation delay affects the frequency of operation or speed of a circuit.

☐ interpret the speed-power product as a measure of performance.

☐ use data sheets to obtain information about a specific device.

☐ explain what the fan-out of a gate means.

☐ describe how basic TTL and CMOS gates operate at a component level.

☐ recognize the difference between TTL totem-pole outputs and TTL open-collector outputs and understand the limitations and uses of each.

☐ connect circuits in a wired-AND configuration.

☐ describe the operation of tristate circuits.

☐ properly terminate unused gate inputs.

☐ compare the characteristics of TTL and CMOS.

☐ handle CMOS devices without risk of damage due to electrostatic discharge.

☐ interface TTL and CMOS devices.

☐ state the advantages of ECL and I^2L technologies.

☐ describe PMOS and NMOS circuits.

13

Integrated Circuit Technologies

BASIC OPERATIONAL CHARACTERISTICS AND PARAMETERS

13–1

When you work with digital ICs, you should be familiar, not only with their logical operation, but also with such operational properties as voltage levels, noise immunity, power dissipation, fan-out, and propagation delays. In this section the practical aspects of these properties are discussed.

dc Supply Voltage

The nominal value of the dc supply voltage for TTL (transistor-transistor logic) and CMOS (complementary metal-oxide semiconductor) devices is +5 V. Although omitted from logic diagrams for simplicity, this voltage is connected to the V_{CC} or V_{DD} pin of an IC package, and ground is connected to the GND pin. Both voltage and ground are distributed internally to all elements within the package, as illustrated in Figure 13–1.

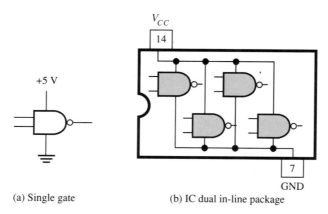

(a) Single gate   (b) IC dual in-line package

FIGURE 13–1

Example of V_{CC} and ground connection and distribution in an IC package.

TTL Logic Levels

Logic levels were discussed briefly in Chapter 1. For TTL circuits the range of input voltages that can represent a valid LOW (logic 0) is from 0 V to 0.8 V. The range of input voltages that can represent a valid HIGH (logic 1) is from 2 V to V_{CC} (usually 5 V), as indicated in Figure 13–2(a).

The range of values between 0.8 V and 2 V is a region of unpredictable performance. When an input voltage is in this range, it can be interpreted as either a HIGH or a LOW by the logic circuit. Therefore, TTL gates cannot be operated reliably when input voltages are in this range.

The range of TTL output voltages is shown in Figure 13–2(b). Notice that the minimum HIGH output voltage, $V_{OH(min)}$, is greater than the minimum HIGH input voltage, $V_{IH(min)}$. Also, the maximum LOW output voltage, $V_{OL(max)}$, is less than the maximum LOW input voltage, $V_{IL(min)}$.

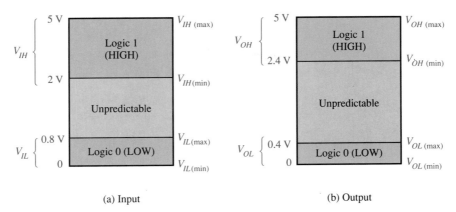

FIGURE 13–2
Input and output logic levels for TTL.

CMOS Logic Levels

The input and output logic levels for high-speed CMOS (HCMOS) are given in Figure 13–3 for $V_{DD} = 5$ V.

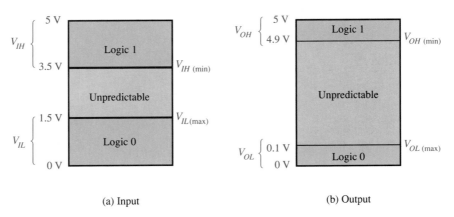

FIGURE 13–3
Input and output logic levels for CMOS.

Noise Immunity

Noise is unwanted voltage that is induced in electrical circuits and can present a threat to the proper operation of the circuit. Wires and other conductors within a system can pick up stray high-frequency electromagnetic radiation from adjacent conductors in which currents are changing rapidly or from many other sources external to the system. Also, power-line voltage fluctuation is a form of low-frequency noise.

In order not to be adversely affected by noise, a logic circuit must have a certain amount of *noise immunity*. This is the ability to tolerate a certain amount of unwanted voltage fluctuation on its inputs without changing its output state. For example, if noise voltage causes the input of a TTL gate to drop below 2 V in the HIGH state, the operation is in the unpredictable region (see Figure 13–2[a]). Thus, the gate may interpret the fluctuation below 2 V as a LOW level, as illustrated in Figure 13–4(a). Similarly, if noise causes a gate input to go above 0.8 V in the LOW state, an uncertain condition is created, as illustrated in part (b).

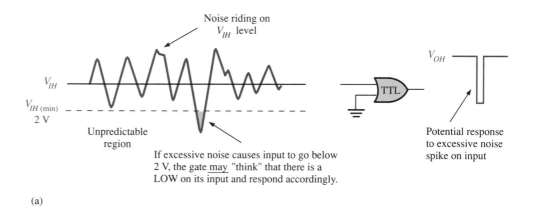

(a)

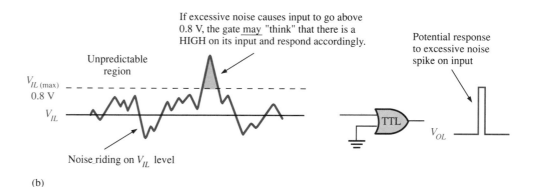

(b)

FIGURE 13–4
Illustration of the effects of input noise on gate operation.

Noise Margin

A measure of a circuit's noise immunity is called the *noise margin,* which is expressed in volts. There are two values of noise margin specified for a given logic circuit: the HIGH-level noise margin (V_{NH}) and the LOW-level noise margin (V_{NL}). These parameters are defined by the following equations:

$$V_{NH} = V_{OH(min)} - V_{IH(min)} \qquad \textbf{(13–1)}$$

$$V_{NL} = V_{IL(max)} - V_{OL(max)} \qquad \textbf{(13–2)}$$

As you can see, V_{NH} is the difference between the lowest possible HIGH output from a driving gate ($V_{OH(min)}$) and the lowest possible HIGH input that the load gate can tolerate ($V_{IH(min)}$). Noise margin V_{NL} is the difference between the maximum possible LOW input that a gate can tolerate ($V_{IL(max)}$) and the maximum possible LOW output of the driving gate ($V_{OL(max)}$). Noise margins are illustrated in Figure 13–5.

FIGURE 13–5
Noise margins.

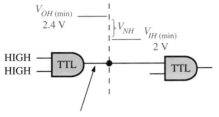

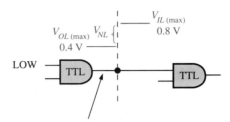

The voltage on this line will never be less than 2.4 V unless noise or improper operation is introduced.

(a) HIGH-level noise margin

The voltage on this line will never exceed 0.4 V unless noise or improper operation is introduced.

(a) LOW-level noise margin

EXAMPLE 13–1

Determine the HIGH- and LOW-level noise margins for TTL and for CMOS by using the information in Figures 13–2 and 13–3.

Solution For TTL:

$$V_{IH(min)} = 2 \text{ V}$$
$$V_{IL(max)} = 0.8 \text{ V}$$
$$V_{OH(min)} = 2.4 \text{ V}$$
$$V_{OL(max)} = 0.4 \text{ V}$$
$$V_{NH} = V_{OH(min)} - V_{IH(min)} = 2.4 \text{ V} - 2 \text{ V} = 0.4 \text{ V}$$
$$V_{NL} = V_{IL(max)} - V_{OL(max)} = 0.8 \text{ V} - 0.4 \text{ V} = 0.4 \text{ V}$$

A TTL gate is immune to up to 0.4 V of noise for both the HIGH and LOW input states.

For CMOS:

$$V_{IH(min)} = 3.5 \text{ V}$$
$$V_{IL(max)} = 1.5 \text{ V}$$
$$V_{OH(min)} = 4.9 \text{ V}$$
$$V_{OL(max)} = 0.1 \text{ V}$$
$$V_{NH} = V_{OH(min)} - V_{IH(min)} = 4.9 \text{ V} - 3.5 \text{ V} = 1.4 \text{ V}$$
$$V_{NL} = V_{IL(max)} - V_{OL(max)} = 1.5 \text{ V} - 0.1 \text{ V} = 1.4 \text{ V}$$

CMOS has a much higher noise margin than TTL and therefore has a higher immunity to noise.

Power Dissipation

A logic gate draws current from the dc supply voltage source as indicated in Figure 13–6. When the gate is in the HIGH output state, an amount of current designated by I_{CCH} is drawn, and in the LOW output state, a different amount of current, I_{CCL}, is drawn. For a TTL gate, I_{CCL} is greater than I_{CCH}.

FIGURE 13–6
Currents from the dc supply.

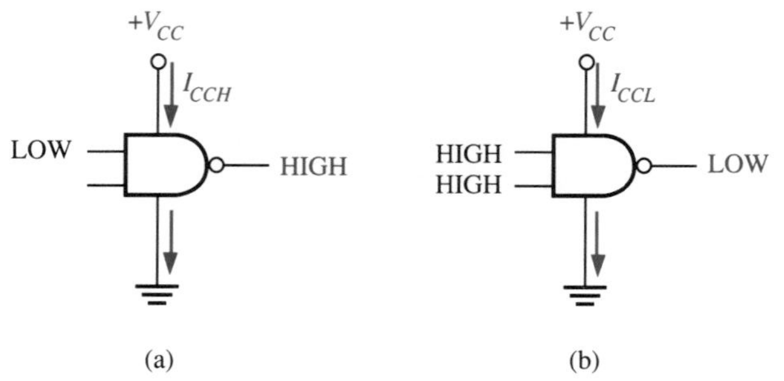

(a)   (b)

As an example, if I_{CCH} is specified as 1.5 mA when V_{CC} is 5 V, and if the gate is in a static (nonchanging) HIGH output state, the power dissipation of the gate is

$$P_D = V_{CC}I_{CCH} = (5 \text{ V})(1.5 \text{ mA}) = 7.5 \text{ mW}$$

When a gate is pulsed, its output switches back and forth between HIGH and LOW, and the amount of supply current varies between I_{CCH} and I_{CCL}. The *average* power dissipation depends on the duty cycle and is usually specified for a duty cycle of 50%. When the duty cycle is 50%, the output is HIGH half the time and LOW the other half. The average supply current is therefore

$$I_{CC} = \frac{(I_{CCH} + I_{CCL})}{2} \tag{13–3}$$

The average power dissipation is

$$P_D = V_{CC}I_{CC} \tag{13–4}$$

EXAMPLE 13–2

A certain gate draws 2 mA when its output is HIGH and 3.6 mA when its output is LOW. What is its average power dissipation if V_{CC} is 5 V and the gate is operated on a 50% duty cycle?

Solution The average I_{CC} is

$$I_{CC} = \frac{I_{CCH} + I_{CCL}}{2} = \frac{2.0 \text{ mA} + 3.6 \text{ mA}}{2} = 2.8 \text{ mA}$$

The average power dissipation is

$$P_D = V_{CC}I_{CC} = (5 \text{ V})(2.8 \text{ mA}) = 14 \text{ mW}$$

Power dissipation in a TTL circuit is essentially constant over its range of operating frequencies. Power dissipation in CMOS, however, is frequency dependent. It is extremely low under static (dc) conditions and increases as the frequency increases. These characteristics are shown in the general curves of Figure 13–7.

For example, the power dissipation of a low-power Schottky (LS) TTL gate is a constant 2 mW. The power dissipation of an HCMOS gate is 0.0000025 mW under static conditions and 0.17 mW at 100 kHz.

FIGURE 13–7

Power-versus-frequency curves for TTL and CMOS.

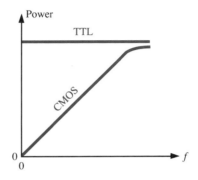

Propagation Delay

When a signal passes (propagates) through a logic circuit, it always experiences a time delay, as illustrated in Figure 13–8. A change in the output level always occurs a short time, called the propagation delay time, later than the change in the input level that caused it. As mentioned in Chapter 3, there are two propagation delay times specified for logic gates:

☐ t_{PHL}: the time between a designated point on the input pulse and the corresponding point on the output pulse when the output is changing from HIGH to LOW

☐ t_{PLH}: the time between a designated point on the input pulse and the corresponding point on the output pulse when the output is changing from LOW to HIGH

FIGURE 13–8

A basic illustration of propagation delay.

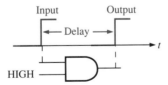

These propagation delay times are illustrated in Figure 13–9, with the 50% points on the pulse edges used as references.

The propagation delay of a gate limits the frequency at which it can be operated. The greater the propagation delay, the lower the maximum frequency. Thus, a *higher-speed* circuit is one that has a smaller propagation delay. For example, a gate with a delay of 3 ns is faster than one with a 10-ns delay.

FIGURE 13–9
Propagation delay times.

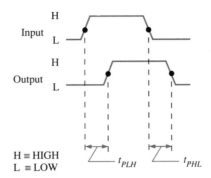

Speed-Power Product

The purpose of this parameter is to give a basis for comparison of logic circuits when *both* propagation delay and power dissipation are important considerations in the selection of the type of logic to be used in a certain application. The unit of speed-power product is the picojoule (pJ).

The lower the speed-power product, the better. The CMOS circuits typically have much lower values for this parameter than do TTL circuits, because of much lower power dissipation. For example, HCMOS has a speed-power product of 1.4 pJ at 100 kHz while LS TTL has a value of 20 pJ.

Loading and Fan-out

When the output of a logic gate is connected to one or more inputs of other gates, a *load* on the driving gate is created, as shown in Figure 13–10.

There is a limit to the number of load gate inputs that a given gate can drive. This limit is called the *fan-out* of the gate.

FIGURE 13–10
Loading a gate output with gate inputs.

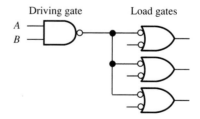

TTL loading A TTL driving gate *sources* current to a load gate input in the HIGH state (I_{IH}) and *sinks* current from the load gate in the LOW state (I_{IL}). This *current sourcing* and *current sinking* concept is illustrated in simplified form in Figure 13–11,

FIGURE 13–11

FIGURE 13–11

Basic illustration of current sourcing and current sinking in logic gates.

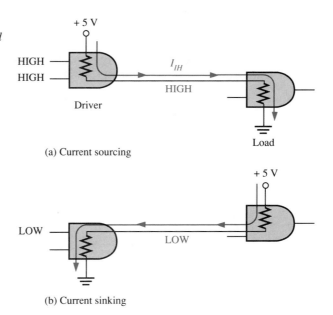

(a) Current sourcing

(b) Current sinking

where the resistors represent the input and output resistance of the gate. This will be covered in more detail when we look at the actual circuitry later.

As more load gates are connected to the driving gate, the loading on the driving gate increases. The total source current increases with each load gate input that is added, as illustrated in Figure 13–12. As this current increases, the internal voltage drop of the driving gate increases, causing the output, V_{OH}, to decrease. If an excessive number of load gate inputs are connected, V_{OH} drops below $V_{OH(\text{min})}$, and the HIGH-level noise margin is reduced, thus compromising the circuit operation. Also, as the total source current increases, the power dissipation of the driving gate increases.

FIGURE 13–12

HIGH-state TTL loading.

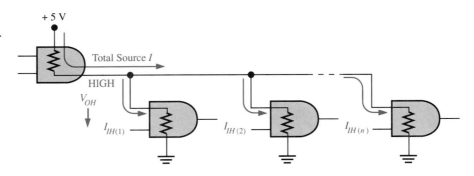

The fan-out is the maximum number of load gate inputs that can be connected without adversely affecting the specified operational characteristics of the gate. For example, low-power Schottky TTL has a fan-out of 20 unit loads. One input of the same logic family as the driving gate is called a *unit load*.

The total sink current also increases with each load gate input that is added, as shown in Figure 13–13. As this current increases, the internal voltage drop of the driving gate increases, causing V_{OL} to increase. If an excessive number of loads are added, V_{OL} exceeds $V_{OL(max)}$, and the LOW-level noise margin is reduced.

FIGURE 13–13
LOW-state TTL loading.

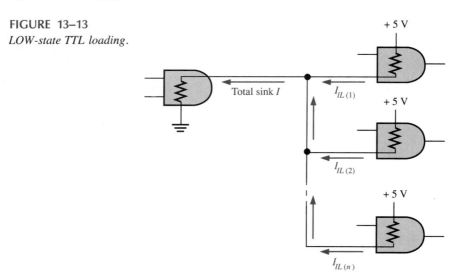

In TTL the current-sinking capability (LOW state) is the limiting factor in determining the fan-out. You will see why later.

CMOS loading   Loading in CMOS differs from that in TTL because the field-effect transistors (*FETs*) used in CMOS logic present a predominantly capacitive load to the driving gate, as illustrated in Figure 13–14. In this case the limitations are the charging and discharging times associated with the output resistance of the driving gate and the input capacitance of the load gates. When the output of the driving gate is HIGH, the input capacitance of the load gate is charging through the output resistance of the driving gate. When the output of the driving gate is LOW, the capacitance is discharging, as indicated in Figure 13–14.

When more load gate inputs are added to the driving gate output, the total capacitance increases because the input capacitances effectively appear in parallel. This increase in capacitance increases the charging and discharging times, thus reducing the maximum frequency at which the gate can be operated. Therefore, the fan-out of a CMOS gate depends on the frequency of operation. The fewer the load gate inputs, the greater the maximum frequency.

**SECTION 13–1
REVIEW**

Answers are found at the end of the chapter.
1.   Define V_{IH}, V_{IL}, V_{OH}, and V_{OL}.
2.   Is it better to have a lower value of noise margin or a higher value?
3.   Gate *A* has a greater propagation delay time than gate *B*. Which gate can operate at a higher frequency?
4.   How does excessive loading affect the noise margin of a gate?

FIGURE 13–14
Capacitive loading of a CMOS gate.

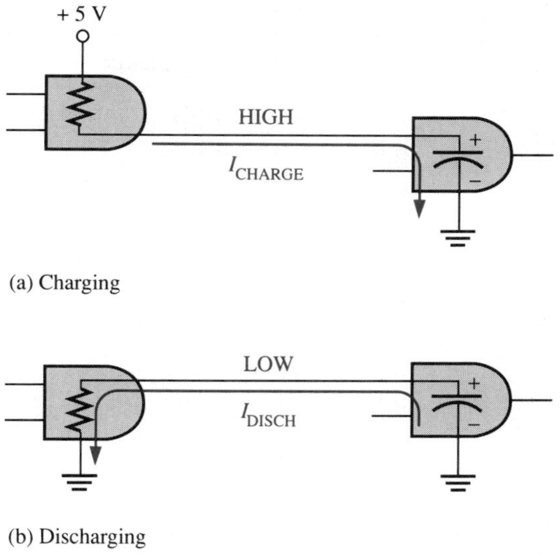
+5 V
HIGH
I_{CHARGE}
(a) Charging
LOW
I_{DISCH}
(b) Discharging

DATA SHEET INTERPRETATION

13–2

In the last section several digital IC parameters were introduced. Values for all these parameters are given in the manufacturer's data sheet for each IC. In this section a precise definition is given for each parameter, and a sample data sheet listing of values is shown for a specific device.

A typical *data sheet* is divided into three main sections: recommended operating conditions, electrical characteristics, and switching characteristics. As an example, Figure 13–15 shows the arrangement of a data sheet for the 5400/7400 quad two-input NAND gate. Additional sample data sheets are given in Appendix A.

Parameter	5400			7400			Units
	Minimum	Typical	Maximum	Minimum	Typical	Maximum	
Supply voltage (V_{CC})	4.5	5.0	5.5	4.75	5.0	5.25	V
Operating free-air temperature range	−55	25	125	0	25	70	°C
HIGH level output current (I_{OH})			−400			−400	μA
LOW level output current (I_{OL})			16			16	mA

(a) Recommended operating conditions

FIGURE 13–15
Data sheet for the 5400/7400 quad two-input NAND gate.

| Parameter | Limits | | | Units | Test Conditions[1] |
	Minimum	Typical[2]	Maximum		
HIGH level input voltage (V_{IH})	2.0			V	
LOW level input voltage (V_{IL})			0.8	V	
HIGH level output voltage (V_{OH})	2.4	3.4		V	V_{CC} = min., I_{OH} = 0.4 mA, V_{IN} = 0.8 V
LOW level output voltage (V_{OL})		0.2	0.4	V	V_{CC} = min., I_{OL} = 16 mA, V_{IN} = 2.0 V
HIGH level input current (I_{IH})			40	μA	V_{CC} = max., V_{IN} = 2.4 V
LOW level input current (I_{IL})			−1.6	mA	V_{CC} = max., V_{IN} = 0.4 V
Short-circuit output current[3] (I_{OS}) 5400	−20		−55	mA	V_{CC} = max.
Short-circuit output current[3] (I_{OS}) 7400	−18		−55	mA	
Total supply current with outputs HIGH (I_{CCH})		4.0	8.0	mA	V_{CC} = max.
Total supply current with outputs LOW (I_{CCL})		12	22	mA	V_{CC} = max.

(b) Electrical characteristics over operating temperature range (unless otherwise noted)

| Parameter | Limits | | | Units | Test Conditions |
	Minimum	Typical	Maximum		
Propagation delay time, LOW-to-HIGH output (t_{PLH})		11	22	ns	V_{CC} = 5.0 V
Propagation delay time, HIGH-to-LOW output (t_{PHL})		7.0	15	ns	C_{LOAD} = 15 pF R_{LOAD} = 400 Ω

(c) Switching characteristics (T_A = 25°C)

NOTES:
[1] For conditions shown as min. or max., use the appropriate value specified under recommended operating conditions for the applicable device type.
[2] Typical limits are at V_{CC} = 5.0 V, 25°C.
[3] Not more than one output should be shorted at a time. Duration of short not to exceed 1 s.

FIGURE 13–15
(*Continued*)

Explanation of Data Sheet Parameters

The following list describes the parameters of the data sheet in Figure 13–15:

1. V_{CC}: The dc voltage that supplies power to the device. Below the specified minimum, reliable operation cannot be guaranteed. Above the specified maximum, damage to the device may occur.

2. I_{OH}: The output current that the gate provides (sources) to a load when the output is at the HIGH level. By convention, the current out of a terminal is assigned a negative value. Figure 13–16(a) illustrates this parameter.

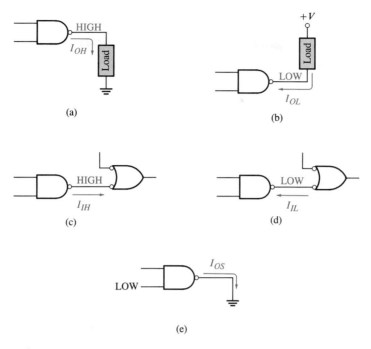

FIGURE 13–16
Illustration of some data sheet parameters.

3. I_{OL}: The output current that the gate sinks when the output is at the LOW level. By convention, the current into a terminal is assigned a positive value. Figure 13–16(b) illustrates this parameter.

4. V_{IH}: The value of input voltage that can be accepted as a HIGH level by the gate. This parameter and the next three were introduced in the discussion of noise margin.

5. V_{IL}: The value of input voltage that can be accepted as a LOW level by the gate.

6. V_{OH}: The value of HIGH level output voltage that the gate produces.

7. V_{OL}: The value of LOW level output voltage that the gate produces.

8. I_{IH}: The value of input current for a HIGH level input voltage. Figure 13–16(c) illustrates this parameter.

9. I_{IL}: The value of input current for a LOW level input voltage. Figure 13–16(d) illustrates this parameter.

10. I_{OS}: The output current when the gate output is shorted to ground and with input conditions that establish a HIGH level output. Figure 13–16(e) illustrates this parameter.

11. I_{CCH}: The total current from the V_{CC} supply when all gate outputs are at the HIGH level.

12. I_{CCL}: The total current from the V_{CC} supply when all gate outputs are at the LOW level.

**SECTION 13–2
REVIEW**

Answers are found at the end of the chapter.
1. Using the data sheet in Figure 13–15, calculate the worst-case noise margins for the 7400 gate.
2. What is the maximum current that a 7400 gate can supply to a load when the output is HIGH?

TTL CIRCUITS

13–3

The internal circuit operation of standard TTL logic gates with **totem-pole** *outputs is covered in this section. Other TTL series, including 74L, 74S, 74LS, and 74ALS, are also presented and are compared in power consumption and speed. Also, the operation of TTL gates with* **open-collector** *outputs and the operation of* **tristate** *gates are covered.*

The Bipolar Junction Transistor

The bipolar junction transistor (*BJT*) is the active switching element used in all TTL circuits. Figure 13–17 shows the symbol for an npn BJT with its three terminals: *base, emitter,* and *collector.* The basic switching operation is as follows:

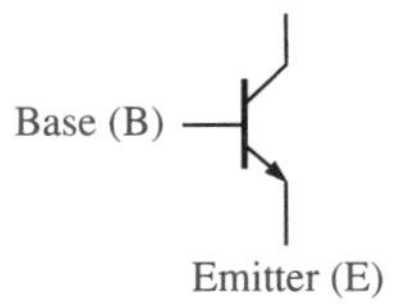

FIGURE 13–17
The symbol for a BJT.

Collector (C)

Base (B)

Emitter (E)

When the base is approximately 0.7 V more positive than the emitter and when sufficient current is provided into the base, the transistor turns on and goes into saturation. In saturation, the transistor ideally acts like a closed switch between the collector and the emitter, as illustrated in Figure 13–18(a).

When the base is less than 0.7 V more positive than the emitter, the transistor turns off and becomes an open switch between the collector and the emitter, as shown in part (b).

FIGURE 13–18
The ideal switching action of the BJT.

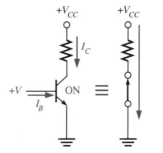

(a) Saturated (ON) transistor
and ideal switch equivalent

(b) OFF transistor and
ideal switch equivalent

To summarize in general terms, a HIGH on the base turns the transistor on and makes it a closed switch. A LOW on the base turns the transistor off and makes it an open switch.

In TTL some BJTs have multiple emitters, as you will see.

TTL Inverter

Transistor-transistor logic (TTL or T^2L) is one of the most widely used integrated circuit technologies. The logic function of an inverter or any type of gate is always the same, regardless of the type of circuit technology that is used. Using TTL logic is only one way to build an inverter or any other logic function. Figure 13–19 shows a standard TTL circuit for an inverter. In this figure Q_1 is the input coupling transistor, and D_1 is the input clamp diode. Transistor Q_2 is called a *phase splitter,* and the combination of Q_3 and Q_4 forms the output circuit often referred to as a *totem-pole* arrangement.

FIGURE 13–19
A standard TTL inverter circuit.

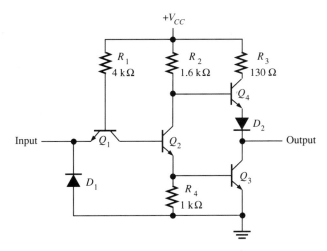

When the input is a HIGH, the base-emitter junction of Q_1 is reverse biased, and the base-collector junction is forward biased. This condition permits current through R_1 and the base-collector junction of Q_1 into the base of Q_2, thus driving Q_2 into saturation. As a result, Q_3 is turned on by Q_2, and its collector voltage, which is the output, is near ground potential. We therefore have a LOW output for a HIGH input. At the same time, the collector of Q_2 is at a sufficiently low voltage level to keep Q_4 off.

When the input is LOW, the base-emitter junction of Q_1 is forward biased, and the base-collector junction is reverse biased. There is current through R_1 and the base-emitter junction of Q_1 to the LOW input. A LOW provides a path to ground for the current. There is no current into the base of Q_2, so it is off. The collector of Q_2 is HIGH, thus turning Q_4 on. A saturated Q_4 provides a low-resistance path from V_{CC} to the output; we therefore have a HIGH on the output for a LOW on the input. At the same time, the emitter of Q_2 is at ground potential, keeping Q_3 off.

The purpose of diode D_1 in the TTL circuit is to prevent negative spikes of voltage on the input from damaging Q_1. Diode D_2 ensures that Q_4 will turn off when Q_2 is on (HIGH input). In this condition the collector voltage of Q_2 is equal to the base-to-emitter voltage, V_{BE}, of Q_3 plus the collector-to-emitter voltage, V_{CE}, of Q_2. Diode D_2 provides an additional V_{BE} equivalent drop in series with the base-emitter junction of Q_4 to ensure its turn-off when Q_2 is on.

The operation of the TTL inverter for the two input states is illustrated in Figure 13–20. In the upper circuit the base of Q_1 is 2.1 V above ground, so Q_2 and Q_3 are on. In the lower circuit the base of Q_1 is about 0.7 V above ground—not enough to turn Q_2 and Q_3 on.

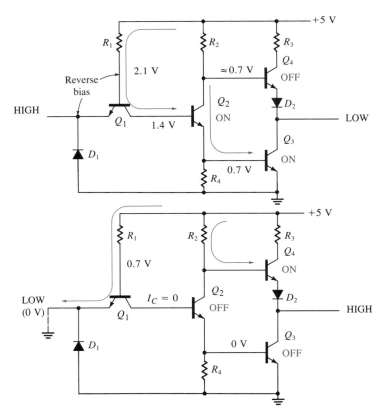

FIGURE 13–20
Operation of a TTL inverter.

TTL NAND Gate

A two-input standard TTL NAND gate is shown in Figure 13–21. Basically, it is the same as the inverter circuit except for the additional input emitter of Q_1. In TTL technology multiple-emitter transistors are used for the input devices. These can be compared to the diode arrangement, as shown in Figure 13–22.

FIGURE 13-21
A standard TTL NAND gate circuit.

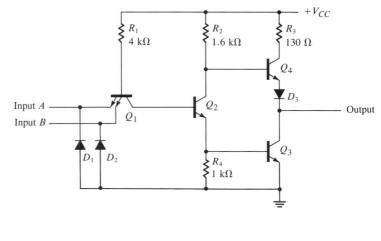

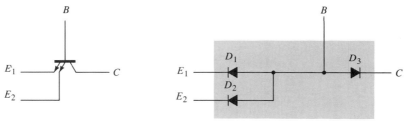

FIGURE 13-22
Diode equivalent of a TTL multiple-emitter transistor.

Perhaps you can understand the operation of this circuit better by visualizing Q_1 in Figure 13-21 replaced by the diode arrangement. A LOW on either input A or input B forward-biases the respective diode and reverse-biases D_3 (Q_1 base-collector junction). This action keeps Q_2 off and results in a HIGH output in the same way as described for the TTL inverter. Of course, a LOW on both inputs will do the same thing.

A HIGH on both inputs reverse-biases both input diodes and forward-biases D_3 (Q_1 base-collector junction). This action turns Q_2 on and results in a LOW output in the same way as described for the TTL inverter.

You should recognize this operation as that of the NAND function: the output is LOW if and only if all inputs are HIGH.

TTL NOR Gate

A two-input standard TTL NOR gate is shown in Figure 13-23. Compare this with the NAND gate circuit, and notice the additional transistors.

Both Q_1 and Q_2 are input transistors. Transistors Q_3 and Q_4 are in parallel and act as a phase splitter. You should recognize Q_5 and Q_6 as the ordinary TTL totem-pole output circuit.

FIGURE 13–23
A standard TTL NOR gate circuit.

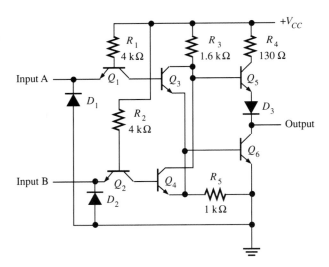

The operation is as follows: When both inputs are LOW, the forward-biased base-emitter junctions of the input transistors pull current away from the phase-splitter transistors, Q_3 and Q_4, keeping them *off*. As a result, Q_5 is on and Q_6 is off, producing a HIGH output.

When input A is LOW and input B is HIGH, Q_3 is off and Q_4 is on. Transistor Q_4 turns on Q_6 and turns off Q_5, producing a LOW output.

When input A is HIGH and input B is LOW, Q_3 is on and Q_4 is off. Transistor Q_3 turns on Q_6 and turns off Q_5, producing a LOW output.

When both inputs are HIGH, both Q_3 and Q_4 are on. Having both on has the same effect as having either one on, turning Q_6 on and Q_5 off. The result is still a LOW output. You should recognize this operation as that of the NOR function: the output is LOW when any of the inputs are HIGH.

TTL AND Gates and OR Gates

Figure 13–24 shows a standard TTL two-input AND gate and a standard two-input OR gate. Compare these to the NAND and NOR gate circuits, and notice the additional circuitry that is shaded. In each case this transistor arrangement provides an inversion so that the NAND becomes AND and the NOR becomes OR.

Open-Collector Gates

The TTL gates described in the previous sections all had the totem-pole output circuit. Another type of output available in TTL integrated circuits is the *open-collector* output. A standard TTL inverter with an open collector is shown in Figure 13–25(a). The other types of gates are also available with open-collector outputs.

Notice that the output is the collector of transistor Q_3 with nothing connected to it, hence the name *open collector*. In order to get the proper HIGH and LOW logic levels out of the circuit, an *external pull-up resistor* must be connected to V_{CC} from the collector of Q_3, as shown in Figure 13–25(b). When Q_3 is off, the output is pulled up to V_{CC} through the external resistor. When Q_3 is on, the output is connected to near-ground through the saturated transistor.

FIGURE 13–24
Standard TTL AND gate and OR gate circuits.

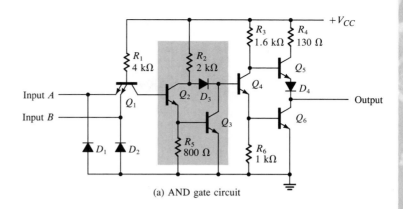

(a) AND gate circuit

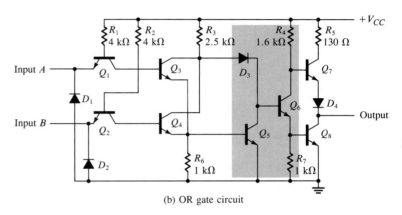

(b) OR gate circuit

FIGURE 13–25
Standard TTL inverter with open-collector output.

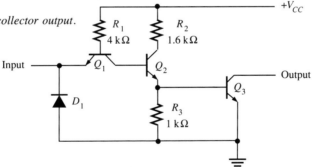

(a) Open-collector inverter circuit

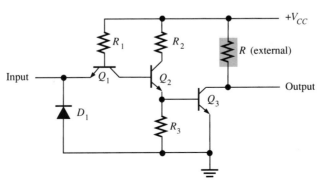

(b) With external pull-up resistor

The ANSI/IEEE standard symbol that designates an open-collector output is shown in Figure 13–25(c) for an inverter.

FIGURE 13–25
(*Continued*)

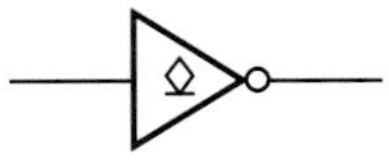

(c) Open-collector symbol in an inverter

Gates with Tristate Outputs

The tristate output combines the advantages of the totem-pole and open-collector circuits. The three output states are HIGH, LOW, and high-impedance (high-Z). When selected for normal logic-level operation, as determined by the state of the Enable input, a tristate circuit operates in the same way as a regular gate. When a tristate circuit is selected for high-Z operation, the output is effectively disconnected from the rest of the circuit. The operation of a tristate circuit is illustrated in Figure 13–26. The inverted triangle (∇) designates a tristate output.

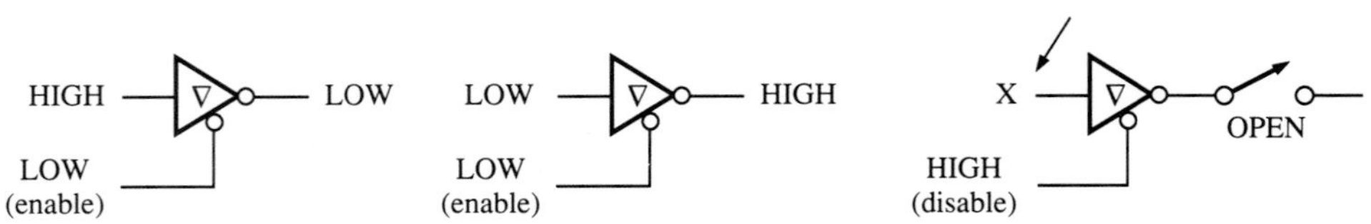

HIGH ———▷o— LOW   LOW ———▷o— HIGH   X ———▷o—o⟋—o OPEN

LOW (enable)   LOW (enable)   HIGH (disable)

don't care

(a) Enabled for normal logic operation   (b) High-Z state

FIGURE 13–26
The three states of a tristate circuit.

Figure 13–27 shows the basic circuit for a TTL tristate inverter. When the Enable input is LOW, Q_2 is off, and the output circuit operates as a normal totem-pole configuration, in which the output state depends on the input state. When the Enable input is HIGH, Q_2 is on. There is thus a LOW on the second emitter of Q_1, causing Q_3 and Q_5 to turn off, and diode D_1 is forward biased, causing Q_4 also to turn off. When both totem-pole transistors are off, they are effectively open, and the output is completely disconnected from the internal circuitry, as illustrated in Figure 13–28.

Other TTL Series

The basic TTL NAND gate circuit was discussed earlier. It is a *current-sinking* type of logic that draws current from the load when in the LOW output state and sources negligible current to the load when in the HIGH output state. Figure 13–29 is a *standard* 54/74 TTL two-input NAND gate, and it will be used for comparison with the other types of TTL circuits.

FIGURE 13–27
Basic tristate inverter circuit.

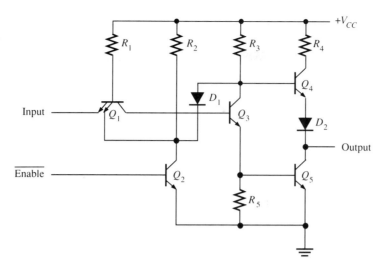

FIGURE 13–28
An equivalent circuit for the tristate totem-pole in the high-Z state.

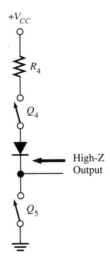

FIGURE 13–29
A 54/74 series standard TTL NAND gate.

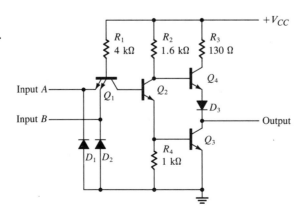

Low-power TTL (54L/74L)   The 54L/74L series of TTL circuits is designed for low power consumption. A typical gate circuit is shown in Figure 13–30. Notice that the circuit's resistor values are considerably higher than those of the standard gate in Figure 13–29. The higher resistance, of course, results in less current and, therefore, less power but increases the switching time of the gate. The typical power dissipation of a standard 54/74 gate is 10 mW, and that of a 54L/74L gate is 1 mW. The saving of power, however, is paid for in loss of speed. A typical standard 54/74 gate has a propagation delay time of 10 ns, compared with 33 ns for a 54L/74L gate.

FIGURE 13–30
A 54L/74L series TTL NAND gate.

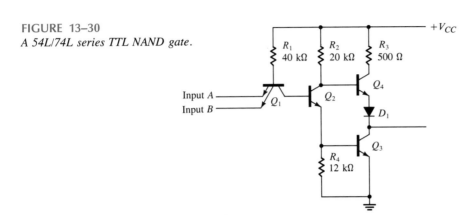

Schottky TTL (54S/74S)   The 54S/74S series of TTL circuits provides a faster switching time by incorporating Schottky diodes to prevent the transistors from going into saturation, thereby decreasing the time for a transistor to turn on or off. Notice the symbol for the Schottky transistor.

Typical propagation delay for a 54S/74S gate is 3 ns, and the power dissipation is 19 mW. Figure 13–31 shows a Schottky gate circuit.

FIGURE 13–31
A 54S/74S series TTL NAND gate.

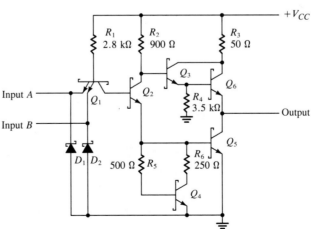

Low-power Schottky TTL (54LS/74LS) This TTL series compromises speed to obtain a lower power dissipation. A NAND gate circuit is shown in Figure 13–32. Notice that the input uses Schottky diodes rather than the conventional transistor input. The typical power dissipation for a gate is 2 mW, and the propagation delay is 9.5 ns.

FIGURE 13–32
A 54LS/74LS series TTL NAND gate.

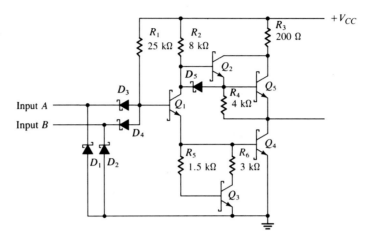

Advanced Schottky and advanced low-power Schottky (AS/ALS) These technologies are advanced versions of the S and LS series. The typical static power dissipation is 8.5 mW for the AS series and 1 mW for the ALS series. The typical propagation delay time is 4 ns for the ALS and 1.5 ns for the AS. One version of the AS series, produced by Fairchild Semiconductor, is called the F series. The label *F* stands for *FAST* (*F*airchild *a*dvanced *S*chottky *T*TL).

SECTION 13–3
REVIEW

Answers are found at the end of the chapter.
1. True or false: An npn BJT is on when the base is more negative than the emitter.
2. In terms of switching action, what do the on and off states of a BJT represent?
3. What are the two major types of output circuits in TTL?
4. Explain how tristate logic differs from normal, two-state logic.

PRACTICAL CONSIDERATIONS IN THE USE OF TTL

13–4

Now that you know something about TTL circuits, we will examine several practical considerations in their use and application.

Current Sinking and Current Sourcing

The concept of current sinking and sourcing was introduced in the first section of this chapter. Now that you are familiar with the totem-pole-output circuit configuration used in TTL, we will take a closer look at the sinking and sourcing action.

Figure 13–33 shows a standard TTL inverter with a totem-pole output connected to the input of another TTL inverter.

When the driving gate is in the HIGH output state, the driver is sourcing current to the load, as shown in Figure 13–33(a). The input to the load gate is like a reverse-biased diode, so there is practically no current required by the load. Actually, since the input is nonideal, there is a maximum of 40 μA from the totem-pole output of the driver into the load gate input.

When the driving gate is in the LOW output state, the driver is sinking current from the load, as shown in part (b). This current is 1.6 mA maximum for standard TTL and is indicated on the data sheet with a negative value because it is *out* of the input.

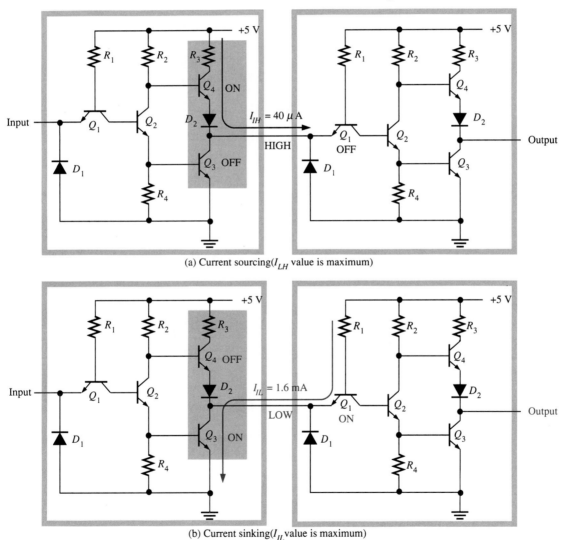

(a) Current sourcing(I_{LH} value is maximum)

(b) Current sinking(I_{IL} value is maximum)

FIGURE 13–33
Current sinking and sourcing action in TTL.

EXAMPLE 13–3

When a standard TTL NAND gate drives five standard TTL inputs, how much current does the driver output source, and how much does it sink?

Solution   Total source current (in HIGH output state):

$$I_{IH(\text{max})} = 40 \text{ μA per input}$$
$$I_T = (5 \text{ inputs})(40 \text{ μA/input}) = 5(40 \text{ μA}) = 200 \text{ μA}$$

Total sink current (in LOW output state):

$$I_{IL(\text{max})} = -1.6 \text{ mA per input}$$
$$I_T = (5 \text{ inputs})(-1.6 \text{ mA/input}) = 5(-1.6 \text{ mA}) = -8.0 \text{ mA}$$

EXAMPLE 13–4

Refer to the data sheet in Figure 13–15, and determine the fan-out of the 7400 NAND gate.

Solution   According to the data sheet, the current parameters are as follows:

$$I_{IH(\text{max})} = 40 \text{ μA}$$
$$I_{IL(\text{max})} = -1.6 \text{ mA}$$
$$I_{OH(\text{max})} = -400 \text{ μA}$$
$$I_{OL(\text{max})} = 16 \text{ mA}$$

Fan-out for the HIGH output state is figured as follows: Current $I_{OH(\text{max})}$ is the maximum current that the gate can source to a load. Each load input requires an $I_{IH(\text{max})}$ of 40 μA. The HIGH-state fan-out is

$$\frac{I_{OH(\text{max})}}{I_{IH(\text{max})}} = \frac{400 \text{ μA}}{40 \text{ μA}} = 10$$

For the LOW output state, fan-out is calculated as follows: $I_{OL(\text{max})}$ is the maximum current that the gate can sink. Each load input produces an $I_{IL(\text{max})}$ of 1.6 mA. The LOW-state fan-out is

$$\frac{I_{OL(\text{max})}}{I_{IL(\text{max})}} = \frac{16 \text{ mA}}{1.6 \text{ mA}} = 10$$

In this case both the HIGH-state fan-out and the LOW-state fan-out are the same.

Using Open-Collector Gates for Wired-AND Operation

The outputs of open-collector gates can be wired together to form what is called a *wired-AND* configuration. Figure 13–34 illustrates how four inverters are connected to produce a four-input negative-AND gate. A single external pull-up resistor, R_p, is required in all wired-AND circuits.

FIGURE 13–34
A wired-AND (actually negative-AND) configuration of four inverters.

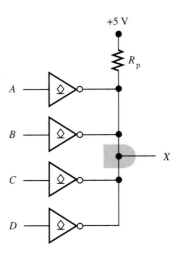

When one (or more) of the inverter inputs is HIGH, the output X is pulled LOW because an output transistor is on and acts as a closed switch to ground, as illustrated in Figure 13–35(a). In this case only one inverter has a HIGH input, but this is sufficient to pull the output LOW through the saturated output transistor Q_1 as indicated.

For the output X to be HIGH, *all* inverter inputs must be LOW so that all the open-collector output transistors are off as indicated in Figure 13–35(b). When this condition exists, the output X is pulled HIGH through the pull-up resistor.

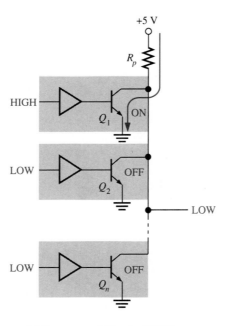

(a) One or more ON, output LOW

(b) All OFF, ouput HIGH

FIGURE 13–35
Open-collector wired-AND operation.

Thus, the output X is HIGH only when *all* the inputs are LOW. Therefore, we have a negative-AND function, as expressed in the following equation:

$$X = \overline{A}\overline{B}\overline{C}\overline{D}$$

EXAMPLE 13–5

Write the output expression for the wired-AND configuration of open-collector AND gates in Figure 13–36.

FIGURE 13–36

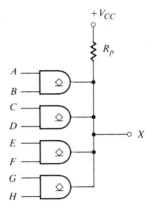

Solution   The output expression is

$$X = ABCDEFGH$$

The wired-AND connection of the four 2-input AND gates creates an 8-input AND gate.

EXAMPLE 13–6

Three open-collector NAND gates are connected in a wired-AND configuration as shown in Figure 13–37.
(a)   Write the logic expression for X.
(b)   Determine the minimum value of R_p if $I_{OL(max)}$ for each gate is 30 mA and $V_{OL(max)}$ is 0.4 V.
Assume that the wired-AND circuit is driving four TTL inputs (-1.6 mA each).

FIGURE 13–37

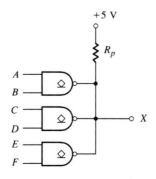

Solutions

(a) $X = \overline{AB} \cdot \overline{CD} \cdot \overline{EF}$

(b) $4(1.6 \text{ mA}) = 6.4 \text{ mA}$

$I_{Rp} = I_{OL(max)} - 6.4 \text{ mA} = 30 \text{ mA} - 6.4 \text{ mA} = 23.6 \text{ mA}$

$R_p = \dfrac{V_{CC} - V_{OL(max)}}{I_{Rp}} = \dfrac{5 \text{ V} - 0.4 \text{ V}}{23.6 \text{ mA}} = 195 \ \Omega$

Connection of Totem-Pole Outputs

Totem-pole outputs cannot be connected together because such a connection might produce excessive current and result in damage to the devices. For example, in Figure 13–38, when Q_1 in device A and Q_2 in device B are both on, the output of device A is effectively shorted to ground through Q_2 of device B.

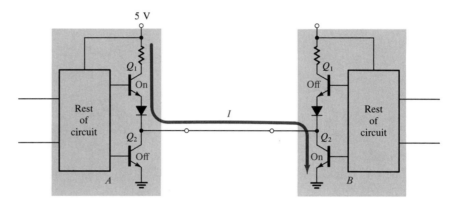

FIGURE 13–38

Totem-pole outputs wired together. Such a connection may cause excessive current through Q_1 of device A *and* Q_2 *of device* B.

Open-Collector Buffer/Drivers

A TTL circuit with a totem-pole output is limited in the amount of current that it can sink in the LOW state ($I_{OL(max)}$) to 16 mA for standard TTL and 20 mA for AS TTL. In many special applications, a gate must drive external devices, such as LEDs, lamps, or relays, that may require more current than that.

Because of their higher voltage and current-handling capability, circuits with open-collector outputs are generally used for driving LEDs, lamps, or relays. However, totem-pole outputs can be used, as long as the output current required by the external device does not exceed the amount that the TTL driver can sink.

With an open-collector TTL gate, the collector of the output transistor is connected to an LED or incandescent lamp as illustrated in Figure 13–39. In part (a) the limiting resistor is used to keep the current below maximum LED current. When the output of the gate is LOW, the output transistor is sinking current, and the LED is on. The LED is off when the output transistor is off and the output is HIGH. A typical open-collector buffer gate can sink up to 40 mA. In part (b) of the figure, the lamp requires no limiting resistor because the filament is resistive. Typically, up to $+30$ V can be used on the open collector, depending on the particular logic family.

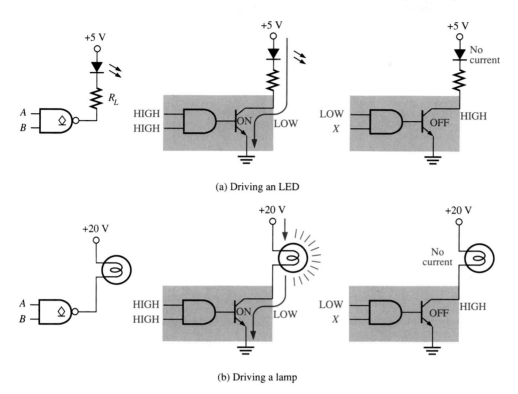

(a) Driving an LED

(b) Driving a lamp

FIGURE 13–39
Some applications of open-collector drivers.

EXAMPLE 13–7

Determine the value of the limiting resistor, R_L, in the open-collector circuit of Figure 13–40 if the LED current is to be 20 mA. Assume a 0.7 V drop across the LED when it is forward biased and a LOW-state output voltage of 0.1 V at the output of the gate.

FIGURE 13–40

+5 V

R_L

A

B

Solution

$$V_{RL} = 5 \text{ V} - 0.7 \text{ V} - 0.1 \text{ V} = 4.2 \text{ V}$$

$$R_L = \frac{V_{RL}}{I} = \frac{4.2 \text{ V}}{20 \text{ mA}} = 210 \ \Omega$$

Unused TTL Inputs

An unconnected input on a TTL gate acts as a HIGH because an open input results in a reverse-biased emitter junction on the input transistor, just as a HIGH level does. This effect is illustrated in Figure 13–41. However, because of noise sensitivity, it is best not to leave unused TTL inputs unconnected (open). There are several alternative ways to handle unused inputs.

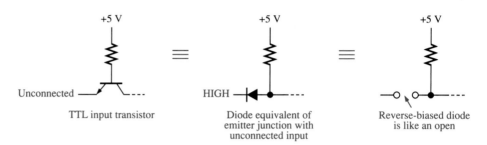

FIGURE 13–41
Comparison of an open TTL input and a HIGH level input.

Tied-together inputs   The most common method for handling unused gate inputs is to connect them to a used input of the same gate. For AND gates and NAND gates, all tied-together inputs count as one unit load in the LOW state; but for OR gates and NOR gates, each input tied to another input counts as a separate unit load in the LOW state. In the HIGH state, each tied-together input counts as a separate load for all types of TTL gates. In Figure 13–42(a) are two examples of the connection of two unused inputs to a used input.

The reason that AND and NAND gates present only a single unit load no matter how many inputs are tied together whereas OR and NOR gates present a unit load for each tied-together input can be seen by looking back at Figures 13–21 and 13–23. The NAND gate uses a multiple-emitter input transistor. No matter how many inputs are LOW, the total LOW-state current is limited by R_1 to a fixed value. The NOR gate uses a separate transistor for each input, so the LOW-state current is the sum of the currents from all the tied-together inputs.

Inputs to V_{CC} or ground   Unused inputs of AND and NAND gates can be connected to V_{CC} through a 1-kΩ resistor. This connection pulls the unused inputs to a HIGH level. Unused inputs of OR and NOR gates can be connected to ground. These methods are illustrated in Figure 13–42(b).

Inputs to unused output   A third method of terminating unused inputs may be appropriate in some cases when an unused gate or inverter is available. The unused gate output must be a constant HIGH for unused AND and NAND inputs and a constant LOW for unused OR and NOR inputs, as illustrated in Figure 13–42(c).

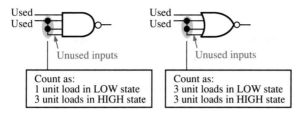

(a) Tied-together inputs

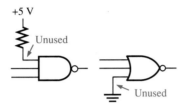

(b) Inputs to V_{CC} or ground

(c) Inputs to unused output

FIGURE 13–42
Methods for handling unused TTL inputs.

**SECTION 13–4
REVIEW**

Answers are found at the end of the chapter.
1. In what output state does a TTL circuit sink current from a load?
2. Why does a TTL circuit source less current into a TTL load than it sinks?
3. Why can TTL circuits with totem-pole outputs not be connected together?
4. What type of TTL circuit must be used for a wired-AND configuration?
5. What type of TTL circuit would you use to drive a lamp?
6. True or false: An unconnected TTL input acts as a LOW.

CMOS CIRCUITS

13–5

Basic internal CMOS circuitry and its operation are presented in this section. The abbreviation **CMOS** *stands for* **c***omplementary* **m***etal-***o***xide* **s***emiconductor. The term* **complementary** *refers to the use of two types of transistors in the output circuit in a configuration similar to the totem-pole in TTL. An* **n***-channel* **MOSFET** *(MOS field-effect transistor) and a* **p***-channel MOSFET are used.*

The MOSFET

Metal-oxide semiconductor field-effect transistors (MOSFETs) are the active switching elements in CMOS circuits. These devices differ greatly in construction and internal operation from BJTs but the switching action is basically the same: they function ideally as open or closed switches, depending on the input.

Figure 13–43(a) shows the symbols for both *n*-channel and *p*-channel MOSFETs. As indicated, the three terminals of a MOSFET are *gate, drain,* and *source.* When the gate voltage of an *n*-channel MOSFET is more positive than the source, the MOSFET is on, and there is a low value of resistance between the drain and the source (R_{ON}). When the gate-to-source voltage is zero, the MOSFET is off, and there is an extremely high resistance between the drain and the source (R_{OFF}). This operation is illustrated in Figure 13–43(b). The *p*-channel MOSFET operates with opposite voltage polarities, as

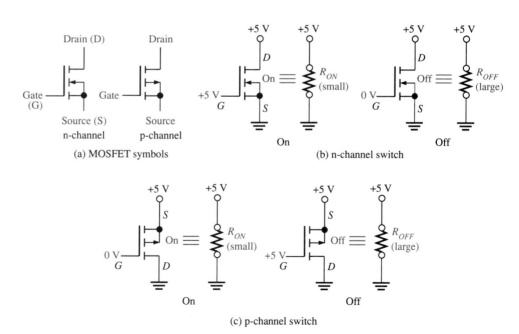

FIGURE 13–43
Basic symbols and switching action of MOSFETs.

shown in part (c). Ideally, R_{ON} and R_{OFF} can be neglected, and the MOSFET can be considered as a closed switch when it is on and an open switch when it is off.

CMOS Inverter

Complementary MOS (CMOS) logic uses the MOSFET as its basic element. It uses both *p*- and *n*-channel enhancement MOSFETs, as shown in the inverter circuit in Figure 13–44.

FIGURE 13–44
A CMOS inverter circuit.

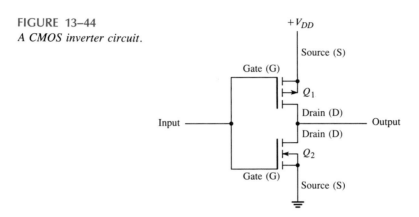

When a HIGH is applied to the input, the *p*-channel MOSFET Q_1 is off, and the *n*-channel MOSFET Q_2 is on. This condition connects the output to ground through the *on* resistance of Q_2, resulting in a LOW output. When a LOW is applied to the input, Q_1 is on and Q_2 is off. This condition connects the output to $+V_{DD}$ through the *on* resistance of Q_1, resulting in a HIGH output. The operation is illustrated in Figure 13–45.

FIGURE 13–45
Operation of a CMOS inverter.

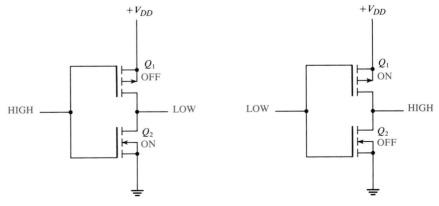

CMOS NAND Gate

Figure 13–46 shows a CMOS NAND gate with two inputs. Notice the arrangement of the complementary pairs (n- and p-channel MOSFETs).

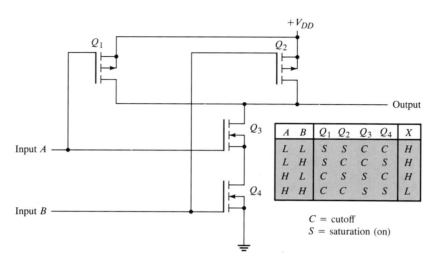

FIGURE 13–46

A CMOS NAND gate circuit.

A	B	Q_1	Q_2	Q_3	Q_4	X
L	L	S	S	C	C	H
L	H	S	C	C	S	H
H	L	C	S	S	C	H
H	H	C	C	S	S	L

C = cutoff
S = saturation (on)

The operation is as follows: When both inputs are LOW, Q_1 and Q_2 are on, and Q_3 and Q_4 are off. The output is pulled HIGH through the *on* resistance of Q_1 and Q_2 in parallel. When input A is LOW and input B is HIGH, Q_1 and Q_4 are on, and Q_2 and Q_3 are off. The output is pulled HIGH through the low *on* resistance of Q_1.

When input A is HIGH and input B is LOW, Q_1 and Q_4 are off, and Q_2 and Q_3 are on. The output is pulled HIGH through the low *on* resistance of Q_2.

Finally, when both inputs are HIGH, Q_1 and Q_2 are off, and Q_3 and Q_4 are on. In this case, the output is pulled LOW through the *on* resistance of Q_3 and Q_4 in series to ground.

CMOS NOR Gate

Figure 13–47 shows a CMOS NOR gate with two inputs. Notice the arrangement of the complementary pairs.

The operation is as follows: When both inputs are LOW, Q_1 and Q_2 are on, and Q_3 and Q_4 are off. As a result, the output is pulled HIGH through the *on* resistance of Q_1 and Q_2 in series.

When input A is LOW and input B is HIGH, Q_1 and Q_4 are on, and Q_2 and Q_3 are off. The output is pulled LOW through the low *on* resistance of Q_4 to ground.

When input A is HIGH and input B is LOW, Q_1 and Q_4 are off, and Q_2 and Q_3 are on. The output is pulled LOW through the *on* resistance of Q_3 to ground.

When both inputs are HIGH, Q_1 and Q_2 are off, and Q_3 and Q_4 are on. The output is pulled LOW through the *on* resistance of Q_3 and Q_4 in parallel to ground.

FIGURE 13–47

A CMOS NOR gate circuit.

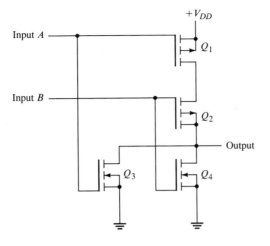

A	B	Q_1	Q_2	Q_3	Q_4	X
L	L	S	S	C	C	H
L	H	S	C	C	S	L
H	L	C	S	S	C	L
H	H	C	C	S	S	L

C = cutoff
S = saturation (on)

Open-Drain Gates

An open-drain gate is the CMOS counterpart of an open-collector TTL gate. An open-drain output circuit is a single *n*-channel MOSFET as shown in Figure 13–48(a). As with open-collector TTL, a pull-up resistor must be used, as shown in part (b), to produce a HIGH output state. Also, like open-collector TTL, open-drain outputs can be connected in a wired-AND configuration.

FIGURE 13–48

Open-drain gates.

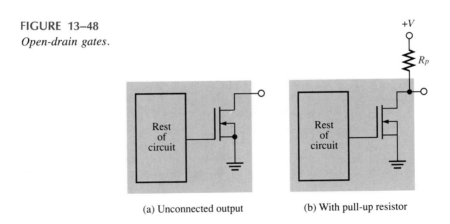

(a) Unconnected output

(b) With pull-up resistor

Tristate Gates

The circuitry in a tristate CMOS gate, as shown in Figure 13–49, allows each of the output transistors Q_1 and Q_2 to be turned off at the same time, thus disconnecting the output from the rest of the circuit.

When the Enable input is LOW, the device is enabled for normal logic operation. When the Enable is HIGH, both Q_1 and Q_2 are off, and the circuit is in the high-Z state.

FIGURE 13–49
A tristate CMOS inverter

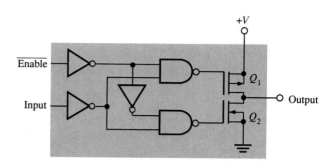

Precautions for Handling CMOS

All CMOS devices are subject to damage from electrostatic discharge (ESD). Therefore, they must be handled with special care, including the following precautions:

1. All CMOS devices are shipped in conductive foam to prevent electrostatic charge buildup. When they are removed from the foam, the pins should not be touched.
2. The devices should be placed with pins down on a grounded surface, such as a metal plate, when removed from protective material. Do not place CMOS devices in polystyrene foam or plastic trays.
3. All tools, test equipment, and metal workbenches should be earth-grounded. A person working with CMOS devices should, in certain environments, have his or her wrist grounded with a length of cable and a large-value resistor. The resistor prevents severe shock should the person come in contact with a voltage source.
4. Do not insert CMOS devices (or any other ICs) into sockets or pc boards with the power on.
5. All unused inputs should be connected to the supply voltage or ground as indicated in Figure 13–50. If left open, an input can acquire electrostatic charge and "float" to unpredicted levels.

FIGURE 13–50
Handling unused CMOS inputs.

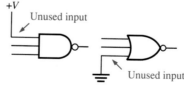

6. After assembly on pc boards, protection should be provided by storing or shipping boards with their connectors in conductive foam. The CMOS input and output pins may also be protected with large-value resistors connected to ground.

SECTION 13–5
REVIEW

Answers are found at the end of the chapter.
1. What type of transistor is used in CMOS logic?
2. What is meant by the term *complementary MOS?*
3. Why must CMOS devices be handled with great care?

COMPARISON OF CMOS AND TTL CHARACTERISTICS

13–6

In this section the main operational and performance characteristics of HCMOS (high-speed CMOS), which is the 74HC series, are compared with those of the various TTL series, particularly low-power Schottky (74LS).

Table 13–1 gives a broad performance comparison of HCMOS and several TTL series. Also, the CMOS 4000 series is listed for comparison. Keep in mind that parameter values for specific devices within a series may vary, so you should always consult the appropriate data sheet. The *maximum clock frequency, f_s,* is the maximum rate at which flip-flops can be clocked reliably. All values are typical and are specified for a supply voltage of $+5$ V. Fan-outs are specified for same-series loads (standard TTL to standard TTL, LS TTL to LS TTL, etc.) and also for LS loads as a basis for comparison among the various devices.

TABLE 13–1

Performance comparison of CMOS and TTL logic families.

Technology	Silicon-Gate CMOS	Metal Gate CMOS	Std. TTL	Low-Power Schottky TTL	Schottky TTL	Advanced Low-Power Schottky TTL	Advanced Schottky TTL
Device series	SN74HC	4000	SN74	SN74LS	SN74AS	SN74ALS	SN74AS
Power dissipation per gate (mW)							
Static	0.0000025	0.001	10	2	19	1	8.5
At 100 kHz	0.17	0.1	10	2	19	1	8.5
Propagation delay time (ns) ($C_L = 15$ pF)	8	105	10	10	3	4	1.5
Maximum clock frequency (MHz) ($C_L = 15$ pF)	40	12	35	40	125	70	200
Speed/Power product (pJ) (at 100 kHz)	1.4	11	100	20	57	4	13
Minimum output drive I_{OL} (mA) ($V_O = 0.4$ V)	4	1.6	16	8	20	8	20
Fan-out							
LS loads	10	4	40	20	50	20	50
Same-series	*	*	10	20	20	20	40
Maximum input current, I_{IL} (mA) ($V_I = 0.4$ V)	±0.001	−0.001	−1.6	−0.4	−2.0	−0.1	−0.5

*Fan out is frequency dependent.

The HCMOS devices can operate at speeds comparable to LS TTL. At higher frequencies the power consumption of HCMOS approaches that of LS TTL. The extremely low static (no-switching) power consumption of CMOS makes it ideal for battery-powered and battery-backup systems in which the current drain on the battery is a major consideration.

An HCMOS circuit has a better noise immunity than any of the TTL series, as indicated in Table 13–2.

TABLE 13–2

Comparison of typical noise margins.

Noise Margin	HCMOS	Std. TTL	LS TTL	AS TTL	ALS TTL
V_{NH}	1.4 V	0.4 V	0.7 V	0.7 V	0.7 V
V_{NL}	0.9 V	0.4 V	0.4 V	0.4 V	0.4 V

SECTION 13–6 REVIEW

Answers are found at the end of the chapter.
1. True or false: A CMOS circuit consumes less power than a TTL circuit.
2. Does the power consumption of CMOS increase or decrease with frequency?
3. True or false: A TTL circuit has better noise immunity than an HCMOS circuit.

INTERFACING LOGIC FAMILIES

13–7

In this section the interfacing of TTL and CMOS logic devices is considered. When two different types of technologies are interfaced, the input and output voltages and currents of each are important parameters. In fact, the difference in these parameters creates the interfacing problem.

Table 13–3 shows typical worst-case values of the input and output parameters for a CMOS family and several TTL families.

TABLE 13–3

Worst-case values of interfacing parameters.

Parameter	74H CMOS	74 TTL	74LS TTL	74AS TTL	74ALS TTL
$V_{IH(min)}$	3.5 V	2 V	2 V	2 V	2 V
$V_{IL(max)}$	1 V	0.8 V	0.8 V	0.8 V	0.8 V
$V_{OH(min)}$	4.9 V	2.4 V	2.7 V	2.7 V	2.7 V
$V_{OL(max)}$	0.1 V	0.4 V	0.4 V	0.4 V	0.4 V
$I_{IH(max)}$	1 μA	40 μA	20 μA	200 μA	20 μA
$I_{IL(max)}$	−1 μA	−1.6 mA	−400 μA	−2 mA	−100 μA
$I_{OH(max)}$	−4 mA	−400 μA	−400 μA	−2 mA	−400 μA
$I_{OL(max)}$	4 mA	16 mA	8 mA	20 mA	4 mA

CMOS-to-TTL Interfacing

This discussion is for the case in which CMOS is driving TTL. Table 13–3 shows that the minimum HIGH output voltage $V_{OH(min)}$ for CMOS is 4.9 V. Since this value exceeds the minimum HIGH input voltage $V_{IH(min)}$ of 2 V that is required by TTL, the CMOS is compatible with TTL in the HIGH state.

A CMOS has a maximum LOW output voltage $V_{OL(max)}$ of 0.1 V. Since this value is less than the maximum LOW input voltage $V_{IL(max)}$ of 0.8 V required by TTL, the CMOS is also compatible with TTL in the LOW state.

In terms of currents, CMOS can sink 4 mA, $I_{OL(max)}$, in the LOW output state with a guaranteed output voltage. When driving *standard* TTL, the CMOS gate must be able to sink 1.6 mA from each TTL input. This limits the fan-out of the CMOS gate to two TTL inputs (2×1.6 mA $= 3.2$ mA), as shown in Figure 13–51(a).

When driving *low-power Schottky* TTL (74LS), the CMOS gate must be able to sink 400 μA from each TTL input. This limits the fan-out of the CMOS gate to ten LS TTL inputs (10×400 μA $= 4$ mA), as shown in Figure 13–51(b).

When driving *advanced Schottky* TTL (74AS), the CMOS gate must be able to sink 2 mA from each TTL input. This limits the fan-out of the CMOS gate to two AS TTL inputs (2×2 mA $= 4$ mA), as shown in Figure 13–51(c).

Finally, when driving *advanced low-power Schottky* TTL (74ALS), the CMOS gate must be able to sink 100 μA from each TTL input. This limits the fan-out of the CMOS gate to forty ALS TTL inputs (40×100 μA $= 4$ mA), as shown in Figure 13–51(d).

FIGURE 13–51

Interfacing of CMOS to TTL.

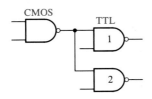

(a) CMOS-to-standard TTL.
Fanout = 2

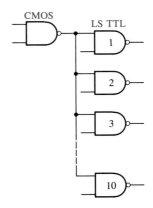

(b) CMOS-to-low power Schottky TTL.
Fanout = 10

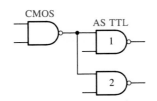

(c) CMOS-to-advanced Schottky TTL.
Fanout = 2

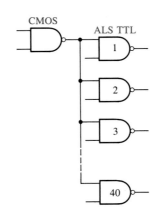

(d) CMOS-to-advanced low power
Schottky TTL.
Fanout = 40

TTL-to-CMOS Interfacing

When TTL is driving CMOS, the interface is not as simple as CMOS-to-TTL. As you can see in Table 13–3, the TTL families have minimum HIGH output voltages $V_{OH(min)}$ of 2.4 V to 2.7 V. The minimum HIGH input voltage required by CMOS is 3.5 V. Thus, the output voltage of TTL is not sufficient to drive CMOS in the HIGH state. In the LOW state, the voltages are compatible, as you can verify in Table 13–3.

For TTL to be properly interfaced to CMOS, a pull-up resistor (R_p) to V_{CC} must be added as shown in Figure 13–52.

FIGURE 13–52
Proper TTL-to-CMOS interfacing.

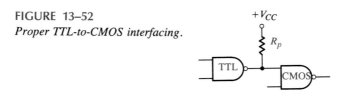

In the LOW state, the TTL driving gate must sink current from the resistor as well as from the CMOS inputs to which it is connected, as indicated in Figure 13–53.

FIGURE 13–53
LOW-state current sinking.

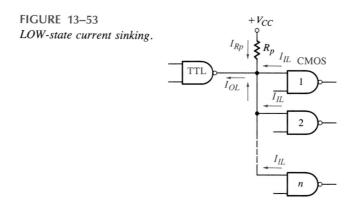

This requirement sets the value of R_p according to the following equation:

$$R_p = \frac{V_{CC} - V_{OL(min)}}{I_{OL(TTL)} + nI_{IL(CMOS)}}$$

(13–5)

where n is the number of CMOS inputs being driven and $I_{OL(TTL)} + nI_{IL(CMOS)} = I_{Rp}$.

EXAMPLE 13–8

A 74LS TTL gate drives four 74HC CMOS gates. The minimum V_{CC} is 4.75 V. Determine the minimum value of pull-up resistor for interfacing these devices.

Solution From Table 13–3, we have $I_{OL} = 8$ mA, $I_{IL} = -1$ μA, and $V_{OL(min)}$ unspecified. Therefore, a value of 0 for $V_{OL(min)}$ is assumed, to ensure that $I_{OL(max)}$ is never exceeded. Thus,

$$R_p = \frac{V_{CC} - V_{OL(min)}}{I_{OL} + 4I_{IL}}$$

$$= \frac{4.75 \text{ V} - 0 \text{ V}}{8 \text{ mA} - 4 \text{ μA}} = \frac{4.75 \text{ V}}{7.996 \text{ mA}}$$

$$= 594 \ \Omega$$

**SECTION 13–7
REVIEW**

Answers are found at the end of the chapter.
1. When is a pull-up resistor required in interfacing TTL and CMOS?
2. True or false: A higher fan-out requires a lower value of R_p.

ECL CIRCUITS

13–8

*The abbreviation **ECL** stands for **emitter-coupled** logic, which, like TTL, is a bipolar technology. The typical ECL circuit, shown in Figure 13–54(a), consists of a differential amplifier input circuit, a bias circuit, and emitter-follower outputs.*

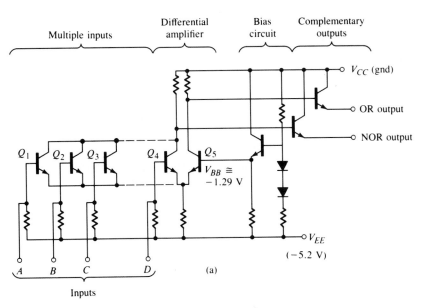

FIGURE 13–54
An ECL OR/NOR gate circuit.

The emitter-follower outputs provide the OR logic function and its NOR complement, as indicated by Figure 13–54(b).

Because of the low output impedance of the emitter-follower and the high input impedance of the differential amplifier input, high fan-out operation is possible. In this type of circuit, saturation is not possible. The lack of saturation results in higher power consumption and limited voltage swing (less than 1 V), but it permits high-frequency switching.

The V_{CC} pin is normally connected to ground, and -5.2 V from the power supply is connected to V_{EE} for best operation. Notice that in Figure 13–54(c) the output varies from a LOW level of -1.75 V to a HIGH level of -0.9 V with respect to ground. In positive logic a 1 is the HIGH level (less negative), and a 0 is the LOW level (more negative).

FIGURE 13–54
(*Continued*)

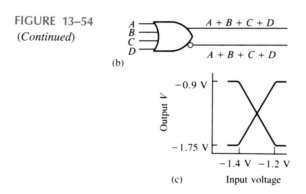

(b)

(c)   Input voltage

ECL Circuit Operation

Beginning with LOWs on all inputs (typically -1.75 V), assume that Q_1 through Q_4 are off because the base-emitter junctions are reverse biased, and that transistor Q_5 is conducting (but not saturated). The bias circuit holds the base of Q_5 at -1.29 V, and, therefore, the emitter of Q_5 is approximately 0.8 V below the base at -2.09 V. The voltage differential from base to emitter of the input transistors Q_1 through Q_4 is -2.09 V $-$ (-1.75 V) $=$ -0.34 V. This is less than the forward-biased voltage of these transistors, and they are therefore off. This condition is shown in Figure 13–55(a).

When any one or all of the inputs are raised to the HIGH level (-0.9 V), that transistor (or transistors) will conduct. When this happens, the voltage at the emitters of Q_1 through Q_5 increases from -2.09 V to -1.7 V (one base-emitter drops below the -0.9-V base). Since the base of Q_5 is held at a constant -1.29 V by the bias circuit, Q_5 turns off. The resulting collector voltages are coupled through the emitter-followers to the output terminals. Because of the differential action of Q_1 through Q_4 with Q_5, Q_5 is off when one or more of the Q_1 through Q_4 transistors conduct, thus providing simultaneous complementary outputs. This operation is shown in Figure 13–55(b).

When all the inputs are returned to the LOW state, Q_1 through Q_4 are again cut off, and Q_5 conducts.

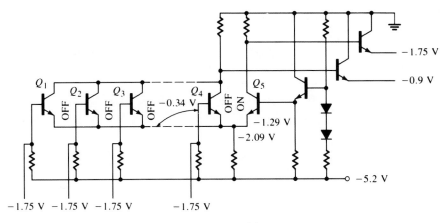

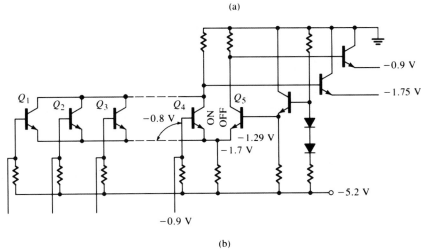

(a)

(b)

FIGURE 13–55
Basic ECL gate operation.

Noise Margin

As you know, the noise margin of a gate is the measure of its immunity to undesired voltage fluctuations (noise). Typical ECL circuits have noise margins from about 0.2 V to 0.25 V. These are less than for TTL and make ECL unreliable in high-noise environments.

Comparison of ECL with Advanced Schottky TTL

Table 13–4 shows a comparison of typical values of key parameters for advanced Schottky (AS) TTL and ECL.

TABLE 13–4

	Propagation Delay Time	Power Dissipation per Gate	Voltage Swing	dc Supply Voltage	Flip-Flop Clock Freq.
74AS	1.5 ns	8.5 mW	3.0 V	+5 V	200 MHz
ECL	1 ns	60 mW	0.85 V	−5.2 V	500 MHz

SECTION 13–8
REVIEW

Answers are found at the end of the chapter.
1. What is the primary advantage of ECL over TTL?
2. Name two disadvantages of ECL compared with TTL.

I²L CIRCUITS

13–9

Integrated injection logic (I²L) is a bipolar technology that allows an extremely high component density on a chip (up to ten times that of TTL). The I²L technology is being used for some complex LSI functions, such as microprocessors, and is simpler to fabricate than either TTL or MOS. It also has a low power requirement and reasonably good switching speeds that are improving all the time.

The basic I²L gate is extremely simple, as indicated in Figure 13–56. Transistor Q_1 acts as a current source and active pull-up, and the multiple-collector transistor Q_2 operates as an inverter.

FIGURE 13–56
An I²L gate.

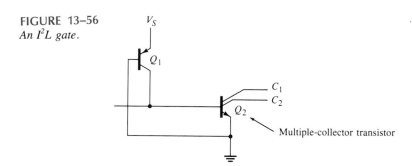

Because the base of Q_1 and the emitter of Q_2 are common and the collector of Q_1 and the base of Q_2 are common, the entire I²L gate (constructed on a silicon chip) takes only the space of a single TTL multiple-emitter transistor.

Transistor Q_1 is called a *current-injector transistor* because when its emitter is connected to an external power source, it can supply current into the base of Q_2. The switching action of Q_2 is accomplished by steering the injector current as follows: A LOW on the base of Q_2 will pull the injector current away from the base of Q_2 and through the low-impedance path(s) provided by the driving gate(s), thus turning Q_2 off.

When the output transistor of the driving gate is off (open), it corresponds to a HIGH input; this condition causes the injector current to be steered into the base of Q_2, turning it on. This action is illustrated in Figure 13–57. Figure 13–58 shows an example of an I^2L implementation of a logic function.

FIGURE 13–57
Basic I^2L operation.

FIGURE 13–58
An I^2L latch.

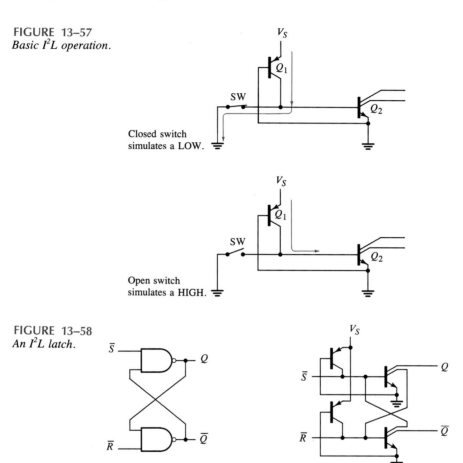

SECTION 13–9
REVIEW

Answer is found at the end of the chapter.
1. What is the main feature of I^2L logic?

PMOS AND NMOS

13–10

The MOS circuits are used largely in LSI functions, such as long shift registers, large memories, and microprocessor products. Such use is a result of the low power consumption and very small chip area required for MOS transistors. Of all the MOS technologies, CMOS is the only one that provides SSI functions at the gate and flip-flop level in a variety comparable to that of TTL.

PMOS

One of the first high-density MOS circuit technologies to be produced was *PMOS*. It utilizes enhancement-mode *p*-channel MOS transistors to form the basic gate building blocks. Figure 13–59 shows a basic PMOS gate that produces the NOR function in positive logic.

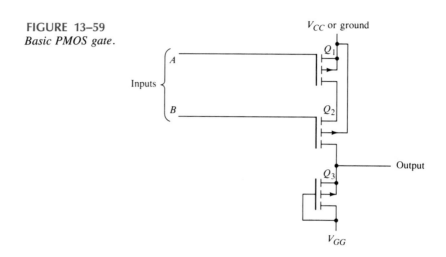

FIGURE 13–59
Basic PMOS gate.

The operation of the PMOS gate is as follows: The supply voltage V_{GG} is a negative voltage, and V_{CC} is a positive voltage or ground (0 V). Transistor Q_3 is permanently biased to create a constant drain-to-source resistance. Its sole purpose is to function as a current-limiting resistor. If a HIGH (V_{CC}) is applied to input *A* or *B*, then Q_1 or Q_2 is off, and the output is pulled down to a voltage near V_{GG}, which represents a LOW. When a LOW voltage (V_{GG}) is applied to both input *A* and input *B*, both Q_1 and Q_2 are turned on. This causes the output to go to a HIGH level (near V_{CC}). Since a LOW output occurs when either or both inputs are HIGH, and a HIGH output occurs only when all inputs are LOW, we have a NOR gate.

NMOS

The *NMOS* devices were developed as processing technology improved, and now most memories and microprocessors use NMOS. The *n*-channel MOS transistor is used in NMOS circuits, as shown in Figure 13–60 for a NAND gate and a NOR gate.

In Figure 13–60(a), Q_3 acts as a resistor to limit current. When a LOW (V_{GG} or ground) is applied to one or both inputs, then at least one of the transistors (Q_1 or Q_2) is off, and the output is pulled up to a HIGH level near V_{CC}. When HIGHs (V_{CC}) are applied to both *A* and *B*, both Q_1 and Q_2 conduct, and the output is LOW. This action, of course, identifies this circuit as a NAND gate.

In Figure 13–60(b), Q_3 again acts as a resistor. A HIGH on either input turns Q_1 or Q_2 on, pulling the output LOW. When both inputs are LOW, both transistors are off, and the output is pulled up to a HIGH level.

FIGURE 13–60
Two NMOS gates.

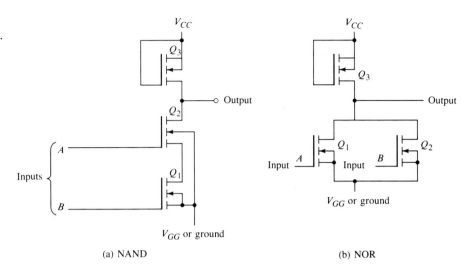

(a) NAND   (b) NOR

Answer is found at the end of the chapter.
1. What is the main feature of NMOS and PMOS technology in integrated circuits?

SUMMARY

☐ *Formulas*

$$V_{NH} = V_{OH(min)} - V_{IH(min)} \qquad \text{High-level noise margin}$$

$$V_{NL} = V_{IL(max)} - V_{OL(max)} \qquad \text{Low-level noise margin}$$

$$I_{CC} = \frac{(I_{CCH} + I_{CCL})}{2} \qquad \text{Average dc supply current}$$

$$P_D = V_{CC}I_{CC} \qquad \text{Power dissipation}$$

$$R_p = \frac{V_{CC} - V_{OL(max)}}{I_{OL}(TTL) + nI_{IL(CMOS)}} \qquad \text{Pull-up resistor for TTL-to-CMOS interface}$$

☐ Totem-pole outputs of TTL cannot be connected together.

☐ Open-collector and open-drain outputs can be connected for wired-AND.

☐ An HCMOS device offers lower power dissipation than any of the TTL series.

☐ The HCMOS operating speed is comparable to that of LS TTL. Some TTL series (S TTL, AS TTL, and ALS TTL) offer greater operating speed.

☐ The $V_{IH(min)}$ of HCMOS is not compatible with the $V_{OH(min)}$ of TTL. When TTL is driving CMOS, use HCT MOS (HCMOS that is TTL compatible), or use pull-up resistors.

☐ The output voltages of HCMOS are TTL compatible.

☐ The output current capability of HCMOS is not as large as for TTL.

☐ An HCMOS device has a smaller fan-out to LS devices than the TTL series.

☐ An HCMOS device has a wide operating supply-voltage range (2 V to 6 V).

☐ A TTL device is not as vulnerable to electrostatic discharge (ESD) as is a CMOS device.

☐ Because of ESD, CMOS devices must be handled with great care.

GLOSSARY

Base   One of the three regions in a bipolar junction transistor.

Bias   The polarity of a voltage, which controls the operation of a semiconductor device.

BJT   Bipolar junction transistor. A semiconductor device used for switching or amplification. A BJT has two junctions, the base-emitter junction and the base-collector junction.

CMOS   Complementary metal-oxide semiconductor. A type of digital integrated circuit that uses MOS field-effect transistors.

Collector   One of the three regions in a bipolar junction transistor.

Current sinking   The action of a circuit in which it accepts current into its output from a load.

Current sourcing   The action of a circuit in which it sends current out of its output and into a load.

Data sheet   A document that specifies parameter values and operating conditions for an integrated circuit or other device.

Diode   A semiconductor device that conducts current in only one direction when properly biased.

Drain   One of the terminals of a field-effect transistor.

ECL   Emitter-coupled logic. A type of IC technology.

Emitter   One of the three regions in a bipolar junction transistor.

Fan-out   The maximum number of equivalent gate inputs that a logic circuit can drive.

FET   Field-effect transistor.

Forward bias   A voltage polarity condition that allows a semiconductor pn junction in a transistor or diode to conduct current.

Gate   A logic circuit. Also, one of the three terminals of a field-effect transistor.

High-Z   The high-impedance state of a tristate circuit, in which the output is effectively disconnected from the rest of the circuit.

I^2L   Integrated injection logic. An IC technology.

Junction   The boundary between an n region and a p region in a BJT.

LED   Light-emitting diode.

MOS   Metal-oxide semiconductor. A type of transistor technology.

Noise immunity   The ability of a circuit to reject unwanted signals.

Noise margin   The difference between the maximum LOW output of a gate and the maximum acceptable LOW input of an equivalent gate. Also, the difference between the minimum HIGH output of a gate and the minimum acceptable HIGH input of an equivalent gate.

NMOS   An n-channel metal-oxide semiconductor.

npn   A type of junction structure of a BJT.

Open-collector   A type of output in a logic circuit in which the collector of the output transistor is left disconnected from any internal circuitry and is available for external connection. Normally used for driving higher-current or higher-voltage loads.

PMOS   A p-channel metal-oxide semiconductor.

pnp   A type of junction structure of a BJT.

Power dissipation   The amount of power required by a circuit.

Propagation delay   The time interval between the occurrence of an input transition and the resulting output transition in a logic circuit.

Reverse bias   A voltage polarity condition that prevents a pn junction of a transistor or diode from conducting current.

Source   One of the terminals of a field-effect transistor.

Totem-pole   A type of output in TTL circuits.

Tristate   A type of output in logic circuits that exhibits three states: HIGH, LOW, and open (disconnected). The open state is called the high-Z state.

Transistor A semiconductor device exhibiting current and/or voltage gain. When used as a switching device, it approximates an open or closed switch.

TTL Transistor-transistor logic. A type of IC technology that uses bipolar junction transistors.

Unit load One input of a gate in the same logic family as the driving gate.

SELF-TEST

1. Describe one major difference between a bipolar IC and an MOS IC.
2. Explain why an open TTL input acts as a HIGH.
3. List five series of TTL circuits.
4. The fan-out of a standard TTL gate is 10 unit loads; that is, the gate can drive 10 other standard TTL gate inputs. How many LS gate inputs can the standard TTL gate drive?
5. If two unused inputs of a standard TTL gate are connected to an input being driven by another standard TTL gate, how many other inputs can be driven by this gate? Assume that the fan-out is 10 unit loads.
6. How much current does a standard TTL gate sink if it is driving seven standard unit loads?
7. If the frequency of operation of a CMOS device is increased, what happens to the power dissipation?
8. Which performs better in a high-noise environment: CMOS or TTL? Why?
9. Why are MOS devices shipped in conductive foam?
10. What is the main advantage of ECL over other IC technologies? What is a second advantage?
11. In what type of application should ECL not be considered? Why?
12. What type of MOS technology is predominant in LSI devices?

PROBLEMS

Section 13–1 Basic Operational Characteristics and Parameters

13–1 A certain logic gate has a $V_{OH(min)} = 2.2$ V, and it is driving a gate with a $V_{IH(min)} = 2.5$ V. Are these gates compatible for HIGH-state operation? Why?

13–2 A certain logic gate has a $V_{OL(max)} = 0.45$ V, and it is driving a gate with a $V_{IL(max)} = 0.75$ V. Are these gates compatible for LOW-state operation? Why?

13–3 A TTL gate has the following actual voltage level values: $V_{IH(min)} = 2.25$ V, $V_{IL(max)} = 0.65$ V. Assuming it is being driven by a gate with $V_{OH(min)} = 2.4$ V and $V_{OL(max)} = 0.4$ V, what are the HIGH- and LOW-level noise margins?

13–4 What is the maximum amplitude of noise spikes that can be tolerated on the inputs in both the HIGH state and the LOW state for the gate in Problem 13–3?

13–5 Voltage specifications for three types of logic gates are given in Table 13–5. Select the gate that you would use in a high-noise industrial environment.

TABLE 13–5

	$V_{OH(min)}$	$V_{OL(max)}$	$V_{IH(min)}$	$V_{IL(max)}$
Gate A	2.4 V	0.4 V	2 V	0.8 V
Gate B	3.5 V	0.2 V	2.5 V	0.6 V
Gate C	4.2 V	0.2 V	3.2 V	0.8 V

13–6 A certain gate draws a dc supply current from a + 5 V source of 2 mA in the LOW state and 3.5 mA in the HIGH state. What is the power dissipation in the LOW state? What is the power dissipation in the HIGH state? Assuming a 50% duty cycle, what is the average power dissipation?

13–7 Each gate in the network of Figure 13–61 has a t_{PLH} of 4 ns. If a positive-going pulse is applied to the input as indicated, how long will it take the output pulse to appear?

FIGURE 13–61

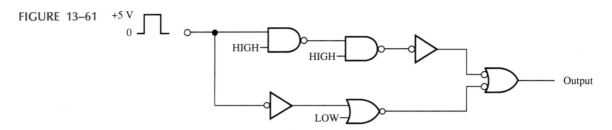

13–8 For a certain gate, t_{PLH} = 3 ns and t_{PHL} = 2 ns. What is the average propagation delay time?

13–9 Table 13–6 lists parameters for three types of gates. Basing your decision on the speed-power product, which one would you select for best performance?

TABLE 13–6

	t_{PLH}	t_{PHL}	P_D
Gate A	1 ns	1.2 ns	15 mW
Gate B	5 ns	4 ns	8 mW
Gate C	10 ns	10 ns	0.5 mW

13–10 Which gate in Table 13–6 would you select if you wanted the gate to operate at the highest possible frequency?

13–11 A standard TTL gate has a fan-out of 10. Are any of the gates in Figure 13–62 overloaded? If so, which ones?

FIGURE 13–62

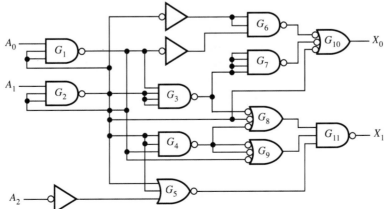

13–12 Which CMOS gate network in Figure 13–63 can operate at the highest frequency?

FIGURE 13–63

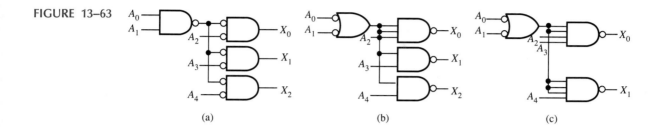

(a) (b) (c)

Section 13–2   Data Sheet Interpretation

13–13 From the data sheet in Appendix A, determine the typical average power dissipation of a 74ALS00A under the conditions given.

13–14 Determine the noise margins of a 74ALS00A for $V_{CC} = 4.5$ V and $I_{OL} = 8$ mA.

13–15 What is the maximum frequency at which a 74AS74 flip-flop can be operated?

Section 13–3   TTL Circuits

13–16 Determine which BJTs in Figure 13–64 are off and which are on.

13–17 Determine the output state of each TTL gate in Figure 13–65.

FIGURE 13–64

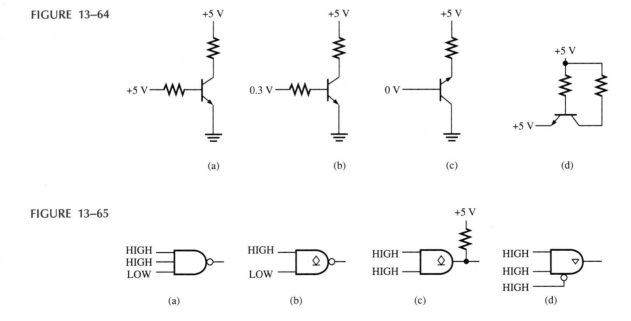

(a) (b) (c) (d)

FIGURE 13–65

(a) (b) (c) (d)

Section 13–4   Practical Considerations in the Use of TTL

13–18 Determine the output level of each TTL gate in Figure 13–66.

FIGURE 13–66

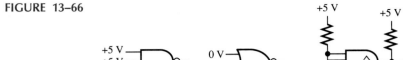
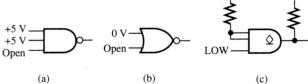

(a)   (b)   (c)

13–19 For each part of Figure 13–67, tell whether each driving gate is sourcing or sinking current. Specify the maximum current out of or into the output of the driving gate or gates in each case. All gates are standard TTL.

13–20 Use open-collector inverters to implement the following logic functions:
(a) $X = \overline{A}\,\overline{B}C$ **(b)** $X = \overline{ABC}\overline{D}$ **(c)** $X = \overline{ABCD}\overline{E}\overline{F}$

13–21 Write the logic expression for each of the circuits in Figure 13–68.

FIGURE 13–67

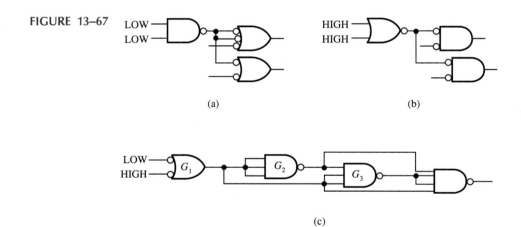

(a)   (b)

(c)

FIGURE 13–68

(a)   (b)   (c)

13–22 Determine the minimum value for the pull-up resistor in each circuit in Figure 13–68 if $I_{OL(max)} = 40$ mA and $V_{OL(max)} = 0.25$ V for each gate. Assume that 10 standard TTL unit loads are being driven from output X and the supply voltage is 5 V.

13–23 A certain relay requires 60 mA. Devise a way to use open-collector NAND gates with $I_{OL(max)} = 40$ mA to drive the relay.

Section 13–5   CMOS Circuits

13–24 Determine the state (on or off) of each MOSFET in Figure 13–69.

13–25 The CMOS gate network in Figure 13–70 is incomplete. Indicate the changes that should be made.

13–26 Devise a circuit, using appropriate CMOS logic gates and/or inverters, with which signals from four different sources can be connected to a common line at different times without interfering with each other.

FIGURE 13–69

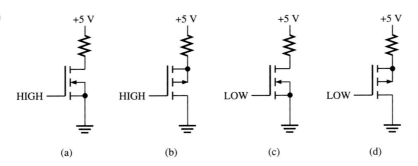

(a)   (b)   (c)   (d)

FIGURE 13–70

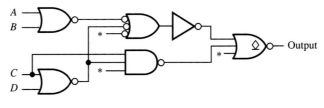

* unused inputs

Section 13–6   Comparison of CMOS and TTL Characteristics

13–27 Determine from Table 13–1 the logic family that is most appropriate for each of the following requirements:

(a) highest speed

(b) lowest static power

(c) lowest power at 100 kHz

(d) optimum compromise between speed and power

(e) highest fan-out into LS loads

(f) highest fan-out into same-series loads

13–28 Determine the amount of current that gate G_1 is sinking when its output is LOW for each circuit in Figure 13–71.

13–29 One of the flip-flops in Figure 13–72 has an erratic output. Which is it and why?

FIGURE 13–71

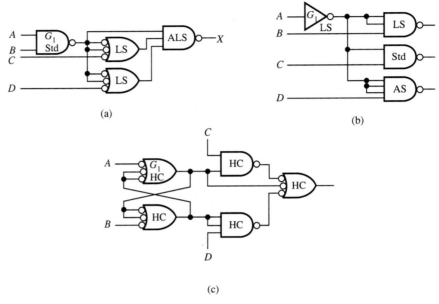

(a)

(b)

(c)

FIGURE 13–72

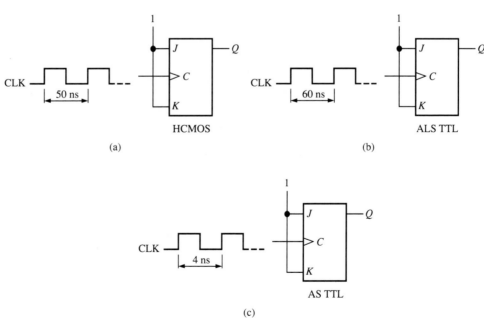

HCMOS

ALS TTL

(a)

(b)

AS TTL

(c)

Section 13–7 Interfacing Logic Families

13–30 The network in Figure 13–73 is a mixture of IC technologies interfaced together. Determine whether the circuit is properly designed from an interface point of view, and if not, make the necessary design corrections.

FIGURE 13–73

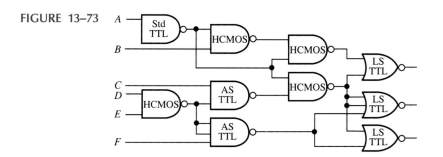

13–31 Repeat Problem 13–30 for the gate network in Figure 13–74.

FIGURE 13–74

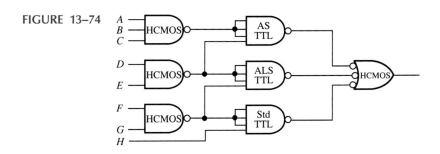

13–32 A standard TTL gate drives five HCMOS gate inputs. Determine the minimum value of pull-up resistor for $V_{CC} = 5$ V. Sketch the circuit.

Section 13–8   ECL Circuits

13–33 What is the basic difference between ECL circuitry and TTL circuitry?

13–34 Select ECL, HCMOS, or the appropriate TTL series for each of the following requirements:

(a) highest speed

(b) lowest power

(c) best compromise between high speed and low power

(d) high-noise environment applications

ANSWERS TO SECTION REVIEWS

Section 13–1   Review

1. V_{IH}: HIGH level input voltage; V_{IL}: LOW level input voltage; V_{OH}: HIGH level output voltage; V_{OL}: LOW level output voltage

2. higher value

3. gate B

4. Excessive loading reduces the noise margin.

Section 13–2   Review

1. $V_{NL} = 0.4$ V, $V_{NH} = 0.4$ V

2. 400 μA

Section 13–3   Review

1. false
2. The on state is a closed switch; the off state is an open switch.
3. totem-pole and open-collector
4. Tristate logic provides a high-impedance state, in which the output is disconnected from the rest of the circuit.

Section 13–4   Review

1. LOW
2. because a TTL load looks like a reverse-biased diode in the HIGH state
3. The transistors cannot handle the current when one output tries to go HIGH and the other is LOW.
4. open-collector
5. open-collector
6. false

Section 13–5   Review

1. MOSFET
2. The output circuit consists of an n-channel and a p-channel MOSFET.
3. because electrostatic discharge can damage them

Section 13–6   Review

1. true
2. It increases.
3. false

Section 13–7   Review

1. TTL to CMOS
2. false

Section 13–8   Review

1. ECL is faster.
2. more power; less noise margin

Section 13–9   Review

1. Fewer components mean higher component densities on a chip.

Section 13–10   Review

1. high density

Data Sheets

TYPES SN54HC00, SN74HC00
QUADRUPLE 2-INPUT POSITIVE-NAND GATES

D2684, DECEMBER 1982 – REVISED MARCH 1984

- Package Options Include Both Plastic and Ceramic Chip Carriers in Addition to Plastic and Ceramic DIPs
- Dependable Texas Instruments Quality and Reliability

description

These devices contain four independent 2-input NAND gates. They perform the Boolean functions $Y = \overline{A \cdot B}$ or $Y = \overline{A} + \overline{B}$ in positive logic.

The SN54HC00 is characterized for operation over the full military temperature range of $-55\,°C$ to $125\,°C$. The SN74HC00 is characterized for operation from $-40\,°C$ to $85\,°C$.

FUNCTION TABLE (each gate)

INPUTS		OUTPUT
A	**B**	**Y**
H	H	L
L	X	H
X	L	H

logic symbol

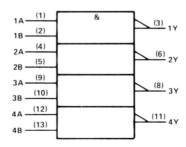

Pin numbers shown are for J and N packages.

SN54HC00 . . . J PACKAGE
SN74HC00 . . . J OR N PACKAGE
(TOP VIEW)

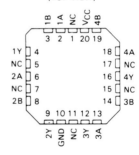

SN54HC00 . . . FH OR FK PACKAGE
SN74HC00 . . . FH OR FN PACKAGE
(TOP VIEW)

NC – No internal connection

maximum ratings, recommended operating conditions, and electrical characteristics

See Table I, page 2-4.

(Courtesy of Texas Instruments Inc.)

TYPES SN54HC00, SN74HC00
QUADRUPLE 2-INPUT POSITIVE-NAND GATES

switching characteristics over recommended operating free-air temperature range (unless otherwise noted), C_L = 50 pF (see Note 1)

PARAMETER	FROM (INPUT)	TO (OUTPUT)	V_{CC}	T_A = 25°C MIN	T_A = 25°C TYP	T_A = 25°C MAX	SN54HC00 MIN	SN54HC00 MAX	SN74HC00 MIN	SN74HC00 MAX	UNIT
t_{pd}	A or B	Y	2 V		45	90		135		115	ns
			4.5 V		9	18		27		23	
			6 V		8	15		23		20	
t_t		Y	2 V		38	75		110		95	ns
			4.5 V		8	15		22		19	
			6 V		6	13		19		16	

C_{pd}	Power dissipation capacitance per gate	No load, T_A = 25°C	20 pF typ

NOTE 1: For load circuit and voltage waveforms, see page 1-14.

(Courtesy of Texas Instruments Inc.)

TYPES SN54ALS00A, SN54AS00, SN74ALS00A, SN74AS00
QUADRUPLE 2-INPUT POSITIVE-NAND GATES

D2661, APRIL 1982–REVISED DECEMBER 1983

- **Package Options Include Both Plastic and Ceramic Chip Carriers in Addition to Plastic and Ceramic DIPs**

- **Dependable Texas Instruments Quality and Reliability**

description

These devices contain four independent 2-input NAND gates. They perform the Boolean functions $Y = \overline{A \cdot B}$ or $Y = \overline{A} + \overline{B}$ in positive logic.

The SN54ALS00A and SN54AS00 are characterized for operation over the full military temperature range of $-55\,°C$ to $125\,°C$. The SN74ALS00A and SN74AS00 are characterized for operation from $0\,°C$ to $70\,°C$.

SN54ALS00A, SN54AS00 . . . J PACKAGE
SN74ALS00A, SN74AS00 . . . N PACKAGE
(TOP VIEW)

```
        ___ ___
1A  [ 1  U  14 ]  VCC
1B  [ 2     13 ]  4B
1Y  [ 3     12 ]  4A
2A  [ 4     11 ]  4Y
2B  [ 5     10 ]  3B
2Y  [ 6      9 ]  3A
GND [ 7      8 ]  3Y
```

FUNCTION TABLE (each gate)

INPUTS		OUTPUT
A	B	Y
H	H	L
L	X	H
X	L	H

logic symbol

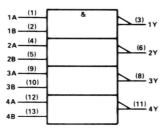

Pin numbers shown are for J and N packages.

SN54ALS00A, SN54AS00 . . . FH PACKAGE
SN74ALS00A, SN74AS00 . . . FN PACKAGE
(TOP VIEW)

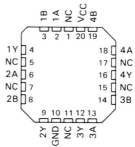

NC – No internal connection

(Courtesy of Texas Instruments Inc.)

TYPES SN54ALS00A, SN74ALS00A
QUADRUPLE 2-INPUT POSITIVE-NAND GATES

absolute maximum ratings over operating free-air temperature range (unless otherwise noted)

Supply voltage, V_{CC} . . . . 7 V
Input voltage . . . . 7 V
Operating free-air temperature range: SN54ALS00A . . . . −55 °C to 125 °C
SN74ALS00A . . . . 0 °C to 70 °C
Storage temperature range . . . . −65 °C to 150 °C

recommended operating conditions

		SN54ALS00A			SN74ALS00A			UNIT
		MIN	NOM	MAX	MIN	NOM	MAX	
V_{CC}	Supply voltage	4.5	5	5.5	4.5	5	5.5	V
V_{IH}	High-level input voltage	2			2			V
V_{IL}	Low-level input voltage			0.8			0.8	V
I_{OH}	High-level output current			−0.4			−0.4	mA
I_{OL}	Low-level output current			4			8	mA
T_A	Operating free-air temperature	−55		125	0		70	°C

electrical characteristics over recommended operating free-air temperature range (unless otherwise noted)

PARAMETER	TEST CONDITIONS		SN54ALS00A			SN74ALS00A			UNIT
			MIN	TYP†	MAX	MIN	TYP†	MAX	
V_{IK}	$V_{CC} = 4.5$ V,	$I_I = -18$ mA			−1.5			−1.5	V
V_{OH}	$V_{CC} = 4.5$ V to 5.5 V,	$I_{OH} = -0.4$ mA	$V_{CC}-2$			$V_{CC}-2$			V
V_{OL}	$V_{CC} = 4.5$ V,	$I_{OL} = 4$ mA		0.25	0.4		0.25	0.4	V
	$V_{CC} = 4.5$ V,	$I_{OL} = 8$ mA					0.35	0.5	
I_I	$V_{CC} = 5.5$ V,	$V_I = 7$ V			0.1			0.1	mA
I_{IH}	$V_{CC} = 5.5$ V,	$V_I = 2.7$ V			20			20	µA
I_{IL}	$V_{CC} = 5.5$ V,	$V_I = 0.4$ V			−0.1			−0.1	mA
I_O‡	$V_{CC} = 5.5$ V,	$V_O = 2.25$ V	−15		−70	−15		−70	mA
I_{CCH}	$V_{CC} = 5.5$ V,	$V_I = 0$ V		0.5	0.85		0.5	0.85	mA
I_{CCL}	$V_{CC} = 5.5$ V,	$V_I = 4.5$ V		1.5	3		1.5	3	mA

†All typical values are at $V_{CC} = 5$ V, $T_A = 25$ °C.
‡The output conditions have been chosen to produce a current that closely approximates one half of the true short-circuit output current, I_{OS}.

switching characteristics (see Note 1)

PARAMETER	FROM (INPUT)	TO (OUTPUT)	$V_{CC} = 4.5$ V to 5.5 V, $C_L = 50$ pF, $R_L = 500$ Ω, $T_A = $ MIN to MAX				UNIT
			SN54ALS00A		SN74ALS00A		
			MIN	MAX	MIN	MAX	
t_{PLH}	A or B	Y	3	14	3	11	ns
t_{PHL}	A or B	Y	2	10	2	8	ns

NOTE 1: For load circuit and voltage waveforms, see page 1-12.

(Courtesy of Texas Instruments Inc.)

TYPES SN54ALS74A, SN54AS74, SN74ALS74A, SN74AS74
DUAL D-TYPE POSITIVE-EDGE-TRIGGERED
FLIP-FLOPS WITH CLEAR AND PRESET

D2661, APRIL 1982—REVISED FEBRUARY 1984

- Package Options Include Both Plastic and Ceramic Chip Carriers in Addition to Plastic and Ceramic DIPs

- Dependable Texas Instruments Quality and Reliability

TYPE	TYPICAL MAXIMUM CLOCK FREQUENCY (C_L = 50 pF)	TYPICAL POWER DISSIPATION PER FLIP-FLOP
'ALS74A	50 MHz	6 mW
'AS74	134 MHz	26 mW

description

These devices contain two independent D-type positive-edge-triggered flip-flops. A low level at the Preset or Clear inputs sets or resets the outputs regardless of the levels of the other inputs. When Preset and Clear are inactive (high), data at the D input meeting the setup time requirements are transferred to the outputs on the positive-going edge of the clock pulse. Clock triggering occurs at a voltage level and is not directly related to the rise time of the clock pulse. Following the hold time interval, data at the D input may be changed without affecting the levels at the outputs.

The SN54ALS74A and SN54AS74 are characterized for operation over the full military temperature range of −55°C to 125°C. The SN74ALS74A and SN74AS74 are characterized for operation from 0°C to 70°C.

SN54ALS74A, SN54AS74 . . . J PACKAGE
SN74ALS74A, SN74AS74 . . . N PACKAGE
(TOP VIEW)

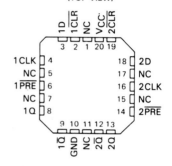

```
1CLR  [ 1   U  14 ] VCC
1D    [ 2      13 ] 2CLR
1CLK  [ 3      12 ] 2D
1PRE  [ 4      11 ] 2CLK
1Q    [ 5      10 ] 2PRE
1Q    [ 6       9 ] 2Q
GND   [ 7       8 ] 2Q
```

SN54ALS74A, SN54AS74 . . . FH PACKAGE
SN74ALS74A, SN74AS74 . . . FN PACKAGE
(TOP VIEW)

```
        1D  1CLR  NC  VCC  2CLR
         3    2    1   20   19
1CLK [ 4              18 ] 2D
NC   [ 5              17 ] NC
1PRE [ 6              16 ] 2CLK
NC   [ 7              15 ] NC
1Q   [ 8              14 ] 2PRE
         9   10  11  12  13
        1Q  GND  NC  2Q  2Q
```

NC—No internal connection

logic symbol

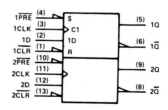

```
1PRE  (4)    ─┤ S        ├─ (5)  1Q
1CLK  (3)    ─┤ C1       │
1D    (2)    ─┤ 1D       ├─ (6)  1Q
1CLR  (1)    ─┤ R        │
2PRE  (10)   ─┤ S        ├─ (9)  2Q
2CLK  (11)   ─┤ C1       │
2D    (12)   ─┤ 1D       ├─ (8)  2Q
2CLR  (13)   ─┤ R        │
```

Pin numbers shown are for J and N packages.

FUNCTION TABLE

INPUTS				OUTPUTS	
PRESET	CLEAR	CLOCK	D	Q	Q̄
L	H	X	X	H	L
H	L	X	X	L	H
L	L	X	X	H*	H*
H	H	↑	H	H	L
H	H	↑	L	L	H
H	H	L	X	Q_0	$\overline{Q}_0$

*The output levels in this configuration are not guaranteed to meet the minimum levels for V_{OH} if the lows at Preset and Clear are near V_{IL} maximum. Furthermore, this configuration is nonstable; that is, it will not persist when either Preset or Clear returns to its inactive (high) level.

absolute maximum ratings over operating free-air temperature range (unless otherwise noted)

Supply voltage, V_{CC} . . . . 7 V
Input voltage . . . . 7 V
Operating free-air temperature range: SN54ALS74A, SN54AS74 . . . . −55°C to 125°C
SN74ALS74A, SN74AS74 . . . . 0°C to 70°C
Storage temperature range . . . . −65°C to 150°C

(Courtesy of Texas Instruments Inc.)

TYPES SN54AS74, SN74AS74
DUAL D-TYPE POSITIVE-EDGE-TRIGGERED
FLIP-FLOPS WITH CLEAR AND PRESET

recommended operating conditions

			SN54AS74			SN74AS74			UNIT
			MIN	NOM	MAX	MIN	NOM	MAX	
V_{CC}	Supply voltage		4.5	5	5.5	4.5	5	5.5	V
V_{IH}	High-level input voltage		2			2			V
V_{IL}	Low-level input voltage				0.8			0.8	V
I_{OH}	High-level output current				−2			−2	mA
I_{OL}	Low-level output current				20			20	mA
f_{clock}	Clock frequency		0		90	0		105	MHz
t_w	Pulse duration	$\overline{PRE}$ or $\overline{CLR}$ low	4			4			ns
		CLK high	4			4			
		CLK low	5.5			5.5			
t_{su}	Setup time	Data	4.5			4.5			ns
	before CLK↑	$\overline{PRE}$ or $\overline{CLR}$ inactive	2			2			
t_h	Hold time, data after CLK↑		0			0			ns
T_A	Operating free-air temperature		−55		125	0		70	°C

electrical characteristics over recommended operating free-air temperature range (unless otherwise noted)

PARAMETER		TEST CONDITIONS		SN54AS74			SN74AS74			UNIT
				MIN	TYP[†]	MAX	MIN	TYP[†]	MAX	
V_{IK}		$V_{CC} = 4.5$ V,	$I_I = −18$ mA			−1.2			−1.2	V
V_{OH}		$V_{CC} = 4.5$ V to 5.5 V,	$I_{OH} = −2$ mA	$V_{CC}−2$			$V_{CC}−2$			V
V_{OL}		$V_{CC} = 4.5$ V,	$I_{OL} = 20$ mA		0.25	0.5		0.25	0.5	V
I_I		$V_{CC} = 5.5$ V,	$V_I = 7$ V			0.1			0.1	mA
I_{IH}	CLK or D	$V_{CC} = 5.5$ V,	$V_I = 2.7$ V			20			20	µA
	$\overline{PRE}$ or $\overline{CLR}$					40			40	
I_{IL}	CLK or D	$V_{CC} = 5.5$ V,	$V_I = 0.4$ V			−0.5			−0.5	mA
	$\overline{PRE}$ or $\overline{CLR}$					−1.8			−1.8	
I_{IO}[‡]		$V_{CC} = 5.5$ V,	$V_O = 2.25$ V	−30		−112	−30		−112	mA
I_{CC}		$V_{CC} = 5.5$ V	See Note 1		10.5	16		10.5	16	mA

[†]All typical values are at $V_{CC} = 5$ V, $T_A = 25$°C.
[‡]The output conditions have been chosen to produce a current that closely approximates one half of the true short-current output current, I_{OS}.
NOTE 1: I_{CC} is measured with D, CLK, and $\overline{PRE}$ grounded, then with D, CLK, and $\overline{CLR}$ grounded.

switching characteristics (see Note 2)

PARAMETER	FROM (INPUT)	TO (OUTPUT)	$V_{CC} = 4.5$ V to 5.5 V, $C_L = 50$ pF, $R_L = 500$ Ω, $T_A = $ MIN to MAX				UNIT
			SN54AS74		SN74AS74		
			MIN	MAX	MIN	MAX	
f_{max}			90		105		MHz
t_{PLH}	$\overline{PRE}$ or $\overline{CLR}$	Q or $\overline{Q}$	3	8.5	3	7.5	ns
t_{PHL}			3.5	11.5	3.5	10.5	
t_{PLH}	CLK	Q or $\overline{Q}$	3.5	9	3.5	8	ns
t_{PHL}			4.5	10.5	4.5	9	

NOTE 2: For load circuit and voltage waveforms, see page 1-12 of the TTL Data Book, Volume 3.

(Courtesy of Texas Instruments Inc.)

TYPES SN54ALS151, SN54AS151, SN74ALS151, SN74AS151
1 OF 8 DATA SELECTORS/MULTIPLEXERS

D2661, APRIL 1982–REVISED FEBRUARY 1984

- **8-Line to 1-Line Multiplexers**
  Can Perform As:
  Boolean Function Generators
  Parallel-to-Serial Converters
  Data Source Selectors

- **Input Clamping Diodes Simplify System Design**

- **Fully Compatible With Most TTL Circuits**

- **Package Options Include Both Plastic and Ceramic Chip Carriers in Addition to Plastic and Ceramic DIPs**

- **Dependable Texas Instruments Quality and Reliability**

description

These monolithic data selectors/multiplexers provide full binary decoding to select one of eight data sources. The strobe input (G) must be at a low logic level to enable the inputs. A high level at the strobe terminal forces the W output high and the Y output low.

The SN54ALS151 and SN54AS151 are characterized for operation over the full military temperature range of −55 °C to 125 °C. The SN74ALS151 and SN74AS151 are characterized for operation from 0 °C to 70 °C.

SN54ALS151, SN54AS151 . . . J PACKAGE
SN74ALS151, SN74AS151 . . . N PACKAGE
(TOP VIEW)

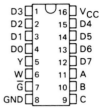

SN54ALS151, SN54AS151 . . . FH PACKAGE
SN74ALS151, SN74AS151 . . . FN PACKAGE
(TOP VIEW)

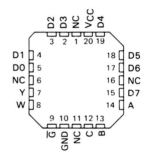

NC – No internal connection

logic symbol

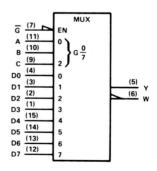

Pin numbers shown are for J and N packages.

FUNCTION TABLE

INPUTS				OUTPUTS	
SELECT			STROBE		
C	B	A	$\overline{G}$	Y	W
X	X	X	H	L	H
L	L	L	L	D0	$\overline{D0}$
L	L	H	L	D1	$\overline{D1}$
L	H	L	L	D2	$\overline{D2}$
L	H	H	L	D3	$\overline{D3}$
H	L	L	L	D4	$\overline{D4}$
H	L	H	L	D5	$\overline{D5}$
H	H	L	L.	D6	$\overline{D6}$
H	H	H	L	D7	$\overline{D7}$

H = high level, L = low level, X = irrelevant
D0, D1 . . . D7 = the level of the D respective input

(Courtesy of Texas Instruments Inc.)

TYPES SN54ALS151, SN54AS151, SN74ALS151, SN74AS151
1 OF 8 DATA SELECTORS/MULTIPLEXERS

logic diagram (positive logic)

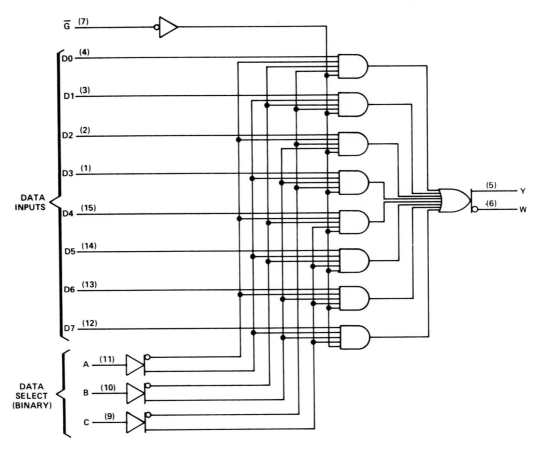

Pin numbers shown are for J and N packages.

absolute maximum ratings over operating free-air temperature range (unless otherwise noted)

Supply voltage, V_{CC} .... 7 V
Input voltage .... 7 V
Operating free-air temperature range: SN54ALS151, SN54AS151 .... −55 °C to 125 °C
                                       SN74ALS151, SN74AS151 .... 0 °C to 70 °C
Storage temperature range .... −65 °C to 150°C

(Courtesy of Texas Instruments Inc.)

TYPES SN54ALS151, SN74ALS151
1 OF 8 DATA SELECTORS/MULTIPLEXERS

recommended operating conditions

		SN54ALS151			SN74ALS151			UNIT
		MIN	NOM	MAX	MIN	NOM	MAX	
V_{CC}	Supply voltage	4.5	5	5.5	4.5	5	5.5	V
V_{IH}	High-level input voltage	2			2			V
V_{IL}	Low-level input voltage			0.8			0.8	V
I_{OH}	High-level output current			−1			−2.6	mA
I_{OL}	Low-level output current			12			24	mA
T_A	Operating free-air temperature	−55		125	0		70	°C

electrical characteristics over recommended operating free-air temperature range (unless otherwise noted)

PARAMETER	TEST CONDITIONS		SN54ALS151			SN74ALS151			UNIT
			MIN	TYP[†]	MAX	MIN	TYP[†]	MAX	
V_{IK}	V_{CC} = 4.5 V,	I_I = −18 mA			−1.5			−1.5	V
V_{OH}	V_{CC} = 4.5 V to 5.5 V,	I_{OH} = −0.4 mA	V_{CC}−2			V_{CC}−2			V
	V_{CC} = 4.5 V,	I_{OH} = −1 mA	2.4	3.3					
	V_{CC} = 4.5 V,	I_{OH} = −2.6 mA				2.4	3.2		
V_{OL}	V_{CC} = 4.5 V,	I_{OL} = 12 mA		0.25	0.4		0.25	0.4	V
	V_{CC} = 4.5 V,	I_{OL} = 24 mA					0.35	0.5	
I_I	V_{CC} = 5.5 V,	V_I = 7 V			0.1			0.1	mA
I_{IH}	V_{CC} = 5.5 V,	V_I = 2.7 V			20			20	μA
I_{IL}	V_{CC} = 5.5 V,	V_I = 0.4 V			−0.1			−0.1	mA
I_O[‡]	V_{CC} = 5.5 V,	V_O = 2.25 V	−30		−112	−30		−112	mA
I_{CC}	V_{CC} = 5.5 V,	Inputs at 4.5 V		7.5	12		7.5	12	mA

[†]All typical values are at V_{CC} = 5 V, T_A = 25 °C.
[‡]The output conditions have been chosen to produce a current that closely approximates one half of the true short-circuit output current, I_{OS}.

switching characteristics (see Note 1)

PARAMETER	FROM (INPUT)	TO (OUTPUT)	V_{CC} = 4.5 V to 5.5 V, C_L = 50 pF, R_L = 500 Ω, T_A = MIN to MAX				UNIT
			SN54ALS151		SN74ALS151		
			MIN	MAX	MIN	MAX	
t_{PLH}	A, B, or C	Y	4	21	4	18	ns
t_{PHL}			8	28	8	24	
t_{PLH}	A, B, or C	W	7	28	7	24	ns
t_{PHL}			7	26	7	23	
t_{PLH}	Any D	Y	3	12	3	10	ns
t_{PHL}			5	18	5	15	
t_{PLH}	Any D	W	3	18	3	15	ns
t_{PHL}			4	18	4	15	
t_{PLH}	$\overline{G}$	Y	4	21	4	18	ns
t_{PHL}			4	23	4	19	
t_{PLH}	$\overline{G}$	W	5	23	5	19	ns
t_{PHL}			5	26	5	23	

NOTE 1: For load circuit and voltage waveforms, see page 1-12.

(Courtesy of Texas Instruments Inc.)

SN54ALS138, SN54AS138, SN74ALS138, SN74AS138
3-LINE TO 8-LINE DECODERS/DEMULTIPLEXERS

D2661, APRIL 1982 REVISED APRIL 1985

- Designed Specifically for High-Speed Memory Decoders and Data Transmission Systems

- Incorporates 3 Enable Inputs to Simplify Cascading and/or Data Reception

- Package Options Include Both Plastic and Ceramic Chip Carriers in Addition to Plastic and Ceramic DIPs

- Dependable Texas Instruments Quality and Reliability

description

The 'ALS138 and 'AS138 circuits are designed to be used in high-performance memory-decoding or data-routing applications requiring very short propagation delay times. In high-performance memory systems this decoder can be used to minimize the effects of system decoding. When employed with high-speed memories utilizing a fast enable circuit, the delay times of this decoder and the enable time of the memory are usually less than the typical access time of the memory. This means that the effective system delay introduced by the Schottky-clamped system decoder is negligible.

The conditions at the binary select inputs and the three enable inputs select one of eight input lines. Two active-low and one active-high enable inputs reduce the need for external gates or inverters when expanding. A 24-line decoder can be implemented without external inverters and a 32-line decoder requires only one inverter. An enable input can be used as a data input for demultiplexing applications.

The SN54ALS138 and SN54AS138 are characterized for operation over the full military temperature range of $-55\,^{\circ}$C to $125\,^{\circ}$C. The SN74ALS138 and SN74AS138 are characterized for operation from $0\,^{\circ}$C to $70\,^{\circ}$C.

SN54ALS138, SN54AS138 . . . J PACKAGE
SN74ALS138, SN74AS138 . . . N PACKAGE
(TOP VIEW)

```
        ┌───┬─∪─┬───┐
    A  ☐│1      16│☐ VCC
    B  ☐│2      15│☐ Y0
    C  ☐│3      14│☐ Y1
  G2A  ☐│4      13│☐ Y2
  G2B  ☐│5      12│☐ Y3
   G1  ☐│6      11│☐ Y4
   Y7  ☐│7      10│☐ Y5
  GND  ☐│8       9│☐ Y6
        └────────────┘
```

SN54ALS138, SN54AS138 . . . FK PACKAGE
SN74ALS138, SN74AS138 . . . FN PACKAGE
(TOP VIEW)

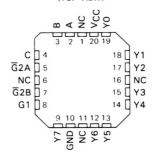

NC – No internal connection

(Courtesy of Texas Instruments Inc.)

SN54ALS138, SN54AS138, SN74ALS138, SN74AS138
3-LINE TO 8-LINE DECODERS/DEMULTIPLEXERS

logic symbols (alternatives)

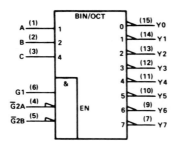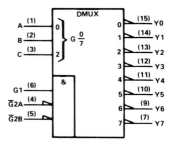

logic diagram (positive logic)

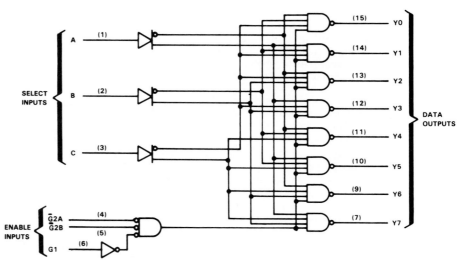

Pin numbers shown are for J and N packages.

(Courtesy of Texas Instruments Inc.)

SN54ALS138, SN54AS138, SN74ALS138, SN74AS138
3-LINE TO 8-LINE DECODERS/DEMULTIPLEXERS

FUNCTION TABLE

ENABLE INPUTS			SELECT INPUTS			OUTPUTS							
G1	$\overline{G2A}$	$\overline{G2B}$	C	B	A	Y0	Y1	Y2	Y3	Y4	Y5	Y6	Y7
X	H	X	X	X	X	H	H	H	H	H	H	H	H
X	X	H	X	X	X	H	H	H	H	H	H	H	H
L	X	X	X	X	X	H	H	H	H	H	H	H	H
H	L	L	L	L	L	L	H	H	H	H	H	H	H
H	L	L	L	L	H	H	L	H	H	H	H	H	H
H	L	L	L	H	L	H	H	L	H	H	H	H	H
H	L	L	L	H	H	H	H	H	L	H	H	H	H
H	L	L	H	L	L	H	H	H	H	L	H	H	H
H	L	L	H	L	H	H	H	H	H	H	L	H	H
H	L	L	H	H	L	H	H	H	H	H	H	L	H
H	L	L	H	H	H	H	H	H	H	H	H	H	L

absolute maximum ratings over operating free-air temperature range (unless otherwise noted)

Supply voltage, V_{CC} . . . . 7 V
Input voltage . . . . 7 V
Operating free-air temperature range; SN54ALS138, SN54AS138 . . . . −55 °C to 125 °C
SN74ALS138, SN74AS138 . . . . 0 °C to 70 °C
Storage temperature range . . . . −65 °C to 150 °C

recommended operating conditions

		SN54ALS138			SN74ALS138			UNIT
		MIN	NOM	MAX	MIN	NOM	MAX	
V_{CC}	Supply voltage	4.5	5	5.5	4.5	5	5.5	V
V_{IH}	High-level input voltage	2			2			V
V_{IL}	Low-level input voltage			0.8			0.8	V
I_{OH}	High-level output current			−0.4			−0.4	mA
I_{OL}	Low-level output current			4			8	mA
T_A	Operating free-air temperature	−55		125	0		70	°C

electrical characteristics over recommended operating free-air temperature range (unless otherwise noted)

PARAMETER	TEST CONDITIONS		SN54ALS138			SN74ALS138			UNIT
			MIN	TYP[†]	MAX	MIN	TYP[†]	MAX	
V_{IK}	$V_{CC} = 4.5$ V,	$I_I = -18$ mA			−1.5			−1.5	V
V_{OH}	$V_{CC} = 4.5$ V to 5.5 V,	$I_{OH} = -0.4$ mA	$V_{CC} - 2$			$V_{CC} - 2$			V
V_{OL}	$V_{CC} = 4.5$ V,	$I_{OL} = 4$ mA		0.25	0.4		0.25	0.4	V
	$V_{CC} = 4.5$ V,	$I_{OL} = 8$ mA					0.35	0.5	
I_I	$V_{CC} = 5.5$ V,	$V_I = 7$ V			0.1			0.1	mA
I_{IH}	$V_{CC} = 5.5$ V,	$V_I = 2.7$ V			20			20	µA
I_{IL}	$V_{CC} = 5.5$ V,	$V_I = 0.4$ V			−0.1			−0.1	mA
I_O [‡]	$V_{CC} = 5.5$ V,	$V_O = 2.25$ V	−30		−112	−30		−112	mA
I_{CC}	$V_{CC} = 5.5$ V			5	10		5	10	mA

[†] All typical values are at $V_{CC} = 5$ V, $T_A = 25$ °C.
[‡] The output conditions have been chosen to produce a current that closely approximates one half of the true short-circuit output current, I_{OS}.

(Courtesy of Texas Instruments Inc.)

TYPES SN54ALS190, SN54ALS191, SN74ALS190, SN74ALS191
SYNCHRONOUS 4-BIT UP/DOWN DECADE AND BINARY COUNTERS

D2661, DECEMBER 1982−REVISED DECEMBER 1983

- Single Down/Up Count Control Line

- Look-Ahead Circuitry Enhances Speed of Cascaded Counters

- Fully Synchronous in Count Modes

- Asynchronously Presettable with Load Control

- Package Options Include Both Plastic and Ceramic Chip Carriers in Addition to Plastic and Ceramic DIPS

- Dependable Texas Instruments Quality and Reliability

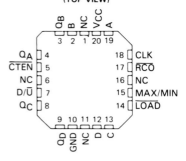

NC — no internal connection.

descriptions

The 'ALS190 and 'ALS191 are synchronous, reversible up/down counters. The 'ALS190 is a 4-bit decade counter and the 'ALS191 is a 4-bit binary counter. Synchronous counting operation is provided by having all flip-flops clocked simultaneously so that the outputs change coincident with each other when so instructed by the steering logic. This mode of operation eliminates the output counting spikes normally associated with asynchronous (ripple clock) counters.

The outputs of the four flip-flops are triggered on a low-to-high-level transition of the clock input if the enable input ($\overline{\text{CTEN}}$) is low. A high at $\overline{\text{CTEN}}$ inhibits counting. The direction of the count is determined by the level of the down/up (D/$\overline{\text{U}}$) input. When D/$\overline{\text{U}}$ is low, the counter counts up and when D/$\overline{\text{U}}$ is high, it counts down.

These counters feature a fully independent clock circuit. Changes at the control inputs ($\overline{\text{CTEN}}$ and D/$\overline{\text{U}}$) that will modify the operating mode have no effect on the contents of the counter until clocking occurs. The function of the counter will be dictated solely by the condition meeting the stable setup and hold times.

These counters are fully programmable; that is, the outputs may each be preset to either level by placing a low on the load input and entering the desired data at the data inputs. The output will change to agree with the data inputs independently of the level of the clock input. This feature allows the counters to be used as modulo-N dividers by simply modifying the count length with the preset inputs.

The CLK, D/$\overline{\text{U}}$, and LOAD inputs are buffered to lower the drive requirement, which significantly reduces the loading on, or current required by, clock drivers, etc., for long parallel words.

Two outputs have been made available to perform the cascading function: ripple clock and maximum/minimum count. The latter output produces a high-level output pulse with a duration approximately equal to one complete cycle of the clock while the count is zero (all outputs low) counting down or maximum (9 or 15) counting up. The ripple clock output produces a low-level output pulse under those same conditions but only while the clock input is low. The counters can be easily cascaded by feeding the ripple clock output to the enable input of the succeeding counter if parallel clocking is used, or to the clock input if parallel enabling is used. The maximum/minimum count output can be used to accomplish look-ahead for high-speed operation.

The SN54ALS190 and SN54ALS191 are characterized for operation over the full military temperature range of −55 °C to 125 °C. The SN74ALS190 and SN74ALS191 are characterized for operation from 0 °C to 70 °C.

(Courtesy of Texas Instruments Inc.)

TYPES SN54ALS191, SN74ALS191
SYNCHRONOUS 4-BIT UP/DOWN BINARY COUNTERS

'ALS191 logic symbol

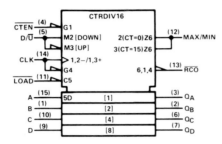

'ALS191 logic diagram (positive logic)

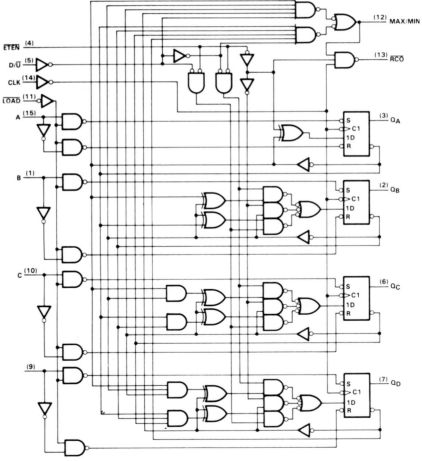

Pin numbers shown are for J and N packages.

(Courtesy of Texas Instruments Inc.)

TYPES SN54ALS190, SN74ALS190
SYNCHRONOUS 4-BIT UP/DOWN BINARY COUNTERS

typical load, count, and inhibit sequences

'ALS190

Illustrated below is the following sequence:

1. Load (preset) to BCD seven.
2. Count up to eight, nine (maximum), zero, one, and two.
3. Inhibit.
4. Count down to one, zero (minimum), nine, eight, and seven.

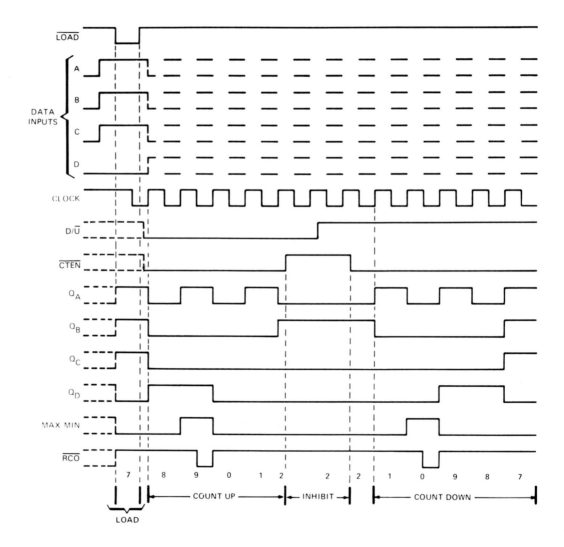

(Courtesy of Texas Instruments Inc.)

TYPES SN54ALS191, SN74ALS191
SYNCHRONOUS 4-BIT UP/DOWN BINARY COUNTERS

typical load, count, and inhibit sequences

'ALS191

Illustrated below is the following sequence:

1. Load (preset) to binary thirteen.
2. Count up to fourteen, fifteen (maximum), zero, one, and two.
3. Inhibit.
4. Count down to one, zero (minimum), fifteen, fourteen, and thirteen.

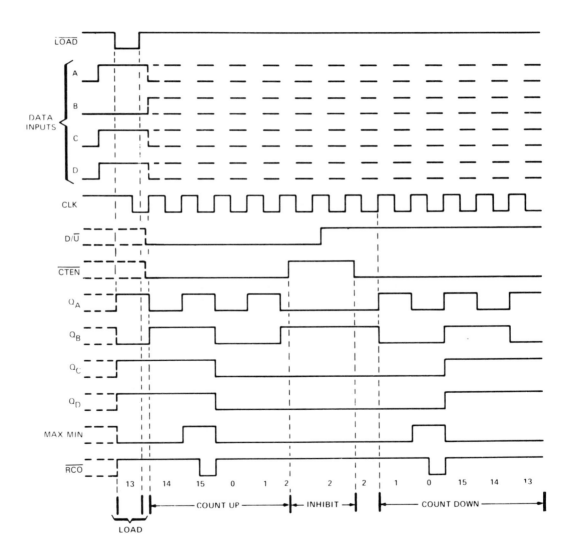

(Courtesy of Texas Instruments Inc.)

TYPES SN54ALS190, SN54ALS191, SN74ALS190, SN74ALS191
SYNCHRONOUS 4-BIT UP/DOWN DECADE AND BINARY COUNTERS

absolute maximum ratings over operating free-air temperature range (unless otherwise noted)

Supply voltage, V_{CC} . . . . 7 V
Input voltage . . . . 7 V
Operating free-air temperature range: SN54ALS190, SN54ALS191 . . . . −55 °C to 125 °C
SN74ALS190, SN74ALS191 . . . . 0 °C to 70 °C
Storage temperature range . . . . −65 °C to 150 °C

recommended operating conditions

				SN54ALS190 SN54ALS191			SN74ALS190 SN74ALS191			UNIT
				MIN	NOM	MAX	MIN	NOM	MAX	
V_{CC}	Supply voltage			4.5	5	5.5	4.5	5	5.5	V
V_{IH}	High-level input voltage			2			2			V
V_{IL}	Low-level input voltage					0.8			0.8	V
I_{OH}	High-level output current					−0.4			−0.4	mA
I_{OL}	Low-level output current					4			8	mA
f_{clock}	Clock frequency	'ALS190		0		20	0		25	MHz
		'ALS191		0		25	0		30	
t_w	Pulse duration	CLK high or low	'ALS190	25			20			ns
			'ALS191	20			16.5			
		LOAD low		25			20			
t_{su}	Setup time	Data before LOAD↑		25			20			ns
		CTEN before CLK↑		25			20			
		D/U before CLK↑		20			20			
		LOAD inactive before CLK↑		20			20			
t_h	Hold time	Data after LOAD↑		5			5			ns
		CTEN after CLK↑		0			0			
		D/U after CLK↑		0			0			
T_A	Operating free-air temperature			−55		125	0		70	°C

electrical characteristics over recommended operating free-air temperature range (unless otherwise noted)

PARAMETER		TEST CONDITIONS		SN54ALS190 SN54ALS191			SN74ALS190 SN74ALS191			UNIT
				MIN	TYP[†]	MAX	MIN	TYP[†]	MAX	
V_{IK}		V_{CC} = 4.5 V,	I_I = −18 mA			−1.5			−1.5	V
V_{OH}		V_{CC} = 4.5 V to 5.5 V,	I_{OH} = −0.4 mA	V_{CC}−2			V_{CC}−2			V
V_{OL}		V_{CC} = 4.5 V,	I_{OL} = 4 mA		0.25	0.4		0.25	0.4	V
		V_{CC} = 4.5 V	I_{OL} = 8 mA					0.35	0.5	
I_I		V_{CC} = 5.5 V,	V_I = 7 V			0.1			0.1	mA
I_{IH}		V_{CC} = 5.5 V,	V_I = 2.7 V			20			20	µA
I_{IL}	CTEN or CLK	V_{CC} = 5.5 V,	V_I = 0.4 V			−0.2			−0.2	mA
	All others					−0.1			−0.1	
I_O[‡]		V_{CC} = 5.5 V,	V_O = 2.25 V	−30		−112	−30		−112	mA
I_{CC}		V_{CC} = 5.5 V,	All inputs at 0 V		12	22		12	22	mA

[†] All typical values are at V_{CC} = 5 V, T_A = 25 °C.
[‡] The output conditions have been chosen to produce a current that closely approximates one half of the true short-circuit output current, I_{OS}.

(Courtesy of Texas Instruments Inc.)

TYPES SN54ALS190, SN54ALS191, SN74ALS190, SN74ALS191
SYNCHRONOUS 4-BIT UP/DOWN DECADE AND BINARY COUNTERS

switching characteristics (see Note 1)

PARAMETER	FROM (INPUT)	TO (OUTPUT)	V_{CC} = 4.5 V to 5.5 V, C_L = 50 pF, R_L = 500 Ω, T_A = MIN to MAX				UNIT
			SN54ALS190 SN54ALS191		SN74ALS190 SN74ALS191		
			MIN	MAX	MIN	MAX	
f_{max}	'ALS190		20		25		MHz
	'ALS191		25		30		
t_{PLH}	$\overline{LOAD}$	Any Q	8	34	8	30	ns
t_{PHL}			8	34	8	30	
t_{PLH}	A, B, C, D	Any Q	4	25	4	21	ns
t_{PHL}			4	25	4	21	
t_{PLH}	CLK	$\overline{RCO}$	5	24	5	20	ns
t_{PHL}			5	24	5	20	
t_{PLH}	CLK	Any Q	3	22	3	18	ns
t_{PHL}			3	22	3	18	
t_{PLH}	CLK	MAX/MIN	8	34	8	31	ns
t_{PHL}			8	34	8	31	
t_{PLH}	$D/\overline{U}$	$\overline{RCO}$	15	42	15	37	ns
t_{PHL}			10	33	10	28	
t_{PLH}	$D/\overline{U}$	MAX/MIN	8	30	8	25	ns
t_{PHL}			8	30	8	25	
t_{PLH}	$\overline{CTEN}$	$\overline{RCO}$	4	21	4	18	ns
t_{PHL}			4	21	4	18	

NOTE 1: For load circuit and voltage waveforms, see page 1-12.

(Courtesy of Texas Instruments Inc.)

MOS
LSI

TMS4016
2048-WORD BY 8-BIT STATIC RAM

FEBRUARY 1981 – REVISED AUGUST 1983

- 2K X 8 Organization, Common I/O
- Single +5-V Supply
- Fully Static Operation (No Clocks, No Refresh)
- JEDEC Standard Pinout
- 24-Pin 600 Mil (15.2 mm) Package Configuration
- Plug-in Compatible with 16K 5 V EPROMs
- 8-Bit Output for Use in Microprocessor-Based Systems
- 3-State Outputs with $\overline{S}$ for OR-ties
- $\overline{G}$ Eliminates Need for External Bus Buffers
- All Inputs and Outputs Fully TTL Compatible
- Fanout to Series 74, Series 74S or Series 74LS TTL Loads
- N-Channel Silicon-Gate Technology
- Power Dissipation Under 385 mW Max
- Guaranteed dc Noise Immunity of 400 mV with Standard TTL Loads
- 4 Performance Ranges:

	ACCESS TIME (MAX)
TMS4016-12	120 ns
TMS4016-15	150 ns
TMS4016-20	200 ns
TMS4016-25	250 ns

TMS4016 . . . NL PACKAGE
(TOP VIEW)

```
        ┌───┬─┬───┐
   A7 ──┤1  U  24├── VCC
   A6 ──┤2     23├── A8
   A5 ──┤3     22├── A9
   A4 ──┤4     21├── W̄
   A3 ──┤5     20├── Ḡ
   A2 ──┤6     19├── A10
   A1 ──┤7     18├── S̄
   A0 ──┤8     17├── DQ8
  DQ1 ──┤9     16├── DQ7
  DQ2 ──┤10    15├── DQ6
  DQ3 ──┤11    14├── DQ5
  VSS ──┤12    13├── DQ4
        └──────────┘
```

PIN NOMENCLATURE	
A0 – A10	Addresses
DQ1 – DQ8	Data In/Data Out
$\overline{G}$	Output Enable
$\overline{S}$	Chip Select
VCC	+5-V Supply
VSS	Ground
$\overline{W}$	Write Enable

description

The TMS4016 static random-access memory is organized as 2048 words of 8 bits each. Fabricated using proven N-channel, silicon-gate MOS technology, the TMS4016 operates at high speeds and draws less power per bit than 4K static RAMs. It is fully compatible with Series 74, 74S, or 74LS TTL. Its static design means that no refresh clocking circuitry is needed and timing requirements are simplified. Access time is equal to cycle time. A chip select control is provided for controlling the flow of data-in and data-out and an output enable function is included in order to eliminate the need for external bus buffers.

Of special importance is that the TMS4016 static RAM has the same standardized pinout as TI's compatible EPROM family. This, along with other compatible features, makes the TMS4016 plug-in compatible with the TMS2516 (or other 16K 5 V EPROMs). Minimal, if any modifications are needed. This allows the microprocessor system designer complete flexibility in partitioning his memory board between read/write and non-volatile storage.

The TMS4016 is offered in the plastic (NL suffix) 24-pin dual-in-line package designed for insertion in mounting hole rows on 600-mil (15.2 mm) centers. It is guaranteed for operation from 0°C to 70°C.

(Courtesy of Texas Instruments Inc.)

TMS4016
2048-WORD BY 8-BIT STATIC RAM

operation

addresses (A0 – A10)

The eleven address inputs select one of the 2048 8-bit words in the RAM. The address-inputs must be stable for the duration of a write cycle. The address inputs can be driven directly from standard Series 54/74 TTL with no external pull-up resistors.

output enable ($\overline{G}$)

The output enable terminal, which can be driven directly from standard TTL circuits, affects only the data-out terminals. When output enable is at a logic high level, the output terminals are disabled to the high-impedance state. Output enable provides greater output control flexibility, simplifying data bus design.

chip select ($\overline{S}$)

The chip-select terminal, which can be driven directly from standard TTL circuits, affects the data-in/data-out terminals. When chip select and output enable are at a logic low level, the D/Q terminals are enabled. When chip select is high, the D/Q terminals are in the floating or high-impedance state and the input is inhibited.

write enable ($\overline{W}$)

The read or write mode is selected through the write enable terminal. A logic high selects the read mode; a logic low selects the write mode. $\overline{W}$ must be high when changing addresses to prevent erroneously writing data into a memory location. The $\overline{W}$ input can be driven directly from standard TTL circuits.

data-in/data-out (DQ1   DQ8)

Data can be written into a selected device when the write enable input is low. The D/Q terminal can be driven directly from standard TTL circuits. The three-state output buffer provides direct TTL compatibility with a fan-out of one Series 74 TTL gate, one Series 74S TTL gate, or five Series 74LS TTL gates. The D/Q terminals are in the high impedance state when chip select ($\overline{S}$) is high, output enable ($\overline{G}$) is high, or whenever a write operation is being performed. Data-out is the same polarity as data-in.

(Courtesy of Texas Instruments Inc.)

logic symbol[†]

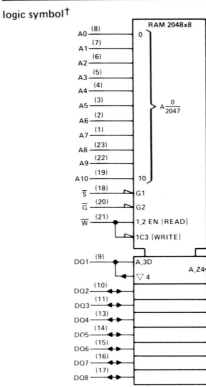

FUNCTION TABLE

$\overline{W}$	$\overline{S}$	$\overline{G}$	DQ1-DQ8	MODE
L	L	X	VALID DATA	WRITE
H	L	L	DATA OUTPUT	READ
X	H	X	HI-Z	DEVICE DISABLED
H	L	H	HI-Z	OUTPUT DISABLED

[†] This symbol is in accordance with IEEE Std 91/ANSI Y32.14 and recent decisions by IEEE and IEC. See explanation on page 10-1.

absolute maximum ratings over operating free-air temperature range (unless otherwise noted)[†]

Supply voltage, V_{CC} (see Note 1) . . . . −0.5 V to 7 V
Input voltage (any input) (see Note 1) . . . . −1 V to 7 V
Continuous power dissipation . . . . 1 W
Operating free-air temperature range . . . . 0 °C to 70 °C
Storage temperature range . . . . −55 °C to 150 °C

[†] Stresses beyond those listed under "Absolute Maximum Ratings" may cause permanent damage to the device This is a stress rating only and functional operation of the device at these or any other conditions beyond those indicated in the "Recommended Operating Conditions" section of this specification is not implied. Exposure to absolute-maximum-rated conditions for extended periods may affect device reliability.

NOTE 1: Voltage values are with respect to the V_{SS} terminal.

recommended operating conditions

PARAMETER	MIN	NOM	MAX	UNIT
Supply voltage, V_{CC}	4.5	5	5.5	V
Supply voltage, V_{SS}		0		V
High-level input voltage, V_{IH}	2		5.5	V
Low-level input voltage, V_{IL} (see Note 2)	−1		0.8	V
Operating free-air temperature, T_A	0		70	°C

NOTE 2: The algebraic convention, where the more negative (less positive) limit is designated as minimum, is used in this data sheet for logic voltage levels only.

(Courtesy of Texas Instruments Inc.)

TMS4016
2048-WORD BY 8-BIT STATIC RAM

electrical characteristics over recommended operating free-air temperature range (unless otherwise noted)

PARAMETER		TEST CONDITIONS		MIN	TYP[†]	MAX	UNIT
V_{OH}	High level voltage	$I_{OH} = -1$ mA,	$V_{CC} = 4.5$ V	2.4			V
V_{OL}	Low level voltage	$I_{OL} = 2.1$ mA,	$V_{CC} = 4.5$ V			0.4	V
I_I	Input current	$V_I = 0$ V to 5.5 V				10	μA
I_{OZ}	Off-state output current	$\overline{S}$ or $\overline{G}$ at 2 V or $\overline{W}$ at 0.8 V, $V_O = 0$ V to 5.5 V				10	μA
I_{CC}	Supply current from V_{CC}	$I_O = 0$ mA, $T_A = 0°C$ (worst case)	$V_{CC} = 5.5$ V,		40	70	mA
C_i	Input capacitance	$V_I = 0$ V,	$f = 1$ MHz			8	pF
C_o	Output capacitance	$V_O = 0$ V,	$f = 1$ MHz			12	pF

[†]All typical values are at $V_{CC} = 5$ V, $T_A = 25°C$.

timing requirements over recommended supply voltage range and operating free-air temperature range

PARAMETER		TMS4016-12		TMS4016-15		TMS4016-20		TMS4016-25		UNIT
		MIN	MAX	MIN	MAX	MIN	MAX	MIN	MAX	
$t_{c(rd)}$	Read cycle time	120		150		200		250		ns
$t_{c(wr)}$	Write cycle time	120		150		200		250		ns
$t_{w(W)}$	Write pulse width	60		80		100		120		ns
$t_{su(A)}$	Address setup time	20		20		20		20		ns
$t_{su(S)}$	Chip select setup time	60		80		100		120		ns
$t_{su(D)}$	Data setup time	50		60		80		100		ns
$t_{h(A)}$	Address hold time	0		0		0		0		ns
$t_{h(D)}$	Data hold time	5		10		10		10		ns

switching characteristics over recommended voltage range, $T_A = 0°C$ to $70°C$ with output loading of Figure 1 (see notes 3 and 4)

PARAMETER		TMS4016-12		TMS4016-15		TMS4016-20		TMS4016-25		UNIT
		MIN	MAX	MIN	MAX	MIN	MAX	MIN	MAX	
$t_{a(A)}$	Access time from address		120		150		200		250	ns
$t_{a(S)}$	Access time from chip select low		60		75		100		120	ns
$t_{a(G)}$	Access time from output enable low		50		60		80		100	ns
$t_{v(A)}$	Output data valid after address change	10		15		15		15		ns
$t_{dis(S)}$	Output disable time after chip select high		40		50		60		80	ns
$t_{dis(G)}$	Output disable time after output enable high		40		50		60		80	ns
$t_{dis(W)}$	Output disable time after write enable low		50		60		60		80	ns
$t_{en(S)}$	Output enable time after chip select low	5		5		10		10		ns
$t_{en(G)}$	Output enable time after output enable low	5		5		10		10		ns
$t_{en(W)}$	Output enable time after write enable high	5		5		10		10		ns

NOTES: 3. $C_L = 100$ pF for all measurements except $t_{dis(W)}$ and $t_{en(W)}$.
$C_L = 5$ pF for $t_{dis(W)}$ and $t_{en(W)}$.
4. t_{dis} and t_{en} parameters are sampled and not 100% tested.

(Courtesy of Texas Instruments Inc.)

B

Error Detection and Correction

ERROR-DETECTION CODES

B–1

The *2-out-of-5* code is sometimes used in communications work. It utilizes five bits to represent the ten decimal digits, so it is a form of BCD code. Each code word has exactly two 1s, a convention that facilitates decoding and provides for better error detection than the single-parity-bit method. If more or less than two 1s appear, an error is indicated.

The *63210 BCD* code is also characterized by having exactly two 1s in each of the five-bit groups. Like the 2-out-of-5 code, it provides reliable error detection and is used in some applications.

The *biquinary* (two-five) code is used in certain counters and is composed of a two-bit group and a five-bit group, each with a single 1. Its weights are 50 43210. The two-bit group, having weights of five and zero, indicates whether the number represented is less than, equal to, or greater than 5. The five-bit group indicates the count above or below 5.

The *ring counter* code has ten bits, one for each decimal digit, and a single 1 makes error detection possible. It is easy to decode but wastes bits and requires more circuitry to implement than the four- or five-bit codes. The name is derived from the fact that the code is generated by a certain type of shift register, a ring counter. Its weights are 9876543210.

Each of these codes is listed in Table B–1. You should realize that this is not an exhaustive coverage of all codes but simply an introduction to some of them.

TABLE B–1

Some codes with error-detection properties.

Decimal	2-out-of-5	63210	5043210	9876543210
0	00011	00110	01 00001	0000000001
1	00101	00011	01 00010	0000000010
2	00110	00101	01 00100	0000000100
3	01001	01001	01 01000	0000001000
4	01010	01010	01 10000	0000010000
5	01100	01100	10 00001	0000100000
6	10001	10001	10 00010	0001000000
7	10010	10010	10 00100	0010000000
8	10100	10100	10 01000	0100000000
9	11000	11000	10 10000	1000000000

HAMMING ERROR-CORRECTION CODE

B–2

This section discusses a method, generally known as the *Hamming code,* that not only provides for the detection of a bit error, but also identifies the bit that is in error so that it can be corrected. The code uses a number of parity bits (dependent on the number of information bits), located at certain positions in the code group.

The Hamming code construction that follows is for *single-error* correction.

Number of Parity Bits

If the number of information bits is designated m, then the number of parity bits, p, is determined by the following relationship:

$$2^p \geq m + p + 1 \tag{B–1}$$

For example, if we have four information bits, then p is found by trial and error with Equation (B–1). Let $p = 2$. Then

$$2^p = 2^2 = 4$$

and

$$m + p + 1 = 4 + 2 + 1 = 7$$

Since 2^p must be equal to or greater than $m + p + 1$, the relationship in Equation (B–1) is *not* satisfied. We have to try again. Let $p = 3$. Then

$$2^p = 2^3 = 8$$

and

$$m + p + 1 = 4 + 3 + 1 = 8$$

This value of p satisfies the relationship of Equation (B–1), so three parity bits are required to provide single-error correction for four information bits. It should be noted here that error detection and correction are provided for *all* bits, both parity and information, in a code group.

Placement of the Parity Bits in the Code

Now that we have found the number of parity bits required in our particular example, we must arrange the bits properly in the code. At this point you should realize that in this example the code is composed of the four information bits and the three parity bits. The left-most bit is designated *bit 1,* the next bit is *bit 2,* and so on as follows:

bit 1,   bit 2,   bit 3,   bit 4,   bit 5,   bit 6,   bit 7

The parity bits are located in the positions that are numbered corresponding to ascending powers of two (1, 2, 4, 8, . . .), as indicated:

$$P_1, \quad P_2, \quad M_1, \quad P_3, \quad M_2, \quad M_3, \quad M_4$$

The symbol P_n designates a particular parity bit, and M_n designates a particular information bit.

Assignment of Parity Bit Values

Finally, we must properly assign a 1 or 0 value to each parity bit. Since each parity bit provides a check on certain other bits in the total code, we must know the value of these others in order to assign the parity bit value. To find the bit values, first number each bit position in *binary;* that is, write the binary number for each decimal position number (as shown in the second two rows of Table B–2). Next, indicate the parity and information bit locations, as shown in the first row of Table B–2. Notice that the binary position number of parity bit P_1 has a 1 for its right-most digit. This parity bit checks all bit positions, including itself, that have *1s in the same location in the binary position numbers*. Therefore, parity bit P_1 checks bit positions 1, 3, 5, and 7.

TABLE B–2
Bit position table for a seven-bit error-correcting code.

Bit designation Bit position Binary position number	P_1 1 001	P_2 2 010	M_1 3 011	P_3 4 100	M_2 5 101	M_3 6 110	M_4 7 111
Information bits (M_n)							
Parity bits (P_n)							

The binary position number for parity bit P_2 has a 1 for its middle bit. It checks all bit positions, including itself, that have 1s in this same position. Therefore, parity bit P_2 checks bit positions 2, 3, 6, and 7.

The binary position number for parity bit P_3 has a 1 for its left-most bit. It checks all bit positions, including itself, that have 1s in this same position. Therefore, parity bit P_3 checks bit positions 4, 5, 6, and 7.

In each case, the parity bit is assigned a value to make the quantity of 1s in the set of bits that it checks odd or even, depending on which is specified. The following examples should make this procedure clear.

EXAMPLE B–1

Determine the single-error-correcting code for the BCD number 1001 (information bits), using even parity.

Solution
Step 1. Find the number of parity bits required. Let $p = 3$. Then

$$2^p = 2^3 = 8$$
$$m + p + 1 = 4 + 3 + 1 = 8$$

Three parity bits are sufficient.

$$\text{Total code bits} = 4 + 3 = 7$$

Step 2. Construct a bit position table, and enter the information bits. Parity bits are determined in the following steps.

Bit designation	P_1	P_2	M_1	P_3	M_2	M_3	M_4
Bit position	1	2	3	4	5	6	7
Binary position number	001	010	011	100	101	110	111
Information bits			1		0	0	1
Parity bits	0	0		1			

Step 3. Determine the parity bits as follows:

Bit P_1 checks bit positions 1, 3, 5, and 7 and must be a 0 for there to be an even number of 1s (2) in this group.

Bit P_2 checks bit positions 2, 3, 6, and 7 and must be a 0 for there to be an even number of 1s (2) in this group.

Bit P_3 checks bit positions 4, 5, 6, and 7 and must be a 1 for there to be an even number of 1s (2) in this group.

Step 4. These parity bits are entered in the table, and the resulting combined code is 0011001.

EXAMPLE B–2

Determine the single-error-correcting code for the information code 10110 for odd parity.

Solution

Step 1. Determine the number of parity bits required. In this case the number of information bits, m, is five. From the previous example we know that $p = 3$ will not work. Try $p = 4$:

$$2^p = 2^4 = 16$$
$$m + p + 1 = 5 + 4 + 1 = 10$$

Four parity bits are sufficient.

$$\text{Total code bits} = 5 + 4 = 9$$

Step 2. Construct a bit position table, and enter the information bits. Parity bits are determined in the following steps.

Bit designation	P_1	P_2	M_1	P_3	M_2	M_3	M_4	P_4	M_5
Bit position	1	2	3	4	5	6	7	8	9
Binary position number	0001	0010	0011	0100	0101	0110	0111	1000	1001
Information bits			1		0	1	1		0
Parity bits	1	0		1				1	

Step 3. Determine the parity bits as follows:

Bit P_1 checks bit positions 1, 3, 5, 7, and 9 and must be a 1 for there to be an odd number of 1s (3) in this group.

Bit P_2 checks bit positions 2, 3, 6, and 7 and must be a 0 for there to be an odd number of 1s (3) in this group.

Bit P_3 checks bit positions 4, 5, 6, and 7 and must be a 1 for there to be an odd number of 1s (3) in this group.

Bit P_4 checks bit positions 8 and 9 and must be a 1 for there to be an odd number of 1s (1) in this group.

Step 4. These parity bits are entered in the table, and the resulting combined code is 101101110.

DETECTING AND CORRECTING AN ERROR

B-3

Now that a method for constructing an error-correcting code has been covered, how do we use it to locate and correct an error? Each parity bit, along with its corresponding group of bits, must be checked for the proper parity. If there are three parity bits in a code word, then three parity checks are made. If there are four parity bits, four checks must be made, and so on. Each parity check will yield a good or a bad result. The total result of all the parity checks indicates the bit, if any, that is in error, as follows:

Step 1. Start with the group checked by P_1.
Step 2. Check the group for proper parity. A 0 represents a good parity check, and 1 represents a bad check.
Step 3. Repeat step 2 for each parity group.
Step 4. The binary number formed by the results of all the parity checks designates the position of the code bit that is in error. This is the *error position code*. The first parity check generates the least significant bit (LSB). If all checks are good, there is no error.

EXAMPLE B–3

Assume that the code word in Example B–1 (0011001) is transmitted and that 0010001 is received. The receiver does not "know" what was transmitted and must look for proper parities to determine if the code is correct. Designate any error that has occurred in transmission if even parity is used.

Solution First, make a bit position table:

Bit designation	P_1	P_2	M_1	P_3	M_2	M_3	M_4
Bit position	1	2	3	4	5	6	7
Binary position number	001	010	011	100	101	110	111
Received code	0	0	1	0	0	0	1

First parity check:
    Bit P_1 checks positions 1, 3, 5, and 7.
    There are two 1s in this group.
    Parity check is good. ———————————————————→ 0 (LSB)
Second parity check:
    Bit P_2 checks positions 2, 3, 6, and 7.
    There are two 1s in this group.
    Parity check is good. ———————————————————→ 0
Third parity check:
    Bit P_3 checks positions 4, 5, 6, and 7.
    There is one 1 in this group.
    Parity check is bad. ————————————————————→ 1 (MSB)
Result:
    The error position code is 100 (binary four). This says that the bit in position 4 is in error. It is a 0 and should be a 1. The corrected code is 0011001, which agrees with the transmitted code.

EXAMPLE B–4

The code 101101010 is received. Correct any errors. There are four parity bits, and odd parity is used.

Solution   First, make a bit position table:

Bit designation Bit position Binary position number	P_1 1 0001	P_2 2 0010	M_1 3 0011	P_3 4 0100	M_2 5 0101	M_3 6 0110	M_4 7 0111	P_4 8 1000	M_5 9 1001
Received code	1	0	1	1	0	1	0	1	0

First parity check:
    Bit P_1 checks positions 1, 3, 5, 7, and 9.
    There are two 1s in this group.
    Parity check is bad. ————————————————————→ 1 (LSB)
Second parity check:
    Bit P_2 checks positions 2, 3, 6, and 7.
    There are two 1s in this group.
    Parity check is bad. ————————————————————→ 1
Third parity check:
    Bit P_3 checks positions 4, 5, 6, and 7.
    There are two 1s in this group.
    Parity check is bad. ————————————————————→ 1
Fourth parity check:
    Bit P_4 checks positions 8 and 9.
    There is one 1 in this group.
    Parity check is good. ———————————————————→ 0 (MSB)
Result:
    The error position code is 0111 (binary seven). This says that the bit in position 7 is in error. The corrected code is therefore 101101110.

C

Conversions

Decimal	BCD(8421)	Octal	Binary	Decimal	BCD(8421)	Octal	Binary	Decimal	BCD(8421)	Octal	Binary
0	0000	0	0	34	00110100	42	100010	68	01101000	104	1000100
1	0001	1	1	35	00110101	43	100011	69	01101001	105	1000101
2	0010	2	10	36	00110110	44	100100	70	01110000	106	1000110
3	0011	3	11	37	00110111	45	100101	71	01110001	107	1000111
4	0100	4	100	38	00111000	46	100110	72	01110010	110	1001000
5	0101	5	101	39	00111001	47	100111	73	01110011	111	1001001
6	0110	6	110	40	01000000	50	101000	74	01110100	112	1001010
7	0111	7	111	41	01000001	51	101001	75	01110101	113	1001011
8	1000	10	1000	42	01000010	52	101010	76	01110110	114	1001100
9	1001	11	1001	43	01000011	53	101011	77	01110111	115	1001101
10	00010000	12	1010	44	01000100	54	101100	78	01111000	116	1001110
11	00010001	13	1011	45	01000101	55	101101	79	01111001	117	1001111
12	00010010	14	1100	46	01000110	56	101110	80	10000000	120	1010000
13	00010011	15	1101	47	01000111	57	101111	81	10000001	121	1010001
14	00010100	16	1110	48	01001000	60	110000	82	10000010	122	1010010
15	00010101	17	1111	49	01001001	61	110001	83	10000011	123	1010011
16	00010110	20	10000	50	01010000	62	110010	84	10000100	124	1010100
17	00010111	21	10001	51	01010001	63	110011	85	10000101	125	1010101
18	00011000	22	10010	52	01010010	64	110100	86	10000110	126	1010110
19	00011001	23	10011	53	01010011	65	110101	87	10000111	127	1010111
20	00100000	24	10100	54	01010100	66	110110	88	10001000	130	1011000
21	00100001	25	10101	55	01010101	67	110111	89	10001001	131	1011001
22	00100010	26	10110	56	01010110	70	111000	90	10010000	132	1011010
23	00100011	27	10111	57	01010111	71	111001	91	10010001	133	1011011
24	00100100	30	11000	58	01011000	72	111010	92	10010010	134	1011100
25	00100101	31	11001	59	01011001	73	111011	93	10010011	135	1011101
26	00100110	32	11010	60	01100000	74	111100	94	10010100	136	1011110
27	00100111	33	11011	61	01100001	75	111101	95	10010101	137	1011111
28	00101000	34	11100	62	01100010	76	111110	96	10010110	140	1100000
29	00101001	35	11101	63	01100011	77	111111	97	10010111	141	1100001
30	00110000	36	11110	64	01100100	100	1000000	98	10011000	142	1100010
31	00110001	37	11111	65	01100101	101	1000001	99	10011001	143	1100011
32	00110010	40	100000	66	01100110	102	1000010				
33	00110011	41	100001	67	01100111	103	1000011				

Powers of Two

2^n	n	2^{-n}
1	0	1.0
2	1	0.5
4	2	0.25
8	3	0.125
16	4	0.062 5
32	5	0.031 25
64	6	0.015 625
128	7	0.007 812 5
256	8	0.003 906 25
512	9	0.001 953 125
1 024	10	0.000 976 562 5
2 048	11	0.000 488 281 25
4 096	12	0.000 244 140 625
8 192	13	0.000 122 070 312 5
16 384	14	0.000 061 035 156 25
32 768	15	0.000 030 517 578 125
65 536	16	0.000 015 258 789 062 5
131 072	17	0.000 007 629 394 531 25
262 144	18	0.000 003 814 697 265 625
524 288	19	0.000 001 907 348 632 812 5
1 048 576	20	0.000 000 953 674 316 406 25
2 097 152	21	0.000 000 476 837 158 203 125
4 194 304	22	0.000 000 238 418 579 101 562 5
8 388 608	23	0.000 000 119 209 289 550 781 25
16 777 216	24	0.000 000 059 604 644 775 390 625
33 554 432	25	0.000 000 029 802 322 387 695 312 5
67 108 864	26	0.000 000 014 901 161 193 847 656 25
134 217 728	27	0.000 000 007 450 580 596 923 828 125
268 435 456	28	0.000 000 003 725 290 298 461 914 062 5
536 870 912	29	0.000 000 001 862 645 149 230 957 031 25
1 073 741 824	30	0.000 000 000 931 322 574 615 478 515 625
2 147 483 648	31	0.000 000 000 465 661 287 307 739 257 812 5
4 294 967 296	32	0.000 000 000 232 830 643 653 869 628 906 25
8 589 934 592	33	0.000 000 000 116 415 321 826 934 814 453 125
17 179 869 184	34	0.000 000 000 058 207 660 913 467 407 226 562 5
34 359 738 368	35	0.000 000 000 029 103 830 456 733 703 613 281 25
68 719 476 736	36	0.000 000 000 014 551 915 228 366 851 806 640 625
137 438 953 472	37	0.000 000 000 007 275 957 614 183 425 903 320 312 5
274 877 906 944	38	0.000 000 000 003 637 978 807 091 712 951 660 156 25
549 755 813 888	39	0.000 000 000 001 818 989 403 545 856 475 830 078 125
1 099 511 627 776	40	0.000 000 000 000 909 494 701 772 928 237 915 039 062 5
2 199 023 255 552	41	0.000 000 000 000 454 747 350 886 464 118 957 519 531 25
4 398 046 511 104	42	0.000 000 000 000 227 373 675 443 232 059 478 759 765 625
8 796 093 022 208	43	0.000 000 000 000 113 686 837 721 616 029 739 379 882 812 5
17 592 186 044 416	44	0.000 000 000 000 056 843 418 860 808 014 869 689 941 406 25
35 184 372 088 832	45	0.000 000 000 000 028 421 709 430 404 007 434 844 970 703 125
70 368 744 177 664	46	0.000 000 000 000 014 210 854 715 202 003 717 422 485 351 562 5
140 737 488 355 328	47	0.000 000 000 000 007 105 427 357 601 001 858 711 242 675 781 25
281 474 976 710 656	48	0.000 000 000 000 003 552 713 678 800 500 929 355 621 337 890 625
562 949 953 421 312	49	0.000 000 000 000 001 776 356 839 400 250 464 677 810 668 945 312 5
1 125 899 906 842 624	50	0.000 000 000 000 000 888 178 419 700 125 232 338 905 334 472 656 25
2 251 799 813 685 248	51	0.000 000 000 000 000 444 089 209 850 062 616 169 452 667 236 328 125
4 503 599 627 370 496	52	0.000 000 000 000 000 222 044 604 925 031 308 084 726 333 618 164 062 5
9 007 199 254 740 992	53	0.000 000 000 000 000 111 022 302 462 515 654 042 363 166 809 082 031 25
18 014 398 509 481 984	54	0.000 000 000 000 000 055 511 151 231 257 827 021 181 583 404 541 015 625
36 028 797 018 963 968	55	0.000 000 000 000 000 027 755 575 615 628 913 510 590 791 702 270 507 812 5
72 057 594 037 927 936	56	0.000 000 000 000 000 013 877 787 807 814 456 755 295 395 851 135 253 906 25
144 115 188 075 855 872	57	0.000 000 000 000 000 006 938 893 903 907 228 377 647 697 925 567 626 953 125
288 230 376 151 711 744	58	0.000 000 000 000 000 003 469 446 951 953 614 188 823 848 962 783 813 476 562 5
576 460 752 303 423 488	59	0.000 000 000 000 000 001 734 723 475 976 807 094 411 924 481 391 906 738 281 25
1 152 921 504 606 846 976	60	0.000 000 000 000 000 000 867 361 737 988 403 547 205 962 240 695 953 369 140 625
2 305 843 009 213 693 952	61	0.000 000 000 000 000 000 433 680 868 994 201 773 602 981 120 347 976 684 570 312 5
4 611 686 018 427 387 904	62	0.000 000 000 000 000 000 216 840 434 497 100 886 801 490 560 173 988 342 285 156 25
9 223 372 036 854 775 808	63	0.000 000 000 000 000 000 108 420 217 248 550 443 400 745 280 086 994 171 142 578 125
18 446 744 073 709 551 616	64	0.000 000 000 000 000 000 054 210 108 624 275 221 700 372 640 043 497 085 571 289 062 5
36 893 488 147 419 103 232	65	0.000 000 000 000 000 000 027 105 054 312 137 610 850 186 320 021 748 542 785 644 531 25
73 786 976 294 838 206 464	66	0.000 000 000 000 000 000 013 552 527 156 088 805 425 093 160 010 874 271 392 822 265 625
147 573 952 589 676 412 928	67	0.000 000 000 000 000 000 006 776 263 578 034 402 712 546 580 005 437 135 696 411 132 812 5
295 147 905 179 352 825 856	68	0.000 000 000 000 000 000 003 388 131 789 017 201 356 273 290 002 718 567 848 205 566 406 25
590 295 810 358 705 651 712	69	0.000 000 000 000 000 000 001 694 065 894 508 600 678 136 645 001 359 283 924 102 783 203 125
1 180 591 620 717 411 303 424	70	0.000 000 000 000 000 000 000 847 032 947 254 300 339 068 322 500 679 641 962 051 391 601 562 5
2 361 183 241 434 822 606 848	71	0.000 000 000 000 000 000 000 423 516 473 627 150 169 534 161 250 339 820 981 025 695 800 781 25
4 722 366 482 869 645 213 696	72	0.000 000 000 000 000 000 000 211 758 236 813 575 084 767 080 625 169 910 490 512 847 900 390 625

Answers and Solutions to Self-Tests

CHAPTER 1

1. 1 and 0
2. HIGH = 0 (negative logic)
3. bit = binary digit
4. negative-going
5. negative-going
6. A *periodic* waveform repeats itself at fixed time intervals. A *nonperiodic* waveform is non-repetitive.
7. *Period* is the time interval for a periodic waveform to repeat, or the time for one cycle.
8. *Frequency* is the rate at which a periodic waveform repeats itself. The unit of frequency is the *hertz*, Hz.
9. Duty cycle increases when pulse width increases with respect to the period.
10. inverter
11. AND: The output is HIGH only when all inputs are HIGH. OR: The output is HIGH when one or more inputs are HIGH.
12. A flip-flop can retain (store) a binary state, but a gate cannot.
13. comparison: determines $>$, $=$, $<$ relationships of two numbers
    arithmetic: adds, subtracts, multiplies, and divides
    encoding: converts information to coded form
    decoding: converts coded data to a familiar form
    counting: counts events; divides frequency
    register: Stores digital data
    multiplexing: puts data from several sources onto a single line in a time sequence
    demultiplexing: reverse of multiplexing
14. dual-in-line package (DIP), a type of integrated circuit package
15. the microprocessor, memory, and I/O interface
16. *Troubleshooting* is the systematic approach to isolating and identifying a fault in a circuit or system.
17. oscilloscope, logic analyzer, signature analyzer, logic probe, pulser

CHAPTER 2

1. 10; ten

2. **(a)** $28 = 2 \times 10 + 8 \times 1$
(b) $389 = 3 \times 100 + 8 \times 10 + 9 \times 1$
(c) $1473 = 1 \times 1000 + 4 \times 100 + 7 \times 10 + 3 \times 1$
(d) $10{,}956 = 1 \times 10{,}000 + 9 \times 100 + 5 \times 10 + 6 \times 1$

3. 100000, 100001, 100010, 100011, 100100, 100101, 100110, 100111, 101000, 101001, 101010.

4. **(a)** $1101 = 1 \times 2^3 + 1 \times 2^2 + 1 \times 2^0 = 8 + 4 + 1 = 13$
(b) $100101 = 1 \times 2^5 + 1 \times 2^2 + 1 \times 2^0 = 32 + 4 + 1 = 37$
(c) $11011101 = 1 \times 2^7 + 1 \times 2^6 + 1 \times 2^4 + 1 \times 2^3 + 1 \times 2^2 + 1 \times 2^0$
$= 128 + 64 + 16 + 8 + 4 + 1 = 221$
(d) $1011.011 = 1 \times 2^3 + 1 \times 2^1 + 1 \times 2^0 + 1 \times 2^{-2} + 1 \times 2^{-3}$
$= 8 + 2 + 1 + 0.25 + 0.125 = 11.375$
(e) $1110.1011 = 1 \times 2^3 + 1 \times 2^2 + 1 \times 2^1 + 1 \times 2^{-1} + 1 \times 2^{-3} + 1 \times 2^{-4}$
$= 8 + 4 + 2 + 0.5 + 0.125 + 0.0625 = 14.6875$

5. **(a)**

```
     8
 2)17
    16
 ───
     1 ──→ 1   (LSB)
     4
 2)8
     8
 ──
     0 ──→ 0
     2
 2)4
     4
 ──
     0 ──→ 0
     1
 2)2
     2
 ──
     0 ──→ 0
     0
 2)1
     0
 ──
     1 ──→ 1
```

$\boxed{10001}$

(b)

```
     51
 2)102
    102
 ───
      0 ──→ 0   (LSB)
     25
 2)51
     50
 ──
      1 ──→ 1
     12
 2)25
     24
 ──
      1 ──→ 1
      6
 2)12
     12
 ──
      0 ──→ 0
      3
 2)6
      6
 ──
      0 ──→ 0
      1
 2)3
      2
 ──
      1 ──→ 1
      0
 2)1
      0
 ──
      1 ──→ 1
```

$\boxed{1100110}$

(c)

$$\begin{array}{r} 277 \\ 2)\overline{555} \\ 554 \\ \hline 1 \end{array} \longrightarrow 1$$
(LSB)

$$\begin{array}{r} 138 \\ 2)\overline{277} \\ 276 \\ \hline 1 \end{array} \longrightarrow 1$$

$$\begin{array}{r} 69 \\ 2)\overline{138} \\ 138 \\ \hline 0 \end{array} \longrightarrow 0$$

$$\begin{array}{r} 34 \\ 2)\overline{69} \\ 68 \\ \hline 1 \end{array} \longrightarrow 1$$

$$\begin{array}{r} 17 \\ 2)\overline{34} \\ 34 \\ \hline 0 \end{array} \longrightarrow 0$$

$$\begin{array}{r} 8 \\ 2)\overline{17} \\ 16 \\ \hline 1 \end{array} \longrightarrow 1$$

$$\begin{array}{r} 4 \\ 2)\overline{8} \\ 8 \\ \hline 0 \end{array} \longrightarrow 0$$

$$\begin{array}{r} 2 \\ 2)\overline{4} \\ 4 \\ \hline 0 \end{array} \longrightarrow 0$$

$$\begin{array}{r} 1 \\ 2)\overline{2} \\ 2 \\ \hline 0 \end{array} \longrightarrow 0$$

$$\begin{array}{r} 0 \\ 2)\overline{1} \\ 0 \\ \hline 1 \end{array} \longrightarrow 1$$

$$\boxed{1000101011}$$

(d)

$$\begin{array}{r} 36 \\ 2)\overline{72} \\ 72 \\ \hline 0 \end{array} \longrightarrow 0$$
(LSB)

$$\begin{array}{r} 18 \\ 2)\overline{36} \\ 36 \\ \hline 0 \end{array} \longrightarrow 0$$

$$\begin{array}{r} 9 \\ 2)\overline{18} \\ 18 \\ \hline 0 \end{array} \longrightarrow 0$$

$$\begin{array}{r} 4 \\ 2)\overline{9} \\ 8 \\ \hline 1 \end{array} \longrightarrow 1$$

$$\begin{array}{r} 2 \\ 2)\overline{4} \\ 4 \\ \hline 0 \end{array} \longrightarrow 0$$

$$\begin{array}{r} 1 \\ 2)\overline{2} \\ 2 \\ \hline 0 \end{array} \longrightarrow 0$$

$$\begin{array}{r} 0 \\ 2)\overline{1} \\ 0 \\ \hline 1 \end{array} \longrightarrow 1$$

$0.66 \times 2 = 1.32 \longrightarrow 1$ (MSB)
$0.32 \times 2 = 0.64 \longrightarrow 0$
$0.64 \times 2 = 1.28 \longrightarrow 1$
$0.28 \times 2 = 0.56 \longrightarrow 0$
$0.56 \times 2 = 1.12 \longrightarrow 1$
$0.12 \times 2 = 0.24 \longrightarrow 0$
$0.24 \times 2 = 0.48 \longrightarrow 0$
$0.48 \times 2 = 0.96 \longrightarrow 0$

$$\boxed{1001000.10101}$$

6. **(a)**
$$\begin{array}{r} 110 \\ +011 \\ \hline 1001 \end{array}$$

(b)
$$\begin{array}{r} 11010 \\ +01111 \\ \hline 101001 \end{array}$$

(c)
$$\begin{array}{r} 110 \\ -010 \\ \hline 100 \end{array}$$

(d)
$$\begin{array}{r} 11011 \\ -10110 \\ \hline 00101 \end{array}$$

(e)
$$\begin{array}{r} 111 \\ \times 101 \\ \hline 111 \\ +1110 \\ \hline 100011 \end{array}$$

(f)
$$\begin{array}{r} 10 \\ 0111)\overline{1110} \\ 111 \\ \hline 0 \end{array}$$

7. **(a)** $110101 \rightarrow 001010$ 1's complement
         $001010 + 1 = 001011$ 2's complement

   (b) $0001101 \rightarrow 1110010$ 1's complement
         $1110010 + 1 = 1110011$ 2's complement

8. **(a)** $1010 = -2$
   (b) $0111 = +7$
   (c) $11010101 = -(1 \times 2^6 + 1 \times 2^4 + 1 \times 2^2 + 1 \times 2^0)$
                $= -(64 + 16 + 4 + 1)$
                $= -85$

9. The magnitude is complemented, and the sign bit is 1.

10. There is overflow when the number of bits in the sum exceeds the number of bits in the largest number. Overflow is possible when both numbers are positive or both are negative.

11. **(a)** **(b)** **(c)**

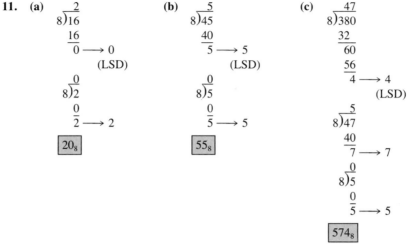

$$20_8$$

$$55_8$$

$$574_8$$

12. **(a)** $7041_8 = 111000100001$
   (b) $65315_8 = 110101011001101$

13. $\underbrace{111}\underbrace{000}\underbrace{010}\underbrace{110}\underbrace{010}$
   $\;\;7\;\;\;\;0\;\;\;\;2\;\;\;\;6\;\;\;\;2\; \rightarrow\; 70{,}262_8$

14. $\underbrace{0011}\underbrace{1101}\underbrace{0001}\underbrace{0000}\underbrace{1011}$
   $\;\;3\;\;\;\;\;D\;\;\;\;\;1\;\;\;\;\;0\;\;\;\;\;B\; \rightarrow\; 3D10B_{16}$

15. $5A34F_{16} = 01011010001101001111$

16. **(a)** $23 = 00100011$ **(b)** $79 = 01111001$
   (c) $718 = 011100011000$ **(d)** $954.61 = 100101010100.01100001$

17. $\underbrace{0100}\underbrace{1001}\underbrace{1000}\underbrace{0110}$
   $\;\;4\;\;\;\;\;9\;\;\;\;\;8\;\;\;\;\;6\; \rightarrow\; 4986$

18. $101011011_2 \rightarrow 111110110$ (Gray)

19. $1100100 \rightarrow 1000111_2$

20. From Table 2–6: H is 1001000, X is 1011000, : is 0111010, = is 0111101, and % is 0100101.

CHAPTER 3

1. **(a)** input HIGH, output LOW   **(b)** input LOW, output HIGH
   (c) input 1, output 0   **(d)** input 0, output 1

2. AND gate with inputs $ABCD$: $X = 1$ when $ABCD = 1111$; $X = 0$ for all other combinations of $ABCD$.

3. OR gate with inputs $ABCD$: $X = 0$ when $ABCD = 0000$; $X = 1$ for all other combinations of $ABCD$.

4. The normally HIGH output goes LOW at $t = 0.8$ ms and back HIGH at $t = 1$ ms, thus creating a 0.2-ms-wide negative-going pulse.

5. The normally HIGH output goes LOW at $t = 0$ ms and back HIGH at $t = 3$ ms, thus creating a 3-ms-wide negative-going pulse.

6. The normally LOW output goes HIGH at $t = 0$ ms and back LOW at $t = 0.8$ ms. It goes HIGH again at $t = 1$ ms and back LOW at $t = 3$ ms, thus creating a 0.8-ms-wide positive pulse, followed 0.2 ms later by a 2-ms-wide positive pulse.

7. t_{PHL}

8. $P_{\text{DISS}} = V_{CC}I_{CC} = (5\text{V})(2\text{ mA}) = 10\text{ mW}$

9. The gate can reliably drive up to 20 inputs to gates in the same family.
10. The 74S is faster than the 74ALS (ALS = advanced low-power Schottky; S = Schottky).
11. CMOS
12. **(a)** 7400: quad two-input NAND   **(b)** 7404: hex inverter
    (c) 7411: triple three-input AND   **(d)** 7420: dual four-input NAND
    (e) 7432: quad two-input OR   **(f)** 7427: triple three-input NOR
13.

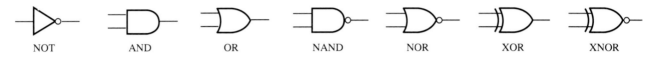

| NOT | AND | OR | NAND | NOR | XOR | XNOR |

FIGURE S–1

14.

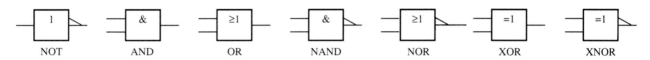

| NOT | AND | OR | NAND | NOR | XOR | XNOR |

FIGURE S–2

CHAPTER 4
1. **(a)** ANDed   **(b)** ORed   **(c)** complemented (inverted)
2. **(a)** three-input AND   **(b)** three-input OR
   (c) three-input NAND   **(d)** three-input NOR
3. **(a)** $A + B = B + A$, commutative
   (b) $AB = BA$, commutative
   (c) $CD = DC$, commutative
   (d) $(A + B) + C = A + (B + C)$, associative
   (e) $A(B + C + D) = AB + AC + AD$, distributive
   (f) $A(BCD) = (ABC)D$, associative
4. **(a)** $B + 1 = 1$ **(b)** $0 + B = B$ **(c)** $B \cdot 1 = B$
   (d) $B \cdot B = B$ **(e)** $C + C = C$ **(f)** $D \cdot D = D$
   (g) $\overline{\overline{C}} = C$ **(h)** $B + BC = B(1 + C)$
   $$= B \cdot 1 = B$$
5. DeMorgan's theorem
6. **(a)** $AB + CD$: two 2-input AND gates and one 2-input OR
   (b) $A + B + CD$: one 2-input AND and one 3-input OR
   (c) $A(B + C + DE)$: two 2-input AND gates and one 3-input OR
7. **(a)** Sum-of-products   **(b)** Product-of-sums
8. **(a)** $ABC + AB\overline{C} = A(BC + B\overline{C}) = A \cdot 1 = A$; correct
   (b) $A + BC + \overline{A}C = A + \overline{A}C + BC = A + C + BC = A + C(1 + B) = A + C$;
   BC not correct; correct answer: $A + C$
   (c) $A(\overline{A}BC + ABCD) = A\overline{A}BC + AABCD = 0 + ABCD = ABCD$; correct
9. $AB(\overline{A} + \overline{B}C) + (C + \overline{D})(A + A\overline{C}) + \overline{A}BC\overline{D}$
   $= AB\overline{A} + AB\overline{B}C + AC + A\overline{C}C + A\overline{D} + A\overline{C}\overline{D} + \overline{A}BC\overline{D}$
   $= 0 + 0 + AC + 0 + A\overline{D}(1 + \overline{C}) + \overline{A}BC\overline{D}$
   $= AC + A\overline{D} + \overline{A}BC\overline{D}$
10. $AC + A\overline{D} + \overline{A}BC\overline{D} = AC(1 + \overline{B}\overline{D}) + A\overline{D}$
    $$= AC + A\overline{D}$$

CHAPTER 5

1. $X = ABCD + EF$
2. **(a)** There are two 3-input AND gates with inputs $A, \overline{B}, C$ and $A, B, \overline{C}$, respectively. The outputs of these two AND gates go into a 2-input NOR gate.
   (b) exclusive-OR
   (c) There are three 4-input AND gates with inputs $A, \overline{B}, \overline{C}, D$ and $A, B, C, \overline{D}$ and $\overline{A}, B, \overline{C}, D$, respectively. The outputs of these AND gates go into a 3-input OR gate.
3. $X = 1$ when $ABC = 010, 101,$ and $111; X = 0$ for other combinations
4. There are three 3-input AND gates with inputs $\overline{A}, B, \overline{C}$ and $A, \overline{B}, C$ and A, B, C, respectively. The outputs of these AND gates go into a 3-input OR.
5. **(a)** $\overline{AB}C + AB\overline{C} = (\overline{\overline{AB}C})(\overline{AB\overline{C}})$
   $= (\overline{A} + B + \overline{C})(\overline{A} + \overline{B} + C)$
   $= \overline{A}\overline{A} + \overline{A}\overline{B} + \overline{A}C + \overline{A}B + B\overline{B} + BC + \overline{A}\overline{C} + \overline{B}\overline{C} + C\overline{C}$
   $= \overline{A} + \overline{A}\overline{B} + \overline{A}C + \overline{A}B + \overline{A}C + BC + \overline{B}\overline{C}$
   $= \overline{A} + BC + \overline{B}\overline{C}$
   (b) no simplification
   (c) no simplification
6. $X = (A + B + C)(D + E)\overline{F}$.
   See Figure S–3.
7. $X = ABC + DE + \overline{F}$.
   See Figure S–4.

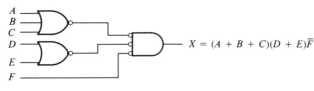

FIGURE S–3

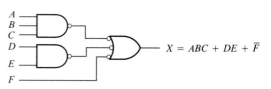

FIGURE S–4

8. $X = AB + \overline{A}\overline{B}$.
   See Figure S–5. This is an XNOR function.

FIGURE S–5

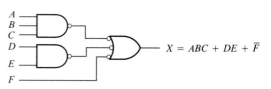

$X = AB + \overline{A}\overline{B}$

CHAPTER 6

1. A full-adder has a carry input, but a half-adder does not.
2. **(a)** $\Sigma = 1, CO = 0$ **(b)** $\Sigma = 0, CO = 1$
   (c) $\Sigma = 1, CO = 0$ **(d)** $\Sigma = 0, CO = 0$
3. **(a)** $\Sigma = 1, CO = 0$ **(b)** $\Sigma = 0, CO = 1$
   (c) $\Sigma = 0, CO = 1$ **(d)** $\Sigma = 1, CO = 1$
4.
   $$\begin{array}{ll} A_1A_0 & 01 \\ +B_1B_0 & +11 \\ \hline CO\Sigma_1\Sigma_0 & 100 \end{array}$$
   $\Sigma_0 = 0, \Sigma_1 = 0, CO = 1$
5. Connect two more 7482s to Figure 6–12 with CO to CI of each.

6. Connect two more 7483As to Figure 6–13 with CO to CI of each.

7. (a) CG = PQ = 1 · 0 = 0; CP = $P + Q$ = 1 + 0 = 1
   (b) CG = PQ = 0 · 0 = 0; CP = $P + Q$ = 0 + 0 = 0
   (c) CG = PQ = 1 · 1 = 1; CP = $P + Q$ = 1 + 1 = 1

8. (a) $P > Q$ = 1, $P < Q$ = 0, $P = Q$ = 0
   (b) < LOW, = HIGH, > LOW

9. Add one more 7485 to Figure 6–26 with $P < Q$ output to < input, $P = Q$ output to = input, and $P > Q$ output to > input.

10. (a) 12 output LOW   (b) 8 output LOW   (c) 2 output LOW

11. (a) 5 output LOW   (b) 9 output LOW
    (c) no output LOW; invalid input

12. (a) b, c (b) a, b, c (c) a, b, c, d, g

13. The complement of BCD 7.

14. (a) 10010011   (b) 01100111

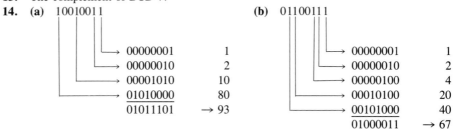

15. two exclusive-OR gates connected as in Figure 6–54.

16. eight exclusive-OR gates connected as in Figure 6–57.

17. (a) $1Y$ = 1, $2Y$ = 0, $3Y$ = 1, $4Y$ = 0
    (b) $1Y$ = 0, $2Y$ = 0, $3Y$ = 0, $4Y$ = 1

18. Output is a square wave (50% duty cycle) beginning with a LOW.

19. See Figure S–6.

FIGURE S–6

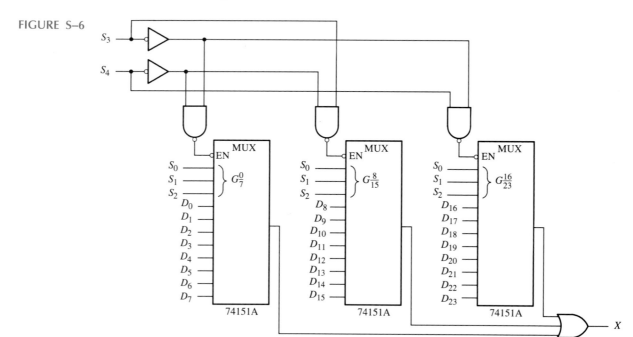

20. See Figure S–7.

FIGURE S–7

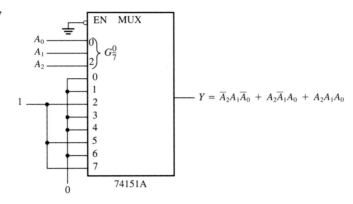

$Y = \overline{A}_2 A_1 \overline{A}_0 + A_2 \overline{A}_1 A_0 + A_2 A_1 A_0$

74151A

21. pin 13, D_{11}
22. (a) $\underline{1}11010$ (b) $\underline{0}1001$ (c) $\underline{1}0111101$
23. (a) $\underline{0}11010$ (b) $\underline{1}1001$ (c) $\underline{0}0111101$

CHAPTER 7

1. SET
2. It forces both Q and $\overline{Q}$ to the LOW state, and when inputs are released, the resulting state of the latch is unpredictable.
3. Output Q goes HIGH when EN goes HIGH the first time; Q goes LOW when D goes LOW.
4. S-R, D, J-K
5. *Positive edge-triggered* flip-flop is sensitive to data inputs only on positive-going edge of clock. *Negative edge-triggered* flip-flop is sensitive to data inputs only on negative-going edge of clock.
6. A J-K has no invalid state.
7. remains 0; remains 0
8. Data go into master on leading edge of clock, then into slave and to output on trailing edge of clock.
9. Data cannot be changed while clock pulse is in active state.
10. The data lock-out flip-flop has a dynamic indicator on the clock input as well as the postponed output symbol.
11. Propagation delay t_{PHL} (clock to Q) is the delay time from triggering edge of clock to HIGH-to-LOW transition of Q.
    Propagation delay t_{PLH} (clock to Q) is the delay time from triggering edge of clock to LOW-to-HIGH transition of Q.
    Propagation delay t_{PHL} (CLR to Q) is the delay time from clear input to Q (always HIGH-to-LOW).
    Propagation delay t_{PLH} (PRE to Q) is the delay time from preset input to Q (always LOW-to-HIGH).
12. $T = t_{H(min)} + t_{L(min)} = 30 \text{ ns} + 37 \text{ ns} = 67 \text{ ns}$
    $f_{max} = 1/T = 1/67 \text{ ns} = 14.9 \text{ MHz}$
13. counting, frequency division, data storage, and data transfer
14. $T = 1/10 \text{ kHz} = 0.1 \text{ ms} = 100 \text{ μs}$; a square wave (50% duty cycle) with a frequency of 10 kHz
15. as timing (clock) sources

CHAPTER 8

1. Each flip-flop in a synchronous counter is clocked simultaneously. Each flip-flop in an asynchronous counter is ripple clocked, resulting in greater delays and lower operating frequency.

2. lower frequency of operation

3. **(a)** $2^2 = 4$ **(b)** $2^4 = 16$ **(c)** $2^5 = 32$
   (d) $2^6 = 64$ **(e)** $2^7 = 128$ **(f)** $2^8 = 256$

4. **(a)** 2   **(b)** 3   **(c)** 4   **(d)** 4
   (e) 5   **(f)** 6   **(g)** 6   **(h)** 8

5. $4(12 \text{ ns}) = 48 \text{ ns}$

6. 12 ns

7. A BCD decade counter has ten states (0 through 9). The Q output of the most significant stage has a frequency that is one-tenth of the clock frequency. See Figure S–8.

FIGURE S–8

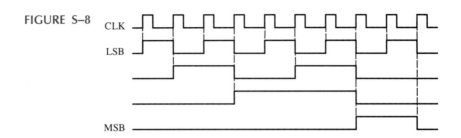

8. 1010, 1011, 1100, 1101, 1110, 1111

9. The counter is parallel-loaded to $D_3 D_2 D_1 D_0 = 0101$.

10. 0001; 1111

11. $J = 0, K = X$ (don't care)

12. **(a)** one flip-flop
    (b) two flip-flops
    (c) modulus-5 counter
    (d) decade counter
    (e) decade counter and flip-flop
    (f) decade counter and two flip-flops
    (g) decade counter and four flip-flops
    (h) two modulus-5 counters and one decade counter
    (i) three decade counters
    (j) four decade counters

13. See Figure S–9.

FIGURE S–9

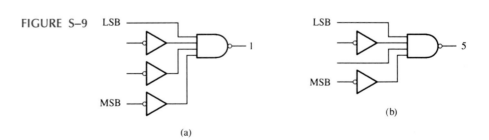

(a)   (b)

FIGURE S–9
(*Continued*)

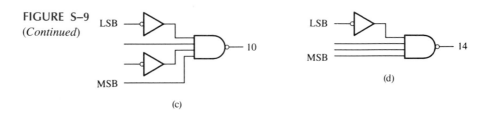

(c)

(d)

14. State 12 is decoded by NAND gate G_2, presetting the counter to 1 by loading the parallel inputs into the 74LS160A and by causing the flip-flop to clear on the next clock pulse via gate G_1.

15. $T = 1/f = 1/10$ kHz $= 0.1$ ms $= 100$ μs
    Time for two bytes: $16(100$ μs$) = 1600$ μs

CHAPTER 9

1. A flip-flop stores a 1 when SET and a 0 when RESET.
2. 16
3. 8
4. Connect the Q output of each flip-flop to the J input of the following one and the $\overline{Q}$ output to the K input. The data input goes to J of the first flip-flop, and the complement of the data input goes to the K input.
5. Take output from Q_2.
6. 01111001 after two clock pulses, 01011110 after four clock pulses, 10110101 after eight clock pulses
7. 00110000
8. five
9. the type of sequence and the way it is generated
10. 0000, 1000, 1100, 1110, 1111, 0111, 0011, 0001
11. 10010000, 01001000, 00100100, 00010010, 00001001, 10000100, 01000010, 00100001, 10010000
12. 24 μs/8 = 3 μs, $T = 3$ μs, $f = 1/3$ μs $= 0.333$ MHz $= 333$ kHz
13. 1010, 0101, 1010, . . .
14. serial-to-parallel and parallel-to-serial conversion
15. 74121
16. to sequentially apply a LOW to each row line to detect a switch closure

CHAPTER 10

1. $512 \times 8 = 4096$ bits
2. read
3. mask ROM, PROM, EPROM; mask ROM, PROM
4. static, dynamic; static
5. four
6. to keep track of the address in memory
7. to temporarily store the address code while it is sent out on the address bus
8. 65,536 bits
9. $1024 \times 4 = 4096$ bits
10. $2^9 = 512$; 9 bits
11. enables the memory to operate
12. provides for a lower power consumption when memory is not in operation
13. 100 ns + 450 ns = 550 ns

14. $(300 \text{ ns})(256) = 76,800 \text{ ns} = 76.8 \text{ } \mu s$
15. The address code is multiplexed in two 7-bit groups (row and column).
16. generates the sequence to address all rows for purposes of refreshing
17. two; word length expansion
18. minor loops
19. error detection
20. A PLD consists of AND and OR gate arrays. A ROM consists of storage cells.

CHAPTER 11

1. Interfacing is the interconnecting of two circuits or systems so that they function properly together.
2. analog: continuous values; digital: discrete values
3. Monotonicity indicates that there are no incorrect step reversals.
4. $100/(2^5 - 1) = 100/31 = 3.23\%$
5. by the number of bits converted
6. Tracking is faster than stairstep-ramp.
7. to disconnect devices from the bus when they are not sending or receiving
8. The Multibus has twenty address lines and the S-100 bus has sixteen.
9. An internal bus connects the various units of a system together. An example is the Multibus. An external bus connects a system to other systems. The GPIB is an example.
10. talker, listener, and controller
11. automated test systems—computer to voltmeters, signal generators, and other test equipment
12. The GPIB provides parallel data interface. The EIA-232-D/RS-232-C provides serial data interface.

CHAPTER 12

1. The CPU controls all microcomputer activities and processes data by arithmetic or logic operations.
2. A program designed to accomplish a specified task that is not associated with system operation.
3. The RAM is a "scratch pad."
4. $2^{24} = 16,777,216$
5. A sixteen-bit data bus permits two bytes to be transferred at a time, thus providing for faster data transfer than an eight-bit data bus.
6. clock generator, bus controller, address latches, and data bus transceiver
7. the external crystal
8. (a) $\overline{\text{MRDC}}$—memory read control
   (b) $\overline{\text{MWTC}}$—memory write control
   (c) $\overline{\text{IORC}}$—I/O read control
   (d) $\overline{\text{IOWC}}$—I/O write control
   All are produced by the bus controller.
9. to signal the bus controller to initiate memory or I/O transfers
10. A positive ALE pulse enables the address latches and places a valid address on the bus lines.
11. 65,535 bytes.
12. $t = 4T$

$$T = \frac{1}{f} = \frac{1}{4.77 \text{ MHz}} = 0.21 \text{ } \mu s$$

$t = 4(0.21 \text{ } \mu s) = 0.84 \text{ } \mu s$

13. The port address must be present on the address bus, and the $\overline{\text{IOWC}}$ signal must be active.
14. Interrupts permit peripheral devices to request service from the CPU.
15. The process of "pointing" the CPU to the beginning address of the appropriate service routine when an interrupt occurs.

16. Instructions are prefetched by the BIU while the EU is executing a previous instruction. This action keeps the EU from waiting while the next instruction is fetched from memory.

17. The instruction pointer is a register that contains the offset address of the next instruction.

18. $C400_{16}$ shifted left $\rightarrow C4000_{16} + 35A1_{16} = C75A1_{16}$

CHAPTER 13

1. A bipolar IC uses bipolar junction transistors. An MOS IC uses MOSFETs.

2. The input *pn* junction acts like a reverse-biased diode when the input is open or when it is HIGH.

3. standard, Schottky (S), low-power Schottky (LS), advanced Schottky (AS), advanced low-power Schottky (ALS).

4. 40 LS loads, from Table 13–1

5. $10 - 3 = 7$ inputs (for HIGH state)

6. $I_{IL(max)} = -1.6 \text{ mA}$
$I_{T(sink)} = 7(-1.6 \text{ mA}) = -11.2 \text{ mA}$

7. Power dissipation increases.

8. CMOS; greater noise immunity.

9. to prevent damage from electrostatic discharge (ESD)

10. It is faster; it has high fan-out.

11. high-noise environment, because ECL has low values of noise margin

12. NMOS and PMOS

Answers to Odd-Numbered Problems

CHAPTER 1

1–1 Geissler, Edison, Brattain, Bardeen, Shockley

1–3 analog: continuous values; digital: discrete values

1–5 (a) 11010001 (b) 00010101

1–7 (a) 0.6 μs (b) 0.45 μs
(c) 2.7 μs (d) 10 V

1–9 250 Hz

1–11 50%

1–13 8 μs; 1 μs

1–15 AND gate

1–17 (a) adder (b) multiplier
(c) multiplexer (d) comparator

1–19 01010000

1–21 for serial transfer to remote location

1–23 LSI

1–25 (a) Begin with pin 1, in upper left, and proceed counterclockwise to pin 24, in upper right.
(b) Pin 1 is middle pin on upper edge. Proceed counterclockwise to pin 28, immediately to the right of pin 1.

1–27 pulse train (continuous sequence of pulses)

CHAPTER 2

2–1 (a) 1 (b) 100 (c) 100,000

2–3 (a) 400; 70; 1 (b) 9000; 300; 50; 6
(c) 100,000; 20,000; 5000; 0; 0; 0

2–5 (a) 3 (b) 4 (c) 7 (d) 8
(e) 9 (f) 12 (g) 11 (h) 15

2–7 (a) 51.75 (b) 42.25 (c) 65.875
(d) 120.625 (e) 92.65625 (f) 113.0625
(g) 90.625 (h) 127.96875

2–9 (a) 5 bits (b) 6 bits (c) 6 bits (d) 7 bits
(e) 7 bits (f) 7 bits (g) 8 bits (h) 8 bits
(i) 9 bits

2–11 (a) 1010 (b) 10001 (c) 11000
(d) 110000 (e) 111101 (f) 1011101
(g) 1111101 (h) 10111010 (i) 100101010

2–13 (a) 100 (b) 100
(c) 1000 (d) 1101
(e) 1110 (f) 11000

2–15 (a) 1001 (b) 1000
(c) 100011 (d) 110110
(e) 10101001 (f) 10110110

2–17 (a) 010 (b) 001 (c) 0101
(d) 00101000 (e) 0001010 (f) 11110

2–19 (a) 10 (b) 001 (c) 0111
(d) 0011 (e) 00100 (f) 01101
(g) 01010000 (h) 11000011

2–21 (a) 11101110 (b) 10011000
(c) 11111010 (d) 10001000

2–23 (a) 0110000 (b) 0011101 (c) 00011000
(d) 1101011 (e) 100111110

2–25 (a) 00011 (b) 00011 (c) 11001
(d) 10001 (e) 11110 (f) 00101

2–27 11111011000

2–29 (a) 10 (b) 23 (c) 46 (d) 52
(e) 67 (f) 367 (g) 115 (h) 532
(i) 4085

2–31 (a) 001011 (b) 101111
(c) 001000001 (d) 011010001
(e) 101100000 (f) 100110101011
(g) 001011010111001 (h) 100101110000000
(i) 001000000010001011 (j) 001000011.100101

2–33 (a) 00111000 (b) 01011001
(c) 101000010100 (d) 010111001000
(e) 0100000100000000 (f) 1111101100010111
(g) 10001010.1001

2–35 (a) 35 (b) 146 (c) 26 (d) 141
(e) 243 (f) 235 (g) 1474 (h) 1792

2–37 (a) 60_{16} (b) $10B_{16}$ (c) $1BA_{16}$

2–39 (a) 00010000 (b) 00010011
(c) 00011000 (d) 00100001
(e) 00100101 (f) 00110110
(g) 01000100 (h) 01010111
(i) 01101001 (j) 10011000
(k) 000100100101 (l) 000101010110

2–41 (a) 000100000100 (b) 000100101000
(c) 000100110010 (d) 000101010000
(e) 000110000110 (f) 001000010000
(g) 001101011001 (h) 010101000111
(i) 0001000001010001 (j) 0010010101100011

2–43 (a) 80 (b) 237 (c) 346 (d) 421
(e) 754 (f) 800 (g) 978 (h) 1683
(i) 9018 (j) 6667

2–45 (a) 00010100 (b) 00010010
(c) 00010111 (d) 00010110
(e) 01010010 (f) 000100001001
(g) 000110010101 (h) 0001001001101001

2–47 The Gray code makes only one bit change at a time when going from one number in the sequence to the next.

2–49 **(a)** 1100 **(b)** 00011 **(c)** 10000011110
2–51 **(a)** 0 **(b)** 6 **(c)** 4
      (d) 13 **(e)** 49 **(f)** 52
2–53 48 65 6C 6C 6F 2E 20 48 6F 77 20 61 72 65 20 79 6F 75 3F

CHAPTER 3

3–1 See Figure P–1.

FIGURE P–1

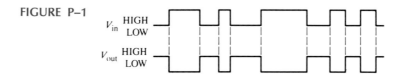

3–3 See Figure P–2.
3–5 See Figure P–3.

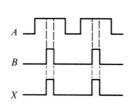

FIGURE P–2

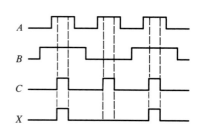

FIGURE P–3

3–7 See Figure P–4.

FIGURE P–4

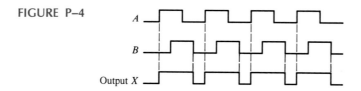

3–9 See Figure P–5.

FIGURE P–5

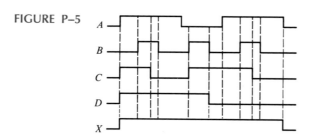

3–11   See Figure P–6.

FIGURE P–6

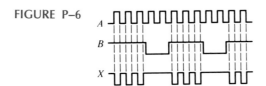

3–13   See Figure P–7.

FIGURE P–7

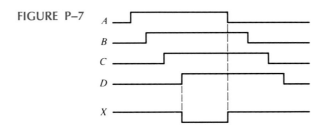

3–15   See Figure P–8.

FIGURE P–8

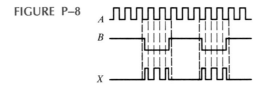

3–17   See Figure P–9.

FIGURE P–9

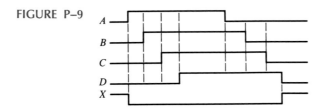

3–19   XOR $= A\overline{B} + \overline{A}B$; OR $= A + B$
3–21   See Figure P–10.

FIGURE P–10

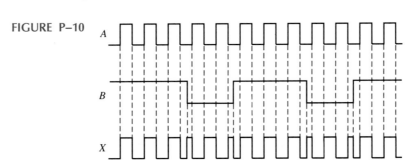

3–23 CMOS

3–25 4.3 ns, 10.5 ns

3–27 20 mW

3–29 NOR

3–31 Add an inverter to the Enable input line of the AND gate.

3–33 See Figure P–11.

FIGURE P–11

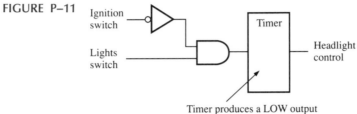

Timer produces a LOW output
15 s after AND gate output
goes HIGH

3–35 The gates in parts (b), (c), (e), are faulty.

3–37 **(a)** defective output (stuck LOW)
(b) Input 2 is open.

3–39 The seatbelt input to the AND gate is open.

CHAPTER 4

4–1 $X = A + B + C + D$

4–3 $Z = \overline{A} + \overline{B} + \overline{C}$

4–5 **(a)** $X = AB$

A	B	X
0	0	0
0	1	0
1	0	0
1	1	1

(b) $X = ABC$

A	B	C	X
0	0	0	0
0	0	1	0
0	1	0	0
0	1	1	0
1	0	0	0
1	0	1	0
1	1	0	0
1	1	1	1

(c) $X = A + B$

A	B	X
0	0	0
0	1	1
1	0	1
1	1	1

(d) $X = A + B + C$

A	B	C	X
0	0	0	0
0	0	1	1
0	1	0	1
0	1	1	1
1	0	0	1
1	0	1	1
1	1	0	1
1	1	1	1

(e) $X = AB + C$

A	B	C	X
0	0	0	0
0	0	1	1
0	1	0	0
0	1	1	1
1	0	0	0
1	0	1	1
1	1	0	1
1	1	1	1

(f) $X = \overline{A} + B$

A	B	X
0	0	1
0	1	1
1	0	0
1	1	1

(g) $X = A\overline{B}\,\overline{C}$

A	B	C	X
0	0	0	0
0	0	1	0
0	1	0	0
0	1	1	0
1	0	0	1
1	0	1	0
1	1	0	0
1	1	1	0

(h) $X = AB + \overline{A}C$

A	B	C	AB	$\overline{A}C$	X
0	0	0	0	0	0
0	0	1	0	1	1
0	1	0	0	0	0
0	1	1	0	1	1
1	0	0	0	0	0
1	0	1	0	0	0
1	1	0	1	0	1
1	1	1	1	0	1

(i) $X = A(B + C)$

A	B	C	B + C	X
0	0	0	0	0
0	0	1	1	0
0	1	0	1	0
0	1	1	1	0
1	0	0	0	0
1	0	1	1	1
1	1	0	1	1
1	1	1	1	1

(j) $X = \overline{A}(\overline{B} + \overline{C})$

A	B	C	$\overline{A}$	$\overline{B} + \overline{C}$	X
0	0	0	1	1	1
0	0	1	1	1	1
0	1	0	1	1	1
0	1	1	1	0	0
1	0	0	0	1	0
1	0	1	0	1	0
1	1	0	0	1	0
1	1	1	0	0	0

4–7 **(a)** $X = AB$ **(b)** $X = \overline{A}$ **(c)** $X = A + B$ **(d)** $X = A + B + C$

4–9 **(a)** $X = A + B$

A	B	X
0	0	0
0	1	1
1	0	1
1	1	1

(b) $X = AB$

A	B	X
0	0	0
0	1	0
1	0	0
1	1	1

(c) $X = AB + BC$

A	B	C	X
0	0	0	0
0	0	1	0
0	1	0	0
0	1	1	1
1	0	0	0
1	0	1	0
1	1	0	1
1	1	1	1

(d) $X = (A + B)C$

A	B	C	X
0	0	0	0
0	0	1	0
0	1	0	0
0	1	1	1
1	0	0	0
1	0	1	1
1	1	0	0
1	1	1	1

(e) $X = (A + B)(\overline{B} + C)$

A	B	C	A + B	$\overline{B}$ + C	X
0	0	0	0	1	0
0	0	1	0	1	0
0	1	0	1	0	0
0	1	1	1	1	1
1	0	0	1	1	1
1	0	1	1	1	1
1	1	0	1	0	0
1	1	1	1	1	1

4–11 reference: Table 4–1
    (a) rule 9   **(b)** rule 8   **(c)** rule 5
    (d) rule 6   **(e)** rule 10   **(f)** rule 11

4–13 **(a)** $\overline{A} + B + \overline{C}D$ **(b)** $\overline{A} + \overline{B} + (\overline{C} + \overline{D})(\overline{E} + \overline{F})$
    (c) $\overline{ABCD} + \overline{A} + \overline{B} + \overline{C} + D$ **(d)** $\overline{A} + B + C + D + A\overline{B}\,\overline{C}D$
    (e) $AB + (\overline{C} + \overline{D})(E + \overline{F}) + ABCD$

4–15 **(a)** $AC + AD + BC + BD$ **(b)** $AD + \overline{B}CD$ **(c)** $ABC + ACD$

4–17 **(a)** $X = AB + \overline{A}\,\overline{B}$, sum-of-products
    (b) $X = AB + AB\overline{C} + ABC\overline{D}$, sum-of-products
    (c) $X = (A + B)(A + C)$, product-of-sums
    (d) $X = (\overline{A} + B)(A + B + C)(\overline{A} + B + C + \overline{D})$, product-of-sums

4–19 $X = ABC + \overline{A}BC + A\overline{B}C + AB\overline{C}$

4–21 See Figure P–12.

4–23 **(a)** A **(b)** AB **(c)** C **(d)** A **(e)** $\overline{A}C + \overline{B}C$

4–25 **(a)** $BD + BE + \overline{D}F$ **(b)** $\overline{A}\overline{B}C + \overline{A}\overline{B}D$
    (c) B **(d)** $AB + CD$ **(e)** ABC

4–27 Networks (b) and (d) are equivalent; $X = A\overline{B} + AC\overline{D}$ for both circuits.

4–29 **(a)** $X = \overline{A}\overline{B} + \overline{B}C$ **(b)** $X = AC$ **(c)** $X = B$
    (d) $X = \overline{C}$

FIGURE P–12

(a) $X = A + B + C$

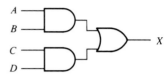

(b) $X = ABC$

(c) $X = AB + C$

(d) $X = AB + CD$

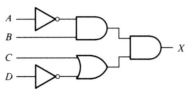

(e) $X = \overline{A}B\,(C + \overline{D})$

4–31 (a) $X = ABC + AB\overline{C} + A\overline{B}C$

(b) $X = A\overline{B}\,\overline{C} + A\overline{B}C + AB\overline{C} + ABC + \overline{A}BC$

(c) $X = A\overline{B}\,\overline{C}D + AB\overline{C}D + A\overline{B}\,\overline{C}\,\overline{D} + AB\overline{C}\,\overline{D} + \overline{A}B\overline{C}D + \overline{A}BC\overline{D}$

(d) $X = A\overline{B}\,\overline{C}\,\overline{D} + A\overline{B}\,\overline{C}D + A\overline{B}C\overline{D} + A\overline{B}CD + \overline{A}B\overline{C}D + \overline{A}BCD + ABCD + \overline{A}BC\overline{D}$
$+ AB\overline{C}D$

4–33 (a) No simplification (b) $X = \overline{AB}\overline{C} + ABC$ (c) $X = B\overline{C} + A\overline{C}D$

(d) $X = \overline{B}C$ (e) $X = \overline{B} + \overline{D}$

4–35 $X = \overline{B} + C$

4–37 $X = \overline{A}\,\overline{B}\overline{C}D + C\overline{D} + BC + A\overline{D}$

CHAPTER 5

5–1 (a) $X = AB$

A	B	X
0	0	0
0	1	0
1	0	0
1	1	1

(b) $X = B$

A	B	X
0	0	0
0	1	1
1	0	0
1	1	1

(c) $X = \overline{A} + B$

A	B	X
0	0	1
0	1	1
1	0	0
1	1	1

(d) $X = A + B$

A	B	X
0	0	0
0	1	1
1	0	1
1	1	1

5–3 $X = A\overline{B} + \overline{A}B$

A	B	X
0	0	0
0	1	1
1	0	1
1	1	0

5–5   **(a)** $X = (AB + C)D + E$ **(b)** $X = \overline{\overline{\overline{(\overline{A} + B)BC}} + D}$

(c) $X = \overline{\overline{(\overline{AB} + \overline{C})}D} + \overline{E}$ **(d)** $X = \overline{\overline{(AB} + \overline{CD})(\overline{EF} + \overline{GH})}$

5–7 $X = 0$ for all combinations of input variables

5–9   See Figure P–13.

FIGURE P–13

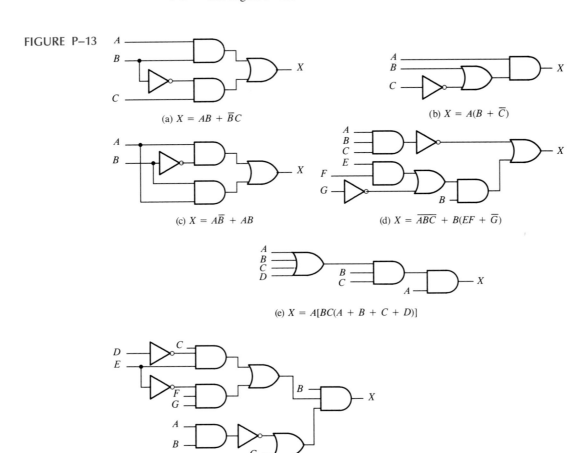

(a) $X = AB + \overline{B}C$

(b) $X = A(B + \overline{C})$

(c) $X = A\overline{B} + AB$

(d) $X = \overline{ABC} + B(EF + \overline{G})$

(e) $X = A[BC(A + B + C + D)]$

(f) $X = B(C\overline{D}E + \overline{E}FG)(\overline{AB} + C)$

5–11   See Figure P–14.

FIGURE P–14 $X = AB + \overline{C}$

5–13 $X =$ lamp on, $A =$ front door switch on, $B =$ back door switch on; see Figure P–15

FIGURE P–15

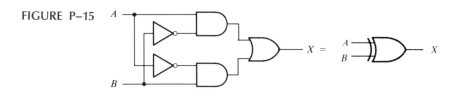

5–15 $X = AB$

5–17 (a) No simplification   (b) No simplification
     (c) $X = A$ (d) $X = \overline{A} + \overline{B} + \overline{C} + EF + \overline{G}$
     (e) $X = ABC$ (f) $X = BC\overline{D}E + \overline{A}B\overline{E}FG + BC\overline{E}FG$

5–19 (a) $X = ABD + CD + E$ (b) $X = \overline{A} + B + D$ (c) $X = ABD + \overline{C}D + \overline{E}$
     (d) $X = \overline{AC} + \overline{AD} + \overline{BC} + \overline{BD} + \overline{EG} + \overline{EH} + \overline{FG} + \overline{FH}$

5–21 See Figure P–16.

FIGURE P–16

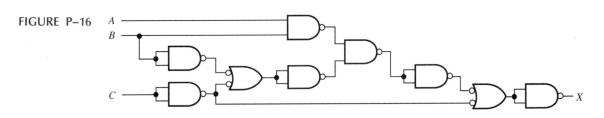

5–23 See Figure P–17.

FIGURE P–17

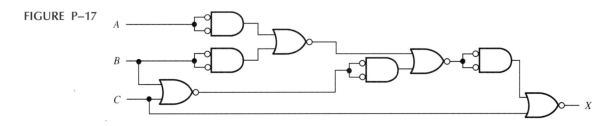

5–25 See Figure P–18.

FIGURE P–18

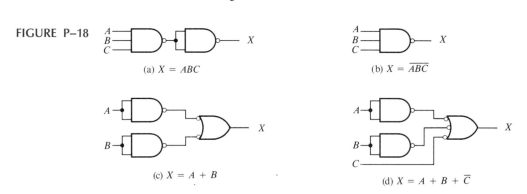

(a) $X = ABC$

(b) $X = \overline{ABC}$

(c) $X = A + B$

(d) $X = A + B + \overline{C}$

FIGURE P–18
(*Continued*)

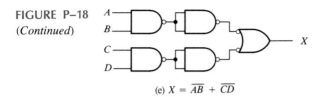

(e) $X = \overline{AB} + \overline{CD}$

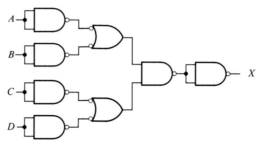

(f) $X = (A + B)(C + D)$

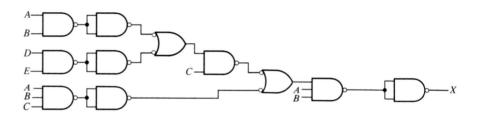

(g) $X = AB[C(\overline{DE} + \overline{AB}) + \overline{BCE}]$

5–27 See Figure P–19.

FIGURE P–19

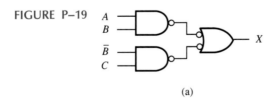

(a)

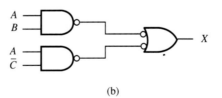

(b)

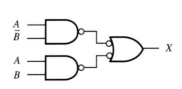

(c)

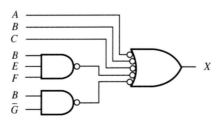

(d)

FIGURE P–19
(Continued)

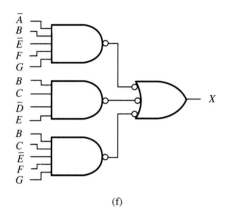

$$X = ABC$$
AND only

(e)   (f)

5–29 $X = A + \overline{B}$; see Figure P–20

5–31 $X = A\overline{B}\overline{C}$; see Figure P–21

5–33 The output pulse width is greater than the specified minimum.

5–35 $X = ABC + D\overline{E}$. Since X is the same as the G_3 output, either G_1 or G_2 has failed, with its output stuck LOW.

5–37 Gate G_5 has a shorted input.

5–39 **(a)** $X = E$. See Figure P–22.
      (b) $X = E$ **(c)** $X = E$

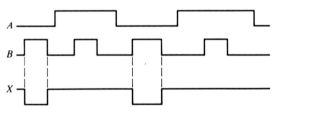

FIGURE P–20

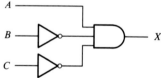

FIGURE P–21

FIGURE P–22

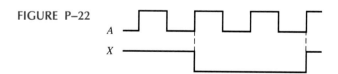

CHAPTER 6

6–1 **(a)** G_1 output $= 0$, G_2 output $= 1$, G_3 output $= 0$, G_4 output $= 1$, G_5 output $= 1$
      (b) G_1 output $= 1$, G_2 output $= 0$, G_3 output $= 1$, G_4 ouput $= 0$, G_5 output $= 1$
      (c) G_1 output $= 1$, G_2 output $= 1$, G_3 output $= 0$, G_4 output $= 0$, G_5 output $= 0$

6–3 no simplification for sum logic; $CO = PQ + QCI + PCI$

6–5 11100

6–7 $\Sigma_0 = 01101110$
     $\Sigma_1 = 10110100$
     $\Sigma_2 = 01101000$
     $\Sigma_3 = 00010100$
     $\Sigma_4 = 10101010$

6–9   225 ns

6–11 $P = Q$ is HIGH when $P_0 = Q_0$ and $P_1 = Q_1$; see Figure P–23

FIGURE P–23

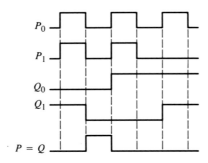

6–13   **(a)** G_1 output = 1   **(b)** G_1 output = 1   **(c)** G_1 output = 1
               G_2 output = 0 G_2 output = 1 G_2 output = 1
               G_3 output = 1 G_3 output = 0 G_3 output = 1
               G_4 output = 0 G_4 output = 0 G_4 output = 1
               G_5 output = 0 G_5 output = 0 G_5 output = 1
               G_6 output = 0 G_6 output = 0 G_6 output = 0
               G_7 output = 1 G_7 output = 0 G_7 output = 0
               G_8 output = 0 G_8 output = 0 G_8 output = 0
               G_9 output = 0 G_9 output = 0 G_9 output = 0
               G_{10} output = 1 G_{10} output = 0 G_{10} output = 0
               G_{11} output = 0 G_{11} output = 0 G_{11} output = 0
               G_{12} output = 0 G_{12} output = 0 G_{12} output = 0
               G_{13} output = 0 G_{13} output = 1 G_{13} output = 0
               G_{14} output = 0 G_{14} output = 0 G_{14} output = 0
               G_{15} output = 0 G_{15} output = 1 G_{15} output = 0
                    $A > B$ $A < B$ $A = B$

6–15   See Figure P–24.

FIGURE P–24

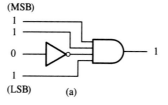

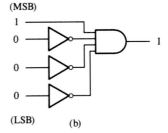

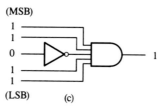

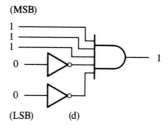

FIGURE P–24
(*Continued*)

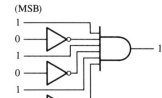

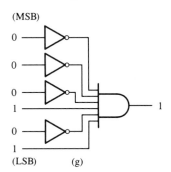

(e)

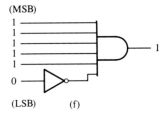

(f)

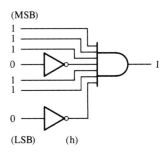

(g)

(h)

6–17 $X = A_3 A_2 \overline{A_1} \overline{A_0} + \overline{A_3} \overline{A_2} \overline{A_1} A_0 + A_3 \overline{A_2} A_1$

6–19 See Figure P–25.

6–21 $A_3 A_2 A_1 A_0 = 1011$, invalid BCD

FIGURE P–25

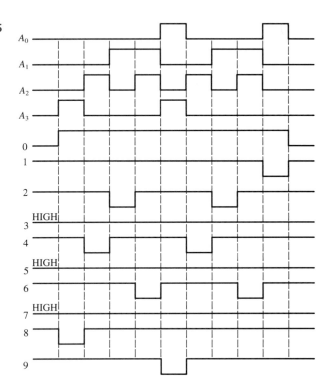

6–23 **(a)** 1111111111   **(b)** 1000010000
      (c) 0000001001   **(d)** 1000000000
      See Figure P–26.

FIGURE P–26

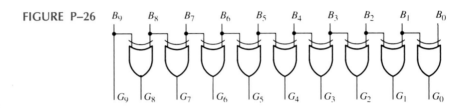

6–25 **(a)** $0010 \rightarrow 0010_2$ **(b)** $1000 \rightarrow 1000_2$
      (c) $00010011 \rightarrow 1101_2$ **(d)** $00100110 \rightarrow 11010_2$
      (e) $00110011 \rightarrow 100001_2$

6–27 See Figure P–27.

FIGURE P–27

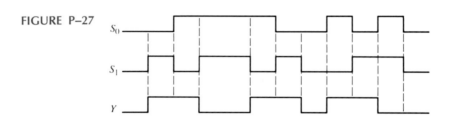

6–29 See Figure P–28.

6–31 $Y = \overline{A_3}\,\overline{A_2}A_1\overline{A_0} + \overline{A_3}\,\overline{A_2}A_1A_0 + \overline{A_3}A_2A_1\overline{A_0} + \overline{A_3}A_2A_1A_0 + A_3\overline{A_2}\,\overline{A_1}\,\overline{A_0} + A_3\overline{A_2}A_1\overline{A_0}$
      $+ A_3\overline{A_2}A_1A_0 + A_3A_2\overline{A_1}A_0 + A_3A_2A_1A_0$

      See Figure P–29.

6–33 See Figure P–30.

6–35 The Σ and CO waveforms are incorrect. Input CI is stuck HIGH.

6–37 Check power; depress each key, and check output; hold down each key while depressing lower-value keys one at a time, and check output.

6–39 **(a)** $\overline{O}$ output of 74139 stuck HIGH or open
      (b) no power; EN input to 74139 open
      (c) f output of 7449 stuck HIGH.
      (d) frequency of the data-select input too low

6–41 **1.** Σ EVEN output of 74180 stuck LOW
      2. error gate faulty
      3. ODD input to 74180 open
      4. output of ODD input inverter open or stuck HIGH

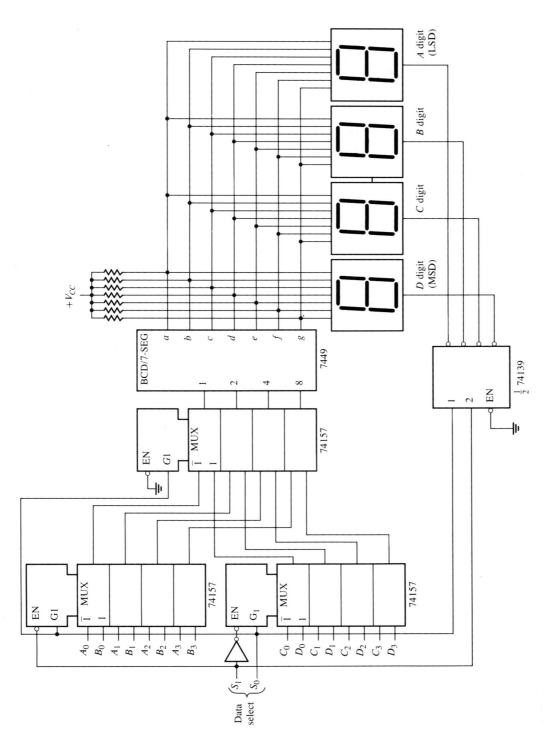

FIGURE P–28

FIGURE P–29

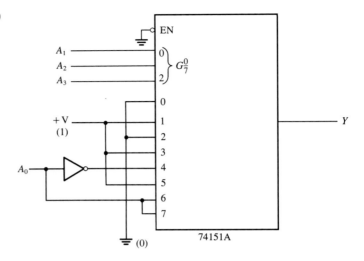

FIGURE P–30

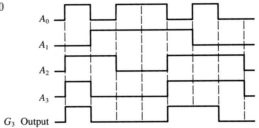

CHAPTER 7

7–1   See Figure P–31.

FIGURE P–31

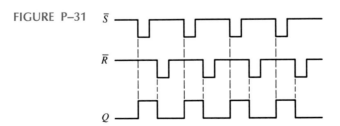

7–3   See Figure P–32.

FIGURE P–32

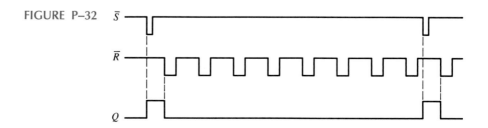

7–5   See Figure P–33.

FIGURE P–33

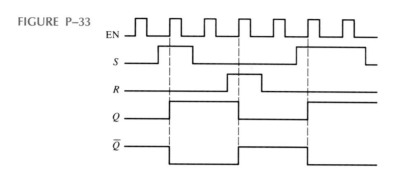

7–7   See Figure P–34.

FIGURE P–34

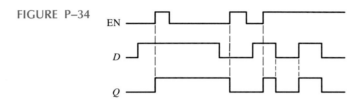

7–9   See Figure P–35.

FIGURE P–35

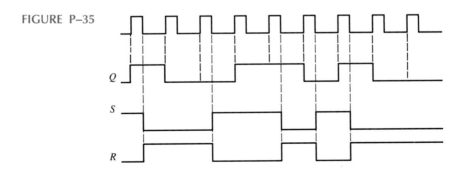

7–11   See Figure P–36.

FIGURE P–36

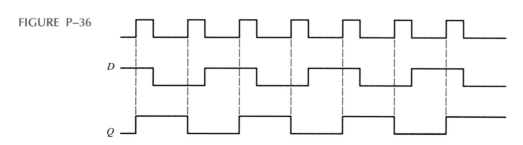

7–13   See Figure P–37.

FIGURE P–37

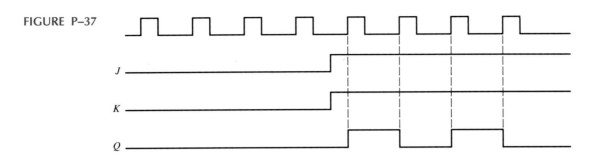

7–15   See Figure P–38.

FIGURE P–38

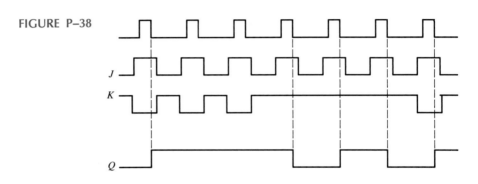

7–17   See Figure P–39.

FIGURE P–39

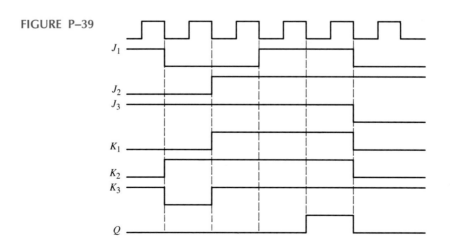

7–19   See Figure P–40.
7–21   See Figure P–41.

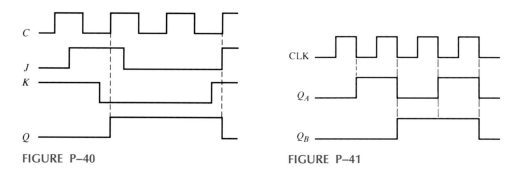

FIGURE P–40

FIGURE P–41

7–23 See Figure P–42.

FIGURE P–42

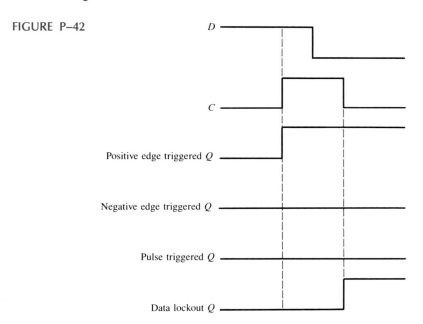

7–25 See Figure P–43.
7–27 150 mA, 750 mW
7–29 divide-by-2; see Figure P–44

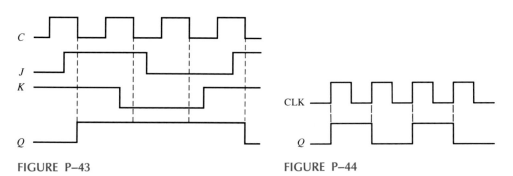

FIGURE P–43

FIGURE P–44

7–31   See Figure P–45.

FIGURE P–45

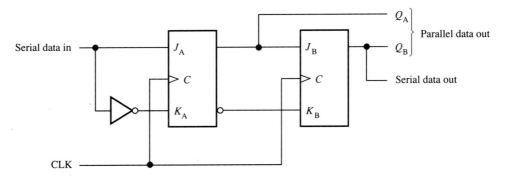

7–33   See Figure P–46.

FIGURE P–46

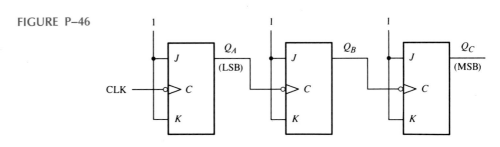

7–35   4.62 μs
7–37   28.57 kHz
7–39 $R = 22$ MΩ, $C = 0.01$ μF
7–41   The wire from pin 6 to pin 10 and the ground wire are reversed on the protoboard.
7–43 $\overline{\text{CLR}}$ shorted to ground
7–45   See Figure P–47. Delays not shown.

FIGURE P–47   (a)

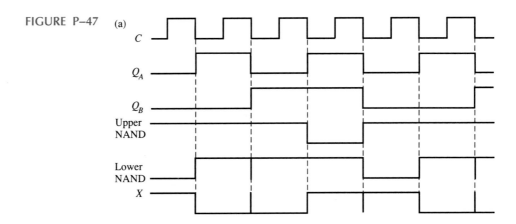

(b) Same as (a)

FIGURE P–47
(*Continued*)

(c)

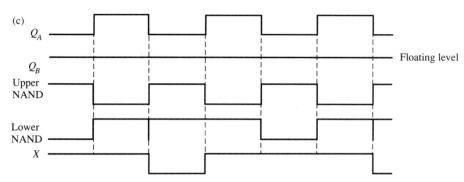

(d) X = LOW if Q_B = 1; $X = \bar{Q}_A$ if Q_B = 0

(e)

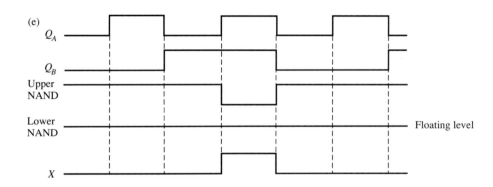

CHAPTER 8

8–1   See Figure P–48.

FIGURE P–48

8–3   Worst-case delay is 24 ns; it occurs when all flip-flops change state from 011 to 100 or from 111 to 000.

8–5   8 ns

8–7   Initially, each flip-flop is RESET.

AT CLK_1: $J_0 = K_0 = 1$ Therefore Q_0 goes to a 1.
               $J_1 = K_1 = 0$ Therefore Q_1 remains a 0.
               $J_2 = K_2 = 0$ Therefore Q_2 remains a 0.
               $J_3 = K_3 = 0$ Therefore Q_3 remains a 0.

At CLK_2: $J_0 = K_0 = 1$ Therefore Q_0 goes to a 0.
               $J_1 = K_1 = 1$ Therefore Q_1 goes to a 1.

$J_2 = K_2 = 0$ Therefore Q_2 remains a 0.
$J_3 = K_3 = 0$ Therefore Q_3 remains a 0.

At CLK_3: $J_0 = K_0 = 1$ Therefore Q_0 goes to a 1.
$J_1 = K_1 = 0$ Therefore Q_1 remains a 1.
$J_2 = K_2 = 0$ Therefore Q_2 remains a 0.
$J_3 = K_3 = 0$ Therefore Q_3 remains a 0.

A continuation of this procedure for the next seven clock pulses will show that the counter progresses through the BCD sequence.

8–9   See Figure P–49.

FIGURE P–49

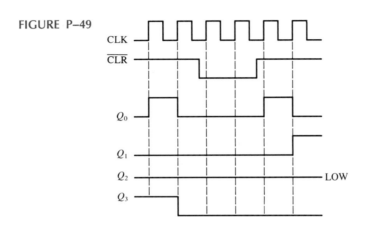

8–11   See Figure P–50.

FIGURE P–50

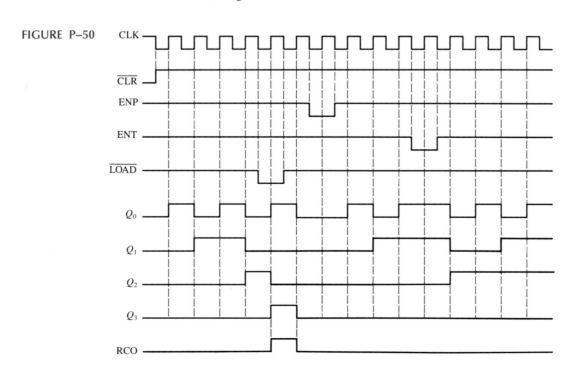

8–13 See Figure P–51.

FIGURE P–51

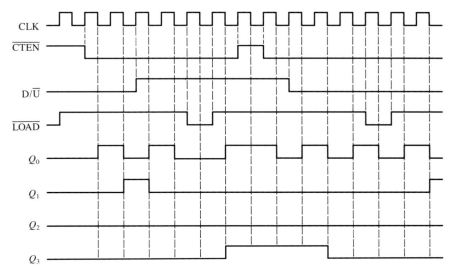

8–15 The sequence is 0000, 1111, 1110, 1101, 1010, 0101. The counter "locks up" in the 1010 and 0101 states and alternates between them.

8–17 See Figure P–52.

FIGURE P–52

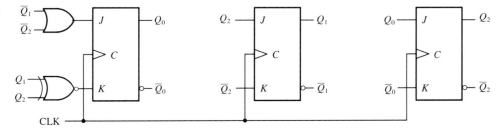

8–19 See Figure P–53.

8–21 See Figure P–54.

8–23 See Figure P–55.

8–25 See Figure P–56.

8–27 CLK_2, output 0; CLK_4, outputs 2, 0; CLK_6, output 4; CLK_8, outputs 6, 4, 0; CLK_{10}, output 8; CLK_{12}, outputs 10, 8; CLK_{14}, output 12; CLK_{16}, outputs 14, 12, 8

8–29 A glitch of the AND gate output occurs on the 111 to 000 transition. Eliminate by AND-ing $\overline{CLK}$ with counter outputs.

8–31 See Figure P–57.

8–33 See Figure P–58.

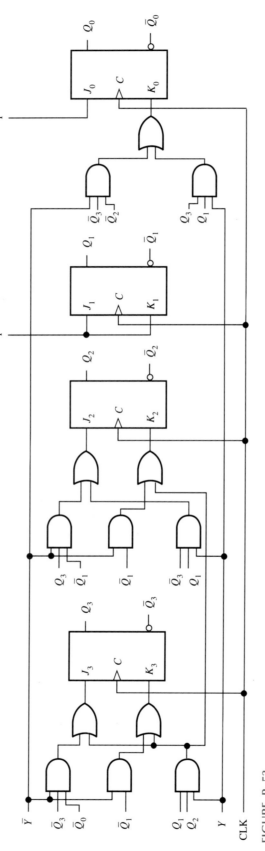

FIGURE P–53

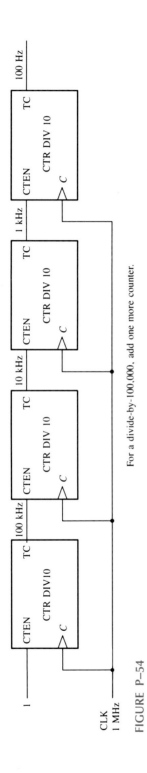

For a divide-by-100,000, add one more counter.

FIGURE P–54

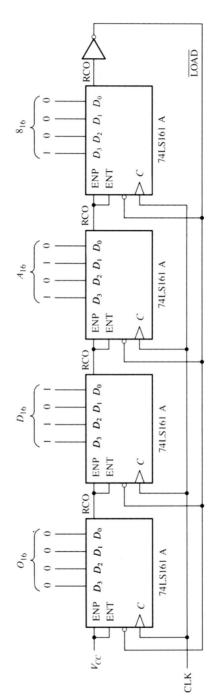

FIGURE P–55

FIGURE P–56

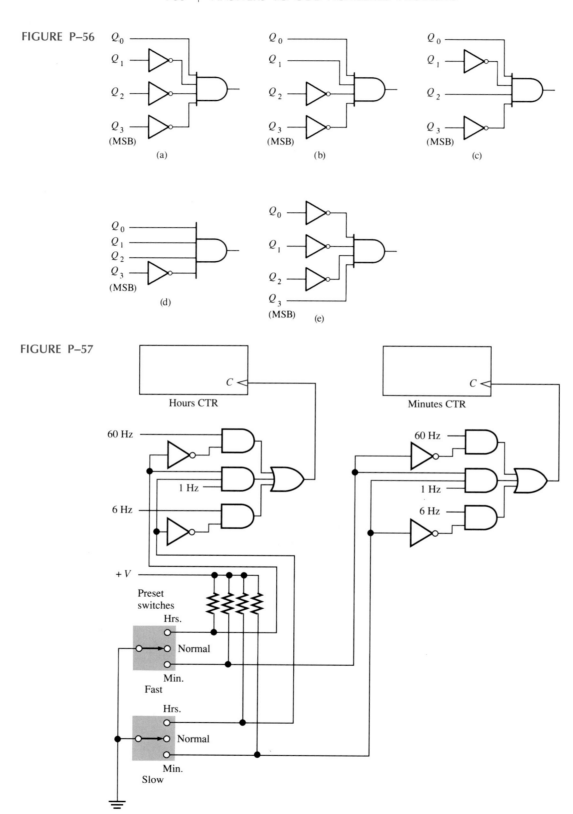

(a)

(b)

(c)

(d)

(e)

FIGURE P–57

FIGURE P–58

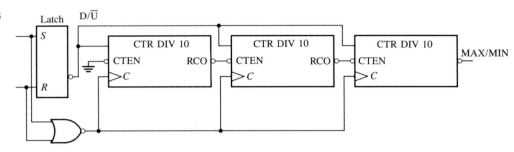

FOR 3000-space counter, add the following:

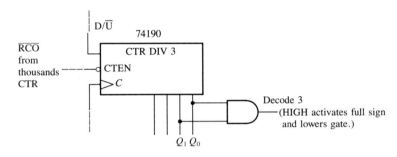

8–35 **(a)** Q_A and Q_B will not change from their initial state.
(b) normal operation except Q_0 floating
(c) Q_0 waveform is normal; Q_1 remains in initial state.
(d) normal operation
(e) The counter will not change from its initial state.

8–37 The K input of FF1 must be connected to ground rather than to the J input. Check for a wiring error.

8–39 upper AND gate input open

8–41 The frequency is not correct. A possible fault is the inverter output stuck HIGH or open.

8–43 flip-flop output stuck HIGH or open; most significant BCD-to-seven-segment decoder input open

8–45 It can be almost anything. Start the process of elimination.

CHAPTER 9

9–1 Shift registers store binary data.

9–3 See Figure P–59.

FIGURE P–59

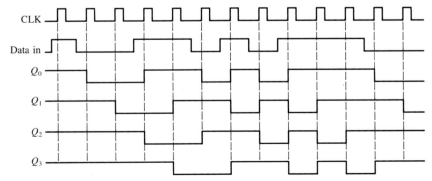

9–5

Initially:	101001111000
CLK_1:	010100111100
CLK_2:	001010011110
CLK_3:	000101001111
CLK_4:	000010100111
CLK_5:	100001010011
CLK_6:	110000101001
CLK_7:	111000010100
CLK_8:	011100001010
CLK_9:	001110000101
CLK_{10}:	000111000010
CLK_{11}:	100011100001
CLK_{12}:	110001110000

9–7 See Figure P–60.

FIGURE P–60

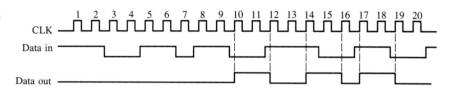

9–9 See Figure P–61.

FIGURE P–61

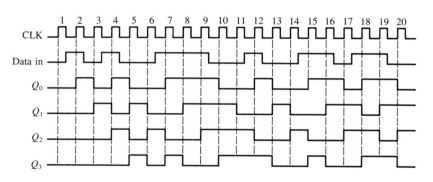

9–11 See Figure P–62.

FIGURE P–62

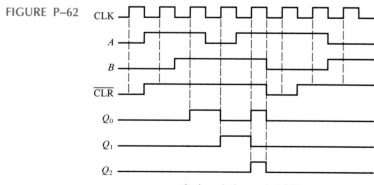

Q_3 through Q_7 remain LOW.

9–13   See Figure P–63.

FIGURE P–63

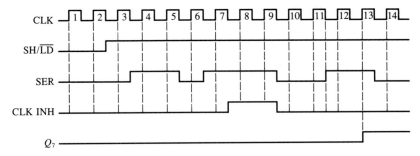

9–15   See Figure P–64.

FIGURE P–64

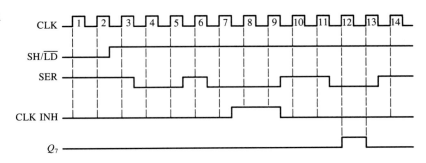

9–17   See Figure P–65.

FIGURE P–65

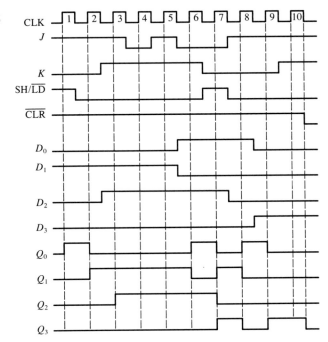

9–19   Initially (76):   01001100
CLK_1:   10011000   left
CLK_2:   01001100   right
CLK_3:   00100110   right
CLK_4:   00010011   right
CLK_5:   00100110   left
CLK_6:   01001100   left
CLK_7:   00100110   right
CLK_8:   01001100   left
CLK_9:   00100110   right
CLK_{10}:   01001100   left
CLK_{11}:   10011000   left

9–21   See Figure P–66.

FIGURE P–66

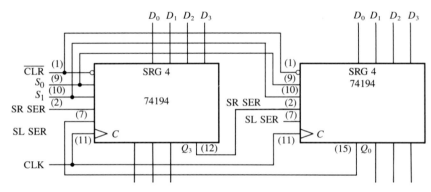

9–23   **(a)** 3   **(b)** 5   **(c)** 7   **(d)** 8
(e) 10   **(f)** 12   **(g)** 18

9–25   See Figure P–67.

9–27   See Figure P–68.

FIGURE P–67

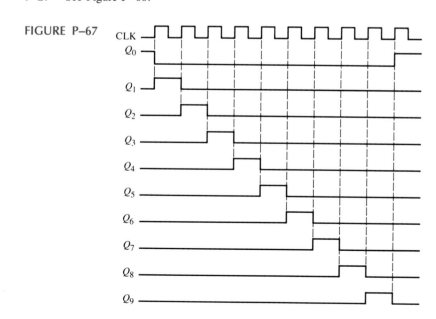

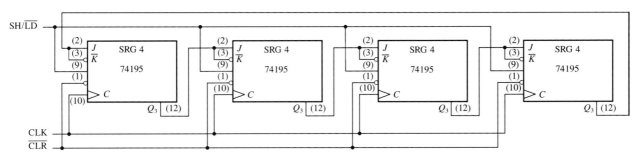

FIGURE P–68

9–29 **1.** Replace divide-by-8 counter with divide-by-16.
2. Add a second eight-bit data-input shift register.
3. Add a second eight-bit data-output register.

9–31 to provide a momentary LOW to parallel-load the ring counter when power is turned on

9–33 An incorrect code may be produced.

9–35 D_3 input open

9–37 **(a)** no clock at switch closure because of faulty NAND (negative-OR) gate or one-shot; open clock (C) input to key code register; open SH/LD input to key code register
(b) diode in third row open; Q_2 output of ring counter open
(c) The NAND (negative-OR) gate input connected to the first column is open or shorted.
(d) the "2" input to the column encoder is open.

9–39 Use an eleven-bit shift register that is alternately loaded with opposite eight-bit patterns but with the same start and stop bits.

CHAPTER 10

10–1 **(a)** ROM   **(b)** RAM

10–3 *Address bus* provides for transfer of address code to memory for accessing a memory location for a read or write operation.
Data bus provides for transfer of data between the microprocessor and the memory or I/O.
Control bus provides for transfer of control signals between microprocessor, memory, and I/O.

10–5

Inputs		Outputs			
A_1	A_0	O_3	O_2	O_1	O_0
0	0	0	1	0	1
0	1	1	0	0	1
1	0	1	1	1	0
1	1	0	0	1	0

10–7 See Figure P–69.

10–9 **(a)** 00010011   **(b)** 01000110   **(c)** 01010101

10–11 blown links: 1–17, 19–23, 25–31, 34, 37, 38, 40–47, 53, 55, 58, 61, 62, 63, 65, 67, 69

FIGURE P–69

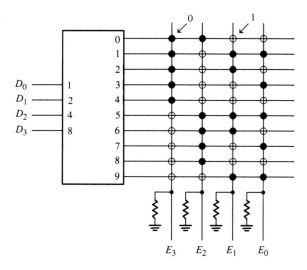

10–13 **(a)** RESET (0) **(b)** RESET (0) **(c)** SET (1)

10–15 64×64

10–17 Use eight 74189s with six address lines. Two of the address lines are decoded to enable the selected memory chips. Four data lines go to each chip.

10–19 8 bits, 64 words; 4 bits, 256K words

10–21 315,392 bytes

10–23 the replication/detection process

10–25 lowest address: $FC0_{16}$
highest address: FFF_{16}

10–27 See Figure P–70.

10–29 See Figure P–71.

10–31 See Figure P–72.

10–33 See Figure P–73.

FIGURE P–70

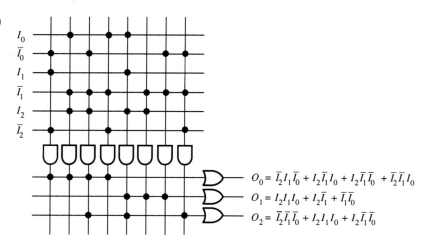

$$O_0 = \bar{I}_2 I_1 \bar{I}_0 + I_2 \bar{I}_1 I_0 + I_2 \bar{I}_1 \bar{I}_0 + \bar{I}_2 \bar{I}_1 I_0$$

$$O_1 = I_2 I_1 I_0 + I_2 \bar{I}_1 + \bar{I}_1 \bar{I}_0$$

$$O_2 = \bar{I}_2 \bar{I}_1 \bar{I}_0 + I_2 I_1 I_0 + I_2 \bar{I}_1 \bar{I}_0$$

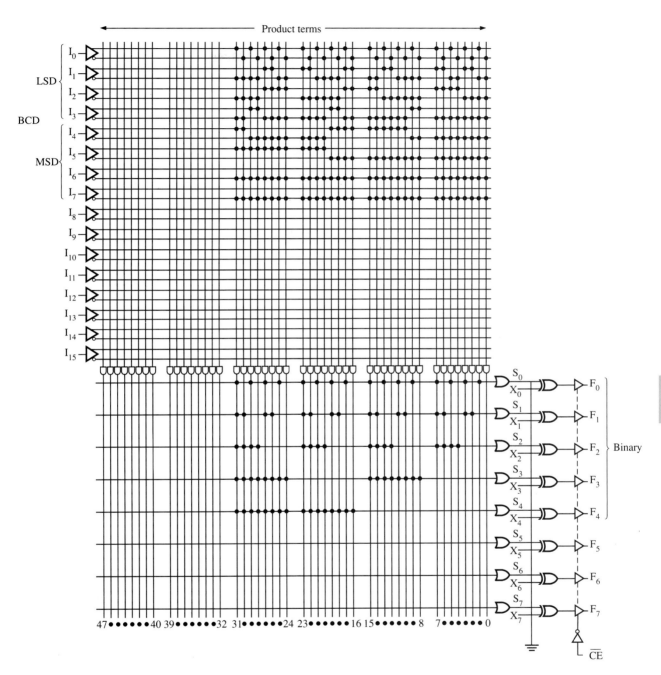

FIGURE P–71

FIGURE P–72

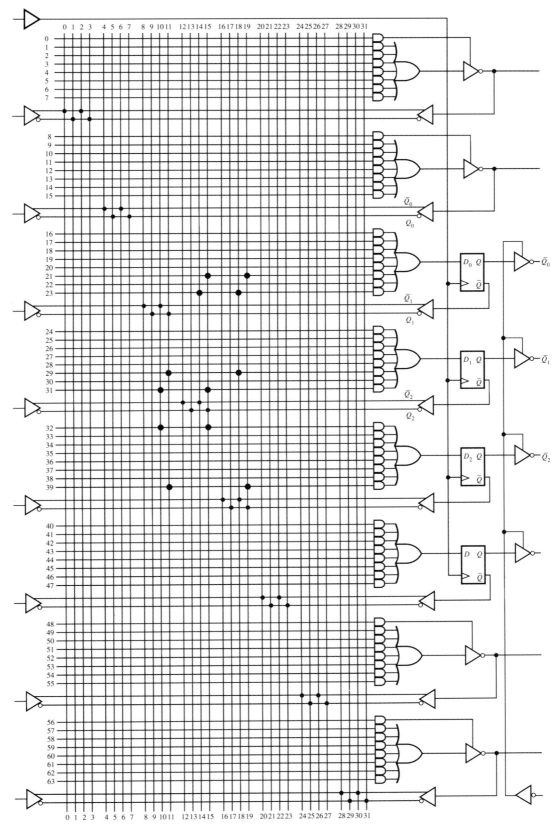

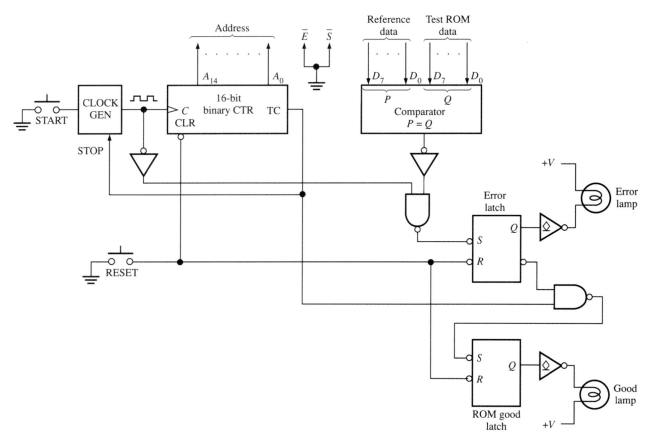

FIGURE P–74

10–35 See Figure P–74 for one approach.

10–37 **(a)** ROM0: 00_{16}, $1F_{16}$
ROM1: 20_{16}, $3F_{16}$
ROM2: 40_{16}, $5F_{16}$
ROM3: 60_{16}, $7F_{16}$

(b) same as Figure 10–84 except that last address is specified as $7E_{16}$ ($7F_{16} - 1$)

(c) See Figure P–75.

(d) A single checksum will not isolate the faulty chip.

10–39 fault in 0_1 waveform

CHAPTER 11

11–1 1000, 1010, 1101, 1110, 1110, 1110, 1100, 1001, 0110, 0100, 0010, 0001, 0001, 0011, 0101, 1000, 1010, 1101, 1110, 1110, 1110, 1100, 1001, 0110, 0100, 0010.

11–3 See Figure P–76.

11–5 330 kΩ

11–7 See Figure P–77.

11–9 **(a)** 14.29% **(b)** 0.098% **(c)** 0.00038%

11–11 000, 001, 100, 110, 101, 100, 011, 010, 001, 001, 100, 111, 111, 111, 111, 111, 111, 111, 110, 100

11–13 0000, 0000, 0000, 1110, 1100, 0111, 0110, 0011, 0010, 1100

FIGURE P–75

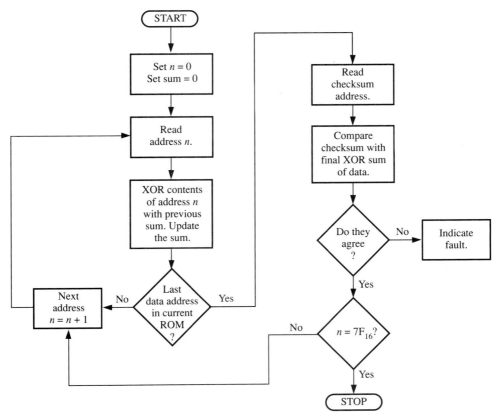

FIGURE P–76

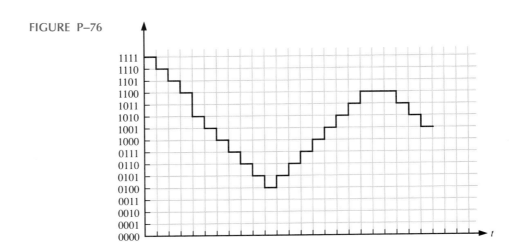

11–15

SAR	Comment
1000	Greater than V_{IN}, reset MSB
0100	Less than V_{IN}, keep the 1
0110	Equal to V_{IN}, keep the 1 (final state)

FIGURE P–77

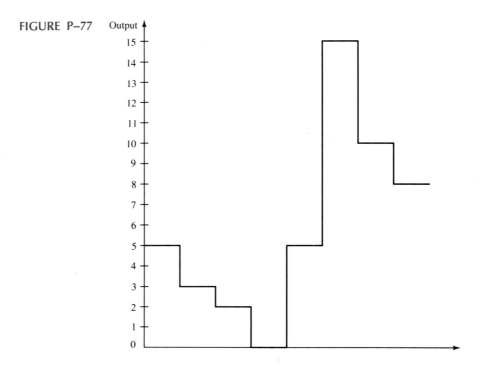

11–17 See Figure P–78.
11–19 See Figure P–79.

FIGURE P–78

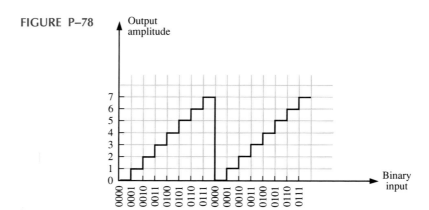

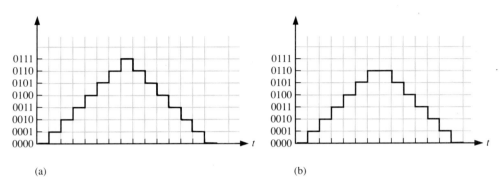

(a)

(b)

FIGURE P–79

11–21 See Figure P–80.

11–23 seven

FIGURE P–80

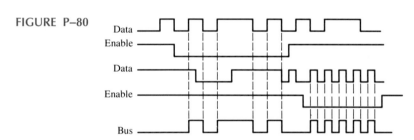

11–25 A controller is sending data to two listeners. The first two bytes of data (3F and 41) go to the listener with address 001A. The second two bytes go to the listener with address 001B. The handshaking signals (DAV, NRFD, and NDAC) indicate the data transfer is successful. See Figure P–81.

FIGURE P–81

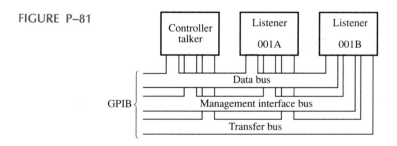

CHAPTER 13

13–1   no; $V_{OH(min)} < V_{IH(min)}$

13–3   0.15 V in HIGH state; 0.25 V in LOW state.

13–5   gate C

13–7   12 ns

13–9   gate C

13–11   yes, G_2

13–13   5.5 mW

13–15   125 MHz

13–17   **(a)**   HIGH   **(b)**   floating
        (c)   HIGH   **(d)**   high-Z

13–19   **(a)**   sourcing; 120 μA
        (b)   sinking; -3.2 mA
        (c) G_1 sourcing; 240 μA
             G_2 sinking; -3.2 mA
             G_3 sourcing; 80 μA

13–21   **(a)** $X = AB\overline{C}D$
        (b) $X = \overline{ABC} \cdot \overline{DE} \cdot \overline{FG}$
        (c) $X = \overline{ABCDEFGH}$

13–23   two open collector gates in parallel

13–25   terminate unused NAND inputs to $+V_{CC}$ and unused NOR inputs to ground.

13–27   **(a)**   AS   **(b)**   HCMOS   **(c)**   4000 CMOS
        (d)   HCMOS   **(e)**   AS   **(f)**   AS

13–29   (c), $f_{CLOCK} > f_{max}$

13–31   Need pull-up resistors on all TTL outputs.

13–33   ECL operates with *nonsaturated* BJTs.

Index

Boldface italic page numbers refer to glossary definitions.